Smart Nanomaterials Technology

Series Editors

Azamal Husen, Wolaita Sodo University, Wolaita, Ethiopia

Mohammad Jawaid, Department of Chemical and Petroleum Engineering, United Arab Emirates University, Al Ain, United Arab Emirates

Nanotechnology is a rapidly growing scientific field and has attracted a great interest over the last few years because of its abundant applications in different fields like biology, physics and chemistry. This science deals with the production of minute particles called nanomaterials having dimensions between 1 and 100 nm which may serve as building blocks for various physical and biological systems. On the other hand, there is the class of smart materials where the material that can stimuli by external factors and results a new kind of functional properties. The combination of these two classes forms a new class of smart nanomaterials, which produces unique functional material properties and a great opportunity to larger span of application. Smart nanomaterials have been employed by researchers to use it effectively in agricultural production, soil improvement, disease management, energy and environment, medical science, pharmaceuticals, engineering, food, animal husbandry and forestry sectors.

This book series in Smart Nanomaterials Technology aims to comprehensively cover topics in the fabrication, synthesis and application of these materials for applications in the following fields:

- Energy Systems—Renewable energy, energy storage (supercapacitors and electrochemical cells), hydrogen storage, photocatalytic water splitting for hydrogen production
- Biomedical—controlled release of drugs, treatment of various diseases, biosensors,
- Agricultural—agricultural production, soil improvement, disease management, animal feed, egg, milk and meat production/processing,
- Forestry—wood preservation, protection, disease management
- Environment—wastewater treatment, separation of hazardous contaminants from wastewater, indoor air filters.

Akil Ahmad · Mohammed B. Alshammari ·
Mohammad Jawaid · Asma Khatoon
Editors

Metal Oxide Based Nanophotocatalyst for Wastewater Purification

Springer

Editors
Akil Ahmad
Department of Chemistry
College of Science and Humanities
Prince Sattam Bin Abdulaziz University
Al Kharj, Saudi Arabia

Mohammed B. Alshammari
Department of Chemistry
College of Science and Humanities
Prince Sattam Bin Abdulaziz University
Al-Kharj, Saudi Arabia

Mohammad Jawaid
Department of Chemical and Petroleum Engineering
United Arab Emirates University (UAEU)
Al Ain, Abu Dhabi, United Arab Emirates

Asma Khatoon
Department of Chemistry
Aligarh Muslim University
Aligarh, Uttar Pradesh, India

ISSN 3004-8273 ISSN 3004-8281 (electronic)
Smart Nanomaterials Technology
ISBN 978-981-96-7682-8 ISBN 978-981-96-7683-5 (eBook)
https://doi.org/10.1007/978-981-96-7683-5

This Springer imprint is published by the registered company Springer Nature Singapore Pte Ltd.
The registered company address is: 152 Beach Road, #21-01/04 Gateway East, Singapore 189721, Singapore

Contents

Introduction and Classification of Nanophotocatalysts

Mohammad Harun-Ur-Rashid, Israt Jahan, Abu Bin Imran, and Md. Abu Bin Hasan Susan

Abstract Nanophotocatalysts are a significant achievement in material science, offering remarkable potential for environmental remediation, energy conversion, and organic synthesis. Materials of this kind take the advantages of the unique features of nanostructures to augment photocatalytic efficiency, addressing limitations of traditional bulk catalysts. Nanophotocatalysts are introduced and classified as a diverse spectrum of materials, which inter alia include: metal nitrides, oxides, sulfides, and hybrid composites, each with distinct mechanisms and applications. The chapter explores the fundamental principles underlying the design and synthesis of nanophotocatalysts, emphasizing their structural and electronic characteristics that drive catalytic activity. It delves into various classification schemes based on composition, morphology, and functionalization, providing a thorough overview of the most recent advances in the area. Advanced characterization techniques and methodologies for assessing photocatalytic performance are discussed, highlighting the importance of nanostructure engineering in optimizing efficiency. The chapter further addresses the emerging trends and future directions in nanophotocatalyst research, such as incorporation of artificial intelligence and machine learning into catalyst design and development of sustainable and scalable manufacturing processes. This ultimately aims to leverage nanophotocatalysts for innovative solutions to global challenges.

M. Harun-Ur-Rashid
Department of Chemistry, International University of Business Agriculture and Technology (IUBAT), Dhaka, Bangladesh
e-mail: mrashid@iubat.edu

I. Jahan
Genetic Engineering and Biotechnology Research Laboratory (GEBRL), Centre for Advanced Research in Sciences (CARS), University of Dhaka, Dhaka, Bangladesh

A. B. Imran
Department of Chemistry, Bangladesh University of Engineering and Technology, Dhaka, Bangladesh

Md. A. B. H. Susan (✉)
Department of Chemistry and Dhaka University Nanotechnology Center (DUNC), University of Dhaka, Dhaka, Bangladesh
e-mail: susan@du.ac.bd

A. Ahmad et al. (eds.), *Metal Oxide Based Nanophotocatalyst for Wastewater Purification*, Smart Nanomaterials Technology,
https://doi.org/10.1007/978-981-96-7683-5_1

Keywords Nanophotocatalysts · Environmental remediation · Energy conversion · Nanostructure engineering · Catalyst design

1 Introduction

Nanophotocatalysis is a transformative field in environmental science, providing innovative solutions to the critical challenge of water pollution [1]. Photocatalysis, which uses light energy to drive chemical reactions, provides a sustainable and efficient solution to environmental restoration. Nanophotocatalysts, especially those based on metal oxides, represent an innovative class of materials designed to maximize the efficiency and efficacy of photocatalytic processes. These materials, when engineered at the nanoscale, exhibit a remarkable ability to degrade organic pollutants, neutralize toxic substances, and purify contaminated water sources with high precision [2, 3]. Their versatility and effectiveness in wastewater purification have garnered significant attention from the scientific community and industry alike, promising a paradigm shift in how we approach environmental cleanup. At the core of this field lies the integration of nanotechnology with photocatalysis. Advancements in nanotechnology have facilitated the creation of diverse engineered nanomaterials, transforming practical applications and driving theoretical research to new horizons. These advancements include the development of structural colored nanomaterials [4, 5], nanocomposite gels [6, 7], smart nanomaterials [8, 9], light-sensitive nanostructures [10, 11], and engineering nanomaterials [12–14]. Such diverse nanomaterials have found extensive applications, increasingly replacing conventional bulk materials in various industrial [15, 16], environmental [17–19], and health [20–23] sectors. Nanotechnology improves photocatalyst performance by manipulating material properties at the molecular and even atomic levels, leading to improvements in surface area, light absorption capacity, and charge carrier mobility [24, 25]. These developments are critical in increasing the reactivity and efficiency of photocatalysts, notably for breaking down persistent organic contaminants and removing heavy metals from wastewater. Metal oxide nanophotocatalysts, particularly TiO_2 and ZnO, have received much interest in this domain because of their affordability, non-toxic characteristics, and robust photocatalytic efficacy [26].

The treatment of wastewater remains a pressing challenge in both industrial and domestic contexts, as traditional methods often struggle to manage emerging contaminants such as pharmaceuticals, pesticides, and microplastics [27]. Nanophotocatalysts present a cutting-edge solution, capable of addressing these issues while enabling the complete breakdown of pollutants into non-toxic by-products [28]. Their ability to efficiently utilize solar energy, combined with high catalytic performance, makes them highly effective for large-scale environmental cleanup. For instance, the use of TiO_2 in degrading organic dyes through photocatalysis has demonstrated notable success, with ongoing research broadening the application of these materials to tackle an ever-growing array of pollutants. Among the various classes of nanophotocatalysts, metal oxides stand out due to their stability, availability, and robust photocatalytic properties. TiO_2 has been extensively studied for its high efficiency under UV light irradiation [29], while ZnO offers complementary benefits, including enhanced visible light absorption and antimicrobial properties [30]. Other metal oxides such as cerium oxide (CeO_2) [31], tungsten trioxide (WO_3) [32], and iron oxide (Fe_2O_3) [33] are gaining attention for their unique redox properties and potential to be tailored for specific environmental applications. The ability to modify and engineer these nanomaterials through doping, heterostructure formation, and surface functionalization further enhances their photocatalytic performance, making them versatile tools in the fight against water pollution.

This chapter presents a groundbreaking approach to understanding and classifying nanophotocatalysts, focusing on their transformative potential in wastewater treatment. It offers a novel framework that systematically categorizes these materials based on composition, morphology, doping strategies, and heterostructure formation, thereby bridging theoretical insights and practical applications. The work delves into emerging trends, emphasizing sustainable synthesis methods such as green and biogenic approaches, while integrating state-of-the-art technologies like artificial intelligence for catalyst design. By addressing the critical parameters influencing photocatalytic efficiency, including bandgap engineering, surface morphology, and dynamics of charge carrier, key advancements that enhance light absorption, stability, and reusability of these materials have been highlighted. This comprehensive effort synthesizes insights from a wide array of seminal studies to map the evolution of nanophotocatalyst technologies and provide a detailed understanding of their applications in pollutant degradation, heavy metal removal, and microbial disinfection. Practical challenges such as scalability, photocatalyst stability, and electron–hole recombination are critically analyzed, and innovative strategies are proposed to overcome these limitations. The discussion extends to the development of multifunctional catalysts capable of addressing complex environmental challenges, offering solutions tailored for both industrial and large-scale applications. Detailed tables and schematic representations are included to compare synthesis methods, key properties, and performance metrics of various nanophotocatalysts, ensuring clarity and accessibility. This chapter establishes a roadmap for future advancements in nanophotocatalysis, positioning it as a cornerstone technology for accomplishing long-term environmental restoration.

2 Fundamentals of Nanophotocatalysis

2.1 Mechanisms of Photocatalytic Processes

The photocatalytic mechanism is central to the functionality of nanophotocatalysts, driving their extensive use in environmental remediation. Interaction of light energy with a semiconductor material, often a metal oxide, generates electron–hole (e^-–h^+) pairs through excitation of electrons from valence band (VB) to the conduction band

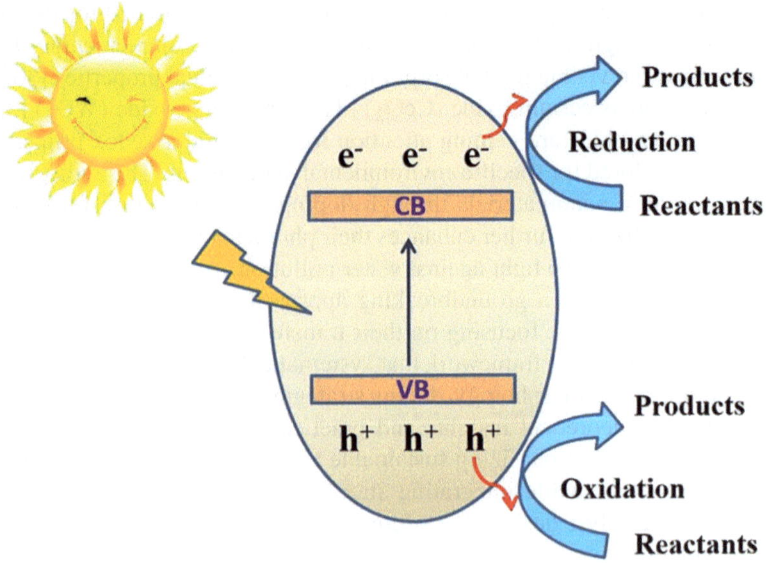

Fig. 1 The fundamental operation of photocatalysts. When photons with sufficient energy (≥ bandgap of the semiconducting material) hit the photocatalyst, electron–hole pairs drive chemical reactions. The figure has been reproduced with the permission from He and Zhang [35]. Copyright 2019 by Springer Nature

(CB) to initiate critical redox reactions. These reactions are vital for degrading pollutants, breaking down organic contaminants, and purifying water during wastewater treatment [34]. Upon absorption of light, the photocatalyst produces reactive species including H_2O_2, hydroxyl radicals (·OH), and superoxide anions (O_2^-·) through complex interactions with water and oxygen molecules on its surface. These reactive species effectively oxidize or reduce pollutants, converting them into non-toxic by-products like water and CO_2. The efficiency of photocatalysis is influenced by the photocatalyst, light source, and characteristics of the pollutants being addressed. Understanding the intricacies of these mechanisms is critical for optimizing nanophotocatalysts for real-world applications. Figure 1 illustrates the basic mechanism of photocatalyst [35].

2.2 *Principles of Light Absorption and Electron Excitation*

The process begins with absorption of light and excitation of electron, which are crucial initial steps [36]. When photons with energy ≥ the bandgap of a nanophotocatalyst strike its surface, electrons in the VB gain sufficient energy to be excited to CB, leaving positively charged holes (h^+) behind. This separation of e^-–h^+ pairs is essential, as these charge carriers drive the redox reactions responsible for breaking down pollutants. The bandgap of a photocatalyst determines the wavelengths of the light capable of triggering electron excitation. For instance, TiO_2, with a bandgap of approximately 3.2 eV, performs efficiently under UV light but has limited activity in the visible light spectrum.

Recent innovations in photocatalyst design have focused on modifying the bandgap. Techniques such as doping with elements like nitrogen, carbon, iron, or copper, and creating heterostructures, have been developed to improve visible light absorption, thereby expanding the applicability of these materials. These modifications are critical in improving the overall photocatalytic efficiency, especially for outdoor applications that rely on solar radiation as the primary source of light [37]. Moreover, the nanoscale dimensions of the photocatalyst allow more photons to interact with the material, further enhancing absorption of light and generation of charge carrier [38].

2.3 *Photocatalytic Redox Reactions at the Nanoscale*

At the nanoscale, photocatalytic efficiency is significantly enhanced by the unique properties of nanomaterials. Electrons in the CB and holes (h^+) in the VB drive reduction and oxidation processes, respectively. In the presence of water, h^+ oxidizes water molecules to produce ·OH, highly reactive species capable of breaking down organic pollutants through oxidation. Meanwhile, the electron reduces molecular oxygen to generate $O_2^-\cdot$, which further contribute to degradation of contaminants [39]. Nanophotocatalysts offer numerous active sites for redox reactions, leading to faster pollutant degradation compared to bulk photocatalysts. Their nanoscale dimensions shorten the distance of charge carriers to reach the surface, reducing electron–hole recombination—a common factor that lowers photocatalytic efficiency. Various nanostructures such as quantum dots, nanotubes, and nanorods have also been developed to ensure sustained photocatalytic activity and enhanced effective degradation [40].

2.4 *Key Parameters Influencing Photocatalytic Efficiency*

The efficiency of a nanophotocatalyst in environmental applications, particularly wastewater purification, is influenced by several key parameters.

2.4.1 Bandgap Energy

As mentioned earlier, the bandgap energy determines capacity of a photocatalyst for absorption of light across various regions of the electromagnetic spectrum. For optimal performance, an ideal nanophotocatalyst should possess a bandgap that facilitates effective absorption of visible light while retaining high photocatalytic efficiency.

2.4.2 Charge Carrier Dynamics

Effective separation and transport of e^{-}–h^{+} pairs are crucial to achieve high photocatalytic efficiency. To minimize e^{-}–h^{+} recombination and improve performance, techniques like introducing heterojunctions or applying surface modifications have worked incredibly well.

2.4.3 Surface Area and Morphology

The interactions between the photocatalyst and pollutants are improved by a larger surface area boosting its effectiveness. Morphological designs, such as nanowires, nanotubes, and nanosheets, further optimize performance by providing additional active sites and enabling more efficient mass transfer.

2.4.4 Catalyst Stability

To ensure effectiveness in wastewater treatment, a photocatalyst must exhibit stability under extended exposure to light and reactive species. Enhancements such as protective coatings and doping with stabilizing agents are commonly employed for nanophotocatalysts to enhance their durability and recyclability over time.

2.4.5 Reactivity with Pollutants

The affinity of the photocatalyst for different pollutants, including heavy metals, organic compounds, and pathogens, is another critical factor. Some nanophotocatalysts are specifically tailored to target certain contaminants, while others are designed to have broad-spectrum activity.

By optimizing these parameters, researchers are continuously advancing nanophotocatalysts to offer higher efficiency and selectivity in wastewater treatment. The combination of material design, nanoscale engineering, and the strategic modification of photocatalytic properties is paving the way for next-generation solutions to environmental challenges.

Table 1 provides a comparative overview of essential parameters that impact the performance of nanophotocatalysts, including bandgap energy, charge carrier dynamics, surface area and morphology, catalyst stability, and reactivity with pollutants.

Table 1 Key parameters influencing the photocatalytic efficiency of nanophotocatalysts

Parameter	Description	Strategies/ Modifications	Benefits	Challenges
Bandgap energy	Determines light absorption capabilities; influences which part of the light spectrum the photocatalyst can utilize	Doping with metals (e.g., Fe, Cu) and nonmetals (e.g., N, S); heterostructure formation	Enhances visible light absorption and photocatalytic activity under sunlight	Narrowed bandgaps increase electron–hole recombination, potentially reducing efficiency
Charge carrier dynamics	Essential for maintaining high photocatalytic activity by preventing electron–hole recombination during redox reactions	Heterojunctions, doping, and surface modifications	Improves charge separation, increasing efficiency of redox reactions	Managing charge carrier recombination remains challenging and may impact overall activity
Surface area and morphology	A higher surface area provides more interaction sites with pollutants; morphology can affect adsorption and mass transfer efficiency	Nanostructures (e.g., nanowires, nanotubes, nanosheets)	Increases pollutant interaction due to larger surface area, enhancing degradation rates	Larger surface areas may lead to faster recombination; increased complexity in synthesis
Catalyst stability	The durability and resilience of nanophotocatalysts, especially under prolonged light and reactive conditions, are vital for long-term applications	Protective coatings, dopants for stability, core–shell structures	Ensures sustained performance over time, reduces degradation risks	Some materials like ZnO are prone to photocorrosion, limiting practical applications
Reactivity with pollutants	Different nanophotocatalysts exhibit varying affinities for organic pollutants, heavy metals, and pathogens, affecting performance in specific applications	Surface functionalization, tailored materials targeting specific contaminants	Allows customized applications for pollutants in wastewater, increases process selectivity	Broad-spectrum reactivity requires balancing specificity and efficiency

3 Classification of Nanophotocatalysts

3.1 Based on Composition

Nanophotocatalysts are typically classified based on their composition, as material properties are key determinants of their photocatalytic efficiency. Due to their capacity to catalyze processes when exposed to light, metal oxide and nonmetal oxide nanophotocatalysts have both been extensively investigated for wastewater treatment, allowing for the efficient degradation of pollutants.

3.1.1 Metal Oxide Nanophotocatalysts

Metal oxide nanophotocatalysts are extensively studied for photocatalytic applications and are valued for their excellent stability, chemical resilience, and robust redox properties. Their capacity to absorb UV or visible light enables efficient degradation of pollutants, making them highly effective in environmental remediation.

TiO_2

TiO_2 is the most widely researched metal oxide nanophotocatalyst, renowned for its outstanding photocatalytic performance, non-toxic nature, and chemical stability. It is highly effective in breaking down organic pollutants and eliminating harmful contaminants from water, making it an ideal choice for wastewater treatment. However, its wide bandgap ($\sim$ 3.2 eV) restricts its efficiency under visible light. To address this, recent advancements have focused on modification of TiO_2 through doping or coupling with other materials to enhance its visible light activity [40].

ZnO

ZnO is a recognized nanophotocatalyst for its strong photocatalytic activity and ability to absorb a wider range of light, including UV and portions of the visible spectrum. Nanostructured ZnO nanostructures, like nanowires and nanorods, offer unique surface characteristics that enhance their effectiveness in photocatalytic applications. Inherent antibacterial properties of ZnO make it particularly valuable for water disinfection and pathogen control. However, its susceptibility to photocorrosion under prolonged light exposure remains a challenge. To overcome this, researchers are exploring surface modifications and the development of hybrid structures to improve its stability and durability [41].

Other Metal Oxides

CeO_2, Fe_2O_3, and WO_3 have emerged as promising nanophotocatalysts for environmental remediation. CeO_2, for example, is known for its ability to store and release oxygen, which enhances its catalytic performance. WO_3 has a relatively small bandgap ($\sim$ 2.7 eV) compared to TiO_2, allowing for better visible light absorption. Fe_2O_3, with its low cost and abundance, has also been studied for its potential in large-scale applications, although it faces challenges with charge carrier recombination. To overcome these limitations, researchers are investigating various doping strategies and heterostructure formations.

3.1.2 Nonmetal Oxide Nanophotocatalysts

While metal oxides dominate the field, nonmetal oxide nanophotocatalysts are gaining interest for photocatalysis under visible light. These materials provide alternative approaches for enhancing photocatalytic performance.

Graphene-Based Nanophotocatalysts

Graphene has shown remarkable potential in enhancing photocatalytic activity when used in conjunction with other materials. Graphene-based nanophotocatalysts have superior electron mobility, which reduces electron–hole recombination. Large surface area of graphene provides more active sites to facilitate photocatalytic reactions, enhancing the performance of other photocatalysts like TiO_2 [42] or ZnO [43].

Carbon Nitride

Carbon nitride (g-C_3N_4) is promising for its ability to absorb visible light due to its narrow bandgap ($\sim$ 2.7 eV). g-C_3N_4 offers a metal-free alternative for photocatalysis, making it highly appealing for green chemistry and environmental sustainability. Moreover, g-C_3N_4 is known for its high stability and ease of synthesis, and through doing and formation of composite, the efficiency can be further enhanced [44].

3.2 Based on Morphology

Nanostructured photocatalysts can be classified based on their morphology, in other words, dimensionality, which affects how they interact with light and pollutants.

3.2.1 Zero Dimensional

Zero-dimensional nanophotocatalysts, such as quantum dots and nanoparticles, have unique optical and electronic properties. Quantum dots, for instance, exhibit size-dependent bandgap tuning, allowing for precise control over their light absorption properties. This makes zero-dimensional materials particularly effective for applications requiring enhancement of light absorption and separation of charge carriers.

3.2.2 One Dimensional

Nanotubes and nanorods offer improved charge transport pathways compared to zero-dimensional materials. Nanotubes, for example, provide high electron mobility along their length, which reduces recombination rates and enhances photocatalytic efficiency. Similar to this, nanorods have a high surface area-to-volume ratio, which makes them suited to the adsorption and degradation of pollutants.

3.2.3 Two Dimensional

Two-dimensional nanophotocatalysts, including nanosheets and nanoflakes, are particularly effective in layered configurations, where e^-–h^+ pairs are better separated to enhance photocatalytic activity. 2D materials such as graphene and g-C_3N_4 nanosheets are highly tunable, allowing for the optimization of their electronic and surface properties.

3.2.4 Three Dimensional

Three-dimensional nanostructures, such as nanocubes and nanoflowers, are characterized by their complex geometries, which offer multiple active sites and enhanced light scattering. Nanocubes and nanoflowers are particularly useful in hybrid photocatalytic systems, where their complex shapes enhance catalytic performance through improved mass transfer and surface reactivity.

3.3 *Based on Doping*

Doping refers to the deliberate introduction of foreign atoms or molecules into a photocatalyst to modify its electronic and optical properties. Doping can narrow down the bandgap, improve charge separation, or introduce new active sites.

3.3.1 Metal-Doped Nanophotocatalysts

Metal doping involves the incorporation of metal ions (e.g., Fe, Cu, Ag) into lattice of nanophotocatalyst to improve its light absorption properties. Metal doping can also introduce localized electronic states that facilitate charge carrier separation, increasing the overall photocatalytic efficiency. For example, when exposed to visible light, Fe-doped TiO_2 has demonstrated higher photocatalytic activity compared to undoped TiO_2 [45].

3.3.2 Nonmetal-Doped Nanophotocatalysts

Nonmetal doping involves the incorporation of elements like nitrogen, sulfur, or carbon to enhance photocatalytic performance. Nonmetal dopants are often used to narrow the bandgap of wide-bandgap materials, enabling them to absorb visible light more efficiently. Nitrogen-doped g-C_3N_4, for instance, exhibit improved photocatalytic activity under visible light due to improved charge carrier dynamics and bandgap narrowing [45].

3.3.3 Co-doped Systems for Enhanced Activity

Co-doping, the simultaneous introduction of both metal and nonmetal dopants, has been shown to further enhance the photocatalytic properties of nanophotocatalysts. Co-doped systems can achieve better absorption of light, higher reactivity, and increased separation of charge carrier and are promising for visible light-driven photocatalysis in complex environmental applications.

3.4 Based on Formation of Heterostructure

Heterostructures combine two or more semiconductor materials to form interfaces that promote charge separation and reduce recombination losses, improving photocatalytic performance.

3.4.1 Type I, II, and Z-Scheme Heterostructures

Type I and II heterostructures refer to band alignment configurations that influence the separation of e^-–h^+ pairs [46]. In Type II heterostructures, the CB of one material aligns with the VB of another, facilitating charge separation across the interface. Z-scheme heterostructures mimic natural photosynthesis by promoting the transfer of electrons between two materials, allowing for stronger redox reactions and higher photocatalytic efficiency.

3.4.2 Core–Shell Structures

Core–shell nanostructures consist of a core material encased in a shell layer, which can provide improved stability, charge separation, or light absorption. These are designed to exploit advantageous features of different materials, such as enhancing light absorption with the core while protecting against photocorrosion with the shell [47].

3.4.3 Hybrid Systems with Organic Components

Hybrid photocatalysts combine inorganic nanomaterials with organic components, such as polymers or organic semiconductors, to enhance photocatalytic performance. These systems offer improved flexibility, light absorption, and reactivity. Organic–inorganic hybrid nanophotocatalysts are particularly useful in flexible or multi-functional applications, such as photocatalytic membranes for water purification [48].

4 Synthesis Techniques for Nanophotocatalysts

The synthesis of nanophotocatalysts helps determining their structural, morphological, and functional properties. Every technique offers distinct benefits, and the selection of a method is contingent upon the intended photocatalytic use. The following are the key synthesis techniques used to fabricate nanophotocatalysts for wastewater purification.

4.1 Sol–Gel Synthesis

Sol–gel synthesis is widely used for producing nanomaterials, particularly metal oxide-based nanophotocatalysts [49]. The method involves the transition of a solution (sol) into a solid (gel) through hydrolysis and condensation of metal alkoxides or metal salts. The method allows precise control over composition, particle size, and morphology of the resulting nanomaterial. This is particularly suitable for producing highly homogeneous nanophotocatalysts with controlled porosity and surface area, which are essential for efficient photocatalysis. In the context of wastewater purification, nanophotocatalysts synthesized via sol–gel methods, such as TiO_2 [50] and ZnO [51], exhibit enhanced photocatalytic activity due to their well-controlled nanostructures. By modifying pH, temperature, and precursor concentration, photocatalytic properties of the synthesized materials can be fine-tuned. It has the ability to incorporate various dopants or form composite materials, such as TiO_2 with graphene or carbon nanotubes. Figure 2 illustrates the Pechini-type sol–gel synthesis method used for preparing Ni(II)-doped $LiMnBO_3/Li_2MnO_3/Li_2B_2O_4$ nanocomposites for use as nanophotocatalysts [52]. This process involves mixing metal ion precursors with a chelating agent, such as EDTA, and a solvent like ethylene glycol to form a polymeric complex. Heating the solution promotes gel formation, followed by calcination, leading to a stable nanocomposite structure. The synthesized nanocomposite is then used for photocatalytic applications, specifically under UV–vis irradiation, to degrade pollutants like dyes in wastewater, benefiting from enhanced stability and photocatalytic efficiency due to the Ni(II) doping.

4.2 Hydrothermal/Solvothermal Methods

Hydrothermal and solvothermal methods involve crystallization of nanomaterials from aqueous or non-aqueous solutions under high temperature and pressure conditions. These techniques allow for the growth of nanophotocatalysts with well-defined crystalline structures and unique morphologies, such as nanorods, nanowires, and nanotubes, which are essential for enhancing photocatalytic performance. Hydrothermal synthesis, conducted in an autoclave, uses water as the solvent and is commonly employed to produce metal oxide nanophotocatalysts like TiO_2 [53], ZnO [54], and Fe_2O_3 [55]. Solvothermal synthesis, on the other hand, uses organic solvents and allows for greater flexibility in the formation of nanocomposites or hybrid materials. By adjusting the solvent composition, temperature, and reaction time, the properties of the nanophotocatalysts can be tailored. Figure 3 illustrates the hydrothermal synthesis process of TiO_2 nanorods, highlighting the formation chemistry, growth mechanisms, and adjustments to surface properties to optimize photocatalytic activity. Two primary hydrothermal approaches have been highlighted—alkaline and acidic conditions—where different solvents (NaOH or HCl) influence the formation of unique TiO_2 morphologies. Key synthesis factors like temperature, time, and pH are shown to directly impact morphology, crystallinity, and specific surface area. This method enables precise tailoring of TiO_2 NRs for enhanced photocatalytic degradation of pollutants and microbial disinfection applications.

4.3 Chemical Vapor Deposition

Chemical vapor deposition (CVD) is an advanced technique to deposit nanophotocatalysts as thin films onto substrates by utilizing vapor-phase precursors. In this process, metal precursors are vaporized and delivered to a reaction chamber to undergo chemical reaction to form a solid material that adheres to the substrate. This method is highly effective for producing uniform coatings of TiO_2 and ZnO [57], on various surfaces. CVD provides precise control over the thickness, composition, and morphology of the deposited nanomaterials, making it a versatile approach for enhancing photocatalytic performance. By carefully adjusting temperature, pressure,

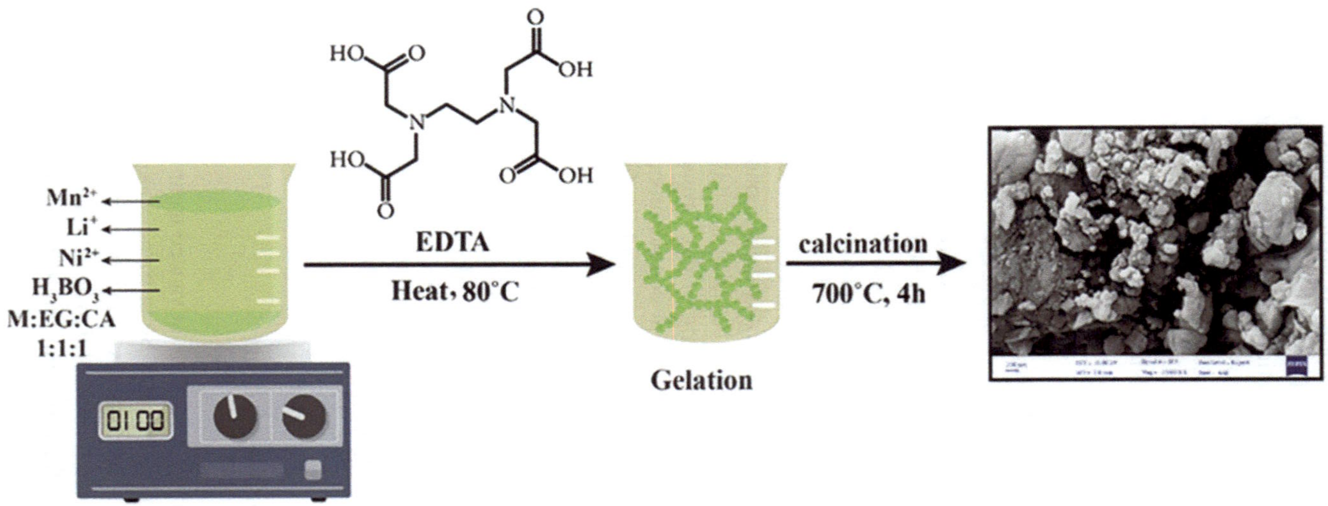

Fig. 2 Pechini-type sol–gel synthesis of Ni(II)-doped $LiMnBO_3/Li_2MnO_3/Li_2B_2O_4$ nanocomposites, illustrating the stepwise process of combining metal ions, chelating agents, and solvents to form a polymeric complex. Calcination of the gel produces a stable, photocatalytically active nanocomposite used for pollutant degradation under UV–vis irradiation. The figure has been reproduced with the permission from Ranjeh et al. [52]. Copyright 2021 by Elsevier

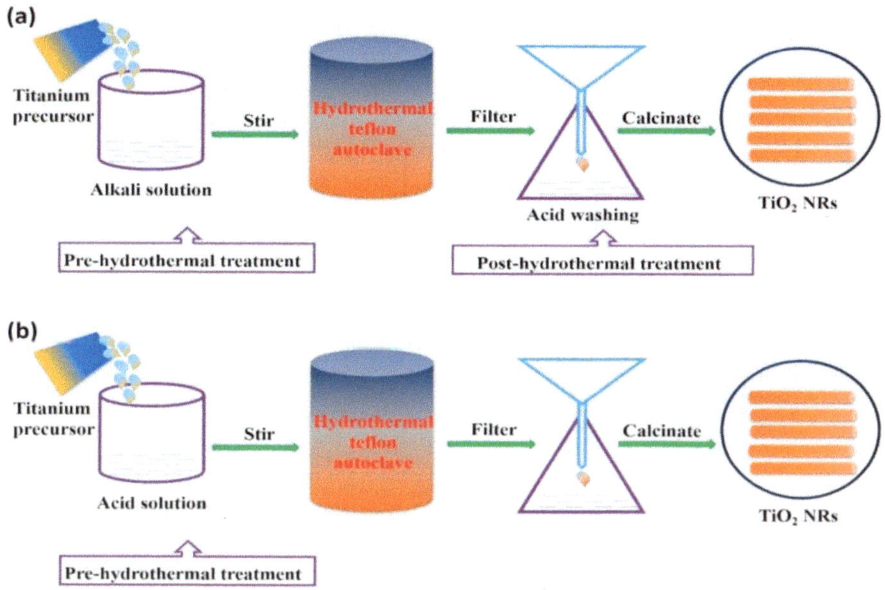

Fig. 3 Hydrothermal synthesis of TiO_2 nanorods, depicting the formation process under alkaline and acidic conditions. The figure has been reproduced with the permission from Gupta et al. [56]. Copyright 2021 by Elsevier

and gas flow rates, the process can produce nanostructures with high surface areas and improved light absorption characteristics, which are essential for maximizing photocatalytic efficiency in applications like wastewater treatment. In addition to metal oxides, CVD can be employed to fabricate graphene-based nanophotocatalysts, which exhibit excellent electron mobility and charge separation properties. These graphene-based materials can be used as supports or active components in hybrid photocatalysts for improved performance.

Figure 4 demonstrates the process of fabricating nanophotocatalyst by CVD method [58]. The figure outlines the steps involved in pretreating the substrate with different alkali metal cations (Li^+, Na^+, K^+) using a spraying method, followed by titanium isopropoxide vapor deposition at 600 °C. The resulting titania/titanate thin films are subsequently used for photocatalytic dye degradation under UV irradiation. The photocatalytic efficiency is shown to vary depending on the type of alkali metal cation, with Na^+ demonstrating the highest efficiency, followed by Li^+ and K^+. The illustration highlights the dye degradation process and the relative contribution of cations for enhancement of the photocatalytic performance.

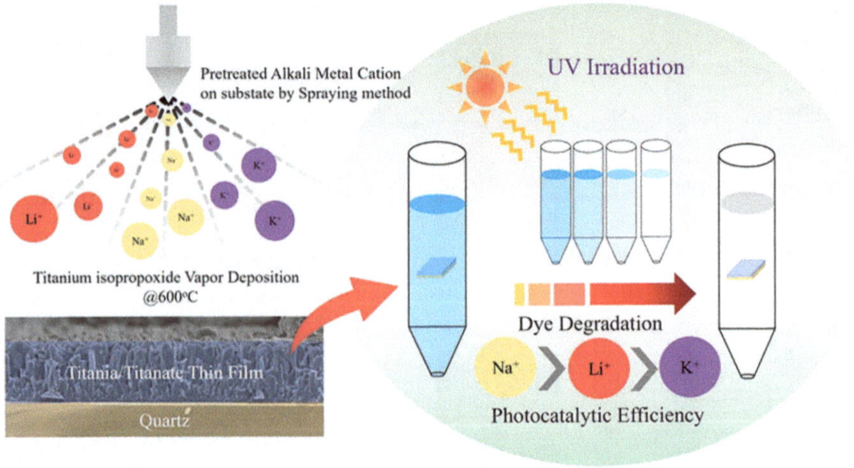

Fig. 4 Schematic representation of the ecofriendly alkali metal cation diffusion process for fabricating mixed-phase titania/titanate thin films via CVD. The process involves pretreatment of the substrate with alkali metal cations (Li^+, Na^+, K^+) followed by titanium isopropoxide vapor deposition at 600 °C. The resulting thin films are used to photocatalytically degrade azo dyes under UV irradiation, with Na^+ pretreated films demonstrating the highest photocatalytic efficiency. The figure has been reproduced with the permission from Chankhunthod et al. [58]. Copyright 2023 by Elsevier

4.4 Electrochemical Methods

Electrochemical methods are highly effective for synthesizing nanophotocatalysts, particularly for creating nanostructured metal oxides. These methods involve applying an electric current to drive deposition of films of metal oxide or the formation of nanostructures on electrodes. Electrochemical deposition is often used to produce nanophotocatalysts with high surface areas and enhanced photocatalytic properties for environmental applications. One common electrochemical method is anodic oxidation, which is used to create highly ordered TiO_2 nanotube arrays [59]. These nanotubes exhibit enhanced photocatalytic activity. TiO_2 nanotube arrays synthesized via electrochemical methods have been used for water splitting, pollutant degradation, and microbial disinfection in wastewater treatment. Electrodeposition is another electrochemical technique used to fabricate thin films of nanophotocatalysts, such as ZnO [60], on various substrates with controlled thickness and morphology of the deposited material.

Figure 5 illustrates schematic of the electrochemical preparation process of ZnO nanoparticles containing nanophotocatalyst [61]. It involves an undivided two-electrode cell setup, where a Zn metal foil serves as the sacrificial anode, and a Pt foil acts as the cathode. The electrolysis process takes place in a solution containing tetrabutylammonium bromide (TBAB) as the stabilizer in various solvent media such as pure ethanol, water, or a mixture of water and ethanol. When a constant current is

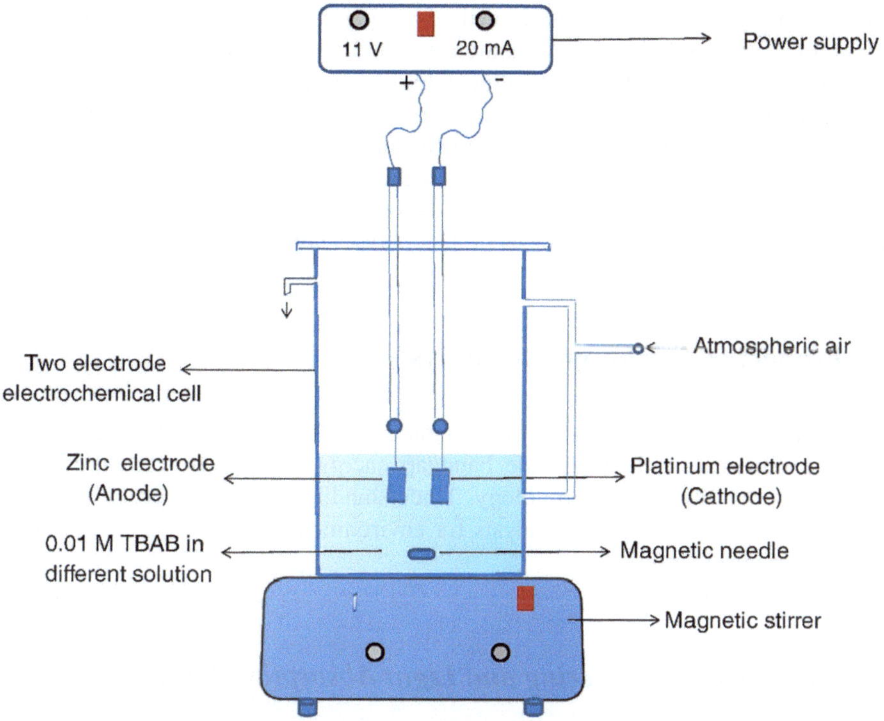

Fig. 5 Electrochemical synthesis of ZnO nanoparticles using a zinc anode and platinum cathode in a TBAB-stabilized solution. The process, conducted in varying solvent media (ethanol, water, or a mixture), influences the morphology of the resulting ZnO nanoclusters. The figure has been reproduced with the permission from Dhayagude et al. [61]. Copyright 2017 by Elsevier

applied, Zn^{2+} ions are generated at the anode, which then combine with oxygen from the atmosphere to form ZnO nanoclusters stabilized by TBAB. The morphology and properties of the resulting ZnO nanoparticles vary depending on the solvent medium used during the electrochemical synthesis.

4.5 Green Synthesis Approaches

Green synthesis approaches have gained significant attention in recent years as environmentally friendly alternatives to conventional synthesis methods. These approaches focus on the use of natural, non-toxic, and renewable resources to synthesize nanophotocatalysts, reducing the environmental impact associated with chemical synthesis [62]. One of the most promising green synthesis methods is the use of plant extracts or biological agents, such as bacteria [63], fungi, and algae, to synthesize metal oxide nanophotocatalysts [64, 65]. These biological systems serve as reducing and capping agents. Green synthesis methods are particularly beneficial for producing nanophotocatalysts that are biocompatible and environmentally benign, making them ideal for sustainable wastewater treatment. Furthermore, green-synthesized nanomaterials often exhibit enhanced surface properties and photocatalytic performance due to the presence of natural organic capping agents.

Table 2 provides a clear comparison of the different synthesis methods, highlighting their unique advantages and applications, making it easier to understand the various techniques used in synthesis of nanophotocatalysts.

5 Properties of Nanophotocatalysts

The photocatalytic performance of nanophotocatalysts is heavily influenced by their intrinsic properties, which include bandgap energy, surface area, charge carrier dynamics, stability, and morphology. Understanding these properties is essential for optimizing nanophotocatalysts for environmental applications, particularly wastewater purification.

5.1 Bandgap Engineering and Light Absorption Properties

Bandgap engineering is a pivotal factor to optimize light absorption properties of nanophotocatalysts [66]. Photocatalysis requires the absorption of photons with energy equal to or greater than the bandgap to excite electrons from the VB to the CB, generating electron–hole pairs that drive redox reactions.

Traditional metal oxides like TiO_2 have relatively wide bandgaps ($\sim$ 3.2 eV), limiting their absorption to UV light, which represents a small portion of the solar spectrum. Various bandgap engineering approaches have been developed. These include doping with metals (e.g., Fe, Cu) or nonmetals (e.g., N, S), forming composites, and designing heterostructures to narrow the bandgap and extend light absorption into the visible spectrum [67]. For instance, nitrogen-doped TiO_2 improves visible light absorption by introducing mid-gap states [68].

However, bandgap engineering must also address the trade-off between enhanced absorption of light and dynamics of charge carrier. While a narrower bandgap increases absorption of light, it may also raise the risk of electron–hole recombination, reducing photocatalytic efficiency. Therefore, optimizing the balance between light absorption and effective charge carrier separation is essential for achieving high photocatalytic performance [69].

Table 2 Comparative analysis of various synthesis techniques for nanophotocatalysts, detailing their description, advantages, commonly synthesized materials, and key photocatalytic applications in wastewater treatment

Synthesis technique	Description	Advantages	Materials typically synthesized	Photocatalytic applications
Sol–gel synthesis	Transition from a solution (sol) to a solid (gel) through hydrolysis and condensation of metal alkoxides or salts	Precise control over composition, particle size, and morphology. Enables incorporation of dopants	Metal oxide nanophotocatalysts like TiO_2, ZnO can form composites such as TiO_2/graphene or TiO_2/carbon nanotubes to improve visible light absorption	Enhanced photocatalytic activity in wastewater purification by degrading organic pollutants and removing heavy metals
Hydrothermal/solvothermal methods	Crystallization of nanomaterials from aqueous or non-aqueous solutions under high temperature and pressure in an autoclave	Produces high-purity, defect-free nanostructures with unique morphologies like nanorods, nanowires, and nanotubes	ZnO, TiO_2, Fe_2O_3. Solvothermal method allows flexibility for forming nanocomposites or hybrid materials	Effective for organic pollutant degradation and heavy metal reduction in wastewater treatment due to enhanced crystallinity and photocatalytic efficiency
Chemical vapor deposition (CVD)	Deposition of thin films of nanophotocatalysts on substrates via vapor-phase precursors	Offers precise control over thickness, composition, and morphology of deposited nanomaterials	Metal oxides like TiO_2, ZnO, and graphene-based nanophotocatalysts	Used for uniform coating on substrates, enhancing light absorption and charge separation in water splitting and pollutant degradation applications

(continued)

Table 2 (continued)

Synthesis technique	Description	Advantages	Materials typically synthesized	Photocatalytic applications
Electrochemical methods	Application of electric current to deposit metal oxide films or form nanostructures on electrodes	Produces nanostructures with high surface areas, enhancing light harvesting and charge transport	TiO_2 nanotube arrays, ZnO thin films. Anodic oxidation is used for ordered TiO_2 nanotube arrays, while electrodeposition allows control over thickness	Widely used in water splitting, pollutant degradation, and microbial disinfection in wastewater treatment
Green synthesis approaches	Synthesis using natural, renewable, and non-toxic resources such as plant extracts, bacteria, fungi, or algae	Environmentally friendly, avoids toxic chemicals, and produces biocompatible nanomaterials	TiO_2 and ZnO nanophotocatalysts synthesized using biological systems. Nanomaterials exhibit enhanced surface properties due to natural organic capping agents	Ideal for sustainable wastewater treatment with applications in organic pollutant degradation and heavy metal removal

5.2 Surface Area, Porosity, and Charge Carrier Dynamics

The surface area and porosity of nanophotocatalysts significantly influence their photocatalytic efficiency [70]. Nanophotocatalysts with their intrinsically high surface area provide more active sites for photocatalytic reactions, which enhances the interaction between the catalyst and the pollutants in wastewater. Porous nanomaterials, in particular, facilitate the diffusion of reactants and products due to high density of active site, leading to improved catalytic activity [71].

For instance, nanostructured TiO_2 with mesoporous architecture exhibits superior photocatalytic activity [72]. The morphology of the nanostructure, such as the presence of nanowires, nanorods, or nanosheets, plays a critical role in enhancing surface reactivity by exposing more reactive facets of the material [73].

Charge carrier dynamics are another vital aspect that determines the efficiency of photocatalytic reactions. Nanophotocatalysts with optimized charge carrier dynamics, such as reduced electron–hole recombination, exhibit higher photocatalytic efficiency. Strategies to improve charge carrier dynamics include doping, heterostructure formation, and surface modifications that promote faster electron transport and reduce recombination losses [74].

5.3 Stability and Recyclability in Photocatalytic Processes

Stability and recyclability are crucial properties of nanophotocatalysts for their practical use in long-term wastewater treatment applications. Photocatalytic materials must be able to withstand prolonged exposure to light, reactive oxygen species (ROS), and aqueous environments without degrading or losing activity. For instance, while TiO_2 is known for its stability, materials like ZnO are prone to photocorrosion under extended irradiation, which can limit their long-term application [75].

Surface coating, doping, and forming core–shell structures have been employed to enhance the stability of nanophotocatalysts. For example, coating ZnO with a thin protective layer of another metal oxide can prevent photocorrosion while maintaining high photocatalytic activity [76]. Furthermore, metal oxide nanophotocatalysts are often engineered to be reusable, maintaining their photocatalytic efficiency across several cycles without experiencing appreciable loss of activity loss. The recyclability is critical for reducing operational costs in large-scale applications [77].

Another key factor in recyclability is the ease of recovering the nanophotocatalyst from the treated water. Nanophotocatalysts that can be easily separated and recovered, such as magnetic nanocomposites, offer an advantage in industrial wastewater treatment, where the recovery of catalysts after use is essential [78].

5.4 Photocatalyst Size and Shape Effects on Performance

The size and shape of nanophotocatalysts have a profound effect on the photocatalytic performance. For example, quantum dots exhibit size-dependent bandgap tuning, allowing precise control over their light absorption properties and photocatalytic behavior [79].

Nanowires and nanotubes offer enhanced charge transport along their length, minimizing electron–hole recombination and increasing photocatalytic efficiency. These structures also provide a larger surface area for pollutant adsorption, further enhancing catalytic activity [80].

The shape of nanophotocatalysts can also influence their performance by determining the exposure of specific crystal facets that are more catalytically active. For example, TiO_2 nanocrystals with exposed (001) facets exhibit higher photocatalytic activity than those with exposed (001) facets due to the higher surface energy of the (001) facets, which promotes redox reactions [81]. Similarly, hierarchical three-dimensional nanostructures, such as nanoflowers or nanocubes, provide multiple active sites and enhance light trapping, resulting in improved photocatalytic performance [82].

Table 3 provides an overview of the critical properties influencing efficiency of nanophotocatalysts in environmental applications. It highlights how factors such as bandgap manipulation, surface morphology, stability, and nanostructure dimensions enhance photocatalytic activity, while addressing challenges associated with each property.

6 Applications of Nanophotocatalysts in Wastewater Treatment

Nanophotocatalysts have become increasingly significant in addressing environmental challenges, particularly in the purification of contaminated water. Their ability to degrade harmful organic pollutants, remove toxic heavy metals, disinfect microbial pathogens, and adapt to various wastewater treatment processes has positioned them as a transformative technology in environmental remediation.

6.1 Degradation of Organic Pollutants

Industrial dyes and pharmaceuticals represent a significant source of environmental contamination, as many are toxic, persistent, and challenging to break down. Nanophotocatalysts, especially those based on metal oxides like TiO_2 and ZnO, have demonstrated a high capability for degrading such pollutants through photocatalytic processes. When exposed to light, nanophotocatalysts generate ROS, such as ·OH radicals and $O_2^{-}\cdot$ anions, which are capable of attacking and breaking down the complex molecular structures of organic compounds.

For example, TiO_2 degrades various dyes such as methylene blue [83] and rhodamine B [84], where it facilitates the complete mineralization of the pollutants into CO_2 and H_2O. Likewise, ZnO has proven effective in degrading pharmaceuticals such as antibiotics [85], offering an advanced approach for treating pharmaceutical wastewater. Importantly, the incorporation of dopants or the formation of hybrid materials significantly enhances the performance of nanophotocatalysts in visible light, further improving their practical application in real-world conditions.

6.2 Removal of Heavy Metals

Traditional methods for removal of heavy metals from wastewater, including precipitation and ion exchange, often face limitations related to inefficiency and high cost. Nanophotocatalysts provide an innovative and more efficient solution for heavy metal removal by utilizing their photocatalytic properties to reduce and immobilize toxic metal ions.

Table 3 Comparative analysis of key properties of nanophotocatalysts, detailing their impact on photocatalytic performance, including bandgap engineering, surface area and charge carrier dynamics, stability and recyclability, and the effects of size and shape on performance

Property	Description	Key factors	Advantages	Challenges	Examples of materials
Bandgap engineering and light absorption	Refers to adjusting the bandgap of nanophotocatalysts to improve light absorption across different wavelengths, especially visible light	Doping with metals/nonmetals, forming heterostructures, bandgap narrowing	Enhanced light absorption in the visible spectrum. Improved photocatalytic efficiency under sunlight	Narrower bandgaps can increase electron–hole recombination, which reduces photocatalytic efficiency	TiO_2 doped with nitrogen or iron for visible light absorption. g-C_3N_4 with bandgap engineering
Surface area, porosity, and charge carrier dynamics	Larger surface area and higher porosity improve photocatalytic reactions by providing more active sites for pollutant degradation and facilitating mass transfer	Nanostructures like nanotubes, nanowires, nanosheets. Morphological control for high surface area	Enhanced pollutant interaction due to higher surface area. Nanostructured materials with mesoporous architecture show improved photocatalytic performance	Increased surface area can lead to faster degradation, but also increases the potential for charge recombination, reducing overall efficiency	Mesoporous TiO_2, ZnO nanowires, g-C_3N_4 nanosheets

(continued)

Table 3 (continued)

Property	Description	Key factors	Advantages	Challenges	Examples of materials
Stability and recyclability	Stability refers to the ability of the nanophotocatalyst to maintain its structure and efficiency over time. Recyclability ensures that it can be reused in multiple cycles	Surface modifications (doping, coatings) to enhance stability, core–shell structures for durability	Longer operational life of the nanophotocatalyst. Reduced operational costs due to recyclability and sustained performance over multiple cycles	Materials like ZnO can suffer from photocorrosion over time, reducing long-term stability. Recovery of nanophotocatalysts from treated water may be challenging	TiO_2 is known for its high stability, while ZnO requires surface modification to prevent degradation
Photocatalyst size and shape effects	The size and shape of nanophotocatalysts affect their surface-to-volume ratio, light absorption, and charge carrier dynamics, which influence overall performance	Zero-dimensional quantum dots, one-dimensional nanorods/nanowires, two-dimensional nanosheets, three-dimensional structures	Quantum confinement effects in quantum dots allow for precise control over photocatalytic properties. Shapes like nanorods and nanosheets improve light trapping and active site exposure	Size and shape manipulation must strike a balance between enhanced light absorption and avoiding rapid recombination of electron–hole pairs. Complex shapes may complicate fabrication processes	Quantum dots, TiO_2 nanorods, ZnO nanosheets, three-dimensional nanoflowers

For instance, photocatalysts like TiO_2 and Fe_2O_3 can facilitate the reduction of Cr(VI) to its less toxic Cr(III) through photocatalytic redox reactions [86, 87]. Additionally, ZnO nanostructures have been shown to adsorb and remove metal ions such as Pb^{2+} and Cd^{2+}, with high efficiency [88]. By optimizing the surface properties and charge carrier dynamics of these nanophotocatalysts, it is possible to enhance the removal rates of heavy metals from the system.

6.3 Photocatalytic Disinfection

Waterborne pathogens, including viruses, bacteria, and fungi, are responsible for a significant number of diseases worldwide, making disinfection a crucial aspect of wastewater treatment. Nanophotocatalysts offer a powerful means of disinfecting contaminated water by generating ROS that disrupt microbial cell membranes, destroy genetic material, and ultimately lead to the inactivation of pathogens.

ZnO, TiO_2, and Ag-doped materials are widely studied for their antimicrobial properties. For instance, ZnO has been shown to be highly efficacious in killing *E. coli* [89] and other harmful bacteria through photocatalytic action, where ROS generated on the nanocatalyst surface attack the bacterial cell walls and internal components, causing the death of cells. Similarly, silver (Ag)-doped TiO_2 enhances the disinfection efficiency due to the synergistic effect of inherent antimicrobial activity of silver and photocatalytic properties of TiO_2 [90].

6.4 Comparison of Nanophotocatalysts in Various Treatment Processes

Different nanophotocatalysts offer distinct advantages depending on the specific application, pollutant type, and treatment process used in wastewater purification. A comparative evaluation of various nanophotocatalysts can highlight their performance in different contexts and help guide their selection for specific treatment systems. Table 4 compares various nanophotocatalysts, highlighting their advantages and challenges in different treatment processes, such as UV/visible light-driven photocatalysis, water disinfection, and pollutant degradation.

Table 4 Comparison of nanophotocatalysts in various treatment processes

Nanophotocatalyst	Advantages	Challenges	Ideal application scenarios
TiO_2	Highly efficient under UV light, excellent stability, strong redox capabilities, cost-effective	Limited absorption under visible light, needs doping for enhanced solar-driven performance	UV light-driven processes, degradation of organic pollutants, removal of heavy metals
ZnO	Superior visible light absorption, strong antimicrobial properties, effective for water disinfection	Prone to photocorrosion, lower long-term stability than TiO_2	Water disinfection, visible light-driven processes, broader light spectrum applications
Metal oxides (e.g., Fe_2O_3)	High stability, strong photocatalytic activity, particularly effective in heavy metal reduction	Charge carrier recombination limits efficiency, requires surface modification	Reduction of heavy metals like Cr(VI) to Cr(III), large-scale wastewater treatment
Nonmetal oxides (e.g., g-C_3N_4)	Excellent visible light activity, metal-free, environmentally friendly, stable	Lower efficiency in UV light, needs modifications for better charge separation	Solar-driven photocatalysis, organic pollutant degradation, sustainable wastewater treatment
Doped photocatalysts (e.g., N-doped TiO_2, Ag-doped ZnO)	Improved visible light absorption due to narrowed bandgap, enhanced photocatalytic efficiency, superior antimicrobial properties (Ag-doped)	Cost of doping elements, potential environmental toxicity of dopants like silver	Sunlight-driven processes, microbial disinfection, environments with mixed pollutants
Heterostructures (e.g., Z-scheme composites)	Enhanced charge separation, reduced electron–hole recombination, efficient for complex treatment processes	Complicated synthesis, requires precise control over material interfaces	Advanced photocatalytic systems, environments with multiple contaminants, high-stability applications

7 Challenges and Future Prospects

Nanophotocatalysts have shown significant promise in wastewater treatment, but several challenges remain. Unlocking the full potential of these materials in environmental restoration will require addressing these constraints while promoting research and commercialization.

7.1 Current Limitations of Nanophotocatalysts

Despite their impressive performance in laboratory settings, nanophotocatalysts face several challenges when transitioning to large-scale, real-world applications. One of the primary concerns is the narrow light absorption range of many photocatalysts. For example, TiO_2 has a wide bandgap, which restricts its absorption to the UV spectrum, representing only about 4% of sunlight. This limits its practical application under visible light for outdoor or solar-driven processes. Another significant issue is electron–hole recombination [91]. Photocatalytic reactions rely on separation of e^-–h^+ pairs, but rapid recombination of these charge carriers can significantly reduce efficiency, limiting the overall degradation or reduction potential of the material.

The stability and durability of some nanophotocatalysts also pose challenges, particularly under harsh environmental conditions or prolonged exposure to light. For instance, ZnO can suffer from photocorrosion under extended UV light exposure, which diminishes its long-term effectiveness. Additionally, some nanophotocatalysts may leach toxic metals or degrade into harmful by-products, raising environmental concerns. Another practical concern is the cost of large-scale synthesis and recovery. Many advanced synthesis methods, such as doping and heterostructure formation, are complex and expensive. Furthermore, recovering nanophotocatalysts after treatment from water can be challenging, making the process less feasible for continuous use in large-scale systems.

7.2 Strategies for Improving Efficiency and Selectivity

Various strategies have been proposed to enhance the efficiency and selectivity of nanophotocatalysts [92, 93]. One promising approach is bandgap engineering, which involves tuning the electronic properties of nanophotocatalysts. Doping with metals (e.g., Fe, Cu) or nonmetals (e.g., N, S) and forming heterostructures, such as Z-scheme photocatalysts, are effective strategies to reduce the bandgap and improve visible light response.

Improving charge carrier dynamics is another critical area of focus. By designing materials that enhance charge separation and suppress electron–hole recombination, researchers aim to increase the photocatalytic efficiency. For example, constructing core–shell nanostructures or using plasmonic NPs can improve charge transfer and reduce recombination rates, leading to higher overall activity [94].

Stability enhancements are essential for ensuring that nanophotocatalysts remain effective over extended periods. Surface modifications, such as coating nanophotocatalysts with protective layers (e.g., SiO_2), can prevent photocorrosion while maintaining high photocatalytic activity. Additionally, developing recyclable photocatalysts, such as magnetic nanocomposites, can facilitate easier recovery and reuse after wastewater treatment, reducing costs and enhancing sustainability.

Improving the selectivity of photocatalysts is crucial, particularly in complex wastewater environments where multiple contaminants may be present. Functionalizing the surface of nanophotocatalysts with specific ligands or designing materials with targeted adsorption sites can enhance their selectivity for certain pollutants, making the treatment process more efficient.

7.3 Future Trends in Nanophotocatalyst Research

Research on nanophotocatalysts is evolving rapidly, and several emerging trends promise to push the boundaries of this field further. One of the most exciting trends is the development of multifunctional nanophotocatalysts, which can perform multiple tasks simultaneously in a single process. This integration of capabilities will make wastewater treatment more efficient and cost-effective. Another trend is the exploration of non-toxic, earth-abundant materials for nanophotocatalysts. Reducing reliance on rare or hazardous materials is essential for sustainable and scalable photocatalytic applications. Materials such as carbon-based photocatalysts (e.g., graphene, carbon nitride) are being extensively studied for their potential to deliver high photocatalytic activity without the environmental drawbacks associated with some metal oxides.

The use of biogenic or "green" synthesis methods is also gaining traction. These methods use plant extracts or microorganisms to produce nanophotocatalysts in an environmentally benign manner. This not only reduces the environmental footprint of the synthesis process but also produces materials with unique surface properties that can enhance photocatalytic performance [95]. Lastly, advances in computational modeling and machine learning are helping to simulate photocatalytic reactions and predict material properties, and researchers can now quickly identify promising candidates for further experimentation, reducing the time and cost associated with trial-and-error approaches [96].

7.4 Potential for Commercialization in Wastewater Treatment

For nanophotocatalysts to transition from research to widespread application, several commercialization challenges must be addressed. One of the primary considerations is scalability—ensuring that the synthesis methods used to produce nanophotocatalysts can be economically scaled up for industrial applications without compromising performance. Techniques such as sol–gel and hydrothermal synthesis, while highly effective at small scales, must be adapted for large-scale production to meet industrial demands [97]. Another key factor is cost-effectiveness. The use of inexpensive, abundant materials and cost-efficient synthesis techniques will be essential for commercial adoption. The longevity and reusability of nanophotocatalysts are critical, as materials that degrade quickly or require frequent replacement may not be commercially viable [93, 98].

Regulatory approval and environmental safety are also important factors. Nanomaterials must be shown to be safe for the environment and public health, particularly in terms of preventing the release of toxic by-products or NPs into the environment during or after use. Robust testing and adherence to environmental regulations will be required for widespread commercial use [99, 100]. Finally, market demand for clean water solutions is driving the push for commercialization of nanophotocatalysts. The growing global water crisis, combined with stricter environmental regulations, is creating a favorable environment for the adoption of advanced water treatment technologies. Governments, industries, and municipalities are likely to invest in nanophotocatalyst technologies as part of their efforts to meet clean water targets and address pollution challenges.

8 Conclusion

The exploration of nanophotocatalysts, particularly metal oxide-based systems, presents a pivotal advancement in the quest for sustainable and efficient wastewater purification technologies. This chapter has systematically classified these photocatalysts based on their composition, morphology, doping strategies, and heterostructure formation, offering a comprehensive overview of their synthesis, properties, and applications. Key nanomaterials, such as TiO_2, ZnO, and novel composites, have been highlighted for their significant photocatalytic activity both under visible light and UV radiation, demonstrating their capability to degrade organic pollutants, remove heavy metals, and disinfect water.

Despite the promising applications, current challenges, including electron–hole recombination, limited visible light absorption, and issues of stability and recyclability, must be addressed to enhance the commercial viability and scalability of these materials. Strategies such as bandgap engineering, heterostructure development, and green synthesis protocols are essential to overcoming these limitations and pushing the boundaries of nanophotocatalytic technology.

Looking forward, the integration of machine learning for catalyst design, the use of non-toxic earth-abundant materials, and the adoption of green synthesis methods are expected to be transformative in advancing the field. As the global demand for clean water intensifies, nanophotocatalysts are set to play an increasingly crucial role in achieving sustainable water treatment solutions. This chapter provides a roadmap for future innovations in the field, ensuring that nanophotocatalysis remains at the forefront of environmental science and material engineering.

Acknowledgements The authors would like to thank the International University of Business Agriculture and Technology (IUBAT) and the University of Dhaka for providing the necessary resources and support for this research. Special thanks to the colleagues and staff of the Department of Chemistry, IUBAT for their invaluable assistance and contributions throughout this study.

Author Contributions Mohammad Harun-Ur-Rashid contributed to conceptualization, methodology, literature review, analysis, and writing—review and editing; Israt Jahan contributed to conceptualization, methodology, data curation, and writing—original draft preparation; Abu Bin Imran contributed to project administration, and writing—review and editing; and Md. Abu Bin Hasan Susan contributed to supervision, project administration, and writing—review and editing. Each author has approved the final version of the manuscript and agrees to be accountable for the content of the work.

Funding This research was not financially supported.

Conflict of Interest The authors declare that they have no conflicts of interest regarding the publication of this paper.

References

1. Tahir MB, Nawaz T, Nabi G, Sagir M, Khan MI, Malik N (2022) Role of nanophotocatalysts for the treatment of hazardous organic and inorganic pollutants in wastewater. Int J Environ Anal Chem 102(2):491–515. https://doi.org/10.1080/03067319.2020.1723570
2. Napoli FSR, Uriarte D, Garrido M, Domini C, Acebal C (2023) Nanophotocatalysis for degradation of organic contaminants. In: Shanker U, Hussain CM, Rani M (eds) Handbook of green and sustainable nanotechnology: fundamentals, developments and applications. Springer International Publishing, Cham, pp 1149–1191. https://doi.org/10.1007/978-3-031-16101-8_43
3. Thakur S, Ojha A, Kansal SK, Gupta NK, Swart HC, Cho J, Kuznetsov A, Sun S, Prakash J (2024) Advances in powder nano-photocatalysts as pollutant removal and as emerging contaminants in water: analysis of pros and cons on health and environment. Adv Powder Mater 3(6):100233. https://doi.org/10.1016/j.apmate.2024.100233
4. Harun-Ur-Rashid M, Imran AB, Seki T, Takeoka Y, Ishii M, Nakamura H (2009) Template synthesis for stimuli-responsive angle independent structural colored smart materials. Trans Mater Res Soc Japan 34(2):333–337. https://doi.org/10.14723/tmrsj.34.333
5. Takeoka Y, Yoshioka S, Teshima M, Takano A, Harun-Ur-Rashid M, Seki T (2013) Structurally coloured secondary particles composed of black and white colloidal particles. Sci Rep 3(1):2371. https://doi.org/10.1038/srep02371
6. Karim MR, Harun-Ur-Rashid M, Imran AB (2023) Effect of sizes of vinyl modified narrow-dispersed silica cross-linker on the mechanical properties of acrylamide based hydrogel. Sci Rep 13(1):5089. https://doi.org/10.1038/s41598-023-32185-4
7. Rezaul Karim M, Harun-Ur-Rashid M, Imran AB (2020) Highly stretchable hydrogel using vinyl modified narrow dispersed silica particles as cross-linker. ChemistrySelect 5(34):10556–10561. https://doi.org/10.1002/slct.202003044
8. Harun-Ur-Rashid M, Imran AB (2022) Organometallic nanomaterials synthesis and sustainable green nanotechnology applications. In: Green nanoarchitectonics. Jenny Stanford Publishing, pp 249–270
9. Imran AB, Harun-Ur-Rashid M, Takeoka Y (2019) Polyrotaxane actuators. In: Asaka K, Okuzaki H (eds) Soft actuators: materials, modeling, applications, and future perspectives. Springer Singapore, Singapore, pp 81–147. https://doi.org/10.1007/978-981-13-6850-9_6
10. Islam MJ, Pal K, Abdullah MM, Harun-Ur-Rashid M (2024) Threshold bandgap of lead halide hybrid perovskites for making stable single crystal by NIR laser trapping. J Mol Struct 1301:137322. https://doi.org/10.1016/j.molstruc.2023.137322
11. Islam MJ, Pal K, Harun-Ur-Rashid M, Kumar A, Hossain MF, Asthana N, Rajendran S, Vallinayagam S, Malik A (2024) Optical maneuvering of photofunctioning hybrid perovskite

for future photonics potential application. J Mol Liq 408:125343. https://doi.org/10.1016/j.molliq.2024.125343

12. Harun-Ur-Rashid M, Imran AB (2024) Emerging trends in engineering polymers: a paradigm shift in material engineering. Recent Progr Mater 06(03):024. https://doi.org/10.21926/rpm.2403024
13. Harun-Ur-Rashid M, Jahan I, Islam MJ, Kumer A, Huda MN, Imran AB, Gouadria S, Alsalhi SA (2024) Global advances and smart innovations in supramolecular polymers. J Mol Struct 1304:137665. https://doi.org/10.1016/j.molstruc.2024.137665
14. Harun-Ur-Rashid M, Naidu Krishna SB, Golla N, Imran AB (2024) Polyphenol oxidases polyphenol oxidases: the future toward global sustainability. In: Verma P, Agrawal K, Shah MP (eds) Function, wastewater remediation, and biosensors. De Gruyter, pp 309–344. https://doi.org/10.1515/9783111033525-012
15. Harun-Ur-Rashid M, Imran AB, Susan MABH (2023a) Fire-resistant polymeric foams and their applications. In: Polymeric foams: applications of polymeric foams (Volume 2). ACS Symposium Series, vol 1440. American Chemical Society, pp 97–121. https://doi.org/10.1021/bk-2023-1440.ch005
16. Harun-Ur-Rashid M, Imran AB, Susan MABH (2023) Green polymer nanocomposites in automotive and packaging industries. Curr Pharm Biotechnol 24(1):145–163. https://doi.org/10.2174/1389201023666220506111027
17. Harun-Ur-Rashid M, Imran AB, Susan MABH (2024a) Emerging 3D printed polymers and composites for water quality preservation. In: Reference module in materials science and materials engineering. Elsevier. https://doi.org/10.1016/B978-0-323-95486-0.00010-7
18. Harun-Ur-Rashid M, Imran AB, Susan MABH (2024b) Magnetic polymer nanocomposites for removal of dyes and metals from wastewater. In: Reference module in materials science and materials engineering. Elsevier. https://doi.org/10.1016/B978-0-323-95486-0.00008-9
19. Huda MN, Harun-Ur-Rashid M, Hosen A, Akter M, Islam MM, Emon SZ, Rahman A, Jashim ZB, Shahrukh S, Ismail M (2024) A potential toxicological risk assessment of heavy metals and pesticides in irrigated rice cultivars near industrial areas of Dhaka, Bangladesh. Environ Monit Assess 196(9):794. https://doi.org/10.1007/s10661-024-12927-1
20. Harun-Ur-Rashid M, Foyez T, Jahan I, Pal K, Imran AB (2022) Rapid diagnosis of COVID-19 via nano-biosensor-implemented biomedical utilization: a systematic review. RSC Adv 12(15):9445–9465
21. Harun-Ur-Rashid M, Jahan I (2024) Innovative hybrid nanostructures: pioneering advances in modern therapy. Front Nanotechnol 6
22. Harun-Ur-Rashid M, Jahan I, Foyez T, Imran AB (2023c) Bio-inspired nanomaterials for micro/nanodevices: a new era in biomedical applications. Micromachines 14(9). https://doi.org/10.3390/mi14091786
23. Harun-Ur-Rashid M, Jahan I, Imran AB (2023d) Recent advancement of nanostructured materials for clinical challenges in vaccinology. In: Pal K (ed) Nanovaccinology: clinical application of nanostructured materials research to translational medicine. Springer International Publishing, Cham, pp 135–160. https://doi.org/10.1007/978-3-031-35395-6_8
24. Harun-Ur-Rashid M, Pal K, Imran AB (2023) Hybrid nanocomposite fabrication of nanocatalyst with enhanced and stable photocatalytic activity. Top Catal. https://doi.org/10.1007/s11244-023-01809-4
25. Qumar U, Hassan JZ, Bhatti RA, Raza A, Nazir G, Nabgan W, Ikram M (2022) Photocatalysis vs adsorption by metal oxide nanoparticles. J Mater Sci Technol 131:122–166. https://doi.org/10.1016/j.jmst.2022.05.020
26. Puri N, Gupta A (2023) Water remediation using titanium and zinc oxide nanomaterials through disinfection and photo catalysis process: a review. Environ Res 227:115786. https://doi.org/10.1016/j.envres.2023.115786
27. Sudarsan JS, Dogra K, Kumar R, Raval NP, Leifels M, Mukherjee S, Trivedi MH, Jain MS, Zang J, Barceló D, Mahlknecht J, Kumar M (2024) Tricks and tracks of prevalence, occurrences, treatment technologies, and challenges of mixtures of emerging contaminants in the environment: with special emphasis on microplastic. J Contam Hydrol 265:104389. https://doi.org/10.1016/j.jconhyd.2024.104389

28. Thennarasu G, Rajendran S, Kalairaj A, Rathore HS, Panda RC, Senthilvelan T (2024) A comprehensive review on the application of semiconductor nanometal oxides photocatalyst for the treatment of wastewater. Clean Technol Environ Policy. https://doi.org/10.1007/s10098-024-02960-6
29. Pelaez M, Nolan NT, Pillai SC, Seery MK, Falaras P, Kontos AG, Dunlop PSM, Hamilton JWJ, Byrne JA, O'Shea K, Entezari MH, Dionysiou DD (2012) A review on the visible light active titanium dioxide photocatalysts for environmental applications. Appl Catal B 125:331–349. https://doi.org/10.1016/j.apcatb.2012.05.036
30. Mohammadipour-Nodoushan R, Shekarriz S, Shariatinia Z, Heydari A, Montazer M (2023) Improved cotton fabrics properties using zinc oxide-based nanomaterials: a review. Int J Biol Macromol 242:124916. https://doi.org/10.1016/j.ijbiomac.2023.124916
31. Xu Y, Zhou Y, Li Y, Liu Y, Ding Z (2024) Advances in cerium dioxide nanomaterials: synthesis strategies, property modulation, and multifunctional applications. J Environ Chem Eng 12(5):113719. https://doi.org/10.1016/j.jece.2024.113719
32. Samuel O, Othman MHD, Kamaludin R, Sinsamphanh O, Abdullah H, Puteh MH, Kurniawan TA (2022) WO3–based photocatalysts: a review on synthesis, performance enhancement and photocatalytic memory for environmental applications. Ceram Int 48(5):5845–5875. https://doi.org/10.1016/j.ceramint.2021.11.158
33. Pourmadadi M, Rahmani E, Shamsabadipour A, Mahtabian S, Ahmadi M, Rahdar A, Díez-Pascual AM (2022) Role of iron oxide (Fe_2O_3) nanocomposites in advanced biomedical applications: a state-of-the-art review. Nanomaterials 12(21). https://doi.org/10.3390/nano12213873
34. Koe WS, Lee JW, Chong WC, Pang YL, Sim LC (2020) An overview of photocatalytic degradation: photocatalysts, mechanisms, and development of photocatalytic membrane. Environ Sci Pollut Res 27(3):2522–2565. https://doi.org/10.1007/s11356-019-07193-5
35. He X, Zhang C (2019) Recent advances in structure design for enhancing photocatalysis. J Mater Sci 54(12):8831–8851. https://doi.org/10.1007/s10853-019-03417-8
36. Gomes Silva C, Juárez R, Marino T, Molinari R, García H (2011) Influence of excitation wavelength (UV or visible light) on the photocatalytic activity of Titania containing gold nanoparticles for the generation of hydrogen or oxygen from water. J Am Chem Soc 133(3):595–602. https://doi.org/10.1021/ja1086358
37. Marcelino RBP, Amorim CC (2019) Towards visible-light photocatalysis for environmental applications: band-gap engineering versus photons absorption—a review. Environ Sci Pollut Res 26(5):4155–4170. https://doi.org/10.1007/s11356-018-3117-5
38. Feng D, Cheng Y, He J, Zheng L, Shao D, Wang W, Wang W, Lu F, Dong H, Liu H, Zheng R, Liu H (2017) Enhanced photocatalytic activities of g-C_3N_4 with large specific surface area via a facile one-step synthesis process. Carbon 125:454–463. https://doi.org/10.1016/j.carbon.2017.09.084
39. Shen M, Ding T, Rackers WH, Tan C, Mahmood K, Lew MD, Sadtler B (2021) Single-molecule colocalization of redox reactions on semiconductor photocatalysts connects surface heterogeneity and charge-carrier separation in bismuth oxybromide. J Am Chem Soc 143(30):11393–11403. https://doi.org/10.1021/jacs.1c02377
40. Zhao Q, Ji M, Qian H, Dai B, Weng L, Gui J, Zhang J, Ouyang M, Zhu H (2014) Controlling structural symmetry of a hybrid nanostructure and its effect on efficient photocatalytic hydrogen evolution. Adv Mater 26(9):1387–1392. https://doi.org/10.1002/adma.201304652
41. Peng Y, Ji J, Chen D (2015) Ultrasound assisted synthesis of ZnO/reduced graphene oxide composites with enhanced photocatalytic activity and anti-photocorrosion. Appl Surf Sci 356:762–768. https://doi.org/10.1016/j.apsusc.2015.08.070
42. Khan H (2024) In situ nitrogen-doped graphene-TiO_2 nano-hybrid as an efficient photocatalyst for pollutant degradation. Environ Sci Pollut Res 31(32):45383–45398. https://doi.org/10.1007/s11356-024-34114-y
43. Ashouri R, Rasekh B, Kasaeian A, Sheikhpour M, Yazdian F, Dehghani Mobarakeh M (2021) Effect of ZnO-based nanophotocatalyst on degradation of aniline. J Mol Model 27(3):92. https://doi.org/10.1007/s00894-021-04710-7

44. Wang Y, Wang X, Antonietti M (2012) Polymeric graphitic carbon nitride as a heterogeneous organocatalyst: from photochemistry to multipurpose catalysis to sustainable chemistry. Angew Chem Int Ed 51(1):68–89. https://doi.org/10.1002/anie.201101182
45. Aritonang AB, Pratiwi E, Warsidah W, Nurdiansyah SI, Risko R (2021) Fe-doped TiO_2/kaolinite as an antibacterial photocatalyst under visible light irradiation. BCREC 16(2). https://doi.org/10.9767/bcrec.16.2.10325.293-301
46. Shabbir A, Sardar S, Mumtaz A (2024) Mechanistic investigations of emerging type-II, Z-scheme and S-scheme heterojunctions for photocatalytic applications—a review. J Alloy Compd 1003:175683. https://doi.org/10.1016/j.jallcom.2024.175683
47. Pan C, Xu J, Wang Y, Li D, Zhu Y (2012) Dramatic activity of C_3N_4/$BiPO_4$ photocatalyst with core/shell structure formed by self-assembly. Adv Func Mater 22(7):1518–1524. https://doi.org/10.1002/adfm.201102306
48. Mendes-Felipe C, Veloso-Fernández A, Vilas-Vilela JL, Ruiz-Rubio L (2022) Hybrid organic–inorganic membranes for photocatalytic water remediation. Catalysts 12(2). https://doi.org/10.3390/catal12020180
49. Tahir MB, Rafique M, Rafique MS, Fatima N, Israr Z (2020) Chapter 6—Metal oxide- and metal sulfide-based nanomaterials as photocatalysts. In: Tahir MB, Rafique M, Rafique MS (eds) Nanotechnology and photocatalysis for environmental applications. Elsevier, pp 77–96. https://doi.org/10.1016/B978-0-12-821192-2.00006-1
50. Rotami M, Hamadanian M, Rahimi-Nasrabadi M, Ganjali MR (2019) Sol–gel preparation of metal and nonmetal-codoped TiO_2–graphene nanophotocatalyst for photodegradation of MO under UV and visible-light irradiation. Ionics 25(4):1869–1878. https://doi.org/10.1007/s11581-019-02861-5
51. Hassanzadeh-Tabrizi SA, Do T-O (2018) Sol–gel synthesis and photocatalytic activity of ZnO–Ag–Sm nanoparticles for water treatment. J Mater Sci: Mater Electron 29(13):10986–10991. https://doi.org/10.1007/s10854-018-9180-3
52. Ranjeh M, Masjedi-Arani M, Amiri O, Salavati-Niasari M (2021) Pechini sol-gel synthesis of Ni(II) doped $LiMnBO_3$/Li_2MnO_3/$Li_2B_2O_4$ nano-photocatalyst under UV–Vis irradiation. Int J Hydrogen Energy 46(17):10324–10336. https://doi.org/10.1016/j.ijhydene.2020.12.149
53. Cong Y, Zhang J, Chen F, Anpo M (2007) Synthesis and characterization of nitrogen-doped TiO_2 nanophotocatalyst with high visible light activity. J Phys Chem C 111(19):6976–6982. https://doi.org/10.1021/jp0685030
54. Sajid MM, Shad NA, Javed Y, Shafique M, Afzal AM, Khan SB, Amin N, Hassan MA, Khan MUH, Tarabi T, Zhai H (2022) Efficient photocatalytic and antimicrobial behaviour of zinc oxide nanoplates prepared by hydrothermal method. J Cluster Sci 33(2):773–783. https://doi.org/10.1007/s10876-021-02013-8
55. Čižmar T, Kojić V, Rukavina M, Brkljačić L, Salamon K, Grčić I, Radetić L, Gajović A (2020) Hydrothermal synthesis of FeOOH and Fe_2O_3 modified self-organizing immobilized TiO_2 nanotubes for photocatalytic degradation of 1H-Benzotriazole. Catalysts 10(12). https://doi.org/10.3390/catal10121371
56. Gupta T, Samriti CJ, Prakash J (2021) Hydrothermal synthesis of TiO_2 nanorods: formation chemistry, growth mechanism, and tailoring of surface properties for photocatalytic activities. Mater Today Chem 20:100428. https://doi.org/10.1016/j.mtchem.2021.100428
57. Luan J, Ma K, Zhang L, Li M, Li Y, Pan B (2010) Research on different preparation methods of new photocatalysts. Curr Org Chem 14(7):683–698. https://doi.org/10.2174/138527210790963449
58. Chankhunthod N, Junploy P, Suthirakun S, Ngamwongwan L, Phromma C, Ruchusartsawat N, Siyasukh A, Yanu P, Kijjanapanich P, Yimklan S, Rujiwatra A, Drummond-Brydson R, Chimupala Y (2023) Ecofriendly alkali metal cations diffusion improves fabrication of mixed-phase titania polymorphs on fixed substrate by chemical vapor deposition (CVD) for photocatalytic degradation of azo dye. Environ Res 239:117347. https://doi.org/10.1016/j.envres.2023.117347
59. Wang Y, Wu Y, Qin Y, Xu G, Hu X, Cui J, Zheng H, Hong Y, Zhang X (2011) Rapid anodic oxidation of highly ordered TiO_2 nanotube arrays. J Alloy Compd 509(14):L157–L160. https://doi.org/10.1016/j.jallcom.2011.01.096

60. Zhang F, Zhang W, Luo X, Feng G, Zhao L (2017) Electrodeposition of ZnO film with enhanced photocatalytic activity towards methylene blue degradation. Int J Electrochem Sci 12(5):3756–3764. https://doi.org/10.20964/2017.05.62
61. Dhayagude AC, Nikam SV, Kapoor S, Joshi SS (2017) Effect of electrolytic media on the photophysical properties and photocatalytic activity of zinc oxide nanoparticles synthesized by simple electrochemical method. J Mol Liq 232:290–303. https://doi.org/10.1016/j.molliq.2017.02.074
62. Bhattacharjee J, Roy S (2024) A review on photocatalysis and nanocatalysts for advanced organic synthesis. Hybrid Adv 6:100268. https://doi.org/10.1016/j.hybadv.2024.100268
63. Prakash J, Krishna SB, Kumar P, Kumar V, Ghosh KS, Swart HC, Bellucci S, Cho J (2022) Recent advances on metal oxide based nano-photocatalysts as potential antibacterial and antiviral agents. Catalysts 12(9). https://doi.org/10.3390/catal12091047
64. Alprol AE, Mansour AT, El-Beltagi HS, Ashour M (2023) Algal extracts for green synthesis of zinc oxide nanoparticles: promising approach for algae bioremediation. Materials 16(7). https://doi.org/10.3390/ma16072819
65. Nande A, Raut S, Michalska-Domanska M, Dhoble SJ (2021) Green synthesis of nanomaterials using plant extract: a review. Curr Pharm Biotechnol 22(13):1794–1811. https://doi.org/10.2174/1389201021666201117121452
66. Gupta NM (2017) Factors affecting the efficiency of a water splitting photocatalyst: a perspective. Renew Sustain Energy Rev 71:585–601. https://doi.org/10.1016/j.rser.2016.12.086
67. Zhu D, Zhou Q (2021) Nitrogen doped g-C_3N_4 with the extremely narrow band gap for excellent photocatalytic activities under visible light. Appl Catal B 281:119474. https://doi.org/10.1016/j.apcatb.2020.119474
68. Ananpattarachai J, Kajitvichyanukul P, Seraphin S (2009) Visible light absorption ability and photocatalytic oxidation activity of various interstitial N-doped TiO_2 prepared from different nitrogen dopants. J Hazard Mater 168(1):253–261. https://doi.org/10.1016/j.jhazmat.2009.02.036
69. Murali A, Sarswat PK, Free ML (2020) Minimizing electron-hole pair recombination through band-gap engineering in novel ZnO-CeO_2-rGO ternary nanocomposite for photoelectrochemical and photocatalytic applications. Environ Sci Pollut Res 27(20):25042–25056. https://doi.org/10.1007/s11356-020-08990-z
70. Wu H, Irani R, Zhang K, Jing L, Dai H, Chung HY, Abdi FF, Ng YH (2021) Unveiling carrier dynamics in periodic porous $BiVO_4$ photocatalyst for enhanced solar water splitting. ACS Energy Lett 6(10):3400–3407. https://doi.org/10.1021/acsenergylett.1c01454
71. Khdary NH, Alkhuraiji WS, Sakthivel TS, Khdary DN, Salam MA, Alshihri S, Al-Mayman SI, Seal S (2020) Synthesis of superior visible-light-driven nanophotocatalyst using high surface area TiO_2 nanoparticles decorated with Cu_xO particles. Catalysts 10(8). https://doi.org/10.3390/catal10080872
72. Xie M, Meng Q, Luan P, Feng Y, Jing L (2014) Synthesis of mesoporous TiO_2-coupled Fe_2O_3 as efficient visible nano-photocatalysts for degrading colorless pollutants. RSC Adv 4(94):52053–52059. https://doi.org/10.1039/C4RA08451A
73. Saito K, Kazama S, Matsubara K, Yui T, Yagi M (2013) Monoclinic $Ag_2Mo_2O_7$ nanowire: a new Ag–Mo–O nanophotocatalyst material. Inorg Chem 52(15):8297–8299. https://doi.org/10.1021/ic401236b
74. Zhang L, Jin Z (2019) Effective electron-hole separation over controllable construction of CdS/Co-Ni-P core/shell nanophotocatalyst for improved photocatalytic hydrogen evolution under visible-light-driven. Catal Surv Asia 23(3):219–230. https://doi.org/10.1007/s10563-019-09274-4
75. Din MI, Najeeb J, Ahmad G (2018) Recent advancements in the architecting schemes of zinc oxide-based photocatalytic assemblies. Sep Purif Rev 47(4):267–287. https://doi.org/10.1080/15422119.2017.1383918
76. Shafiee A, Rabiee N, Ahmadi S, Baneshi M, Khatami M, Iravani S, Varma RS (2022) Core-shell nanophotocatalysts: review of materials and applications. ACS Appl Nano Mater 5(1):55–86. https://doi.org/10.1021/acsanm.1c03714

77. Wu C, Yan H, Zhong J, Xie J, Wang D, Shi Y, Zhang L, Zhu J, Chen Q (2019) Tape casting and phase inversion derived easily recyclable PVDF@g-$C_3N_4/Bi_2W_{0.98}Nb_{0.02}O_6$ porous film for high performance and long-term photocatalytic pollutants degradation. Appl Surf Sci 496:143731. https://doi.org/10.1016/j.apsusc.2019.143731
78. Ahmad I, Aftab MA, Fatima A, Mekkey SD, Melhi S, Ikram S (2024) A comprehensive review on the advancement of transition metals incorporated on functional magnetic nanocomposites for the catalytic reduction and photocatalytic degradation of organic pollutants. Coord Chem Rev 514:215904. https://doi.org/10.1016/j.ccr.2024.215904
79. Grigioni I, Bernareggi M, Sinibaldi G, Dozzi MV, Selli E (2016) Size-dependent performance of CdSe quantum dots in the photocatalytic evolution of hydrogen under visible light irradiation. Appl Catal A 518:176–180. https://doi.org/10.1016/j.apcata.2015.09.021
80. Weng B, Liu S, Zhang N, Tang Z-R, Xu Y-J (2014) A simple yet efficient visible-light-driven CdS nanowires-carbon nanotube 1D–1D nanocomposite photocatalyst. J Catal 309:146–155. https://doi.org/10.1016/j.jcat.2013.09.013
81. Song J, Yan L, Duan J, Jing C (2017) TiO_2 crystal facet-dependent antimony adsorption and photocatalytic oxidation. J Colloid Interface Sci 496:522–530. https://doi.org/10.1016/j.jcis.2017.02.054
82. Hou T-F, Boppella R, Shanmugasundaram A, Kim DH, Lee D-W (2017) Hierarchically self-assembled ZnO architectures: establishing light trapping networks for effective photoelectrochemical water splitting. Int J Hydrogen Energy 42(22):15126–15139. https://doi.org/10.1016/j.ijhydene.2017.04.121
83. Arbuj SS, Hawaldar RR, Mulik UP, Wani BN, Amalnerkar DP, Waghmode SB (2010) Preparation, characterization and photocatalytic activity of TiO_2 towards methylene blue degradation. Mater Sci Eng, B 168(1):90–94. https://doi.org/10.1016/j.mseb.2009.11.010
84. Zhang J, Yan S, Fu L, Wang F, Yuan M, Luo G, Xu Q, Wang X, Li C (2011) Photocatalytic degradation of rhodamine b on anatase, rutile, and brookite TiO_2. Chin J Catal 32(6):983–991. https://doi.org/10.1016/S1872-2067(10)60222-7
85. Elmolla ES, Chaudhuri M (2010) Degradation of amoxicillin, ampicillin and cloxacillin antibiotics in aqueous solution by the UV/ZnO photocatalytic process. J Hazard Mater 173(1):445–449. https://doi.org/10.1016/j.jhazmat.2009.08.104
86. Gao W, Yan J, Qian L, Han L, Chen M (2018) Surface catalyzing action of hematite (α-Fe_2O_3) on reduction of Cr(VI) to Cr(III) by citrate. Environ Technol Innov 9:82–90. https://doi.org/10.1016/j.eti.2017.11.007
87. Zhao B, Zhang K, Huang Y, Wang Q, Xu H, Wang Y, Li J, Song T, Xia W, Liu J (2021) A novel visible light-driven TiO_2 photocatalytic reduction for hexavalent chromium wastewater and mechanism. Water Sci Technol 83(9):2135–2145. https://doi.org/10.2166/wst.2021.116
88. Hemmami H, Zeghoud S, Ben Amor I, Alnazza Alhamad A, Tliba A, Alsalme A, Cornu D, Bechelany M, Barhoum A (2024) Green synthesis of CaO nanoparticles from chicken eggshells: antibacterial, antifungal, and heavy metal (Pb^{2+}, Cr^{2+}, Cd^{2+} and Hg^{2+}) adsorption properties. Front Environ Sci 12. https://doi.org/10.3389/fenvs.2024.1450485
89. Jiang Y, Zhang L, Wen D, Ding Y (2016) Role of physical and chemical interactions in the antibacterial behavior of ZnO nanoparticles against E. coli. Mater Sci Eng, C 69:1361–1366. https://doi.org/10.1016/j.msec.2016.08.044
90. Jiménez E, Hamdan-Partida A, Padilla-Godínez FJ, Arellano-Lara D, Gómez-López E, López-Goerne TM (2022) Spectroscopic analysis and microbicidal effect of Ag/TiO_2-SiO_2 bionanocatalysts. IEEE Trans Nanobiosci 21(2):246–255. https://doi.org/10.1109/TNB.2021.3122084
91. An L, Onishi H (2015) Electron-hole recombination controlled by metal doping sites in $NaTaO_3$ photocatalysts. ACS Catal 5(6):3196–3206. https://doi.org/10.1021/acscatal.5b00484
92. Ferreira VRA, Santos PRM, Silva CIQ, Azenha MA (2021) Latest developments on TiO_2-based photocatalysis: a special focus on selectivity and hollowness for enhanced photonic efficiency. Appl Catal A 623:118243. https://doi.org/10.1016/j.apcata.2021.118243

93. Xiong L, Tang J (2021) Strategies and challenges on selectivity of photocatalytic oxidation of organic substances. Adv Energy Mater 11(8):2003216. https://doi.org/10.1002/aenm.202003216
94. Yao K, Zhong H, Liu Z, Xiong M, Leng S, Zhang J, Xu Y-x, Wang W, Zhou L, Huang H, Jen AKY (2019) Plasmonic metal nanoparticles with core-bishell structure for high-performance organic and perovskite solar cells. ACS Nano 13(5):5397–5409. https://doi.org/10.1021/acsnano.9b00135
95. Khatami M, Iravani S (2021) Green and eco-friendly synthesis of nanophotocatalysts: an overview. Comments Inorg Chem 41(3):133–187. https://doi.org/10.1080/02603594.2021.1895127
96. Gusarov S (2024) Advances in computational methods for modeling photocatalytic reactions: a review of recent developments. Materials 17(9). https://doi.org/10.3390/ma17092119
97. Petronella F, Truppi A, Dell'Edera M, Agostiano A, Curri ML, Comparelli R (2019) Scalable synthesis of mesoporous TiO_2 for environmental photocatalytic applications. Materials 12(11). https://doi.org/10.3390/ma12111853
98. Ji Z, Zhang Y, Qi X, Wang Y, Xia X, Pei Y (2021) Low-cost and facile fabrication of recyclable and reusable waste-based geopolymer for visible-light photocatalysis degradation. J Clean Prod 310:127434. https://doi.org/10.1016/j.jclepro.2021.127434
99. Markowska-Szczupak A, Endo-Kimura M, Paszkiewicz O, Kowalska E (2020) Are Titania photocatalysts and titanium implants safe? Review on the toxicity of titanium compounds. Nanomaterials 10(10). https://doi.org/10.3390/nano10102065
100. Razavi M, Khandan A (2017) 14—Safety, regulatory issues, long-term biotoxicity, and the processing environment. In: Razavi M, Thakor A (eds) Nanobiomaterials science, development and evaluation. Woodhead Publishing, pp 261–279. https://doi.org/10.1016/B978-0-08-100963-5.00014-8

Synthesis of Nanophotocatalysts

Geetha Devi

Abstract Nanomaterials play a significant role in various fields due to their versatile surface functional properties and diversified applications. Nano catalysts are the main constituents of nano catalysis with dimensions less than 100 nm, and they exist in the form of metal oxides or inorganic materials like semiconductors. The improved selectivity and efficiency of nano catalysts find potential application in resolving various complex industrial issues by offering sustainable solutions. Nano photo catalysts are receiving significant attention due to their extraordinary physical properties, high thermal and chemical stability, low environmental impact, strong oxidation capability, and bountiful potential applications in solar energy systems. Nano photo catalysts are an excellent support material for pollution-control applications. This chapter deals with the various methods of synthesis of nano photocatalysts and their application in wastewater treatment. The advantages and limitations of the nano photocatalysts are presented in this chapter.

Keywords Copper oxide · Nanoparticles · Nano photo catalysts · Sol–gel method · Titanium dioxide · Zinc oxide

1 Introduction

A catalyst is defined as a substance that speeds up the chemical reaction without taking part in the reaction. A photo catalyst is a material that absorbs light or photons and converts light energy into chemical energy. A catalytic process which alters the chemical reaction in presence of light or solar irradiation is described as Photocatalysis. Photo catalysis process is employed in wastewater treatment applications for the degradation of organic pollutants in presence of sunlight containing UV rays. Photocatalysis is also employed in the production and generation of hydrocarbons, water

G. Devi (✉)
College of Engineering, National University of Science and Technology, Muscat, Oman
e-mail: geethadevi@nu.edu.om

A. Ahmad et al. (eds.), *Metal Oxide Based Nanophotocatalyst for Wastewater Purification*, Smart Nanomaterials Technology,
https://doi.org/10.1007/978-981-96-7683-5_2

splitting, drug delivery, and self-cleaning coatings [53]. The exposure of a semiconductor or a photocatalysts with sunlight will result in generation of electron–hole pair. The difference in energy between the conduction band and valence band is known as the band gap. Since band gap is a basic property of a photo catalyst, the material that acts as a photo catalyst should produce a good band gap. One of the requirements of a photo catalyst in water treatment applications is to have better stability and high surface area with uniform surface morphology. Majority of the oxides of chromium, zinc, titanium, and vanadium naturally have the above-mentioned properties. Hence, these can preferably be used as potential photocatalysts in the wastewater treatment.

1.1 Types of Photocatalytic Reactions

There are two types of photocatalytic reactions based on the existence of physical state of reactants.

(i) Homogeneous photocatalysis: If the photocatalyst or semiconductor and the reaction medium are in the similar phase (gas, solid, or liquid) is known as homogeneous photocatalysis. Ozone and photo Fenton processes are the most commonly used homogeneous photocatalysis.
(ii) Heterogeneous photocatalysis: If the photocatalyst or semiconductor and the reaction medium are in different phases is called homogeneous photocatalysis. Alumina and oxides of iron are the two main types of heterogeneous catalysts applied in the production of ammonia by Haber process. Another example of heterogeneous photocatalysis is the hydrogenation of ethylene (C_2H_4) by Platinum (Pt), a metal catalyst.

1.2 Different Types of Photocatalysts

Variety of semiconductor materials are employed as photocatalysts; the common types are titanium dioxide (TiO_2), Iron Oxide (Fe_2O_3), Zinc Oxide (ZnO), Copper oxide (Cu_2O), Zinc Sulphide (ZnS), Cadmium Sulphide (CdS), etc. Nano photocatalysts have excellent surface-to-volume ratio, which offers great driving force for diffusion, mainly at elevated temperatures. Semiconducting oxides such as ZnO and TiO_2 in bulk (macro) form are extensively employed in various industries. When they are in a nanoscale form, they exhibit remarkable surface properties that contribute to the development of new materials. The band gap energy details of TiO_2 and ZnO nanoparticles are presented in Table 1.

Table 1 Band gap energy details of TiO_2 and ZnO nanoparticles

Semiconductors	Band gap energy (eV)
TiO_2	3.030
ZnO	3.36

1.3 Different Methods of Synthesis of Photocatalyst:

Photocatalysts are synthesized using different techniques. Sol–gel is the conventional method of preparation of photocatalysts. Due to the enhanced surface area, preparation of nanomaterials is divided into the chemical and physical methods. However, these synthesis methods are not appropriate for medical and pharmaceutical applications due to their harmful effects. Green synthesis technique has received great deal of attention in the preparation of nanomaterials owing to the simple, nontoxic, ecofriendly, and cost-effective nature [11, 32]. The extensively developed methods for the synthesis of photocatalysts are sol–gel, solvothermal, co-precipitation, sonochemical, chemical vapor deposition, hydrothermal, etc. Among these, the Sol–gel method is one of the most popular methods employed in the preparation of photocatalyst by solidification from their precursor solution. This section presents an overview of some of the most important techniques in the preparation of photocatalysts.

1.3.1 Sol–gel Method

Sol–gel Method is one of the most popular wet-chemical methods for the synthesis of photocatalyst by solidification process from their precursor solution. The three-step synthesis method consists of hydrolysis and polymerization followed by drying of the gel [17]. The metal salts, metal oxides, or alkoxides are used as the starting material for the synthesis process. The solution form of the precursor materials is hydrolyzed to develop a colloidal sol, which is subjected to polymerization process to form gel-like structure. Finally, the resulting gel will be calcined to form xerogel or aerogel. The thermal treatment of the gel will result in photocatalysts powders [8]. The attractions of sol–gel process include less reaction time, simple process, excellent stability, low temperature process, high product purity, and homogeneous. Additionally, the photocatalytic activity of the photocatalyst prepared via sol–gel process can be enhanced by the addition of dopants [27].

1.3.2 Solvothermal Method

Solvothermal process is an extension of hydrothermal method; however, the solvents used are either non-aqueous or organic. The reaction temperature was chosen based on the boiling point of the selected solvent in the synthesis process. The four-step preparation process consists of nucleation, dissolution, recrystallization, and finally growth in a specific direction [17]. Compared to hydrothermal technique, the particle geometry, size distribution, crystallization, and the agglomeration properties of the particles can be better controlled in solvothermal method [49]. The solvent plays a crucial role in the synthesis process as the solvent properties affect the reactivity, solubility, and diffusive nature of reactants. The crystal structure and morphology are also greatly influenced by the choice of solvent.

1.3.3 Co-precipitation Method

Co-precipitation is a technique used to synthesize metal oxide nanophotocatalysts for the application of photocatalytic degradation studies [62]. This process is a very complicated process equipped with higher-end machines, as it requires careful monitoring of the process parameters. The starting materials used in the synthesis process are metal salts of chloride, nitrite, etc. In this process, the metal salts are dissolved in aqueous solution, followed by precipitation of metal hydroxide. The resulting salts undergo washing and calcination processes to get the required metal oxide as photocatalyst. If more than one metal salt precursor solutions are used, then the final step of the photocatalyst product will be a mixture of metal oxides. The particle size distribution can be controlled by adjusting the speed of agitation and temperature. Highly dispersed particles with 60%–80% metal loading can be achieved by employing this process [78]. Highly dispersed, homogeneous, and well-mixed metal oxides can be synthesized using this technique. The main drawback of this process requires further separation process by filtration, and the powder form of the metal oxide is produced. Though Co-precipitation and sol–gel processes are commonly used techniques, but it involves expensive equipment.

1.3.4 Chemical Vapor Deposition (CVD) Method

The production of high-purity, high-performance photocatalyst are carried out using Chemical vapor deposition (CVD) process. The CVD process involves coating or grafting techniques, and the process necessitates ultrahigh vacuum and high vapor pressure for its deposition on substrate. The precursor material is deposited in gaseous phase under inert atmosphere at high temperature and pressure on the surface of substrate [80]. The volatile precursors are deposited onto the substrate surface in the form of thin film. The byproducts are removed by passing the inert gases to the system. Chemical vapor deposition (CVD) method is very expensive compared to other methods, but the outstanding properties of the synthesized photocatalyst may counterbalance the initial investment [47].

2 Synthesis of Titanium Dioxide Nano Photocatalysts (TiO_2NPs)

TiO_2 is a bright white pigment, which is chemically stable and relatively cheap material [28]. The augmented stability, nontoxic and biocompatible nature of TiO_2NPs is suitable for applications as cosmetic agents like sunscreens; dental care products; powders, paints and coatings; plastics; and as food coloring agents [72]. The plant-mediated synthesis of TiO_2NPs from *Sargassum myriocystum* and *Spirulina platensis* by green synthesis techniques is receiving considerable attention. Aravind et al. [9] synthesized TiO_2NPs using the chemical as well as the green synthesis routine. Cubic, square, and spherical particles with diameter range between 50 and 90 nm with a negative surface charge. Vasanth [74] have biosynthesized anatase phase TiO_2NPs using *S. platensis,* and the particles are characterized by XRD analysis. Both methods are environmentally safe and cost-effective, although the second type is more efficient than the first. The *Pseudo kirchneriella subcapitata,* a freshwater algae species was examined by [29] for testing the toxic effect of TiO_2NPs having varying diameters of 10 nm, 30 nm, and 300 nm. Although TiO_2NPs intimidated the algal growth, the aggregation and sedimentation were predominant. Sorption capabilities of TiO_2NPs were also revealed, the mitigation of cadmium toxicity, while the dose-dependent combined toxic effect of TiO_2NPs and Cd (II) was established. This indicates the importance of more comprehensive nano eco toxicological analyses to assess the environmental impacts. Sol–Gel process was employed in the preparation of TiO_2NPs nanoparticles by reacting ethanol, titanium (IV) iso propoxide, and nitric acid at room temperature under continuous stirring for 45 min [31]. The modified sol–gel method was used to synthesize the nanoparticles at ambient temperature. The reaction was performed using Titanium (IV) isopropoxide and ethanol under continuous stirring followed by centrifugation at 5000 RPM for 10 min. The resulting product was calcined in a muffle furnace at 400 °C for four hours followed by drying [64]. Facile green synthesis of TiO_2NPs from the extracts of Jasmine flower was used as a capping agent. The synthesis was performed by boiling 100 ml water and 50 g of jasmine flower for a duration of 30 min followed by filtration. 20 ml of the extract was mixed with 50 ml of titanium tetra isopropoxide under vigorous stirring for 3 h. After 3 h, the resulting mixture centrifugal separation followed by drying at 110 °C for 5 h to get titanium dioxide nanoparticles in powder form [6, 39]. The band gap of the TiO_2 (anatase grade) is 3.2 eV and 3.0 eV for rutile [52]. Owing to the outstanding surface area and interconnected pores in transition metal oxide form of TiO_2NPs is widely employed in photocatalysis [1]. Raghadhiita et al. [59] investigated the preparation of TiO_2NPs by liquid phase synthesis method by the hydrolysis of titanium isopropoxide and nitric acid followed by dilution with distilled water to form a solution of titanyl nitrate. The resulting solution was mixed with glycine which acts as a capping agent. The product was calcined to obtain crystals of TiO_2NPs. The powdered materials were characterized using various characterization tools, and the final product looks like anatase grade of TiO_2NPs and the yield of TiO_2 was 90%.

Synthesis of TiO_2 using two different methods namely Hydrothermal (HT) sol–gel and Flame Spray Pyrolysis (FSP) was carried out by Aboulouard et al. [2]. In HT method, the reaction was performed by reacting titanium tetra isopropoxide and acetic acid and the reaction mixture was continuously stirred to get the homogenous consistency. Nitric acid was added to the resulting colloidal solution and then autoclaved at 239 °C followed by concentrating the product. The final product was characterized using various tools, and it was confirmed that the size of TiO_2 nanoparticles was around 13.25 nm, and it was a pure anatase phase material. The synthesis of TiO_2 nanoparticles using solvothermal method was investigated by [60]. The synthesis was performed by reacting titanium tetra isopropoxide and ethanol by vigorous stirring and hydrolyzed using H_2SO_4. The resulting mixture was stored in an air-tight container and solvothermal growth was performed at 90 °C. The as prepared product was centrifuged and washed with Millipore water to remove sulphate ions and then annealed between at a temperature range between 400 and 500 °C followed by characterization. The size of the TiO_2NPs was around 10 nm. Buraso et al. [14] investigated a simple precipitation method of TiO_2NPs by hydrolysis and calcination at a temperature of 400 °C to 700 °C. The average particle size was found to be 27 nm upon increasing the calcination temperature from 400 °C to 700 °C. TiO_2NPs were also synthesized by in situ synthesized method on an activated carbon fibre substrate by Pratyush et al. [55]. Titanium tetra isopropoxide and ethanol were used as precursor materials and the mixture was impregnated on the substrate. This was hydrothermally treated at 180 °C for 24 h. The synthesized NPs were characterized, and particles size of 400 nm was identified. Sharma et al. [67] successfully synthesized TiO_2 nanoparticles by sol–gel method with a crystalline size of 7.5 nm. The electrochemical performance of the synthesized TiO_2NPs were confirmed its great potential in photovoltaic applications. The excellent catalytic and optical properties of TiO_2NPs are suitable for the transparent protective coatings and sunscreens, which are characterized by a highly transparent appearance. The large surface-to-volume ratio of the nano particles is suitable for photo catalytic purification of water and air by decomposing the organic pollutants present in wastewater. Table 2 represents the various synthesis methods of TiO_2NPs.

3 Synthesis Methods of Copper Oxide Nanoparticles (CuONPs)

Copper oxide nanoparticles (CuONPs) are monoclinic semiconducting compound exhibits unique properties such as electron correlation effects, high-temperature superconductivity, and spin dynamics, making it suitable for a range of applications. CuONPs are applied in gas sensors, batteries, solar energy systems, fungicides, water purifiers, antimicrobials, algaecides, and anti-fouling agents. Bhattacharya et al. [13] synthesized rod-shaped CuONPs using *Anabaena cylindrica* extract, for the disinfection of drinking water. Renuga et al. [61] utilized *Brassica oleracea* algal extract

Table 2 Methods of synthesis of and application of TiO_2NPs

Synthesis method	Application	References
Biosynthesis from Bengal gram	Electrochemical applications	[37]
Sol–gel method	Antimicrobial activity, Methyl Orange dye degradation, Antibacterial activity of *Escherichia coli*	[4, 46, 51]
Microwave assisted hydrothermal route	Electro catalytic activity studies	[15]
Bio synthesis using mycorrhizal sorghum roots	Photo degradation studies of Methyl Orange Dye	[22]
Ionic liquid-assisted hydrothermal method	Photocatalytic degradation of methylene blue dye	[48]
Electro synthesis	Photo degradation of congo red dye without UV radiation	[36]
Surfactant-assisted method	Photocatalytic degradation of methylene blue	[34]
Biosynthesis method using fenugreek seeds	Antimicrobial activity studies	[71]
Sol–gel and solvothermal methods	Due sensitized solar cell applications	[18]
Electrochemical method	Antibacterial activity against gram-negative (*Escherichia coli)* and gram-positive (*Staphylococcus aureus)* bacterial strains	[26]
Green synthesis using papaya leaves	Photocatalytic degradation of Ragnorak Online (RO) dye	[38]
Wet-chemical method using Chitin	Photocatalytic degradation of Methylene Blue	[54]
Solvothermal method	Degradation of Acid Blue (AB 5) dye	[16]

to produce CuONPs, where the distinct functional groups of hydroxyl, carbonyl, and amine in polyols and polysaccharides played significant roles in nanoparticle reduction, capping, and stabilization. Araya-Castro et al. [10] employed protein fractions from *Macrocystis pyrifera* extract for CuONP synthesis. Using UV/VIS diode arrays and TEM-EDS analysis, they observed spherical nanoparticles formed by low and high-molecular-weight proteins. Similarly, the biosynthesis of spherical CuONPs, averaging 22.6 nm in size, using di terpenoids from *Bifurcaria bifurcata.* Khanehzaei et al. [40] utilized Kappaphycus alvarezii seaweed extracts, in which sulphated polysaccharides acted as stabilizers, forming CuONPs with a mean size of 53 nm. The simple ecofriendly and low-cost biosynthesis of CuONPs using *Ephedra Alata* aqueous plant extract was carried out by Afrah Atri et al. [5] and the plant extract acted as reducing and capping agents. The resulting CuONPs were employed in the photocatalytic degradation studies of dye molecules. The outcome of the study shows that the biosynthesized CuONPs are excellent material for the photo degradation of dyes. Among various types of metal oxide nanoparticles, CuONPs have excellent band gap of 1.2–2.1 eV with monoclinic nature are suitable for photo degradation of pollutants present in industrial wastewater. Among the various metal and metal oxide nanoparticles, copper oxide (CuO) nanoparticles showed excellent band gap with p-type semiconductors and monoclinic structures, which has tremendous potential in environmental pollution control [24]. CuONPs with varying geometry and structure are synthesized in the form of nanospheres [76], nanosheets [44], nanorods [58], nanoflowers [77], nano rings [79], etc.

4 Synthesis Methods of Magnesium Oxide Nanoparticles (MgONPs)

Magnesium Oxide Nanoparticles are an essential nutrient required for plant growth, which offers exceptional driving force for the photosynthesis and exhibits abundant reactivity with phytochemicals in algae and plants during nanoparticles synthesis stage [35]. Magnesium oxide nanoparticles (MgONPs) are approved by the Food and Drug administration due to their exceptional bactericidal properties, which include stability at high temperatures and low flammability, results in a long-lasting antibacterial efficiency [41]. A study by [57] reported the ecofriendly green synthesis of MgONPs from *Sargassum wighitii* that acted as excellent capping and reducing agents. UV spectroscopic analysis for all the samples displayed maximum absorbance value of 322 nm, and X-ray diffraction analysis confirmed the crystalline nature with face-centered cubic structure. The EDX analysis confirmed the purity of the synthesized nanoparticles with compositions of magnesium and oxygen. The FTIR analysis established the functional groups present in the nanoparticle surface. The attractive stability, antimicrobial activity, ecofriendly nature, and biodegradability, of the MgONPs are appropriate for variety of applications particularly in plant growth and also protection from the microbes.

5 Synthesis Methods of Zinc Oxide Nanoparticles (ZnONPs)

Zinc oxide nanoparticles (ZnONPs), an n-type semiconducting metal oxide have been extensive used in biomedical, cosmetics, rubber, and electronic systems. Priyadharshini et al. [56] prepared ZnONPs using extract of Gracilaria edulis,the solution containing Zn^{2+} turns to reddish-brown, signifying formation of ZnONPs. When the nanoparticles were coated with quinine to form the rod-shaped nanoparticles, the cellular toxicity of the nanoparticles against the PC3 human malignant cells was raised because the large surface area and better cell penetration and adhesion enabled the nanoparticles to be toxic to the cells. The synthesis of ZnONPs was made by [21] through a process known as reactive milling that utilized agar derived from

Gracilaria gracilis as the template. The prepared nanoparticles showed hexagonal wurtzite structures (18–50 nm) with strong photocatalytic activities and exhibited considerable efficiency in the degradation of aqueous phenol solutions after calcination process. Shariati et al. [65] synthesized indium-doped ZnONPs, using Chlorella vulgaris and *Scenedesmus quadricauda.* These nanoparticles attained 200 mg/L, and these inhibited the algal growth in water by 100%. Shrasti et al. [68] successfully synthesized ZnONPs using *Chlorosarcinopsis* sp. *algae* for water treatment application and as biodiesel. With 10 mg/L ZnONPs they observed 87.2% nitrogen, 97.5% organic carbon and 82.21% phosphorus removal in the experiments. The treatment of leather industry effluent using green synthesized ZnONPs using *Oedogonium* sp. and the result showed considerable reduction in TDS, chlorides, Cr, Cd, and Pb within 45 days. The ZnO NPs were also synthesized by modified sol–gel technique using Zinc acetate dihydrate [$Zn(CH_3COO)_2 2H_2O$] and sodium hydroxide (NaOH) for the development of chitosan-zinc oxide nanocomposite thin films for refinery wastewater treatment and the optimum processing conditions were established. From the experimental studies, the best performance was achieved at pH 8.0, 60 min of contact time with ten layers of thin-film deposition, and a stirring speed of 100 rpm. The research study illustrated that the nanocomposite thin films developed from chitosan and ZnO NPs has a promising pollutant removal capacity in the refinery wastewater treatment [23]. Table 3 represents the synthesis methods of ZnO NPs and their applications.

6 Synthesis Methods of Iron Oxide Nanoparticles (FeONPs)

The synthesis of iron oxide nanoparticles (FeONPs) using seaweeds is one of the green, rapid, and economical approaches. The reaction involves the reduction of ferric chloride ($FeCl_3$) to FeONPs, with the seaweed extracts whereby the sulphated polysaccharides acted as reducing and stabilizing agents. Phytogenic synthesis of FeONPs has been confirmed by the presence of surface plasmon band at 402–415 nm and particle size of 10–19.5 nm of *P. pavonica* and 21.6–27.4 nm of *S. acinarium.* The amino and carboxyl functional groups of polysaccharides aid in the synthesis of forming cubic crystalline particles of ~ 18.4 nm synthesized FeONPs from Sargassum muticum extract [20] produced the FeONPs using Penikeitleria pavonica and Sargassum acinarium. Among various micropollutants, the adsorption capacities of FeONPs for pyrene and benzo(a)pyrene were estimated by Hassan et al. [30] at 2.8 mg/g and 0.029 mg/g, respectively. Further, synthesized green iron nanoparticles with size ranging from 5 to 10 nm were also used and seen to remove 88.2% of Total Petroleum Hydrocarbon (TPH) within 12–32 h. Subsequently, [19] further investigated the preparation of FeONPs from *P. pavonica C. sinuosa* and *P. fascia* could mitigate marine pollution. Instead, the nanoparticles assured extra nitrogen and phosphorus and inhibited the algae's growth. In conclusion, the biosynthesis of FeONPs using brown seaweed extracts is therefore a sustainable

Table 3 Methods of synthesis of and application of ZnO_2NPs

Synthesis method	Application	References
Simple homogenous precipitation without surfactant	Anti-fungal activity against *Candida albicans*	[66]
Sol–gel method	Photocatalytic degradation	[25]
Microwave-assisted method	Dye removal applications	[42]
Organic precursor method	Drug carrier for anti-cancer activity	[7]
Hydrothermal method	Efficiency of solar still	[63]
Simple combustion method	Photo degradation of Methyl orange dye under UV radiation	[12]
Chemical co-precipitation method	Photo degradation of methylene blue under UV irradiation	[33]
Hydrothermal method	Photo degradation of Rhodamine B dye	[43]
Wet-chemical method	Removal of phenol from synthetic industrial wastewater	[70]
Co-precipitation method	Photo degradation studies of Congo red dye	[3]
Chemical method	Enhanced oil recovery	[69]
Synthesis using leaf extract of *Costus igneus*	Antibacterial activity	[75]
Sol–gel method	Photocatalytic degradation of methylene blue under visible light	[73]
Green synthesis using radish root	Effective wound dressing agents	[45]
Ultrasound-assisted synthesis	Photocatalytic activity towards degradation of Nitrogen oxides, (NO_x) in air	[50]

way to produce and have potential application in environmental remediation, water purification, heavy metals adsorption, pollutant elimination, and nutrient recovery.

7 Conclusion and Future Perspective

Due to the abundant availability and multidirectional applications, nanophotocatlysts have received widespread attention in various fields. This chapter portrays the general overview of catalysis and preparation methods of different types of photocatalysts using wet-chemical processes with particular emphasis on various applications are presented. In particular, the commonly employed chemical methodologies including sol–gel, hydrothermal, solvothermal precipitation/co-precipitation, chemical vapor deposition techniques, etc. for the photocatalysts preparation are explained in this chapter. Additionally, the sustainable synthesis methods presented in this chapter support the scale-up and commercial production process and also, an important step towards the circular economy perceptions by aligning with the goals and targets in the

2023 agenda of the United Nations Sustainable Development Goals (UNSDG-2030). The application of nanophotocatalysts in water purification will serve the society by saving a considerable amount of fresh water every day with reusing facility and to produce a clean environment. The various synthesis techniques of nanophotocatalysts offer a flexible approach for developing a diverse range of nanomaterials and microstructures with control of bulk properties of the end products.

References

1. Abisharani JM, Devikala S, Kumar RD et al (2019) Green synthesis of TiO_2 nanoparticles using *Cucurbita pepo* seeds extract. Mater Today Proc 14:302–307
2. Aboulouard A, Gultekin B, Can M et al (2020) Dye sensitized solar cells based on titanium dioxide nanoparticles synthesized by flame spray pyrolysis and hydrothermal sol-gel methods: a comparative study on photovoltaic performances. J Mat Res Tech 9(2):1569–1577
3. Adam RE, Pozina G, Willander M (2018) Synthesis of ZnO nanoparticles by co-precipitation method for solar driven photodegradation of Congo red dye at different pH. Photon Nanostruct Fundamentals Appl 32:11–18
4. Adawiya H, Riyad A, Ghadah K, Zainab (2018) Synthesis and photocatalytic activity for TiO_2 nanoparticles as air purification. MATEC Web Conf 162:05006
5. Afrah A, Mosaab E, Amel B, Imen et al (2023) Green synthesis of copper oxide nanoparticles using Ephedra Alata plant extract and a study of their antifungal, antibacterial activity and photocatalytic performance under sunlight. Heliyon 9:1–14
6. Ahmad W, Jaiswal KK, Soni S (2020) Green synthesis of titanium dioxide (TiO_2) nanoparticles by using Mentha arvensis leaves extract and its antimicrobial properties. Inorg Nano-Metal Chem 50(10):1032–1038
7. Al-Ajmi MF, Hussain A, Ahmed F (2016) Novel synthesis of ZnO nanoparticles and their enhanced anticancer activity: role of ZnO as a drug carrier. Ceram Int 42(3):4462–4469
8. Araoyinbo AO, Abdullah S et al (2018) Phase study of titanium dioxide nanoparticle prepared via sol-gel process. IOP Conference Series: Mat Sci Eng 343:012011
9. Aravind M, Amalanathan M, Sony MM (2021) Synthesis of TiO_2 nanoparticles by chemical and green synthesis methods and their multifaceted properties. SN Appl Sci 3:409–413
10. Araya-Castro K et al (2020) Green synthesis of copper oxide nanoparticles using protein fractions from an aqueous extract of Brown Algae Macrocystis pyrifera. Processes 9:78–85
11. Azar BE, Ramazani A, Fardood ST, Morsali A (2020) Green synthesis and characterization of $ZnAl_2O_4$@ ZnO nanocomposite and its environmental applications in rapid dye degradation. Optik 208:164129
12. Bhatia S, Verma N (2017) Photocatalytic activity of ZnO nanoparticles with optimization of defects. Mater Res Bull 95:468–476
13. Bhattacharya P et al (2019) Disinfection of drinking water via algae mediated green synthesized copper oxide nanoparticles and its toxicity evaluation. J Environ Chem Eng 7:102867
14. Buraso W, Lachom V, Siriya P, Laokul P (2018) Synthesis of TiO_2 nanoparticles via a simple precipitation method and photocatalytic performance. Mat Res Express 5(11):115003
15. Cabello G, Davoglio RA, Pereira EC (2017) Microwave-assisted synthesis of anatase-TiO_2 nanoparticles with catalytic activity in oxygen reduction. J Electroanaly Chem 794:36–42
16. Darbandi M, Shaabani B, Schneider et al (2019) TiO_2 nanoparticles with superior hydrogen evolution and pollutant degradation performance. Int J Hydrogen Energy 44(44):24162–24173
17. Dmitry B, Abduladheem TJ, Supat C, Wanich S et al (2021) Nanomaterial by sol-gel method: synthesis and application. Adv Mat Sci Eng:1–21
18. Dubey RS, Katta Venkata K, Shyam S (2019) Experimental studies of TiO_2 nanoparticles synthesized by sol-gel and solvothermal routes for DSSCs application. Results Phys 14:102390

19. El-Sheekh MM et al (2022) The efficient role of algae as green factories for nanotechnology and their vital applications. Microbiol Res 263:127111
20. El-Sheekh MM et al (2023) Green synthesis, characterization applications of iron oxide nanoparticles for anti-algal and wastewater bioremediation using three brown algae. Int J Phytoremediation 23:1538–1552
21. Francavilla M et al (2014) Efficient and simple reactive milling preparation of photocatalytically active porous ZnO nanostructures using biomass derived polysaccharides. Green Chem 16:2876–2885
22. Fulekar J, Dutta DP, Pathak B, Fulekar MH (2018) Novel microbial and root mediated green synthesis of TiO_2 nanoparticles and its application in wastewater remediation. J Chem Techn Biotech 93(3):736–743
23. Geetha DM, Chandra S (2024) Performance evaluation of chitosan- zinc oxide nanocomposite thin films in refinery wastewater treatment. Indian J Chem Tech 31:921–928
24. Getu KW (2020) Photocatalytic and antibacterial activity of CuO nanoparticles biosynthesized using Verbascum thapsus leaves extract. Optik 204:164230
25. Giahi M, Toutounchi S (2016) Synthesis of ZnO nanoparticles and its application in photocatalytic degradation of LABS by the trial-and-error and Taguchi methods. Russ J Appl Chem 89:823–829
26. Haider AJ, Al–Anbari RH, Kadhim GR, Salame CT (2017) Exploring potential environmental applications of TiO_2 nanoparticles. Energy Procedia 119:332–345
27. Hariharan C (2006) Photocatalytic degradation of organic contaminants in water by ZnO nanoparticles: revisited. Appl Catal A 304:55–61
28. Hartmann NB, Engelbrekt C, Zhang J, Ulstrup J, Kusk KO, Baun A (2013) The challenges of testing metal and metal oxide nanoparticles in algal bioassays: titanium dioxide and gold nanoparticles as case studies. Nanotoxicology 7:1082–1094
29. Hartmann NB, Von der Kammer F, Hofmann T et al (2010) Algal testing of titanium dioxide nanoparticles—testing considerations, inhibitory effects and modification of cadmium bioavailability. Toxicology 269:190–197
30. Hassaan MA, El Nemr A, Ragab S (2020) Green synthesis and application of metal and metal oxide nanoparticles. In: Handbook of nanomaterials and nanocomposites for energy and environmental applications. Springer, Cham, Switzerland, pp 1–27
31. Hebatallah AJ, Geetha D, Muhannad (2023) Development of polyaniline—TiO_2 nano composite films and its application in corrosion inhibition of oil pipelines. J Ind Chem Soc 100:100826
32. Helan V, Prince JJ, Al-Dhabi NA et al (2016) Neem leaves mediated preparation of NiO nanoparticles and its magnetization, coercivity and antibacterial analysis. Results Phys 6:712–718
33. Hitkari G, Singh S, Pandey G (2017) Structural, optical and photocatalytic study of ZnO and ZnO–ZnS synthesized by chemical method. Nano-Struct Nano-Objects 12:1–9
34. Hou C, Hu B, Zhu J (2018) Photocatalytic degradation of methylene blue over TiO_2 pretreated with varying concentrations of NaOH. Catalysts 8:575–563
35. Jeevanandam J, Hii YS, San CY (2021) Biosynthesized metal nanoparticles in bioremediation. In: Rhizomicrobiome dynamics in bioremediation. CRC Press, Boca Raton, FL, USA, pp 126–161
36. Kardanzadeh M, Kazeminezhad I, Mosivand S (2018) Electro-synthesis and characterization of TiO_2 nanoparticles and their application in removal of Congo red from water without UV radiation. Ceramics Int 44(5):5652–5659
37. Kashale AA et al (2016) Bio mediated green synthesis of TiO_2 nanoparticles for lithium ion battery application. Compos B Eng 99:297–304
38. Kaur H, Kaur S, Singh J, Rawat M, Kumar S (2019) Expanding horizon: green synthesis of TiO_2 nanoparticles using Carica papaya leaves for photocatalysis application. Mat Res Express 6(9):095034
39. Khalil AT, Ovais M, Ullah I, Ali M, Shinwari ZK, Maaza M (2020) Physical properties, biological applications and biocompatibility studies on biosynthesized single phase cobalt oxide (Co_3O_4) nanoparticles via Sageretiathea (Osbeck.). Arabian J Chem 13(1):606–619

40. Khanehzaei H et al (2014) Synthesis and characterization of Cu@ Cu_2O core shell nanoparticles prepared in seaweed Kappaphycus alvarezii Media. Int J Electrochem Sci 9:8189–8198
41. Kumar H et al (2020) Flower-based green synthesis of metallic nanoparticles: applications beyond fragrance. Nanomaterials 10:766–772
42. Kumar V, Gohain M, Som S, Kumar V et al (2016) Microwave assisted synthesis of ZnO nanoparticles for lighting and dye removal application. Physica B 480:36–41
43. Kumaresan N, Ramamurthi K, Babu RR, Sethuraman K, Babu SM (2017) Hydrothermally grown ZnO nanoparticles for effective photocatalytic activity. Appl Surf Sci 418:138–146
44. Kwang SJ, Jong DK (2009) Facile and large-scale route to the fabrication of CuO nanosheets from a lamellar mesophase and their reversible self-assembly. Langmuir 25(11):6028–6603
45. Liu D, Liu L, Yao L et al (2020) Synthesis of ZnO nanoparticles using radish root extract for effective wound dressing agents for diabetic foot ulcers in nursing care. J Drug Deliv Sci Tech 55:101364
46. Lusvardi G, Barani C, Giubertoni F, Paganelli G (2017) Synthesis and characterization of TiO_2 nanoparticles for the reduction of water pollutants. Materials 10(10):1208
47. Mahato TH, Prasad GK, Singh B et al (2009) Nanocrystalline zinc oxide for the decontamination of sarin. J Haz Mat 165(1–3):928–932
48. Manjunath K, Yadav LSR, Jayalakshmi T et al (2018) Ionic liquid assisted hydrothermal synthesis of TiO_2 nanoparticles: photocatalytic and antibacterial activity. J Mat Res Technol 7(1):7–13
49. Mayekar V, Dhar S, Radha (2013) To study the role of temperature and sodium hydroxide concentration in the synthesis of zinc oxide nanoparticles. J Sci Res Publ 3(11):2250–3153
50. Meroni D, Gasparini C, Di Michele A et al (2020) Ultrasound-assisted synthesis of ZnO photocatalysts for gas phase pollutant remediation: role of the synthetic parameters and of promotion with WO_3. Ultrason Sonochem 66:105119
51. Mónica AV, Jorge ER (2017) Amorphous TiO_2 nanoparticles: synthesis and antibacterial capacity. J Non-Cryst Solids 459(1):192–205
52. Nabi G, Raza W, Tahir MB (2020) Green synthesis of TiO_2 nanoparticle using cinnamon powder extract and the study of optical properties. J Inorg Organo Met Polym Mater 30(4):1425–1429
53. Nadeem A, Chandra K, Amir K et al (2023) Self-cleaning and fully polymer-based super-moistureresistant gas barrier coating films with 2D polymers for flexible electronic devices and packaging applications. J Mat Chem C 11:15907–15917
54. Nwankwo U, Bucher R, Ekwealor et al (2019) Synthesis and characterizations of rutile-TiO_2 nanoparticles derived from chitin for potential photocatalytic applications. Vacuum 161:49–54
55. Pratyush J, Arun K, Nishith V, Raju KG (2019) In-situ synthesis of TiO_2 nanoparticles in ACF: photocatalytic degradation under continuous flow. Sol Energy 189(1):35–44
56. Priyadharshini RI et al (2014) Microwave-mediated extracellular synthesis of metallic silver and zinc oxide nanoparticles using macro-algae (Gracilaria edulis) extracts and its anticancer activity against human PC3 cell lines. Appl Biochem Biotechnol 174:2777–2790
57. Pugazhendhi A et al (2019) Anticancer, antimicrobial and photocatalytic activities of green synthesized magnesium oxide nanoparticles (MgONPs) using aqueous extract of Sargassum wightii. J Photochem Photobiol B Biol 190:86–97
58. Qi L, Yongye L, Hongjiang L, Jianming H, Zheng X (2006) Solution phase synthesis of CuO nanorods. Mat Chem Phys 98(2–3):519–522
59. Ragadhita RISTI, Nandiyanto ABD, Maulana AC et al (2019) Techno-economic analysis for the production of titanium dioxide nanoparticle produced by liquid-phase synthesis method. J Eng Sci Tech 14(3):1639–1652
60. Ramakrishnan VM, Natarajan M, Santhanam A, Asokan V, Velauthapillai D (2018) Size controlled synthesis of TiO_2 nanoparticles by modified solvothermal method towards effective photo catalytic and photovoltaic applications. Mat Res Bulletin 97:351–360
61. Renuga D et al (2020) Synthesis and characterization of copper oxide nanoparticles using Brassica oleracea var. italic extract for its antifungal application. Mater Res Express 7:045007
62. Ristic M, Musi S (2005) Sol–gel synthesis and characterization of nanocrystalline ZnO powders. J Alloys Compounds 397(1–2)

63. Salehi G, Abazari R, Mahjoub AR (2018) Visible-light-induced graphitic–C_3N_4@ nickel–aluminum layered double hydroxide nanocomposites with enhanced photocatalytic activity for removal of dyes in water. Inorg Chem 57(14):8681–8691
64. Sangeetha BM, Noura AM, Geetha D, Susmita D, Anjali KP (2021) Fabrication, characterization and application of polymer based nanomaterials in the removal of pollutants from industrial effluents. Environ Technol Innov 23:101748
65. Shariati M et al (2018) Inhibitory effects of functionalized indium doped ZnO nanoparticles on algal growth for preservation of adobe mud and earthen-made artworks under humid conditions. Int Biodeterior Biodegrad 127:209–216
66. Sharma RK, Ghose R (2015) Synthesis of zinc oxide nanoparticles by homogeneous precipitation method and its application in antifungal activity against Candida albicans. Ceram Int 41(1):967–975
67. Sharma S, Sharma RK, Gaur K et al (2019) Fueling a hot debate on the application of TiO2 nanoparticles in sunscreen. Materials (Basel). 12(14):2317
68. Shrasti V, Anwesha K, Monika PR (2021) Influence of microalgae-ZnO nanoparticle association on sewage wastewater towards efficient nutrient removal and improved biodiesel application: an integrated approach. J Water Process Eng 39:101711
69. Soleimani H, Baig MK, Yahya, (2018) Synthesis of ZnO nanoparticles for oil–water interfacial tension reduction in enhanced oil recovery. Appl Phys A 124:128
70. Sridhar R, Uma U, Ramanane R et al (2018) ZnO nanoparticles—synthesis, characterization and its application for phenol removal from synthetic and pharmaceutical industry wastewater. Envir Nanotech Monit Man 10:388–393
71. Subhapriya S, Gomathipriya P (2018) Green synthesis of titanium dioxide (TiO_2) nanoparticles by Trigonella foenum-graecum extract and its antimicrobial properties. Microb Pathog 116:215–220
72. Trouiller B, Reliene R, Westbrook A, Solaimani P, Schiestl RH (2009) Titanium dioxide nanoparticles induce DNA damage and genetic instability in vivo in mice. Cancer Res 69:8784–8789
73. Upadhyay GK, Rajput JK, Pathak TK et al (2019) Synthesis of ZnO: TiO_2 nanocomposites for photocatalyst application in visible light. Vacuum 160:154–163
74. Vasanth V (2022) Synthesis of titanium dioxide nanoparticles using Spirulina platensis algae extract. Pharma Innov J SP-11:266–269
75. Vinotha V, Iswarya A, Thaya R, Govindarajan M et al (2019) Synthesis of ZnO nanoparticles using insulin-rich leaf extract: anti-diabetic, antibiofilm and antioxidant properties. J Photochem Photobiol, B 197:111541
76. Yanyan X, Dairong C, Xiuling J (2005) Fabrication of CuO pricky microspheres with tunable size by a simple solution route. J Phys Chem B 109(28):13561–13566
77. Yuanlie Y, Junyan Z (2009) Solution-phase synthesis of rose-like CuO. Mate Letters 63(21):1840–1843
78. Yue S, Yan Z, Shi Y, Ran G (2013) Synthesis of zinc oxide nanotubes within ultrathin anodic aluminum oxide membrane by sol-gel method. Mater Lett 98:246–249
79. Yueli L, Lei L, Jinchai L, Chunxu P (2007) From copper nanocrystalline to CuO nanoneedle array: synthesis, growth mechanism, and properties. J Phys Chem C 111(13):5050–5056
80. Zavar S (2017) A novel three component synthesis of 2-amino4H-chromenes derivatives using nano ZnO catalyst. Arabian J Chem 10:S67–S70

Characterization of Nano Photocatalysts

R. Senthil Kumar and M. Geetha Devi

Abstract Nanomaterials have excellent functional properties with enhanced surface-to-volume ratio and, therefore suitable for diversified applications. In order to study and understand the physical and chemical properties of nanocatalysts, various characterization tools are employed through imaging and analysis by spectroscopy and microscopy. Therefore, this chapter will provide an overview of commonly employed characterization tools in determining the morphological, chemical, physical, and structural properties of nanoparticles and nanostructured materials. The techniques for analyzing the topology of nanocatalysts, including Field Emission Scanning Electron Microscopy (FESEM), Transmission Electron Microscopy (TEM), Energy Dispersive X-ray spectroscopy (EDS), Fourier Transform Infrared Spectroscopy (FTIR), Dynamic Light Scattering (DLS), Atomic Force Microscopy (AFM), UV–Visible spectroscopy, Zeta potential measurements, X-Ray Florescence (XRF) are described. Additionally, the microstructural investigation using X-ray diffraction (XRD) is also included in this chapter.

Keywords Characterization · Dynamic light scattering · Microscopy · Nanomaterials · Particle size analyzer

1 Introduction

The physical properties, chemical composition, functional groups, and band structure of nano materials and photocatalysts are important in assessing their performance in specific applications [1]. Based on the physico-chemical characteristics, nanostructured materials exist in different sizes, surface structural characteristics, constitution,

R. Senthil Kumar
College of Engineering and Technology, University of Technology and Applied Sciences, Suhar, Sultanate of Oman

M. Geetha Devi (✉)
College of Engineering, National University of Science and Technology, Muscat, Oman
e-mail: geethadevi@nu.edu.om

A. Ahmad et al. (eds.), *Metal Oxide Based Nanophotocatalyst for Wastewater Purification*, Smart Nanomaterials Technology,
https://doi.org/10.1007/978-981-96-7683-5_3

solubility, molecular weight, and stability. These properties play a significant role in differentiating nanomaterials. The material scientists and researchers have conducted extensive studies on the application of nanoparticles and nanostructured materials in various fields [2, 3]. The thermal, structural, optical, and electrical properties of nanomaterials are analyzed using these testing methods are essential for the detailed study of elemental analysis, molecular interactions, structural study, surface characterization, and morphology [4]. Some of the common characterization techniques in current practice within the field of nanomaterials are explained. The size and morphology of the nanocatalysts solely depend on several factors such as concentration of reactants, amount of surfactants and condition of solvents, reaction time, and temperature during the synthesis process.

2 Characterization Techniques of Nanocatalysts

Characterization encompasses various practices required to discover material properties and microscopic structures with material analysis, such as thermal analysis, mechanical testing, and density calculation. The characterization tools employed in the analysis of nanomaterials and nanoparticles are vital to assess its purity, morphology, elemental composition, particle size, presence of functional groups, crystallographic structure, surface charge, microstructural characteristics, and bond stretching, etc. The material characterization and analyses techniques of the nanostructured materials are UV–Vis Spectroscopy, TEM, EDX, DLS, XRD, SEM, AFM, FTIR, XRF, and Zeta potential measurement, etc. Brief description of the characterization tools along with the working principle is presented in this section.

2.1 Ultraviolet–Visible Spectroscopy (UV–Vis Spectroscopy)

UV–Vis Spectroscopic techniques are used to analyze inorganic, organic, and biological materials by passing light through the sample and measuring the absorbance values at different wavelengths. UV spectroscopic analysis is employed to determine the relationship between the particle size distributions of quantum-sized nano crystals and absorbance spectra. It works by Beer-Lambert Law. At a particular wavelength, the absorbance of a sample is directly proportional to the concentration of the absorbing substance, and the path length. The absorbance is calculated using Eq. (1)

$$A = abc \tag{1}$$

where A is the absorbance, a is the absorptivity, b is the concentration, and c is the path length.

This characterization tool is employed in identifying, characterizing, and studying variety of nanoparticle formulations in suspension form. Figure 1 shows the

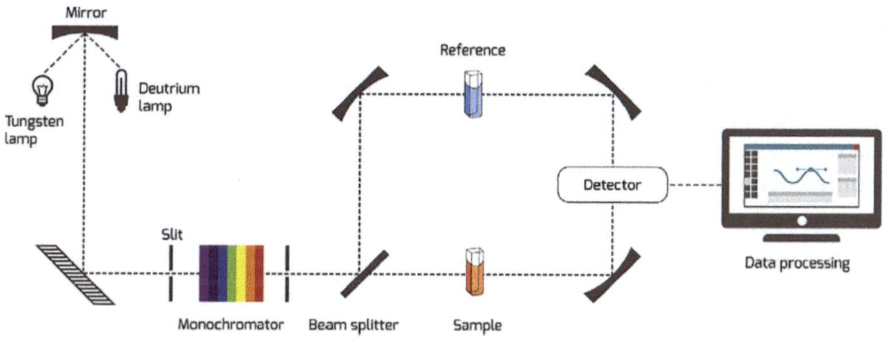

Fig. 1 Schematic representation of the UV–Vis spectroscopy

schematic representation of UV–Vis Spectroscopy. In order to understand the various stages in the development of nanoparticles, UV spectroscopy is the ideally used. This characterization tool can be employed for the quick analysis of surface property, the size and shape of colloidal particles in the dispersed phase. The amount of precursor metal ions consumed during the preparation of metal nanoparticles can be detected quickly by UV spectroscopic analysis [4]. The applications of UV—visible spectrophotometer include pharmaceutical, environmental, quantitative, food, and beverage analyses. A UV–Vis spectrophotometer is a powerful tool that measures the absorbance of liquid samples at different wavelengths and confirms a reliable measurement of the concentration for wide range of substances in various fields by utilizing monochromatic light. It receives incident monochromatic light to illuminate a liquid sample with light across the UV to the visible wavelength range (typically 190 to 900 nm). The UV spectrophotometer measures the absorbed, transmitted, or reflected light by the liquid sample at different wavelengths.

2.2 *Fourier Transforms Infrared Spectroscopy (FTIR)*

In FTIR analysis, infrared radiations are passed through the sample surface. When an Infrared radiation passes, the sample absorbs some of the radiation while the remaining passes through it. The spectrum rising from absorption and transmission modes generates a molecular fingerprint of sample. It is preferred in materials characterization owing to its enhanced sensitivity, speed, accuracy, high precision, easy operation, and the sample will be intact or non-destructive after testing [5]. The change in material composition is predicted by the shift in the characteristic peaks due to the absorption bands in the FTIR spectrum. The analysis using FTIR technique is suitable for the identification and characterization of unknown materials, sensing contaminants, and also for detecting the material composition and the purity. Figure 2 illustrates the working principle of FTIR. In FTIR analysis, the infrared radiation passes through the sample and it is partially absorbed by the sample and some are transmitted. The resulting spectrum develops a molecular fingerprint of the sample. Figure 3 represents the schematic of FTIR.

The various functional groups, broad spectrum and bond stretching vibrations, etc., on the nanoparticle surface are determined using FTIR analysis operated in non-destructive mode, and analysis is conducted at room temperature. The existence of different functional groups present on the silica nanoparticle is displayed in Fig. 4.

Fig. 2 Working principle of Fourier transform infrared spectroscopy [6]

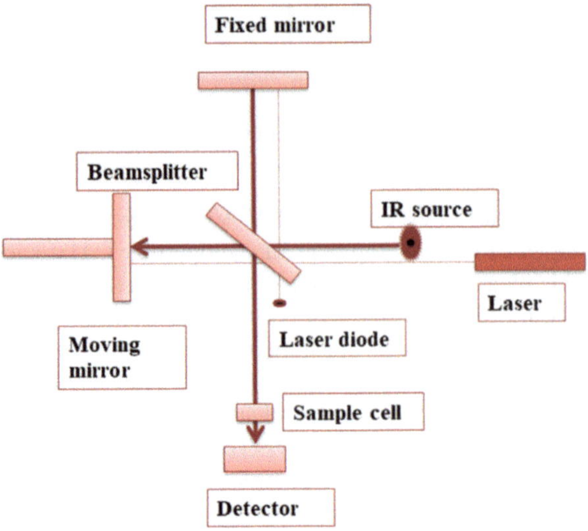

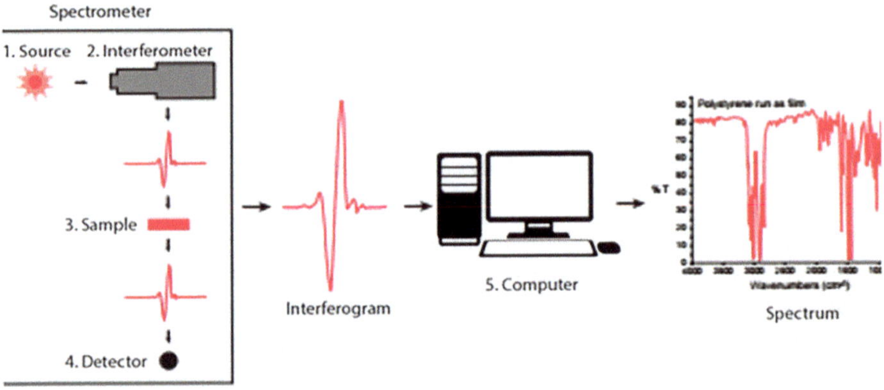

Fig. 3 Schematic of Fourier transform infrared spectroscopy [6]

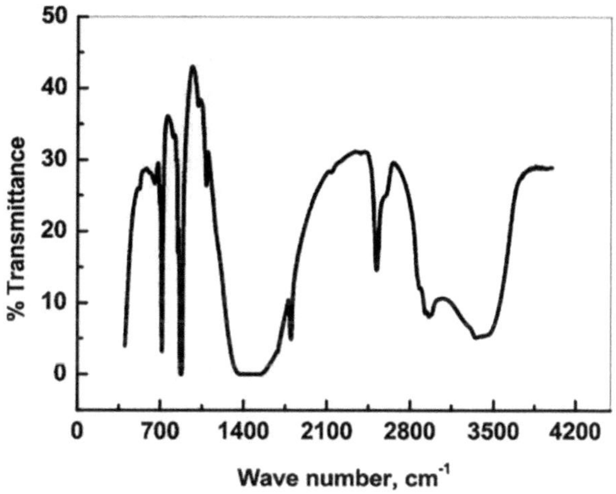

Fig. 4 FTIR spectra of silica nanoparticles [7]

2.3 *Scanning Electron Microscopy (SEM)*

Scanning Electron Microscopy is used to study the surface morphological and microstructural characteristics of solid samples by scanning through high-energy electron beams. The image is developed by passing electron beams and magnified using electromagnetic fields. The conventional light microscope uses light waves to form a magnified image, whereas SEM utilizes electron beams and interacts with the sample surface.

A narrow beam of electrons from an electron gun hits the sample surface and scans the surface by releasing secondary electrons along with other radiations from the specimen surface. The intensity of secondary electrons emitted will vary according to the shape and chemical composition of the powder sample. The detector collects the secondary electrons and produces electronic signals. SEM utilizes both back-scattered and secondary electrons. High-resolution images are produced by SEM at a resolution of 5 nm using the secondary electrons. The SEM samples are prepared by dropping the liquid sample on a stub and dried overnight at ambient conditions to remove the moisture content. Then the sample surface will be coated with platinum and then placed inside the SEM for characterization. Figure 5 shows the image of scanning electron microscope. The electron gun shoots out a bunch of large number of electron beams from the top of the device. The types of electron guns used are Thermionic guns, which heat a filament until electrons stream away. These electrons are directed towards the specimen through lenses to maximize efficiency. The more number of electrons are allowed to pass through then superior will be the view of the sample. The SEM characterization utilizes vacuum chamber. When electron beam hit the sample surface, it radiates X-rays with three types of electrons namely primary and secondary back-scattered electrons, and Auger electrons are generated. The SEM utilizes primary back-scatter electrons and secondary electrons. This information is translated onto a screen which allows three-dimensional images to be represented clearly. In addition, SEM is more user-friendly and performs much faster measurements with less maintenance costs.

The surface morphological analysis of the nanoparticles will provide the minute features and characteristics of the particle distribution, aggregation state, yield, and size, etc. Figure 6 shows the SEM image of the silica nanoparticles at a magnification of 30,000× and a voltage of 15.0 keV [7].

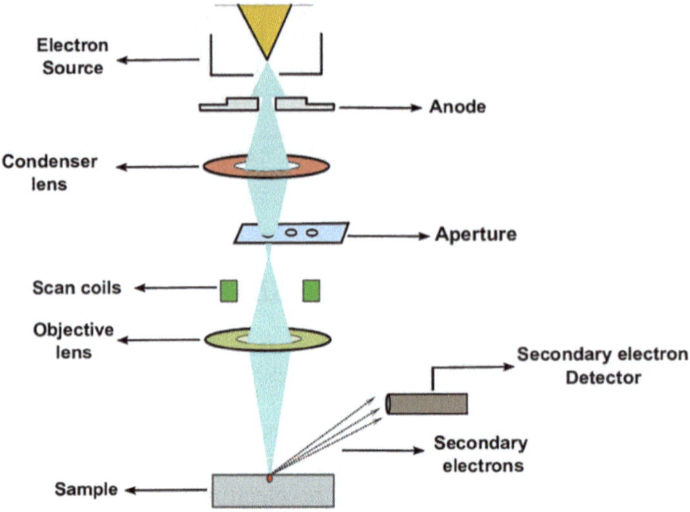

Fig. 5 Working principle of SEM [8]

Fig. 6 SEM image of silica nanoparticles [7]

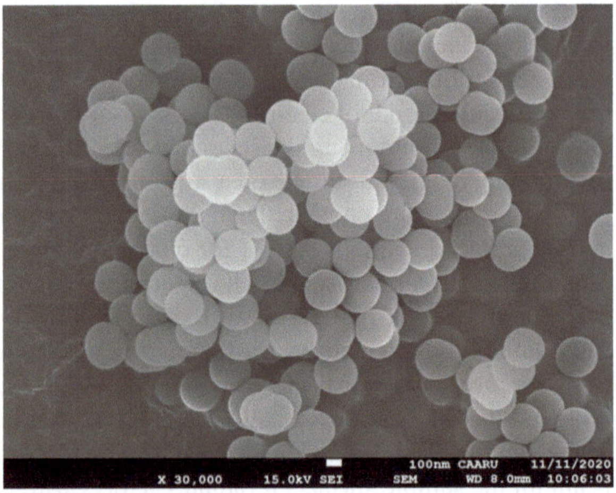

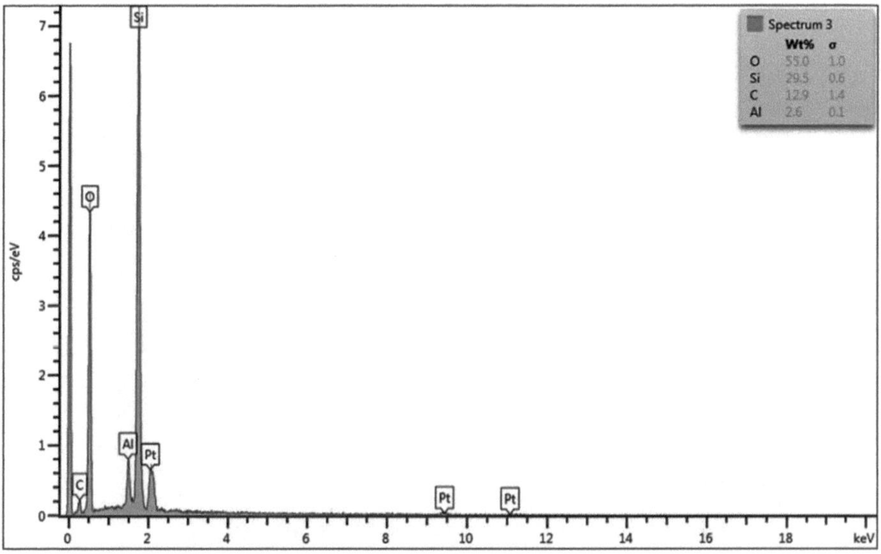

Fig. 7 EDX spectra of silica nanoparticles [7]

2.4 Energy Dispersive X-Ray Analysis (EDX)

Energy Dispersive X-Ray spectroscopy (EDS or EDX) is the analytical tool used to investigate the individual elemental conformation and chemical characterization of the sample. EDX is a quantitative analysis, which is aligned with SEM. Gold, palladium, and silver nanoparticles are easily analyzed using EDX. However, low atomic number elements are difficult to detect by EDX. EDX provided the elemental composition and purity of the silica nanoparticles as shown in Fig. 7.

2.5 X-Ray Diffraction Analysis (XRD)

XRD analysis is used to test the crystallinity of the powder sample. X-ray diffractogram can be used to find the crystal structure and diameter of the nanoparticles. X-Ray diffraction is technique used to determine the purity and degree of crystallinity of a material as a function of time. The differences were correlated to tensile strengths and elasticity data. Also, to determine the crystalline species in a material, strain, preferred orientation, crystallographic structure, and grain size of crystalline materials. When X-Ray diffraction is used in a glancing angle mode, crystalline structure as a function of depth can be obtained [9]. The phase and crystal structure of the synthesized particles are characterized using high-performance X-ray Diffraction (XRD) and the typical Diffractometer is displayed in Fig. 8.

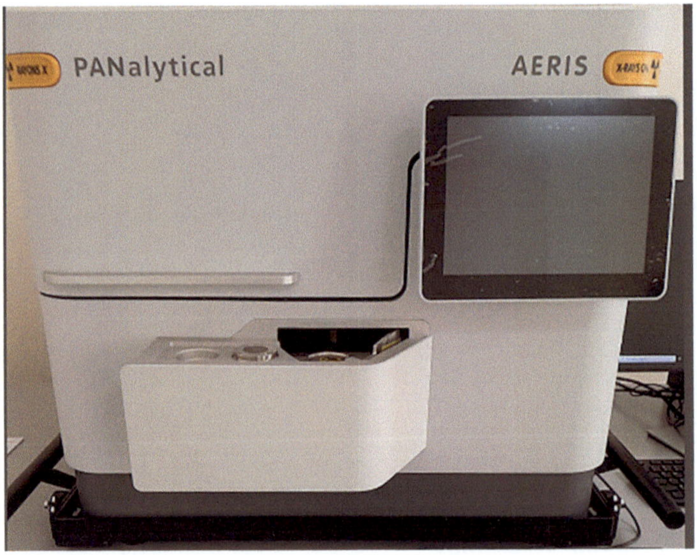

Fig. 8 X-ray diffractometer (XRD)

In XRD analysis, an X-Ray source is used to pass through the specimen to be analyzed and to cause the elements in the specimen to emit their characteristic X-Rays. Figure 9 shows the working principle of XRD. A wavelength dispersive detector was used to measure the peaks of the emitted X-Rays for qualitative/ quantitative analysis of the elements and their quantification. The X-Ray diffraction techniques are also applied in the simultaneous determination of elemental composition and film thickness.

The phase identification of the SiO_2 nanoparticles is judged through XRD analysis with a diffraction angle 2θ ranges between 10° and 90°. The powdered silica particles were used for the determination of average bulk composition. A single peak observed

Fig. 9 Working principle of XRD

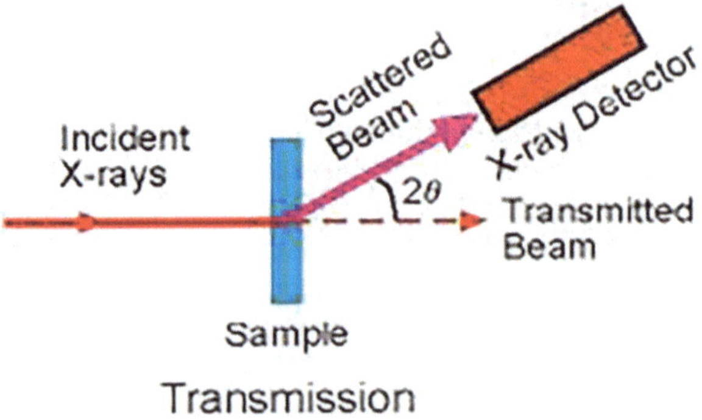

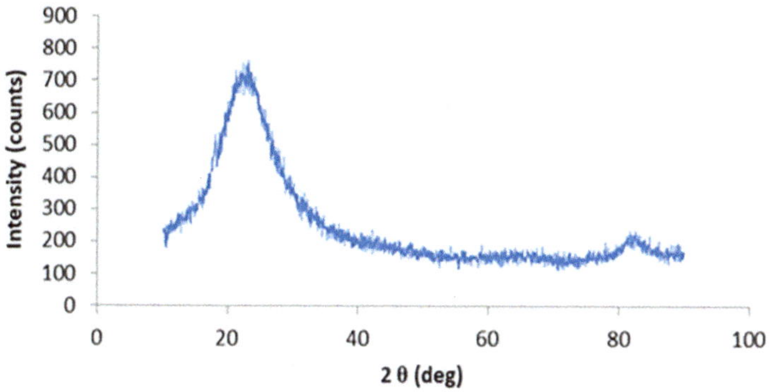

Fig. 10 XRD spectral analysis of silica nanoparticles [7]

in the spectrum demonstrates that particles are amorphous in nature as shown in Fig. 10 [7].

2.6 Dynamic Light Scattering (DLS)

The particle size distribution and characterization of colloidal solutions of nanoparticles are carried out using Dynamic light scattering (DLS) or Photon correlation spectroscopy (PCS).

The DLS measurement is carried out by passing a laser beam through the sample, which hits the colloidal solution, and the particles will undergo Brownian motion which results in light scattering. The scattered light intensity is modulated and analyzed as a function of time. The diameters of the particles are calculated from the time dependence of the scattering intensity measurement. DLS measurements also provide information about the extent of aggregation of nanoparticles. The hydrodynamic diameters mainly depend on the shape and particle size of the macromolecules [10].

Figure 11 represents the distribution of mean particle size of silica nanoparticles synthesized by Stober's protocol. Figure 11 indicates the uniform particle size distribution with an average particle size of 335 nm. It is observed that a single peak demonstrates the particles are of uniform size, which is one of the desirable properties of nanoparticles in wastewater treatment [11].

2.7 Zeta Potential Measurements

The extent of nanoparticle formation and characterization of colloidal solutions are determined based on the surface charge and this measurement will predict the success

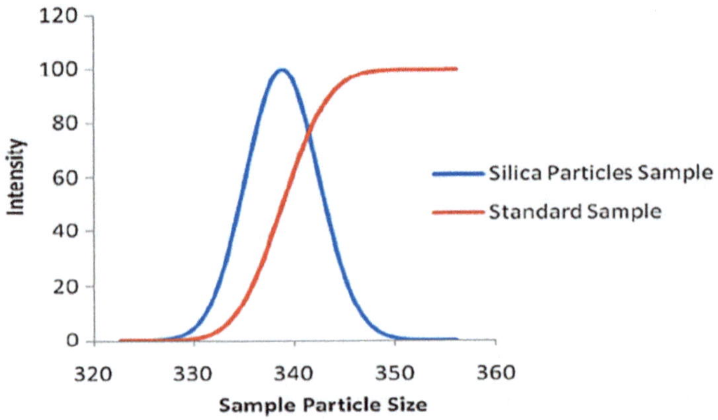

Fig. 11 Particle size analysis using DLS [11]

Fig. 12 Zeta potential of SiO_2 nanoparticles [7]

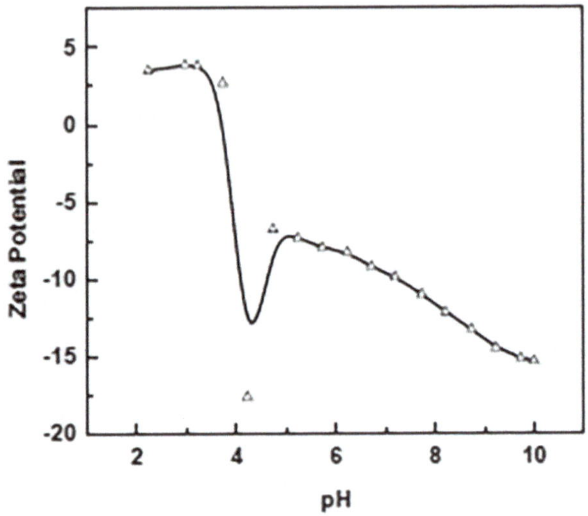

of synthesis. Figure 12 shows the zeta potential measurement of SiO_2 particles at different time intervals. The zeta potential test demonstrates the surface charge and accomplishment of silica nanoparticle formation.

2.8 X-Ray Fluorescence Spectrometers (XRF)

XRF is a non-destructive analytical tool used to find the material chemistry and the composition analysis of a test sample by quantifying the fluorescent X-ray released when it is excited by a primary X-Ray source. It also measures the coating thickness,

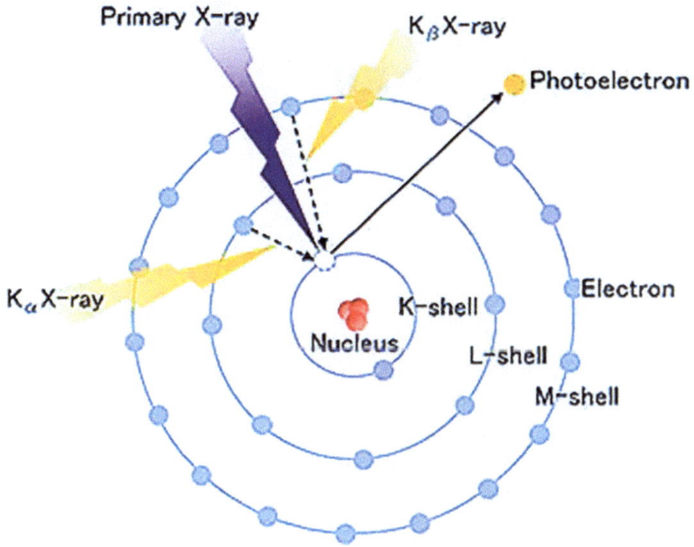

Fig. 13 Working principle of XRF (https://wpo-altertechnology.com/)

composition, and development of surface coatings. XRF can be preferred for the analysis of certain elements from beryllium (Be) to uranium (U) from 100 wt% to sub-ppm levels of concentration.

In XRF analysis, the test specimen is excited with the primary X-rays and the electrons from the inner electron shells are knocked. The detector evaluates the fluorescence radiation. The working and generation of the X-ray fluorescence radiation is shown in Fig. 13.

2.9 *Atomic Force Microscopy (AFM)*

Atomic Force Microscopy is a powerful and versatile characterization tool in the category of microscopy technology mainly employed in the surface topological analysis of samples at a nanoscale. It gives three-dimensional topographical image and also provides surface roughness measurements at atomic resolution with angstrom-scale precision. AFM analysis exhibits the topological image and surface roughness. Additionally, it also provides information about the length, width, and height of the synthesized nanoparticles and other physical properties (such as morphology and surface texture). The surface roughness and textural features of polymer nanocomposites and various materials can be studied by AFM topological investigation. It is, moreover, a non-destructive technique and has a high three-dimensional spatial resolution [12]. Figure 14a, b represents the AFM image in two and three-dimensional view.

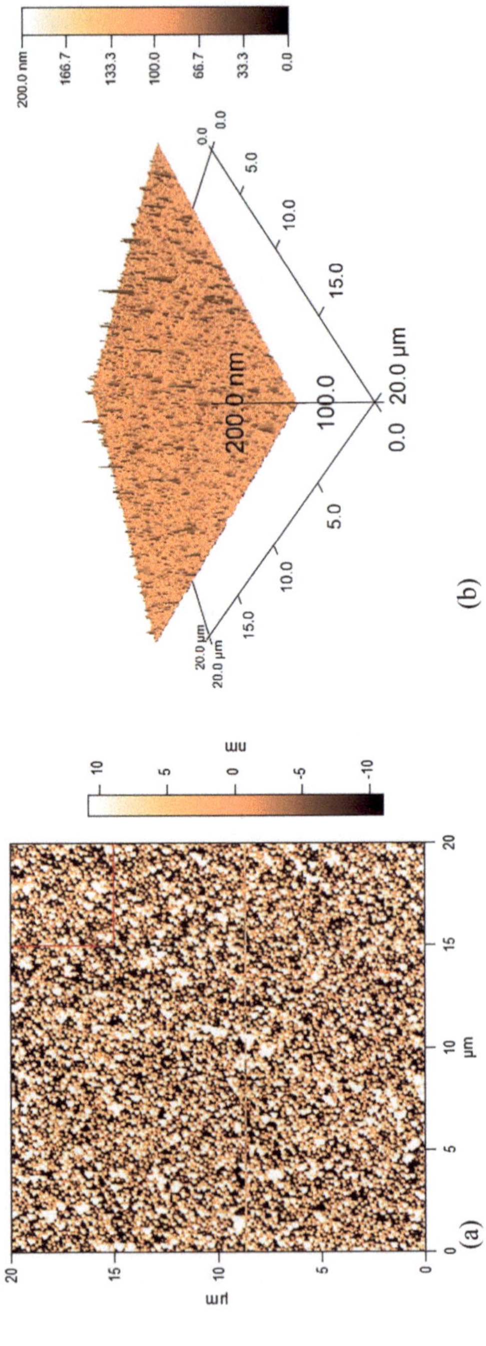

Fig. 14 **a** AFM 2d image of pure silica nanoparticles **b** AFM 3d image of pure silica nanoparticles

2.10 Transmission Electron Microscopy (TEM)

The high-resolution Transmission Electron Microscopy imaging is used for studying nanoscale properties of materials to determine the nanoparticle size, geometry, mono dispersity, shape, aggregation state, detects and localizes nanoparticles in matrices, and also allows study growth kinetics. Due to the high resolution, it can easily assess crystal structures and crystal defects of individual atoms [13, 14]. When the electron beam interacts with the sample, it diffraction in x-axis at certain angles, whereas it is transmitted at other angles and the transmitted electrons are used for imaging purposes. The working of TEM is illustrated in Fig. 15.

The commonly employed nanoparticle characterizations are summarized in Table 1.

Fig. 15 The working principle of TEM

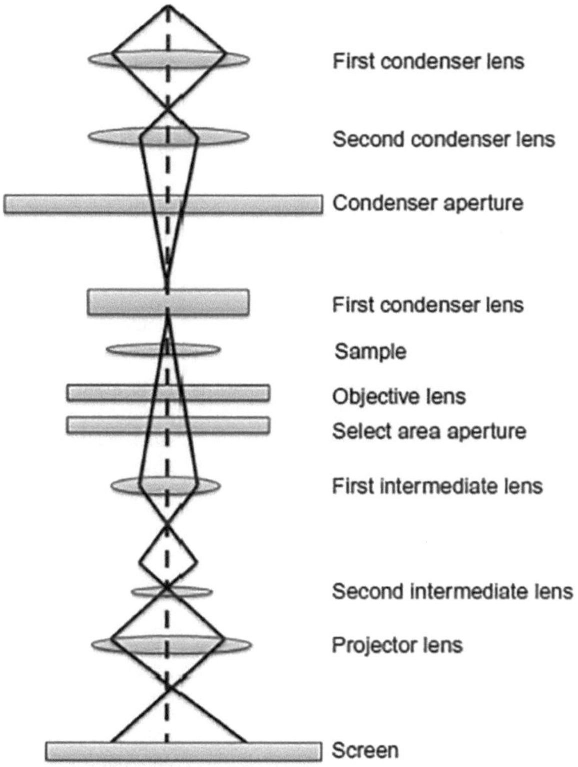

Table 1 Various characterization tools employed in the nanoparticle synthesis process

Type	Characterization technique	Application	References
Nanoparticle formation	UV–Vis spectroscopy	Provide information about nanoparticle size, structural properties, stabilization, and particle aggregation	[15]
Surface morphology and particle size	Transmission electron microscopy	Determine the shape, size (10^{-10} m), surface morphology, and allographic nature of the nanoparticles	[16]
	Atomic-level resolution Transmission electron microscopy	Provide information on microstructures, defects, arrangement of atoms in the nanoparticles, etc.	[17]
	Scanning electron microscopy	Determine the morphology by direct visualization through electron beams	[18]
Surface topology	Atomic force microscopy	Finding the topology, rms value of roughness, thickness and two dimensional (2D) and three dimensional (3D) images	[19]
Size analysis	Photon Correlation Spectroscopy / Dynamic light scattering	Determine the particle size distribution by the Brownian motion of nanoparticles	[20]
Surface charge	Zeta potential	Determine the surface charge and colloidal stability of the nanoparticles	[21]
Functional group	Fourier transform infrared spectroscopy (FTIR)	Determine the different functional groups attached to the nanoparticles surface and to understand the absorption, emission, photoconductivity, of solids and liquids	[22]
	Thermo gravimetric analysis (TGA)	Endorse the development of surfactant or polymer coating and to estimate the surface binding efficiency of magnetic nanoparticles	[17]
Crystallinity	X-Ray diffraction (XRD)	Identify amorphous/crystalline phases of powder samples and to measure composition analysis of nanoparticles	[19]
	Magnetometry	Determine the magnetic property of the nanoparticles	[17]
Other characterization tools	Chromatographic techniques	Separation of nanoparticles based on their affinity towards the mobile phase	[23]
	Energy dispersive X-ray spectra	Identify the elemental composition of the nanoparticles	[22]

(continued)

Table 1 (continued)

Type	Characterization technique	Application	References
	Field flow flotation	Nanoparticle separation based on magnetic susceptibility	[24]
	Centrifugal separation	Separation of nanoparticle suspension by centrifugal force	[25]
	X-Ray fluorescence spectroscopy (XRF)	Detect and measure the composition of powder or liquid samples	[26]

3 Conclusion and Future Perspective

Nanotechnology has emerged as one of the prominent field of current cutting-edge research and move for advanced materials development and shift towards multifunctional applications ranging from environmental remediation to smart materials. The development and application of novel nanomaterials employing innovative technologies has motivated to the discovery of more advanced sophisticated equipment's for the characterization and fine tuning of nanomaterial properties. The progress of nanomaterials formation and their application to tackle global issues requires advancements in nanoparticle characterization. Characterization and analysis of materials are very important to recognize the controlled production of nanoparticles and its application. The important characterization tools employed in the testing and analysis of various features and properties are presented in this chapter. Nanotechnology-oriented approach is needed for the testing and characterization of nanomaterials in a faster, simpler, effective and employing novel techniques for materials characterization to better understand structure of materials and composition.

References

1. Bignozzi CA et al (2013) Nanostructured photo electrodes based on WO3: applications to photooxidation of aqueous electrolytes. Chem Soc Rev 42:2228
2. Mabrouk M, Das DB, Salem ZA, Beherei HH (2021) Nanomaterials for biomedical applications: production, characterisations, recent trends and difficulties. Molecules 26(4):1077. https://doi.org/10.3390/molecules26041077
3. Harish et al (2022) Review on nanoparticles and nanostructured materials: bio imaging, bio sensing, drug delivery, tissue engineering, antimicrobial, and agro-food applications. Nanomaterials 12(3):457
4. Praveen KJ, Sanjay MR, Suchart S (2024) Advanced characterization techniques for nanostructured materials in biomedical applications. Adv Industrial Eng Polym Res 7:122e143
5. Huang X, Beck MJ (2015) Determining the oxidation state of small, hydroxylated metaloxide nanoparticles with infrared absorption spectroscopy. Chem Mater 27:2965

6. Taha M et al (2013) Use of Fourier transform infrared spectroscopy (FTIR) spectroscopy for rapid and accurate identification of Yeasts isolated from human and animals. Int J Veterin Sci Med 1(1):15
7. Amna WYAR, Devi MG, Chandrasekhar G (2022) Synthesis and application of silica and calcium carbonate nanoparticles in the reduction of organics from refinery wastewater. J Ind Chem Soc 99(7):100519
8. Naresh M, Sounkalo D, Nadine P (2014) Scanning electron microscope image signal-to-noise ratio monitoring for micro-nano manipulation. J Scann Microsc 3:11
9. Zaid HJ et al (2023) A review study summarizes the main characterization techniques of nano composite photocatalysts and their applications in photo degradation of organic pollutants. Environ Nanotechnol Monit Manage 19:100765
10. Stetefeld J, McKenna SA, Patel TR (2016) Dynamic light scattering: a practical guide and applications in biomedical sciences. Biophys Rev 8(4):409
11. Ismail I, Balachandran S, Devi MG (2019) Synthesis, characterization and application of nanoparticles in wastewater treatment. Ind Chem Eng 61:77
12. Qin et al (2023) Progress in the applications of atomic force microscope (AFM) for mineralogical research. Micron 170:103460
13. Joshi M, Bhattacharya A, Ali W (2008) Characterization techniques for nanotechnology application in textiles. Ind J Fibre Text Res 33:304
14. Stefanos et al (2018) Characterization techniques for nanoparticles: comparison and complementarity upon studying nanoparticle properties. Nanoscale 10:12871
15. Gupta V, Gupta AR, Kant V (2013) Synthesis, characterization and biomedical application of nanoparticles. Sci Int 5:167
16. Kowalczyk B, Lagzi I, Grzybowski BA (2011) Nanoseparations: strategies for size and/or shape-selective purification of nanoparticles. Curr. Opin. Colloid. Interf. Sci. 16(2):135
17. Faraji M, Yamini Y, Rezaee M (2010) Magnetic nanoparticles: synthesis, stabilization, functionalization, characterization, and applications. J Iran Chem Soc 7(1):1
18. Prashanth et al (2011) Synthesis of plant-mediated silver nanoparticles using medicinal plant extract and evaluation of its antimicrobial activities. Int J Eng Sci Tech 3(8):6235
19. Chauhan RPS, Gupta C, Prakash D (2012) Methodological advancements in green nanotechnology and their applications in biological synthesis of herbal nanoparticles. Int J Bioassays 1(7):6
20. Cookson (2012) The preparation of palladium nanoparticles. Platin Met Rev 56:83
21. Otsuka H, Nagasaki Y, Kataoka K (2003) PEGylated nanoparticles for biological and pharmaceutical applications. Adv Drug Deliv Rev 55(3):403
22. Prasad SG, Abhijit D, Udayan D (2011) Structural and optical investigations of radiation damage in transparent PET polymer films. Int J Spectroscopy 201(1):1
23. Tiede K et al (2008) Detection and characterization of engineered nanoparticles in food and the environment. Food Addit Contami A 25(7):795
24. Vickrey TM, Garcia-Ramirez JA (1980) Magnetic field-flow fractionation: theoretical basis. Sep Sci Techn 15(6):1297
25. Mavrocordatos et al (2007) Strategies and advances in the characterization of environmental colloids by electron microscopy. In: Environmental colloids and particles: behaviour, structure and characterization. Chichester, UK, 6, pp 345
26. López-Serrano A, Olivas RM, Landaluze JS, Cámara C (2014) Nanoparticles: a global vision. Characterization, separation, and quantification methods. Potential environmental and health impact. Analyt Methods 6(1):38

Zinc Oxide (ZnO) Based Catalyst for Wastewater Purification

Thi Thu Hien Chu, Minh Thuy Pham, Nguyen Minh Viet, Nguyen Minh Phuong, Pham Huynh Thanh Trang, and Pham Thi Huong

Abstract The worldwide growth of the global technology industry has significantly contributed to the accumulation of harmful pollutants in the environment. Zinc oxide (ZnO), a semiconductor material, has attracted considerable attention as a photocatalyst due to its eco-friendly nature, biocompatibility, wide availability, and durability in removing hazardous contaminants. This chapter provides a thorough analysis of the results of recent advancements in ZnO-based heterostructures as photocatalysts for treating organically polluted wastewater. Moreover, it analyzes substantial progress made in addressing the problems related to pure ZnO. Considerable work has been done to accomplish metal/non-metal doping, create a heterojunction, and integrate ZnO with a small band gap material. These efforts aim to increase photocatalytic performance and extend ZnO's absorption spectrum from the ultraviolet to the visible range. The mechanisms of charge separation and transfer, strategies for improving photocatalytic activity, and the enhanced light-harvesting capabilities of ZnO are thoroughly discussed. Futhermore, this chapter emphasises the role of ZnO-based photocatalysts in wastewater treatment and identifies promising directions for future research. It is crucial to choose plentiful and non-toxic materials to ensure both economic effectiveness and environmental safety. Moreover, given the abundant availability of solar energy, future studies should prioritize the use of natural sunlight for illumination instead of artificially generated solar light. This chapter of the book explores prospective improvements and compelling future strategies for utilizing modified ZnO to serve as photocatalyst.

Keywords Zinc oxide · Photocatalyst · Pollutants · Application

---

T. T. H. Chu · M. T. Pham
Department of Chemistry, Faculty of Building Materials, Ha Noi University of Civil Engineering (HUCE), Hanoi, Vietnam

N. M. Viet (✉) · N. M. Phuong · P. H. T. Trang
VNU Key Laboratory of Advanced Material for Green Growth, Faculty of Chemistry, VNU University of Science, Hanoi, Vietnam
e-mail: nguyenminhviet@hus.edu.vn

P. T. Huong
Department of Materials Science and Engineering, Gachon University, Seongnam, South Korea

A. Ahmad et al. (eds.), *Metal Oxide Based Nanophotocatalyst for Wastewater Purification*, Smart Nanomaterials Technology,
https://doi.org/10.1007/978-981-96-7683-5_4

Abbreviations

ZnO	Zinc oxide
h^+	The generation of holes
e^-	Electrons
·OH	Hydroxyl
$\cdot O_2^-$	Superoxide radical anions
$HO_2\cdot$	Hydroperoxyl radicals
CBD	Chemical bath deposition
COD	Chemical oxygen demand
CVD	Chemical vapor deposition
PEC	Photoelectrocatalytic
TOC	Both total organic carbon
XRD	X-ray difraction

1 Introduction

Rapid urbanization and industrialization worldwide have exacerbated environmental challenges, leading to severe air and water pollution that threatens global health. Industries such as chemical plants, refineries, and plastic manufacturers utilize hazardous and organic pollutants in their production processes, often discharging them directly into natural water bodies [1, 2]. These contaminants disrupt the decomposition processes of bacteria, insects, and microscopic flora in aquatic ecosystems. Highly toxic and persistent pollutants, including azo dyes, aromatic compounds, and pharmaceuticals, alter the biological oxygen demand in water, posing a significant threat to aquatic life. Exposure to these pollutants can cause allergic dermatitis, skin inflammation, and, in some cases, cancer, further endangering both human health and marine organisms [3–5]. Given these concerns, wastewater remediation has become a critical environmental priority. Various chemical, physical, and biological methods are employed to eliminate hazardous contaminants, including chemical oxidation, photocatalytic treatment, anaerobic processes, activated sludge, flocculation, coagulation, adsorption, reverse osmosis, sedimentation, and ultrafiltration [6]. Among these, photocatalytic treatment stands out as an efficient and cost-effective approach, offering high degradation efficiency and rapid processing, making it a promising solution for wastewater purification [7, 8].

The photocatalytic mechanism is a notable rusting technology that has attracted considerable attention due to its sustainable method for tackling challenges such as water splitting, waste photodegradation, and gaseous pollutant purification [9–12]. The photocatalytic reaction arises from the redox mechanism that produces the generation of holes (h^+) and electrons (e^-) on the catalyst's surface. The interaction between these charge carriers leads to the formation of reactive radicals, including hydroxyl (·OH) and superoxide ($\cdot O_2^-$), along with environmentally benign byproducts [13, 14]. The increasing interest in photocatalysis applications is due to harnessing solar ultraviolet and visible light for environmental remediation. In countries like Malaysia, where abundant sunlight is available, leveraging solar energy for photocatalysis is both cost-effective and sustainable. Solar radiation comprises approximately 42% visible light and 5% ultraviolet light, making it an abundant and renewable resource. However, the use of semiconductor catalysts in wastewater treatment still faces several technological challenges. This includes the development of catalysts with an extended photoactivity range in both the visible and solar light spectra. Additionally, a limited understanding of the theoretical framework governing key operational elements and their interactions hinders process optimization. Gaining deeper insights into these parameters is essential for designing and implementing efficient photocatalytic systems for sustainable wastewater treatment. Further technical obstacles include the absence of predictive analytical techniques and difficulties in identifying intermediates formed during the degradation of various organic pollutants. Moreover, most research has mostly concentrated on single-constituent model solutions, creating a substantial obstacle to practical application, since actual wastewaters are generally comprised of mixes. This chapter aims to enhance the understanding of recent advancements in the photocatalytic treatment of organic dyes using ZnO-based catalysts.

Photoelectrocatalytic (PEC) technologies have emerged as promising solutions for removing organic pollutants from wastewater. This review examines the latest development in PEC methods, with particular emphasis on ZnO-based heterostructures, and evaluates their effectiveness compared to traditional photocatalytic approaches. We explore various synthesis techniques for developing ZnO-based heterostructures as efficient working electrodes in PEC applications. Special attention is given to modification strategies aimed at mitigating ZnO photocorrosion and expanding its optical response range from ultraviolet to visible light through metal and non-metal doping, as well as heterostructure formation. Therefore, this chapter reveals several critical areas requiring further research, including improved catalyst immobilization techniques, long-term stability enhancement, reusability optimization, toxicity studies, and the use of freely available natural sunlight that can be utilized for water purification or wastewater treatment in the near future.

2 Properties of ZnO: Structural, Optical, and Electrical

2.1 ZnO Structure

The structure of ZnO is characterized by alternating layers of tetrahedrally coordinated Zn^{2+} and O_2^- ions aligned along the c–axis. Its hexagonal wurtzite configuration is the most thermodynamically stable and predominant due to its ionic nature, which is situated precisely at the interface between covalent and ionic materials [15]. The bonding in ZnO, similar to most group II–VI compounds, is primarily ionic, with zinc ions (Zn^{2+}) possessing a radius of 0.007 nm and oxygen ions (O_2^-) having a radius of 0.140 nm. This characteristic favors the formation of the wurtzite structure over the zinc blende configuration while also contributing to its enhanced piezoelectric properties. The wurtzite structure is the primary phase of ZnO, however, two additional forms, zincblende and rock salt, also exist. The wurtzite structure dominates ZnO crystallization due to its thermodynamic stability, while metastable phases like the rock salt structure emerge only under specific conditions. For instance, the rock salt configuration of ZnO requires extreme pressures of approximately 10 GPa, making it exceptionally rare [16]. This inherent tendency for wurtzite crystallization explains why the majority of photocatalytic research focuses on this structural arrangement.

2.2 *Optical and Electrical Characteristics of ZnO*

ZnO exhibits favorable electro-optical characteristics and significant electrochemical stability, with a 3.37 eV of bandgap and a noteworthy exciton binding energy of 60 meV at room temperature [17, 18]. ZnO exhibits exceptional stability in chemical, thermal, and high-energy radiation environments [17–19]. The n-type ZnO semiconductor features a broad straight band gap, high electron mobility, and significant breakdown field strength [19–21]. Table 1 outlines the wurtzite properties. Thus, ZnO's exceptional electrical properties have enabled its application in high-performance electronic devices, particularly field emission systems. When modified, ZnO serves as an effective transparent conducting electrode in various optoelectronic applications. The material exhibits distinct optical behavior, where the recombination of electron–hole pairs results in UV/Blue light emission [20, 21]. This characteristic has made ZnO especially valuable in short-wavelength optoelectronic applications, specifically in devices requiring ultraviolet and blue light emission capabilities [17].

Table 1 The effect of manufacturing methods on the properties of zinc oxide

Fabrication	Morphology	Particle size, nm	References
Sol–gel	Spherical shape	20	[22]
Chemical vapor deposition	Nanorod shape	90–564	[23]
Microemulsion	Nanorod	D_{EBS}: 80 D_{DBS}: 300	[24]
Coprecipitation	Crystal shape	20–40	[25]
Sonochemical	Flakes shape	200–400	[26]
Wet chemical	Nanodise	300–500	[27]

3 Principle of ZnO Photocatalysis

Heterogeneous photocatalysis on ZnO typically comprises four essential stages:

(1) Generation of charge carriers;
(2) Trapping of charge carriers;
(3) Recombination of charge carriers;
(4) Photocatalytic degradation of organic pollutants [28].

3.1 Generation of Charge Carriers

Upon exposure to UV light with energy equal to or exceeding its band gap energy, ZnO facilitates the promotion of an electron from the valence band to the conduction band. Figure 1 depicts the process of electron–hole pair formation induced by the irradiation of ZnO particles with adequate light energy (hv). The wavelength for UV light energy generally correlates to $\lambda < 387$ nm. The excitation generates a positive hole in the valence band, hence creating an electron–hole pair (e^-–h^+).

$$ZnO + hv(UV) \rightarrow ZnO\left(h^+ + e^-\right) \quad (1)$$

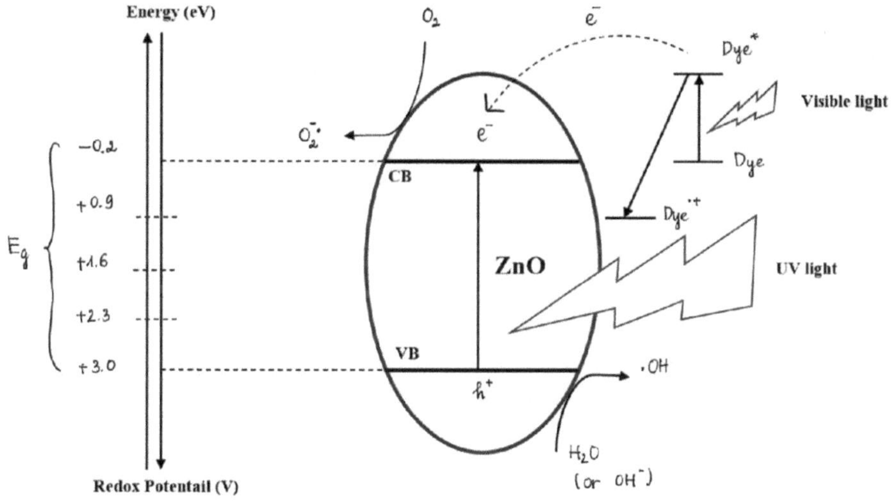

Fig. 1 Schematic representation depicting the production of oxidizing agents in ZnO photocatalysis

3.2 Trapping of Charge Carriers

Recombination is prevented by electron and hole scavengers capturing the e–h+ pair. The positive hole is an effective oxidizer capable of rapidly oxidizing adsorbate contaminants or reacting with electron donors like water or hydroxyl ions (·OH) to generate hydroxyl radicals (OH^-), which are also important oxidizing agents.

$$h^+ + H_2O \rightarrow \cdot OH + H^+ \tag{2}$$

$$h^+ + OH^- \rightarrow \cdot OH \tag{3}$$

Additionally, it is crucial for the electron trapped in the conduction band to be captured by an electron acceptor to prevent recombination with the trapped hole. Molecular oxygen (O_2) acts as an effective electron acceptor. When O_2 is reduced by electrons, reactive superoxide radical anions ($O_2^{\cdot -}$) are formed. These, along with other oxidative species, lead to the production of hydrogen peroxide (H_2O_2) and hydroperoxyl radicals ($HO_2\cdot$). The supplementary OH radicals are produced by the subsequent processes:

$$O_2 + e^- \rightarrow O_2^{\cdot -} \tag{4}$$

$$O_2^{\cdot -} + H^+ \rightarrow HO_2^{\cdot -} \tag{5}$$

$$O_2^{\cdot -} + HO_2^{\cdot -} \rightarrow HO_2^{\cdot} + O^- \tag{6}$$

$$2HO_2^{\cdot} \rightarrow H_2O_2 + O_2 \tag{7}$$

$$H_2O_2 + O_2^{\cdot -} \rightarrow OH^- + \cdot OH + O_2 \tag{8}$$

$$H_2O_2 + e^- \rightarrow OH^- + \cdot OH \tag{9}$$

$$H_2O_2 + hv \rightarrow 2 \cdot OH \tag{10}$$

3.3 Recombination of Charge Carriers

In competition with charge transfer to adsorbed pollutants, electron–hole pair recombination may also occur. This recombination may occur within the photocatalyst's volume or on its surface, accompanied by the production of heat as a by-product.

Prior literature has documented the standard durations for essential steps in the photocatalysis mechanism [29, 30]. Charge carrier entrapment transpires within a duration of around one nanosecond (10–100 ns). Interfacial charge transfer of holes happens slowly (~ 100 ns), whereas the interfacial charge transfer of electrons to the electron acceptor takes the longest (ms).

$$h^+ + e^- \rightarrow ZnO + heat \tag{11}$$

3.4 *Photocatalytic Degradation of Organic Pollutants*

The principal photoreaction (1)–(11) demonstrated the essential function of electron–hole pairs in photocatalytic destruction. The radicals $\cdot OH$, $HO_2\cdot$, and $O_2^{\cdot-}$, along with photogenerated holes (h^+), are extremely reactive intermediates that continuously engage in the reaction system, resulting in entire mineralization of organic contaminants.

In contrast, under visible light with wavelengths exceeding 420 nm, the breakdown of dye pollutants transpires via specific mechanistic pathways referred to as photosensitized decomposition. In photosensitized decomposition, the dye contaminant adhered to the surface is initially activated by visible light illumination. The excited state dye pollutant can inject electrons into the conduction band of ZnO, which are then caught by surface-adsorbed O_2. The accumulation of injected electrons in the conduction band of ZnO makes them especially susceptible to recombination with cationic radicals. Consequently, electron trapping and electron transfer are vital processes for minimizing the radical recombination. The cationic dye radicals produced from the injection of electrons exhibit diminished stability compared to the ground-state dye. As a result, unstable cationic dye radicals can easily decompose through interactions with $\cdot OH$, $HO_2\cdot$, and $O_2^{\cdot-}$ radicals as detailed [23, 31, 32]:

$$Dye + hv \rightarrow Dye^{\cdot} \tag{12}$$

$$Dye^{\cdot} + ZnO \rightarrow Dye^{\cdot+} + ZnO(e^-) \tag{13}$$

$$ZnO(e^-) + O_2 \rightarrow ZnO + O_2^{\cdot-} \tag{14}$$

$$H^+ + O_2^{\cdot-} \rightarrow HO_2^{\cdot} \tag{15}$$

$$2HO_2^{\cdot} \rightarrow O_2 + H_2O_2 \tag{16}$$

$$O_2^{\cdot-} + e^- \rightarrow 2H^+ + H_2O_2 \tag{17}$$

$$H_2O_2 + e^- \rightarrow OH^{\cdot} + OH^- \quad (18)$$

$$Dye^{\cdot+} + \left(O_2^{\cdot-} \text{ or } HO_2^{\cdot} \text{ or } OH^{\cdot}\right) \rightarrow \text{intermediates} \rightarrow \text{final products} \quad (19)$$

4 Bare ZnO as Photoelectrocatalysts

The distinctive characteristics of ZnO nanostructures, arising through their nanoscale size, surface effects, and quantum processes, set them apart from their bulk equivalents. A variety of ZnO nanostructures have been synthesized using multiple production techniques, such as electrochemical deposition, hydrothermal and solvothermal methods, chemical vapor deposition (CVD), physical vapor deposition, and sol–gel processes. Furthermore, its significant influence on shape, dimensions, crystallinity, and photoelectrocatalytic efficiency in the destruction of diverse organic pollutants is acknowledged. Table 1 illustrates the influence of different production procedures on the characteristics of ZnO.

Wang and colleagues [33] explored the synthesis of ZnO nanostructures with two distinct morphologies—propeller and flower shapes—using chemical vapor deposition on silicon substrates. Photoluminescence studies revealed that the flower-shaped structures produced stronger, more defined UV emission peaks at 378 nm, while showing weaker, broader green emission at 495 nm compared to their propeller-shaped counterparts. This spectral pattern indicated better crystallinity and fewer structural defects in the flower-shaped ZnO. However, when tested for photocatalytic performance, the propeller-shaped structures proved more effective at degrading Rhodamine B under UV illumination than the flower-shaped structures.

A study by Shinde and colleagues [34] investigated the fabrication of ZnO electrodes through spray pyrolysis, revealing wurtzite crystal structure formation at various temperatures. Their analysis of substrate temperature effects on film formation showed distinct patterns—lower temperatures produced thicker films due to incomplete dissolution and powder-like characteristics, while higher temperatures resulted in thinner films from rapid precursor evaporation. Films synthesized at 400 °C achieved optimal thickness around 160 nm.

Microscopic analysis revealed distinctive flower-like structures measuring 1–2 μm in diameter. These formations consisted of tapered, sword-shaped nanorods, creating a high surface area morphology ideal for photocatalytic applications. The researchers evaluated this ZnO electrode's effectiveness in removing three organic pollutants from wastewater: salicylic acid (SA), toluene, and 4-chlorophenol (4-CP).

Treatment results showed significant reductions in both total organic carbon (TOC) and chemical oxygen demand (COD). Specifically:

- Toluene: TOC decreased from 35.89 to 12.15 mg/L; COD reduced from 23.92 to 8.27 mg/L;

- Salicylic acid: TOC dropped from 65.17 to 6.26 mg/L; COD fell from 62.57 to 9.19 mg/L;
- 4-Chlorophenol: TOC declined from 48 to 13.72 mg/L; COD decreased from 52 to 12.20 mg/L.

These results suggest that ZnO flower structures show promise as immobilized photocatalysts for water treatment and organic pollutant removal. This development represents an important step towards broader implementation of visible light heterogeneous photocatalysis for environmental remediation.

5 Doped ZnO

The limited visible light sensitivity and inefficient charge carrier separation in prostine ZnO pose significant challenges for photocatalytic applications. Doping ZnO nanomaterials with foreign ions has emerged as an effective strategy to address these limitations.

5.1 Doping with Metal

Metal ion doping is widely recognized as a method to enhance ZnO's photocatalytic performance by extending its photo-response into the visible spectrum [35]. In specific instances, metal dopants can serve as electron or hole scavengers, increasing charge carrier lifetime and reducing recombination probability [36]. This process occurs through specific mechanisms:

$$M^{n+} + e^-_{cb} \rightarrow M^{(n-1)+} \text{(as electron trap)}$$
$$M^{n+} + h^+_{vb} \rightarrow M^{(n+1)+} \text{(as hole trap)}$$

When $M^{n+}/M^{(n-1)+}$ pairs exhibit a potential less negative than ZnO's conduction band edge, while maintaining an energy level less positive than ZnO's valence band edge, both electron and hole trapping occur at the surface. This surface-level interaction influences charge carrier longevity, and efficient metal ion doping at the ZnO surface can significantly improve photocatalytic device performance.

However, research has revealed that metal ion doping can also produce adverse effects. The effectiveness of metal-doped ZnO photocatalysts depends heavily on several factors, including the specific metal ions used, dopant concentration, fabrication methods, and operating conditions. At high doping levels, metal ions may act as recombination centers, hindering electron and hole transport to the interface. Moreover, there exists a critical threshold for metal ion dopant concentration; exceeding this optimal level leads to decreased photocatalytic activity due to increased electron–hole pair recombination [37, 38].

5.1.1 Doped ZnO with Alkali and Alkaline Earth Metals

Alkali and alkaline earth metal ions are widely recognized for their excellent electrical and thermal conductivity properties. Introducing ions such as Li^+, Na^+, and K^+ into ZnO has been shown to improve its photocatalytic efficiency [39–41]. Group I elements serve as effective acceptor dopants in ZnO. Li^+ incorporation at Zn^{2+} lattice sites creates an exceptionally shallow acceptor level at 0.09 eV. Na-doping facilitates hole capture at interstitial sites, which promotes the transfer of holes to adsorbed OH^- ions, generating hydroxyl radicals (·OH). These shallow traps help reduce charge carrier recombination. While Li^+ and Na^+ ions possess suitable ionic sizes for Zn^{2+} site substitution, excessive K^+ doping has been shown to be detrimental due to K_2O formation [42].

Mg^{2+} doping offers unique advantages among Group II elements, primarily by reducing zinc interstitials and oxygen vacancies. This effect stems from the enhanced ionic character of Mg-O bonds compared to Zn–O bonds [43], attributed to magnesium's higher oxygen affinity ($E^0 = -2.37$ V) versus zinc ($E^0 = -0.76$ V) in ZnO [44]. By decreasing oxygen vacancies—a primary carrier source—Mg^{2+} dopants reduce carrier electrons in ZnO films, resulting in a lower Fermi level. Mg^{2+} doping induces quantization effects within ZnO's bandgap, though these effects depend on the synthesis method. Notable bandgap energy increases have been observed even in larger particles (> 10 nm) [45], although some highly doped materials show no blue shift [46, 47]. This evidence suggests that bandgap modifications arise primarily from Mg^{2+}-induced changes in electrical energy levels rather than size quantization effects.

The impact of Mg^{2+} doping on ZnO's band structure is particularly interesting:

- The conduction band, composed of Zn 4*s* orbitals, is significantly affected by Mg^{2+} substitution at Zn^{2+} sites
- The valence band, consisting of O 2*p* orbitals, remains largely unaltered
- Increasing Mg^{2+} doping concentration enhances the influence of 3*s*-like states at the conduction band minimum.
- The higher energy of these 3*s*-like states compared to Zn 4*s* levels [48, 49] shifts the conduction band minimum to higher energies, resulting in an expanded bandgap with increased Mg^{2+} concentration.

The increase in optical band gap energy from Mg^{2+} doping can also be attributed to the Moss-Burstein effect, triggered by electrons from oxygen vacancies. As carrier density increases, the Fermi level shifts into the conduction band, resulting in band gap expansion [49]. According to the semiconductor transition metal hypothesis, band gap width increases when impurity concentrations remain below the Mott transition threshold, but decreases when exceeding it. While Mg 3*s* orbitals enhance carrier transport efficiency and increase the reduction potential of excited electrons, excessive Mg^{2+} doping leads to detrimental formation of cubic magnesium oxide [49], magnesium hydroxide [46], and amorphous products [48], compromising Mg-ZnO performance.

5.1.2 Transition Metal Ion Doping

Transition metal ion doping has emerged as a particularly promising approach due to three key advantages:

(i) Isomorphic compatibility with zinc;
(ii) Multiple oxidation states that enhance charge carrier transfer properties;
(iii) Distinctive donor/acceptor characteristics in zinc oxide that influence electronic band structures.

The incorporation of transition metal dopants alters the electrical environment through zinc atom substitution in the ZnO lattice, with variations based on ionic radius. These dopants can affect both the coordination environment of ZnO and its electronic band energy structure through localized *d–d* transition states. The enhanced visible light sensitivity stems from "*sp-d*" exchange interactions between the localized "d" electrons of substituted metal ions and conduction band electrons.

The incorporation of transition metals modifies ZnO's band structure through "*s–d*" and "*p–d*" exchange interactions, causing the valence band edge to shift positively and the conduction band edge to shift negatively [50]. This results in extended bandgap absorption at longer wavelengths. While most transition metal doping induces visible light response, specific ions such as Cu^{2+}, Mn^{2+}, Zr^{4+}, and Ag^{+} can cause a blue shift in bandgap response under certain experimental conditions. These modifications can produce both beneficial and detrimental effects on photocatalytic performance [51–54].

The enhanced performance of doped ZnO catalysts can be attributed to multiple factors: increased surface area, reduced particle size, better accessibility of active sites, modified defect and oxygen vacancy density, broader light absorption spectrum and intensity, decreased recombination pathways. However, optimizing these characteristics often involves complex trade-offs between material properties and experimental conditions. The effects vary significantly across different dopants and even for the same dopant under different conditions.

For instance, Fe^{3+} ions demonstrate unique behavior due to their half-filled electronic structure, functioning as shallow traps [55]—a property not observed with Mn^{+} ions [52]. The Fe^{3+} ions capture electrons and holes, transforming into Fe^{2+} and Fe^{4+}, which are much less stable than the Fe^{3+} state. To regain stability, the localized electrons and holes are transferred to adsorbed O_2 and OH^{-}, respectively, promoting the generation of free radicals [56].

Qi et al. [44] efficiently synthesized transition metal (Cu, Co, Mn, Fe, and Co) doped ZnO nanocrystals exhibiting increased photocatalytic activity by a simple solvothermal technique utilizing ultrasonic acid. The generated metal-doped ZnO and pristine ZnO and display a wurtzite crystal structure, with average diameters between 200 and 400 nm. Unlike pristine ZnO, metal-doped ZnO demonstrates increased visible light absorption. Metal-doped ZnO displays increased photocatalytic efficacy under simulated sunlight irradiation, with Cu-doped ZnO exhibiting the greatest activity. A photocatalytic mechanism is proposed based on DFT simulations. Cu^{2+}

in the doped ZnO nanoparticles function as efficiency traps, augmenting the separation efficiency of photogenerated carriers and inhibiting their recombination, hence enhancing photocatalytic activity.

In contrast, research by Tsuzuki et al. [36] revealed that Mn-doping can inhibit ZnO's photocatalytic performance in Rhodamine B (RhB) degradation. As dopant concentrations increased from 0 to 5%, degradation efficiency decreased significantly from 100% to approximately 70%. The mechanistic explanation for this reduction lies in the behavior of substitutional Mn^{3+} or Mn^{4+} ions in ZnO's interstitial positions. These ions divert effective charge carriers from the solid solution interface, primarily acting as charge carrier recombination centers, ultimately diminishing photocatalytic activity [57, 58]. Donkova et al. [57] conducted comparative studies examining both Mn and Cu doping effects on methylene blue (MB) decomposition under UV radiation. It was shown that undoped ZnO was more beneficial for MB decomposition than Mn-doped ZnO. A similar tendency was noted in their Cu-doped ZnO catalyst. According to their research, transition metal doping in ZnO may occasionally have a negative impact on photocatalytic activity.

5.1.3 Rare Earth-Doped ZnO

Rare earth metals are particularly efficient dopants for improving the degradation of organic pollutants. Photocatalysts infused with rare earth elements have recently garnered significant attention. The performance of photocatalysts can be improved by leveraging the unique redox and photocatalytic capabilities of rare earth elements. Studies have shown that incorporating rare earth dopants into zinc oxide (ZnO) can significantly enhance its photocatalytic activity compared to undoped ZnO. This enhancement is attributed to the specialized electronic and structural properties imparted by the rare earth dopants, which optimize the surface characteristics and electron transfer efficiency of the photocatalyst material.

Incorporating rare earth metals into ZnO structures is an effective approach to modify electrical properties and improve absorption within the visible light spectrum. Variations in band gap energy are linked to the variations of ionized impurities, the creation of localized impurity level bands by rare earth metals, and charge transfer between the valence/conduction bands of ZnO and the $4f$ and $5d$ states of rare earth metals [59]. The incorporation of rare earth metals into ZnO structures is found to substantially suppress photoinduced electron–hole recombination, which is attributed to the rare earth metal's ability to trap electrons. Although doping ZnO with rare earth metals has advantages, there are drawbacks as well. For example, the concentration of rare earth ions within the ZnO lattice may not be sufficiently saturated because of variations in charge and ionic size, and the rare earth ion energy levels may not be properly aligned with the valence and conduction bands of ZnO [59].

Conclusions are derived from the electrical arrangement of rare earth ions [60]. Eu^{3+} attains a stable electronic configuration ($4f^7$) by capturing conduction band electrons, thereby complicating its de-trapping ability. In contrast, Er^{3+} (f^{11}) and

Dy^{3+} (f^9), upon electron capture, achieve complete electronic configurations (f^{12} and f^{10}, respectively), making the release of electrons from the coupled system difficult. Conversely, Ho^{3+} (f^{10}) converts to Ho^{2+} (f^{11}) through electron capture, facilitating the easy release of the solitary unpaired electron. The restricted electrons migrate to the d orbital of Gd^{3+} or La^{3+}, thereby enabling their liberation to adsorbed O_2. Thus, La-ZnO exhibits enhanced efficacy in the degradation of methyl orange relative to other dopants, including Eu Ho, and Dy ions. Nonetheless, the benefits of Lewis acidity derived from unoccupied *f*-orbitals are still inadequately exploited in these dopants [60]. The intrinsic properties of ZnO and the impacts of the dopant, such as its concentration, location, and dispersion within the matrix, influence how well the electrical configuration in the trapping mechanism works.

La-doped ZnO nanoparticles have been investigated by multiple research groups, who observed an enhancement in electron transfer at the interface, leading to reduced electron–hole pair recombination and a red shift towards longer wavelengths. When La (III) ions were incorporated into the ZnO lattice, the band gap was reduced from 3.15 to 3.02 eV, 2.94 eV, and 3.09 eV for ZnO doped with 0.5%, 1.0%, and 1.5% weight percent of La, respectively. Upon exposure to visible light for 3 h, 1% by weight of La-doped ZnO exhibits exceptional photocatalytic degradation capabilities, reaching a 99% elimination of paracetamol and an 85% reduction in TOC, whereas pure ZnO shows no photocatalytic activity [35].

5.2 Doping with Non-metal

Doping ZnO with nonmetals offers advantages due to their ability to easily replace oxygen atoms in the crystal lattice or occupy vacant positions within the structure, thanks to their better compatibility with the ZnO matrix compared to metal dopants. Non-metal ions enhance photocatalytic activity more effectively than metal ions because they are less likely to create recombination centers. When nonmetals replace oxygen sites, they disrupt the valence band, leading to reduced scattering of conduction band electrons [61]. According to theory, nitrogen enhances light absorption in the visible spectrum by generating isolated N 2*p* energy levels just above the highest point of the valence band (Fig. 2).

Liu et al. [62] investigated the photocatalytic degradation of RhB utilizing C-doped ZnO flowers under visible light. The researchers noticed that because carbon was substituted at oxygen sites in the catalyst, modified ZnO showed better absorption in the UV–Vis spectrum. Modified ZnO demonstrated effective degradation of RhB, indicating superior photocatalytic activity compared to commercial ZnO. Modified ZnO exhibits higher photocatalytic efficiency due to its increased surface area, which provides more active sites and reaction centers for pollutant molecule interaction and degradation.

In non-metallic doped ZnO, nitrogen and carbon serve as the primary non-metal dopants. Carbon doping in anion-doped ZnO is theoretically proposed to provide a potential benefit. In the ZnO matrix, substitutional carbon produced new states (C

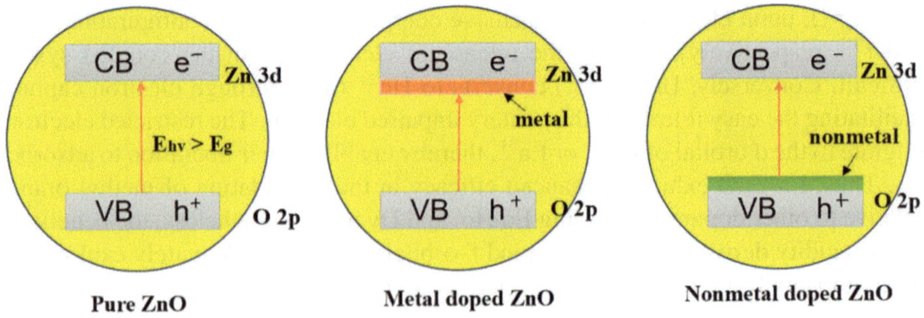

Fig. 2 Bandgap energy of both pure and doped ZnO

2*p*) near the ZnO (O 2*p*) valence band edge [63]. Consequently, the band gap shrank when the valence band edge rose to an energy level greater than that of pure ZnO. The C-doped ZnO catalyst has to narrow its bandgap in order to show a positive response in the visible light spectrum. Zhang et al. [58] observed a substantial reduction in band gap energy following carbon doping. Zhang found that the 30% C-doped sample displayed the greatest total visible light absorption, exhibiting approximately three-fold the absorption of the undoped ZnO nanoparticles. The increased carbon concentration indicates a transition in morphology from nanorods to nanospheres, while the band gap consistently decreases from 3.19 to 2.72 eV with carbon doping up to 40%. In general, photogenerated charge carriers with a nanorod structure exhibit a longer lifespan. Zhang et al. found that as the carbon doping level increased, the photoluminescence (PL) intensity decreased gradually, indicating a lower recombination rate in the nanospheres compared to the nanorods Due to these factors, nanospheres with a higher carbon doping percentage (59%) exhibited improved photocatalytic activity compared to the 10% carbon-doped nanorods.

Because of its similar size to oxygen and reduced energy requirement for the replacement process, nitrogen is the dopant of choice among nonmetals like fluorine, sulfur, and phosphorus. Additionally, in the pristine oxide, the electron density surrounding the doped nitrogen is decreased due to nitrogen's lower electronegativity and energy of ionization when compared to oxygen [64, 65]. Furthermore, a decrease in Zn-N bond ionicity may also result in a decrease in ZnO's bandgap [66]. By trapping through surface-adsorbed oxygen and group hydroxyl, respectively, the electron–hole pairs produced between N 2*p* states and CB Zn 3*d* levels are separated. First principles calculations suggest that nitrogen doping will result in the emergence of an extra carrier, namely a hole, at the peak of the valence band. N-ZnO's conductivity changes from n-type to p-type as a result of charge carriers creating a narrow-deep acceptor level within the bandgap due to the mutual exclusion effect among the holes. P-type semiconductor demonstrate enhanced stability against photocorrosion relative to n-type materials [67]. Chen et al. [68] looked into the photocatalytic degradation of 1.10–4 mol/L Methyl Orange (MO) using N-doped ZnO, which is made by thermally breaking down zinc nitrate in air. The findings showed an increase in absorption strength and a shift towards longer wavelengths in

N-doped ZnO. During exposure to visible light, N-doped ZnO's photocatalytic efficiency significantly outperformed that of pure ZnO. The improvement was linked to N-doped ZnO, which exhibited a reduced energy bandgap and was responsive to visible light. As a result, a greater number of electrons and holes were generated, enabling enhanced photocatalytic breakdown of MO.

5.3 Co-dopants

The addition of co-dopants can reduce the rate at which electron–hole pairs recombine. Co-doped ZnO has been employed for the elimination of several dyes, including Methylene Blue, Napthol Blue Black, Methyl Orange, Reactive Red, and Acid Black 1. Compared to pure ZnO and single-dopant ZnO networks, co-doped ZnO exhibits noticeably higher photodegradation efficacy. The co-dopants concurrently capture electrons generated by photosynthesis from the conduction band of zinc oxide, therefore diminishing the electron–hole recombination rate. Consequently, the photoinduced synthesis of electron–hole pairs will persist, resulting in the production of substantial quantities of highly active superoxide radical anions ($O_2^{\cdot -}$) and hydroxyl radicals ($\cdot OH$), which will improve the dyes' photodecomposition efficiency.

6 ZnO-Based Binary Heterostructure

Semiconductor heterostructures have emerged as a more effective approach to enhance photocatalytic performance compared to using single materials alone. These structures are created by combining semiconductors with different bandgap, whether n-type or p-type, resulting in specific band alignments. The interface between semiconductors creates an internal electric potential that facilitates photogenerated charge carrier separation and transport. To develop robust and efficient photocatalysts, researchers must optimize multiple factors including surface characteristics, crystal architecture, material compatibility, and fabrication methods [69].

When light interacts with a binary heterostructure system, both semiconductor components generate electron–hole pairs. The charge transfer mechanism involves electrons moving from semiconductors with higher conduction bands to those with lower ones, while holes transfer in the opposite direction—from materials with lower valence bands to those with higher ones. This spatial separation of charges reduces recombination, thereby improving photocatalytic efficiency. In p–n heterojunction systems specifically, illumination triggers electron migration from the p-type material's higher conduction band to the n-type material's lower one, while holes transfer from the n-type's lower valence band to the p-type's higher valence band. This charge redistribution generates a stable electric field at the p–n interface throughout the composite structure.

Utilizing the spray pyrolysis approach, Sapkal et al. [70] created nanocrystalline films of ZnO, TiO_2, and ZnO/TiO_2 on glass covered with FTO. The ZnO/TiO_2 thin films showed a red shift, with the energy bandgap estimated to be around 2.99 eV. When evaluated for methylene blue (MB) degradation, the ZnO/TiO_2 thin films achieved nearly complete mineralization of MB, surpassing the performance of individual ZnO and TiO_2 films.

Lachhed et al.'s work [31] confirmed that ZnO undergoes non-structural deformation as a result of forming a heterojunction with Fe_2O_3. This signifies that the composite is devoid of Fe^{3+} inclusion. This indicates that Fe_2O_3 exists as a distinct phase by exclusively establishing a local heterojunction. Electrons were transported from the dye's excited state to the photocatalyst's conduction band at the interface, according to the photocatalytic study. This can make it more likely that the material will absorb visible light through indirect photocatalytic activities. This action demonstrated the shortcomings of the dyes used as reference points in assessments of photocatalyst quality. Using DRS and related Tauc plots, the composite's bandgap energy was determined to be 2.29 eV. The enhanced photocatalytic efficiency of the composite, in contrast to bare ZnO (3.2 eV), was confirmed by electrochemical characterization methods.

Numerous researchers have effectively produced ZnO/CdS heterostructures with various morphologies using multiple methods. Using the coprecipitation approach, Saxena et al. showed that ZnO-CdS nanostructures significantly outperformed pure ZnO and CdS in terms of photodegradation efficiency, reaching a maximum of 99% for MB dye in 110 min [71]. The maximum RhB dye degradation efficiency of ZnO/CdS heterostructures produced using the chemical precipitation approach was 98.16% in 80 min, according to Gurububelli et al. [72]. The heterostructure's creation successfully altered the bandgap energy (2.9796 eV). CdS-ZnO core–shell nanowires were created by Tso et al. using the RF sputtering method. A maximum rate of 9618 mmol/g/h of photocatalytic hydrogen generation was observed [69]. Huo and Chen employed a microwave synthesis method to fabricate a CdS/ZnO system that markedly enhanced hydrogen yield [73]. The significant increase in the H_2 generation rate is attributed to the lowered bandgap of 2.62 eV. Revathi et al. used the microwave method to prepare a ZnO/CdS catalyst that enhances photocatalytic hydrogen generation [74]. Senasu et al. developed a ZnO/CdS photocatalyst that exhibits excellent visible light absorption and remarkable charge carrier separation efficiency [75].

7 ZnO-Based Ternary and Quaternary Heterostructure

Researchers have addressed the limitations of single and binary coupled heterostructures by developing ternary and quaternary ZnO-based structures as catalysts. This has improved optical characteristics, charge transfer efficiency at the interface, and degradation efficacy of various wastewater contaminants under visible light exposure. This section summarizes the photoelectrocatalytic efficiency of ZnO-based ternary and quaternary heterostructured semiconductor catalysts. The mechanisms of reactive oxygen species (ROS) formation were investigated to clarify the roles of hydroxyl radical (·OH), photogenerated holes (h^+), and superoxide anion radical ($\cdot O_2^-$) in the photocatalytic process. Results showed that holes were essential for pollutant degradation in the ternary system. This document outlines the pollutant degradation procedure.

$$\text{Photocatalyst} + h\nu \rightarrow h^+ + e^-$$
$$e^- + O_2 \rightarrow \cdot O^{2-}$$
$$O_2 + 2H_2O \rightarrow 2H_2O_2$$
$$H_2O_2 + 2e^- \rightarrow 2 \cdot OH$$
$$\text{Pollutants} + \cdot O^{2-} \rightarrow \text{oxidation products}$$
$$\text{Pollutants} + h^+ \rightarrow \text{oxidation products}$$
$$\text{Pollutants} + \cdot OH \rightarrow \text{oxidation products}$$

By creating a ternary heterojunction with CeO_2, Fe_2O_3, and ZnO, Lei et al. [32] showed that they could reduce electron–hole recombination, increase absorption of visible light, improve thermal stability, and increase photocatalytic efficiency. The direct ion-exchange technique was used to create the heterojunction on the polyimide substrate. The enhancement in visible light absorption efficiency is confirmed by the notable redshift observed, which has an estimated bandgap energy of 1.6 eV, in contrast to binary heterojunctions.

Li et al. [76] also confirmed the improved visible light performance of the Fe_2O_3/ZnO/$ZnFe_2O_4$ heterojunction. The formation of $ZnFe_2O_4$, known for its strong ability to absorb visible light, is greatly influenced by the calcination temperature, as shown by the interaction between Fe_2O_3 and ZnO. As the calcination temperature increases, the synthesis of $ZnFe_2O_4$ also increases. With a bandgap energy of approximately 1.9 eV, $ZnFe_2O_4$ is key to effectively absorbing visible light and facilitating efficient separation of electron–hole pairs. The proposed mechanism indicates that, during degradation, electrons move from $ZnFe_2O_4$ to Fe_2O_3 and then to ZnO, while holes transfer from Fe_2O_3 to $ZnFe_2O_4$.

Studies have demonstrated the superior performance of ternary heterostructures over their binary and single-component counterparts. For example, Nandi and Das found that ZnO/CdS/CuS systems achieved higher photodegradation rates for RhB and MO dyes compared to ZnO/CdS or pure ZnO. This enhancement was attributed to reduced charge carrier recombination and a favorable bandgap of 2.97 eV, which improved charge separation and transfer. Similarly, Zhou and colleagues reported [77] enhanced photodegradation and H_2 generation from spherical CdS/ZnS/ZnO structures, citing improvements in surface area, pore volume, and visible light absorption. Research by Liu et al. [78] revealed that CdS/g-C_3N_4/ZnO outperformed related binary systems and pure ZnO. Notable achievements in hydrogen evolution include Wang et al.'s [79] CuS-ZnO/rGO/CdS system, which achieved 1073 μmol/g/h under visible light. Sun et al. [80] found that optimized CdS/ZnS/ZnO (CZZ-4) structures showed 2.28 and 93.54 times higher PEC performance than ZnO/CdS and pure ZnO, respectively. Additionally, $ZnO/CdS/MoS_2$ heterostructures developed by Jia et al. [81] demonstrated remarkable H_2 generation rates of 10,247.4 μmol/g/h, exceeding binary ZnO/CdS by three times and pure CdS by thirty times.

Advancing to quaternary systems, Zheng and colleagues [78] developed a ZnO/ZnSe/CdSe/MoS_2 p–n junction core–shell structure on FTO glass using combined chemical bath deposition (CBD) and hydrothermal methods. Structural analysis revealed minimal MoS_2 diffraction peaks due to low concentration and crystallinity on the ZnO/ZnSe/CdSe nanorod surface. XRD confirmed successful ZnSe/CdSe deposition on the ZnO core through ion replacement, while electron microscopy showed well-aligned vertical nanorod arrays. The quaternary structure demonstrated optimal PEC performance among tested samples, efficiently degrading amoxicillin at various concentrations (5–40 ppm) within 21–96 min. The system maintained robust performance through multiple cycles, achieving 85% amoxicillin degradation in the third run. XRD analysis confirmed structural stability throughout repeated use.

8 Conclusion and Future Work

This study offers an in-depth review of the progress made in ZnO-based heterostructures over the past decade and their application as photoelectrocatalysts for pollutant removal in wastewater. This review outlines different synthesis modification methods and chemical additives that were created to reduce charge carrier recombination losses and expand ZnO's spectral response into the visible spectrum through co-doping, rare earth doping, metal doping, non-metal doping, and semiconductor coupling with ZnO.

Conclusions drawn from the current literature are provided, accompanied by recommendations for future research domains:

(i) By combining several semiconductor materials to create heterojunctions, which have a tunable bandgap and improved sensitivity to visible light, ZnO's limitations can be lessened. This is a contemporary and efficient method of lowering the rate at which electron–hole pairs recombine, prolonging the life of photocarriers, and dramatically lowering specific pollutants.

(ii) By using solar light rather than UV radiation, doping bare ZnO with a metal, non-metal, or transition metal enhances its optical and physical characteristics as well as its catalytic effectiveness. In the future, the utilization of plentiful and non-toxic materials will be deemed essential for ensuring the safety and economic viability of the process. Numerous noble metals, including Co, Ni, and Ag, were examined to determine how they related to the ZnO photoelectrocatalyst.

(iii) In the forthcoming period, it is imperative to devise a cost-efficient method for the immobilization of photocatalysts for solid–liquid separation.

(iv) Future projects should concentrate on using natural sunshine as a source of illumination rather than artificially simulated solar light, given the abundance of solar energy. This would be exciting and efficient.
(v) In real-world applications, stability, durability, catalytic efficiency, and photoelectrode reusability or recyclability should be given priority when subjected to a small amount of bias.

References

1. Rajput RB, Shaikh R, Sawant J, Kale RB (2022) Recent developments in ZnO-based heterostructures as photoelectrocatalysts for wastewater treatment: a review. Environ Adv 9:100264
2. Tamtam MR, Koutavarapu R, Shim J (2023) InVO4 nanosheets decorated with $ZnWO_4$ nanorods: a novel composite and its enhanced photocatalytic performance under solar light. Environ Res 227:115735
3. Danish M, Ahmad T, Nadhari WNAW, Ahmad M, Khanday WA, Ziyang L, Pin Z. Optimization of banana trunk-activated carbon production for methylene blue-contaminated water treatment. Appl Water Sci
4. Huang C, Li J (2023) A synergistic effect between ZnO/CdS S-scheme heterojunction and GO cocatalyst for boosting photocatalytic performance. Opt Mater 139:113726
5. Koutavarapu R, Jang WY, Rao MC, Arumugam M, Shim J (2022) Novel $BiVO_4$-nanosheet-supported MoS_2-nanoflake-heterostructure with synergistic enhanced photocatalytic removal of tetracycline under visible light irradiation. Chemosphere 305:135465
6. Adeel M, Saeed M, Khan I, Muneer M, Akram N (2021) Synthesis and characterization of Co–ZnO and evaluation of its photocatalytic activity for photodegradation of Methyl Orange. ACS Omega 6:1426–1435
7. Koutavarapu R, Peera SG, Lee TG, Myla CR, Lee D-Y, Shim J, Balasingam SK (1959) Recent trends in graphitic carbon nitride-based binary and ternary heterostructured electrodes for photoelectrochemical water splitting. Processes 2021:9
8. Wang R, Hao Q, Feng J, Wang G-C, Ding H, Chen D, Ni B (2019) Enhanced separation of photogenerated charge carriers and catalytic properties of ZnO-MnO_2 composites by microwave and photothermal effect. J Alloys Compd 786:418–427
9. Singh D, Verma S, Gautam RK, Krishna V (2015) Removal of tartrazine by activated carbon biosorbents of Lantana camara: kinetics, equilibrium modeling and spectroscopic analysis. J Environ Chem Eng 3:2161–2171
10. Hong E, Kim D, Kim JH (2014) Heterostructured metal sulfide (ZnS–CuS–CdS) photocatalyst for high electron utilization in hydrogen production from solar water splitting. J Ind Eng Chem 20:3869–3874
11. Ikehata K, Naghashkar NJ, Gamal El-Din M (2), Degradation of aqueous pharmaceuticals by ozonation and advanced oxidation processes: a review, Ozone Sci Eng 28:353–414
12. He Y, Zhang L, Fan M, Wang X, Walbridge ML, Nong Q, Wu Y, Zhao L (2015) Zscheme SnO_{2-x}/g-C_3N_4 composite as an efficient photocatalyst for dye degradation and photocatalytic CO_2 reduction. Sol Energy Mater. Solid Cells 137:175–184
13. Lee SJ, Jung HJ, Koutavarapu R, Lee SH, Arumugam M, Kim JH, Choi MY (2019) ZnO supported Au/Pd bimetallic nanocomposites for plasmon improved photocatalytic activity for methylene blue degradation under visible light irradiation. Appl Surf Sci 496:143665
14. Zhang Y, He S, Guo W, Hu Y, Huang J, Mulcahy JR, Wei WD (2018) Surfaceplasmon driven hot electron photochemistry. Chem Rev 118:2927–2954
15. Patil AB, Patil KR, Pardeshi SK (2010) Ecofriendly synthesis and solar photocatalytic activity of S-doped ZnO. J Hazard Mater 183:315e323

16. Özgür Ü, Alivov YI, Liu C, Teke A, Reshchikov MA, Doğan S, Avrutin VC, Cho SJ, Morkoç AH (2005) A comprehensivem review of ZnO materials and devices. J Appl Phys 98:1–103
17. Lu JG, Ye ZZ, Huang JY, Zhu LP, Zhao BH (2006) ZnO quantum dots synthesized by a vapor phase transport process. Appl Phys Lett 88:063110
18. Tan ST, Chen BJ, Sun XW, Fan WJ (2005) Blueshift of optical band gap in ZnO thin films grown by metal-organic chemical-vapor deposition. J Appl Phys 98(1):013505
19. Liu K, Sakurai M, Aono M (2010) ZnO-based ultraviolet photodetectors. Sensors 10(9):8604e8634
20. Rodnyi PA, Khodyuk IV (2011) Optical and luminescence properties of zinc oxide. Opt Spectrosc 111(5):776e785
21. Umar A (2009) Growth of Comb-like ZnO nanostructures for dye sensitized solar cells applications. Nanoscale Res Lett 4(9):1004e1008; Daneshvar N, Salari D, Khataee AR (2004) Photocatalytic degradation of azo dye acid red 14 in water on ZnO as an alternative catalyst to TiO_2. J Photochem Photobio A: Chem 162:317–322
22. Ba-Abbad MM, Kadhum AAH, Mohamad AB, Takriff MS, Sopian K (2013) Optimization of process parameters using D-optimal design for synthesis of ZnO nanoparticles via sol–gel technique. J Ind Eng Chem 19:99–105
23. Laurenti M, Garino N, Porro S, Fontana M, Gerbaldi C (2015) Zinc oxide nanostructures by chemical vapour deposition as anodes for Li-ion batteries. J Alloy Compd 640:321–326
24. Lim SK, Hwang S-H, Kim S, Park H (2011) Preparation of ZnO nanorods by microemulsion synthesis and their application as a CO gas sensor. Sens Actuators B Chem 160:94–98
25. Kumar VR, Wariar PRS, Prasad VS, Koshy J (2011) A novel approach for the synthesis of nanocrystalline zinc oxide powders by room temperature co-precipitation method. Mater Lett 65
26. Ghosh S, Majumder D, Sen A, Roy S (2014) Facile sonochemical synthesis of zinc oxide nanoflakes at room temperature. Mater Lett 130
27. Samanta PK, Mishra S (2013) Wet chemical growth and optical property of ZnO nanodiscs. Opt—Int J Light Electron Opt 124:2871–2873
28. Nishio J, Tokumura M, Znad HT, Kawase Y (2006) Photocatalytic decolorization of azo-dye with zinc oxide powder in an external UV light irradiation slurry photoreactor. J Hazard Mater 138:106–115
29. Mills A, Le Hunte S (1997) An overview of semiconductor photocatalysis. J Photochem Photobio A: Chem 108:1–35
30. Hoffmann MR, Martin ST, Choi WY, Bahnemann DW (1995) Environmental applications of semiconductor photocatalysis. Chem Rev 95:69–96
31. Lachheb H, Ajala F, Hamrouni A, Houas A, Parrino F, Palmisano L (2017) Electron transfer in ZnO–Fe_2O_3 aqueous slurry systems and its effects on visible light photocatalytic activity. Catal Sci Technol 7:4041–4047
32. Lei Y, Huo J, Liao H (2018) Fabrication and catalytic mechanism study of CeO_2–Fe_2O_3–ZnO mixed oxides on double surfaces of polyimide substrate using ion-exchange technique. Mater Sci Semicond Process 74:154–164
33. Wang SL, Zhu HW, Tang WH, Li PG (2012) Propeller-shaped ZnO nanostructures obtained by chemical vapor deposition: photoluminescence and photocatalytic properties. J Nanomater 2012:594290
34. Shinde SS, Shinde PS, Bhosale CH, Rajpure KY (2011) Zinc oxide mediated heterogeneous photocatalytic degradation of organic species under solar radiation. J Photochem Photobiol B Biol 104:425–433
35. Thi VH, Lee BK (2017) Mater Res Bull 96:171–182
36. Tsuzuki T, Smith Z, Parker A, He RL, Wang XG (2009) Photocatalytic activity of manganese-doped ZnO nanocrystalline powders. J Aust Ceramic Soc 45:58–62
37. Jackson D, Hargreaves SJ (2009) Oxide materials in photocatalytic processes. metal oxide catalysis. Wiley-VCH, Weinheim
38. Malato S, Fernández-Ibáñez P, Maldonado MI, Blanco J, Gernjak W (2009) Decontamination and disinfection of water by solar photocatalysis: recent overview and trends. Catal Today 147:1–59

39. Benhebal H, Chaib M, Leonard A, Lambert SD, Crine M (2012) Photodegradation of phenol and benzoic acid by solegel-synthesized alkali metal-doped ZnO. Mater Sci Semicond Process 15:264e269
40. Benhebal H, Chaib M, Malengreaux C, Lambert SD, Leonard A, Crine M, Heinrichs B (2014) Visible-light photo-activity of alkali metal doped ZnO. J Taiwan Inst Chem Eng 45:249e253
41. Lee WC, Canciani GE, Alwhshe BOS, Chen Q (2016) Enhanced photoelectrochemical water oxidation by ZnxMyO (M ¼ Ni, Co, K, Na) nanorod arrays. Int J Hydrogen Energy 41:123e131
42. Li D, Huang JF, Cao LY, OuYang HB, Li JY, Yao CY (2014) Microwave hydrothermal synthesis of K^+ doped ZnO nanoparticles with enhanced photocatalytic properties under visible-light. Mater Lett 118:17e20
43. Kim GH, Jeong WH, Ahn BD, Shin HS, Kim HJ, Kim HJ, Ryu MK, Park KB, Seon JB, Lee SY (2010) Appl Phys Lett 96:163506
44. Ku CJ, Duan ZQ, Reyes PI, Lu YC, Xu Y, Hsueh CL, Garfunkel E (2011) Appl Phys Lett 98:123511
45. Etacheri V, Roshan R, Kumar V (2012) Mg-Doped ZnO nanoparticles for efficient sunlight-driven photocatalysis. ACS Appl Mater Interfaces 4:2717e2725
46. Nouri H, Yangjeh AH (2014) Microwave-assisted method for preparation of Zn1-xMgxO nanostructures and their activities for photodegradation of methylene blue. Adv Powder Technol 25:1016e1025
47. Labhane PK, Sonawane SH, Sonawane GH, Patil SP, Huse VR (2018) Influence of Mg doping on ZnO nanoparticles decorated on graphene oxide (GO) crumpled paper like sheet and its high photo catalytic performance under sunlight. J Phys Chem Solids 114:71e82
48. Boshta M, Helal MOA, Ghoneim D, Mohsen NA, Zaghlool RA (2010) The photocatalytic activity of sprayed $Zn_{1-x}Mg_xO$ thinfilms. Surf Coat Technol 205:271e274
49. Qiu X, Li L, Zheng J, Liu J, Sun X, Li G (2008) Origin of the enhanced photocatalytic activities of semiconductors: a case study of ZnO doped with Mg2þ. J Phys Chem C 112:12242e12248
50. Jayanthi K, Chawla S, Joshi AG, Joshi AG, Khan ZH, Kotnala RK (2010) Fabrication of luminescent, magnetic hollow core nanospheres and nanotubes of Cr-doped ZnO by inclusive coprecipitation method. J Phys Chem C 114:18429e18434
51. Barick KC, Singh S, Aslam M, Bahadur D (2010) Porosity and photocatalytic studies of transition metal doped ZnO nanoclusters, Microporous Mesoporous Mater 134:195e202
52. Rekha K, Nirmala M, Nair MG, Anukalliani A (2010) Structural, optical, photocatalytic and antibacterial activity of zinc oxide and manganese doped zinc oxide nanoparticles. Physica B 405:3180e3185
53. Selvam NCS, Vijaya JJ, Kennedy LJ (2012) Effects of morphology and Zr doping on structural, optical, and photocatalytic properties of ZnO nanostructures. Ind Eng Chem Res 51:16333e16345
54. Jongnavakit P, Amornpitoksuk P, Suwanboon S, Ndiege N (2012) Preparation and photocatalytic activity of Cu doped ZnO thin films prepared by the solegel method. Appl Surf Sci 258:8192e8198
55. Xiao S, Zhao L, Lian J (2014) Enhanced photocatalytic performance of supported Fe doped ZnO nanorod arrays prepared by wet chemical method. Catal Lett 144:347e354
56. Kumar SG, Rao KSRK (2017) Comparison of modification strategies towards enhanced charge carrier separation and photocatalytic degradation activity of metal oxide semiconductors (TiO2, WO3 and ZnO). Appl Surf Sci 391:124e148
57. Zhang X, Dong S, Zhou X, Yan L, Chen G, Dong S, Zhou D (2015) A facile onepot synthesis of Er-Al co-doped ZnO nanoparticles with enhanced photocatalytic performance under visible light. Mater Lett 143:312e314
58. Zhang X, Qin J, Hao R, Wang L, Shen X, Yu R, Limpanart S, Ma M, Liu R (2015) Carbon-doped ZnO nanostructures: facile synthesis and visible light photocatalytic applications. J Phys Chem C 119:20544–20554
59. Iqbal J, Liu X, Zhu H, Wu ZB, Zhang Y, Yu D, Yu R (2009) Acta Mater 57:4790–4796
60. Okte AN (2014) Characterization and photocatalytic activity of Ln (La, Eu, Gd, Dy and Ho) loaded ZnO nanocatalysts. Appl Catal A: Gen 475:27e39

61. Ahmad S, Kharkwal M, Govind, Nagarajan R (2011) Application of $KZnF_3$ as a single source precursor for the synthesis of nanocrystals of ZnO_2:F and ZnO:F; synthesis, characterization, optical, and photocatalytic properties. J Phys Chem C 115:10131e10139
62. Liu SW, Li C, Yu JG, Xiang QJ (2011) Improved visible-light photocatalytic activity of porous carbon self-doped ZnO nanosheet-assembled flowers. Cryst Eng Comm 13:2533–2541
63. Cho SH, Jang JW, Lee JS, Lee KH (2010) Carbon-doped ZnO nanostructures synthesized using vitamin C for visible light photocatalysis. Cryst Eng Comm 12:3929–3925
64. Rajbongshi BM, Ramchiary A, Samdarshi SK (2014) Influence of N-doping on photocatalytic activity of ZnO nanoparticles under visible light irradiation. Mater Lett 134:111e114
65. Asahi R, Morikawa T, Irie H, Ohwaki T (2014) Nitrogen-doped titanium dioxide as visiblelight-sensitive photocatalyst: designs, developments, and prospects. Chem Rev 114:9824e9852
66. Lin HF, Liao SC, Hung SW (2005) The dc thermal plasma synthesis of ZnO nanoparticles for visible-light photocatalyst. J Photochem Photobiol A Chem 174:82e87
67. Shifu C, Wei Z, Sujuan Z, Wei L (2009) Preparation, characterization and photocatalytic activity of N-containing ZnO powder. Chem Eng J 148:263e269
68. Chen SF, Zhao W, Zhang SJ, Liu W (2009) Preparation, characterization and photocatalytic activity of N-containing ZnO powder. Chem Eng J 148:263–269
69. Tso S, Li W-S, Wu B-H, Chen L-J (2018) Enhanced H_2 production in water splitting with CdS-ZnO core-shell nanowires. Nano Energy 43:270–277
70. Sapkal RT, Shinde SS, Waghmode TR, Govindwar SP, Rajpure KY, Bhosale CH (2012) Photo-corrosion inhibition and photoactivity enhancement with tailored zinc oxide thin films. J Photochem Photobiol B Biol 110:15–21
71. Saxena N, Sondhi H, Sharma R, Joshi M, Amirthapandian S, Rajput P, Sinha OP, Krishna R (2022) Equimolar ZnO-CdS nanocomposite for enhanced photocatalytic performance. Chem Phys Impact 5:100119
72. Gurugubelli TR, Ravikumar RVSSN, Koutavarapu R (2022) Enhanced photocatalytic activity of ZnO–CdS composite nanostructures towards the degradation of rhodamine B under solar light. Catalysts 12, 84
73. Huo S, Chen C (2021) One-step synthesis CdS/single crystal ZnO nanorod heterostructures with high photocatalytic H_2 production ability. Inorg Chem Commun 132:108841
74. Revathi M, Pricilla A, Saravanan SP (2021) Design and Fabrication of ZnO/CdS heterostructured nanocomposites for enhanced hydrogen evolution from solar water splitting. Inorg Chem Commun 134:109056
75. Senasu T, Chankhanittha T, Hemavibool K, Nanan S (2021) Visible-light-responsive photocatalyst based on ZnO/CdS nanocomposite for photodegradation of reactive red azo dye and ofloxacin antibiotic. Mater Sci Semicond Process 123:105558
76. Li X, Jin B, Huang J, Zhang Q, Peng R, Chu S (2018) Fe_2O_3/ZnO/$ZnFe_2O_4$ composites for the efficient photocatalytic degradation of organic dyes under visible light. Solid State Sci 80:6–14
77. Zhou Q, Li L, Xin Z, Yu Y, Wang L, Zhang W (2020) Visible light response and heterostructure of composite CdS@ZnS–ZnO to enhance its photocatalytic activity. J Alloys Compd 813:152190
78. Zheng Z, Li X, Li L, Tang Y (2019) Photoelectrocatalytic degradation of amoxicillin over quaternary ZnO/ZnSe/CdSe/MoS_2 hierarchical nanorods. Int J Hydrog Energy 44:20826–20838
79. Wang X, Li Q, Xu H, Gan L, Ji X, Liu H, Zhang R (2020) CuS-modified ZnO rod/reduced graphene oxide/CdS heterostructure for efficient visible-light photocatalytic hydrogen generation. Int J Hydrogen Energy 45:28394–28403
80. Sun D, Shi J-W, Ma D, Zou Y, Sun G, Mao S, Sun L, Cheng Y (2020) CdS/ZnS/ZnO ternary heterostructure nanofibers fabricated by electrospinning for excellent photocatalytic hydrogen evolution without co-catalyst. Chin J Catal 41:1421–1429
81. Jia Y, Wang Z, Qiao X-Q, Huang L, Gan S, Hou D, Zhao J, Sun C, Li D-S (2021) A synergistic effect between S-scheme heterojunction and Noble-metal free cocatalyst to promote the hydrogen evolution of ZnO/CdS/MoS_2 photocatalyst. Chem Eng J 424:130368

Cerium Oxide: Synthesis and Application for Wastewater Purification

Mohamed F. Cheira, Abdelaal S. A. Ahmed, and Emad A. Elshehy

Abstract Rising worldwide populations and changing weather patterns have intensified worries about a possible water crisis throughout the last 20 years. As a major health concern, the presence of heavy metals in wastewater demands prompt action to mitigate its effects. To tackle this critical concern, a multitude of cutting-edge technologies have been put into action. It is worth mentioning that nanotechnology has gained the attention of academics worldwide as a promising subject within this science. Researchers have looked at nano cerium oxide as a potential water filtration material. The redox characteristics and oxidation storage capacity of cerium oxide are highlighted in this chapter as being the inherent traits that make it efficient in technologies like filtration, adsorption, and Advanced Oxidation Processes (AOP). Additional alterations to ceria that improve its qualities include adding nonmetal and metal atoms to its framework and dispersing it on a suitable support. Presenting this study gives a glimpse into the technical trends shaping the future of creating environmentally beneficial ceria-based products. Current breakthroughs in the application of cerium oxide for wastewater treatment are summarized and examined in this chapter, which sheds light on their efficacy and potential. It explains the dangers of commonly detected heavy metals and stresses the need for effective filtering methods. Many different things can benefit greatly from cerium oxide and solid solutions. Its usage in water treatment techniques has become widespread.

Keywords Wastewater treatment · Ceria · APO · Synthesis nanoparticles · Organic and inorganic contaminations

M. F. Cheira · E. A. Elshehy
Nuclear Materials Authority, Maadi, Cairo 11728, Egypt

A. S. A. Ahmed (✉)
Chemistry Department, Faculty of Science, Al-Azhar University, Assiut 71524, Egypt
e-mail: abdelaalsaiyd@azhar.edu.eg

A. Ahmad et al. (eds.), *Metal Oxide Based Nanophotocatalyst for Wastewater Purification*, Smart Nanomaterials Technology,
https://doi.org/10.1007/978-981-96-7683-5_5

1 Introduction

As the global population rises, it is becoming increasingly challenging to meet the raising demand for affordable water [113, 125]. As a result of global warming, coastal regions are experiencing increased droughts and salt intrusion, and water contamination is spreading unregulated [242]. Over 650 million people lack access to safe drinking water. It has been reported that every two minutes, a child younger than five dies from a preventable cause related to water contamination and poor sanitation. Water supply contains numerous dangerous toxins, some naturally occurring and others manufactured by humans. [38] The three most fundamental methods for treating water are purification, disinfection, and precipitation. Among these are harmful bacteria in water, toxic algae that release cyanotoxins, heavy metals, and metalloids in groundwater, and a plethora of other contaminants [162]. Moreover, advanced methods for eliminating the most hazardous pollutants, even those undetectable by the naked eye, can induce cancer or other chronic illnesses in the future. Contaminants include pharmaceuticals, personal care products, traditional pesticides, and volatile organic compounds such as industrial solvents. Water purification, pollution control, and public health initiatives can potentially boost the economic and social capabilities of nations. Another goal of technological improvement is to reduce the operational and maintenance expenses of outdated sewage and water treatment plants [64, 233]. Due to the growing financial incentives for dependable and affordable access to clean water and stricter regulatory mandates for water pollution control, there is a growing need for more cost-effective desalination and wastewater reuse systems. Recently, there has been an abundance of pollution in the environment due to increasing global industrialization and manufacturing. This pollution includes volatile organic compounds (VOCs), particulate matter, sulfur oxide (SO_x), nitrogen oxide (NO_x), sludge, toxic effluents, and greenhouse gases (GHGs). As a result of gravity or rainfall [275], the air pollutants produced by these enterprises eventually settle to the ground, where they combine with dirt or water from the surface, and posing a threat to the ecosystem [84].

Furthermore, natural water supplies are depleted due to the enormous amounts of water utilized for cooking, cleaning, rinsing, processing, cutting, washing, and other industrial purposes. Acute toxicity to aquatic organisms and retardation of light penetration are two of the many negative impacts of industrial effluent on water bodies, which contribute to their reputation as a harmful environmental problem [3, 88, 152]. Additionally, the central nervous system, heart, liver, reproductive system, and kidneys are among the human organs most vulnerable to the harmful effects of contaminants that make it into the food chain. In terms of environmental safety, these effluents are very concerning due to their high chemical oxygen demand (COD), biochemical oxygen demand (BOD), excluded particles, pathogenic load, and strong hue [29]. There are many methods used to eliminate this effluent such as filtration, ozonization, flocculation and coagulation, electrical desalination, rectification, photodegradation, reverse osmosis, ultrasonication, as well as sorption process [43, 62, 82, 105, 108, 118, 119, 190, 250, 254]. To be effective, the remediating chemical must be safe for humans and other nonpathogenic organisms to use, even in small amounts. Nanoparticles (NPs) are an attractive adsorbent for removing toxic contaminants from wastewater [8, 38, 64, 111, 137]. Because of their remarkable properties and enormous surface area, NPs are the best adsorbents in terms of adsorption [56, 57, 60, 147]. Organic pollutants, dangerous nuclides, and heavy metals created some harmful contaminants [104, 164]. As these contaminants build up in aquatic creatures, they go up the food chain and eventually into human bodies [179]. Therefore, traditional wastewater treatment procedures are not effective toward these contaminants [37, 191]. As a result, numerous contemporary techniques, including chemical oxidation, adsorption, biodegradation, and electrochemical combustion, have been developed to treat wastewater [12, 31, 105, 110, 266]. But the majority of them are not very good at dealing with these pollutants. Therefore, much work has been done to developing suitable strategy. One of the trails is AOP which is a widely used technique for wastewater treatment [245]. The active hydroxyl radicals are generated by AOP, which include photocatalytic processes, ozone oxidation, and reactions similar to Fenton's. These species then attack the organic contaminants [150]. Due to their high efficiency, low cost, safety, and environmental friendliness, the photocatalytic AOPs are widely used to remove organic contaminants from wastewater [271]. Semiconductors are one type of photocatalyst that is now being employed for water treatment. This is because these materials are chemically stable and easy to produce [79, 83, 89]. In the last decades, CeO_2 displayed promising ability as photocatalysts [210]. CeO_2 NPs were chosen because of its most crucial characteristic, which is its ability to transition between two distinct oxidation states. Therefore, CeO_2 NPs creates oxygen vacancies and localized electrons on the surface nanoparticles, allowing oxygen and compounds containing oxygen to diffuse more efficiently [165].

The adsorbent is also safer to release into the environment than other alternatives since pure cerium has a high toxicity threshold. CeO_2 NPs can be utilized as oxidizing agent, a catalyst in a variety of reactions, a colorant for glass and ceramics, and a polishing powder [223]. According to the literature, a range of stabilizing chemicals can change the size, coverage, and adsorption characteristics of nanoparticles. [182, 202]. The relative efficacy of surface functionalization of CeO_2 NPs was assessed using different surfactants in wastewater treatment. Great research on wastewater treatment using CeO_2 as a doping; however, these approaches did not prioritize adsorbent repair [26]. The utilization of residual nanoparticles from pollutant adsorption to disinfect wastewater is a significant topic of concern, the biocidal capabilities of CeO_2 NPs [146, 154, 170] gave rise to this application [194]. To confirm the function of the surface coating in dye molecule adsorption, comparisons with bare CeO_2 NPs have also been made. Recently, chromium and arsenate have been eliminated from water by using CeO_2 NPs and CeO_2 nanocomposites. [219, 278]. Both nanoparticles and nanocomposites have shown promising results in the treatment of water.

To comprehend several environmental processes, it is essential to have a molecular-level understanding of the ecological process [11]. After the adsorption–desorption process on the NPs, it is essential to assess the environmental impact following the procedure. The specific adsorption behavior of the resultant NPs is determined by the conditions under which the NPs are synthesized. This is because the redox characteristics of the NPs can be altered, and their dual reversible reaction can lead to modifications in the overall adsorption process. This chapter discusses the structure and synthesis, and how ceria works to photodegrade contaminants in wastewater. Additionally, doping ceria with elements and metal oxides significantly improve the effectiveness of photocatalytic degradation.

2 CeO_2 Structure

Cerium is an element with the atomic number 58 and belongs to the lanthanide family. CeO_2 has a fluorite structure as presented in Fig. 1 [236]. Cerium atoms has a face-centered crystal (FCC) lattice with a lattice constant of 0.541 nm [48]. CeO_2 often contains cerium in two oxidation states, + 3 and + 4, so it may pair with oxygen atoms more effectively. The removal of hydrocarbons, CO, and NO_x from industrial exhausts is aided by these various oxidation states [59]. CeO_2 can mimic some enzymes, including catalase, oxidase peroxidase, superoxide dismutase, and decrease damage caused by oxidative stress because it can change valency states [25, 132]. The ease of the Ce^{4+} to Ce^{3+} transition, and surface properties like porosity and surface area are a few of the variables that influence the oxygen storage capacity and how the doped composite impacts it. In addition, metals (e.g., Au, Pt, and Pd), and supporters (activated carbon, and graphene) are doped into ceria [109]. The dopants primarily utilized are those with varying ionic radii, as opposed to $Ce^{3+}/^{4+}$. Cavity flaws are created, improving ceria's reduction of power and oxygen storage capacity [287, 288]. Adding earth metals and highly alkaline earth metals to the mix further boosts the catalytic activity of CeO_2 [148]. When noble metals are doped, their relatively weak metal–oxygen interaction results in comparatively increased catalytic activity [183]. Similarly, nanocomposites doped with CeO_2 exhibited a lattice structure, with the doped element either filling a void or substituting for cerium itself [22].

3 Synthesis of CeO_2 NPs

There are a lot of variables that influence nanoparticle characteristics, including crystalline size, and shape. The synthesis process displayed a significant impact on these characteristics. In this section, the most popular approaches to synthesis CeO_2 NPs will be presented.

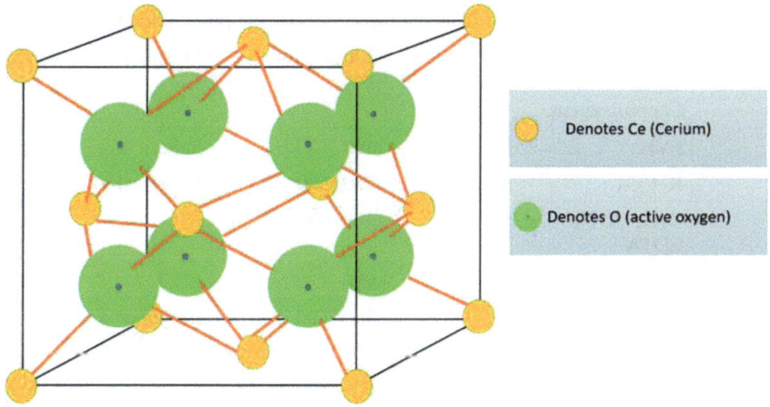

Fig. 1 The FCC lattice of CeO_2. Reproduced from Ref. [236] with permission, Copyright 2020, Royal Society of Chemistry

3.1 Co-precipitation Technique

This process is usually based on the precipitation reaction to achieve a homogenous solution of cations with a uniform composition. Changing various parameters such as precursors, the pH, and temperature affect CeO_2 NPs characteristics [216]. From the lists of the conditions and results of CeO_2 NPs in Table 1, the size and morphology of the prepared CeO_2 NPs remarkably increased by increasing the temperature. [72]. This mainly assigned to reduces oxygen concentration by boosting the temperature which leads to increasing the average particle size [40].

3.2 Hydrothermal Technique

Hydrothermal synthesis happened in an autoclave with a stainless-steel container. The solution was in its aqueous state, and in case of other solvent, the procedure is called the solvothermal method. Typically, the temperature used in hydrothermal synthesis should be greater than the boiling point of water (≈120–200 °C) [252]. From the conditions and the properties of the prepared CeO_2 NPs by hydrothermal technique in Table 2, increasing the solvent and hydrothermal temperature increased the particle size of the prepared CeO_2 NPs [158].

Table 1 The list of conditions for synthesizing CeO_2 NPs by the precipitation method

Compound name	Surfactant/ Solvent/ Capping agent	Particle size (nm)	Temp./Time	Morphology	References
CeO_2	PVP	19–33	600 °C/2 h	Spherical	[72]
CeO_2	K_2CO_3	20	220–600 °C/3 h	Spherical	[40]
CeO_2	$(NH_4)_2CO_3$/ EDTA	8	400–850 °C/2 h	–	[23]
CeO_2	NH_4OH	10	600 °C/2 h	–	[258]
CeO_2	NH_3	53.1–68.6	500 °C/4 h	Irregular	[1]
CeO_2	NH_4OH	7.4–18.8	30–90 °C/2 h	Hexagonal	[40]
CeO_2: Gd_2O_3	$(NH_4)_2CO_3$	22.6–33.6	600–800 °C/2 h	–	[253]
$Ce_{0.8}Gd_{0.2}O$	Ethyl alcohol	3.1	25 °C	–	[87]
CeO_2: Pd	NaOH	7.6–7.7	450–600 °C	Spherical	[144]
CeO_2: Fe	Cetyl trimethyl ammonium bromide (CTAB)	40–80	500 °C/3 h	Rod-shaped	[70]
CeO_2: Al_2O_3	NH_4OH	30–70	800 °C	Spherical	[107]

Table 2 List of conditions for synthesizing CeO_2 NPs by the hydrothermal method

Compound name	Surfactant/ Solvent/ Capping agent	Particle size (nm)	Autoclave temp./Time	Morphology	References
CeO_2	Ammonia	6–15	250 °C/6 h	Quasi spherical/ spheres	[252]
CeO_2: Sm	Tri-sodium phosphate	43.47–58.3	180 °C/15 h	Octahedral and grow-up rod arm on octahedral phases	[212]
CeO_2	NH_4OH	16	180 °C/5 h	Spheres	[158]
CeO_2	NH_4OH	3	180 °C/5 h	–	[158]
CeO_2: Gd	Tri-sodium phosphate	57.4–58.3	180 °C/15 h	Square shape	[122]
CeO_2	NaOH	8.4	180 °C/24 h	Regular octahedron	[153]
CeO_2	NaOH	5–54	70–180 °C/ 4 h	Amorphous	[209]
CeO_2	–	8–16	150 °C/1 h	Rods	[176]
CeO_2: Cu	Polyvinyl pyrrolidone (PVP)	8–15	180 °C/24 h	Hollow spherical	[296]
CeO_2: Co	PVP	13–19	160–200 °C/ 12 h	Spherical	[209]

Table 3 List of conditions for synthesizing CeO_2 NPs by the sol–gel method

Compound name	Surfactant/ Solvent/ Capping agent	Particle size (nm)	Temp./Time	Morphology	References
CeO_2: Al_2O_3	HCl/Al powder	90	800 °C	Spherical	[106]
CeO_2: Fe_2O_3	Propylene oxide	10–50	600 °C/1 h	–	[30]
CeO_2	NH_4OH	9–13	400–600 °C/ 2 h	Spherical	[207]
CeO_2: Ni/ Al_2O_3	Citric acid	22.4–52.5	550 °C/6 h	Irregular	[7]
CeO_2	Methanol	8–30	400 °C/2 h	Elliptical/ Spherical	[86]
CeO_2	Cassavas starch	8–16	200–500 °C/ 1 h	Spherical	[80]
CeO_2: TiO_2	HNO_3/Ethanol	9.4–31.4	200–400 °C	–	[159]
CeO_2: Mn	Citric acid/ glycerol	14–22	600 °C	Spherical	[251]
CeO_2: Zn	Gelatin solution	15–30	600 °C/2 h	Spherical	[240]

3.3 *Sol–Gel Technique*

This process was developed in the 1960s. The principle of this technique is gradually bringing the compound from a dissolved state back into a solid form. Compared to other methods, this method is easy and inexpensive. In this procedure, a chelating agent is introduced dropwise to a homogenous solution, and the resulting gel is then removed followed dried. Lastly, the gel is subjected to calcination over a variety of temperatures. As in Table 3, the size of the prepared NPs increased by raising the calcination temperature [71, 292].

3.4 *Combustion Technique*

The combustion synthesis is a short synthesis time, a high reaction temperature, and a low cost [16]. This process involves adding a precursor to deionized water, mixing fuel or complexing agents, and finally allowing the mixture to evaporate on a hot plate, forming a gel. A deluge of massive gases is released, facilitating the complete process, and the resulting powder undergoes additional calcination at varying temperatures. The list of studies that described the combustion method as a means of synthesizing CeO_2 NPs by adding various fuels are in Table 4.

Table 4 List of conditions for synthesizing CeO_2 NPs by the combustion process

Compound name	Surfactant/ Solvent/ Capping agent/Fuel	Particle size (nm)	Temp./ Time	Morphology	References
CeO_2	Glycine	41	–	Spongy/ Porous	[259]
CeO_2	Citric acid	10	600 °C/ 3 h	Spherical	[168]
CeO_2	Glycine	12.6	550 °C/ 12 h	–	[181]
CeO_2	EDTA	42	450 °C/ 3–4 min	Spherical	[218]
$(SmO_{1.5})_{0.2}(CeO_2)_{0.8}$	Glycine	8.2	550 °C/ 12 h	–	[155]
$(GdO_{1.5})_{0.2}(CeO_2)_{0.8}$	Glycine	10.3	550 °C/ 12 h	–	[181]
CeO_2–$CeAlO_3$	Urea/Glycine	10	900 °C	Ring	[15]
CeO_2–$CeAlO_3$	Urea	28	1100 °C	Spherical	[15]
CuO-CeO_2	Urea	11.9–24.4	550 °C/ 1 h	–	[21]
$CuFe_2O_4$/CeO_2	Urea–nitrate	10–15	600 °C/ 5 h	Irregular particles	[231]

3.5 Ball Milling Technique

This process is called high-energy ball milling. This method produces nanoparticles in the form of a powder. The cylindrical devices used in the milling method are filled with material to be ground and small, hard balls. This process produces NPs by of mechanical attrition, which entails the impact of processing materials with metal balls. Balls spun at high speeds inside cylindrical devices crush materials into nanocrystallites by combining the forces of kinetics and gravity as they fall on the substance to be treated [222]. Table 5 shows the precursors used in the ball milling method of synthesis and the conditions of CeO_2 NPs.

3.6 Green Synthesis Technique

One approach that is gentler on the natural environment is the green production procedure. The goal of this method is to avoid handling hazardous and unnecessary waste. This approach is energy efficient because it does not require any external

Table 5 List of conditions for synthesis CeO_2 NPs by the ball milling procedure

Compound name	Particle size (nm)	Milling time (h)	Morphology	References
CeO_2	10	30	Spherical	[284]
CeO_2	80–120	5–55	Amorphous	[237]
CeO_2	25	30	–	[32]
CeO_2	45–102	24	–	[101]
CeO_2: Fe	200–400	24	Spherical	[249]
CeO_2/ZnO	20.45–29.78	20	Hexagonal	[244]
Ni–CeO_2	6–6.6	10	Spherical	[295]
CeO_2: Fe	30–47	2.5	Spherical	[28]

Table 6 List of conditions for green synthesis of CeO_2 NPs by green method

Compound name	Surfactant/ Solvent/ Capping agent	Particle size (nm)	Temp./Time	Morphology	References
CeO_2	*Prosopis jujliflora*	15	800 °C/2 h	Spherical	[17]
CeO_2	Olive	6	500 °C/2 h	Spherical	[172]
CeO_2	*Acalypha indica*	8–54	400 °C	Elliptical/ Spherical	[120]
CeO_2	*Gloriosa superba*	5	400 °C/2 h	Spherical	[14]
CeO_2	Walnut	12	250 °C/4 h	Spherical	[294]
CeO_2	*Rubia cordifola*	26	500 °C/4 h	Hexagonal	[238]
CeO_2	Honey	23	200–800 °C/ 2 h	Spherical	[51]
CeO_2	Aloe Vera	2–3	–	Spherical	[61]
CeO_2	*Hibiscus sabdarifffa*	3.9	500 °C/2 h	Amorphous	[248]
CeO_2	*Origanum majo rana*	20	450 °C/4 h	Spherical	[18]

processes, like high heat or pressure, to be carried out. In this process, the biological component is used as capping agents [9, 235]. The various approaches used to synthesize CeO_2 NPs by green synthesis are in Table 6.

3.7 Spray Pyrolysis Technique

The aerosol process known as spray pyrolysis involves atomizing a solution and heating the droplets to form solid particles [68]. It is not necessary to have a spotless

Table 7 List of conditions for synthesis CeO_2 NPs by spray pyrolysis

Compound name	Surfactant/Solvent/ Capping agent/Fuel	Particle size (nm)	Morphology	References
CeO_2	Ethanol and water	4.2–20	–	[269]
CeO_2	Double distilled water	10–15	Spherical	[204]
CeO_2	Deionized water	8–50	Large grain	[298]
CeO_2	Water and Ethanol	100–1000	Spherical	[124]
CeO_2–TiO_2	Ethanol	10–13	Spherical	[33]
CeO_2	–	5–10	Hollow sphere	[260]
CuO–CeO_2	Ethanol	28.2–40.6	Irregular	[161]
CeO_2: Mn	Deionized water	29–37	Hexagonal	[205]
CeO_2: Gd	Distilled water	3–22	Ovoid and spherical	[94]
CeO_2: Fe	Ethanol	5.94–6.39	Spherical and equiaxed	[35]

or highly vacuumed atmosphere. This method involves spraying the solution over a surface that has been heated. A thin-film deposition procedure that is both easy and inexpensive. It does not cost much to use the precursor [184]. Table 7 displays the synthesis conditions and outputs of the spray pyrolysis for CeO_2 NPs.

4 Characteristics of CeO_2

CeO_2 can have internal and external defects due to its geometric and electrical characteristics. The substitution of other ions primarily causes extrinsic flaws, whereas intrinsic flaws originate from crystal thermal disorders, a consequence of redox processes between the solid and its surroundings. Oxygen vacancies or lattice defects are formed when crystals undergo lattice strain due to the loss of oxygen and/or its electrons [131]. The distortion can be caused or altered by several factors, including electric fields, temperature, oxygen partial pressure, surface stress, other ion doping, etc. By taking the place of more positively charged elements in the cation sub-lattice, cerium oxide may be able to endure severe oxygen deprivation. At low partial pressures of oxygen and high temperatures, it is also known to emit a substantial amount of oxygen. Oxygen vacancy flaws improve the electrical and chemical characteristics of CeO_2. Typically, Ce^{4+} ions transform to Ce^{3+} ions when two electrons are trapped at two cerium sites, indicating that the lattice is decreased as a result of less oxygen.

The reduction of CeO_2 requiring Ce^{3+} ions and narrow-band 4*f* electron production, is thought to play a role in heterogeneous catalysis involving CeO_2-based catalysts. The extra electrons move to nearby Ce ions rather than the nearest cation sites when oxygen vacancies form on CeO_2 surfaces. Oxygen vacancies make crystal surfaces more reactive, which improves catalytic activity by increasing the mobility of oxygen atoms within the crystal.

4.1 *Amendments to CeO_2 NPs*

There are several ways to alter CeO_2 distinctive characteristics. One method involves doping the CeO_2 lattice with smaller and/or less valence metal ions. New stable materials with unique physical and chemical characteristics can be created using this process, which also supplies an opportunity to improve the operation of metal oxides. Combining ceria with metal or metal oxide molecules results in blend oxides [132]. Many factors influence how ceria changes its oxygen storage behavior when doped with metal cations. These include the dopant's nature and composition, the constancy of the resulting construction, the crystal structure change, the easiness of the Ce^{4+}–Ce^{3+} transition, the concentration of carriers (oxygen vacancies), and surface properties (surface area, porosity, etc.). A different approach would be to spread the ceria or similar substances on appropriate support substances, such as graphene [109], activated carbon [160], etc. Ceria can at times facilitate the metals scattering such as Pd, Pt, Au, etc. [98, 265]. Transition, rare, alkaline earth, and noble metals are the most common modifications to cereal-based materials that improve their activity and thermal stability. Rare earth cations and other dopants with ionic radii somewhat diverse from $Ce^{3+/4+}$ create solid solutions that disrupt the lattice structure of CeO_2. Consequently, abundant oxygen cavity defects would enhance ceria's redox characteristics and oxygen storage capacity [199]. Incorporating alkali and alkaline earth metals, which are highly mobile, into ceria-based oxides improves their catalytic activity [148]. Combined with CeO_2, noble metals have enhanced catalytic activity due to weak metal–oxygen connections [183, 286].

5 Photocatalytic Degradation Mechanism by CeO_2 NPs

Because of their uneven energy levels, the photocatalytic activity of any semiconductor is position-dependent [273]. Figure 2 displayed how ceria works toward photocatalytic destruction of organic contaminants [136]. Ceria has a valence band (VB) of around 2.5 eV and a conduction band (CB) of about 0.4 eV, the energy difference of about 2.9 eV [179, 274]. Sunlight bombarding a semiconductor causes electrons (e^-) in the VB to transfer to the CB, leaving holes (h^+) in the VB [255]. By reacting the produced electrons and holes with redox processes, a reactive oxygen species (ROS) are formed [140]. By combining oxygen molecules, electrons create superoxide ions, and by pairing with hydroxide ions, holes create hydroxide radicals [239]. The chemical formulas for these reactive species are as follows:

$$CeO_2 \rightarrow CeO_2(e^- + h^+) \quad \text{in light}$$

$$e^- + O_2 \rightarrow O_2^{\cdot -}$$

$$O_2^{\cdot -} + H^+ \rightarrow \cdot OOH$$

$$2 \cdot OOH \rightarrow O_2 + H_2O_2$$

$$e^- + H_2O_2 \rightarrow OH^- + \cdot OH$$

$$h^+ + OH^- \rightarrow \cdot OH$$

These active species are employed to break down the pollutants into harmless H_2O and CO_2 as in the following equation [69, 276].

$$\text{Organic pollutant} + \text{radicals} \rightarrow CO_2 + H_2O$$

The photocatalytic effectiveness is weak due to its relatively high bandgap, ceria cannot effectively utilize visible solar radiation, leading to electron–hole recombination [180]. Therefore, CeO_2 is doped to reduce the bandgap and enhance its photocatalytic efficacy, hence overcoming the limitation.

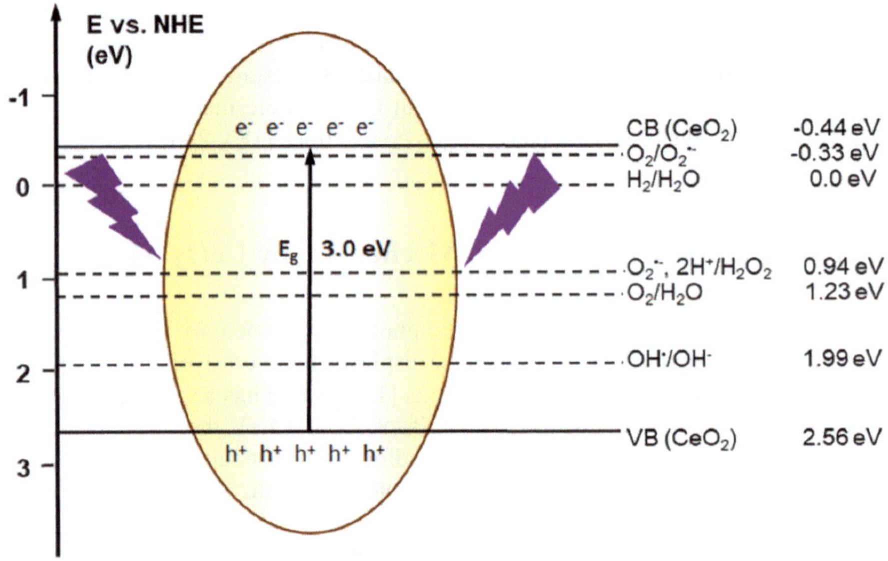

Fig. 2 Scheme for photocatalytic degradation of organic contaminants by CeO_2. (Adapted from Ref. [136] with permission from MDPI)

6 CeO_2-Based Nanocomposites for Water Treatment by Advanced Oxidation (AO) Process

An essential subset of water treatment technologies known as AOPs makes use of strong but transient species like sulfate radicals [78] or hydroxyl radicals ($^{\bullet}OH$) [201]. It can make these radicals in water by combining various energy sources, like solar, electrical, or Waves of sound. Also, these radicals can be obtained by utilizing chemicals such as hydrogen peroxide, ozone, with or without a catalyst. Usually, many water treatment processes can benefit from AOPs include those dealing with industrial effluents (e.g., from textile mills, distilleries, oil fields, pulp and paper mills, metal plating wastes, and agrochemical facilities), unsafe effluents (e.g., from hospitals and slaughterhouses), municipal wastewater treatment plant pathogen removal, and heavy metal removal (e.g., from arsenic and chromium).

In AOPs, the use of heterogeneous and homogeneous catalysts has shown promising results. Some advantages of heterogeneous catalysts over homogeneous catalysts include their resistance to severe working conditions, ease of separation from exhausted catalysts, and the fact that treated water does not need additional treatment to eliminate dissolved metals [135, 188]. Additionally, they work well throughout a wider pH range, from 2 to 9. In AOPs, various catalysts derived from hydrotalcite-like compounds [171], transition metal oxides [129, 187], zeolites [246], and metal-exchanged clays [130] displayed promising activity.

6.1 *Chemical AOPs*

Chemical AOPs involve using a wide variety of chemical reagent combinations to create the transient species. The most often utilized oxidant is hydrogen peroxide, while ozone and air are additionally employed.

6.2 *Catalytic Wet Peroxide Oxidation (CWPO) Process*

Producing $^{\bullet}OH$ from H_2O_2 using heterogeneous catalysts is known as Catalytic Wet Peroxide Oxidation (CWPO). Several metal oxides including iron, chromium, vanadium, copper, cobalt, nickel, and others, can break down H_2O_2 into hydroxyl radicals ($\cdot OH$) [134]. Recent reports have also mentioned graphitic nanocomposites, carbon nanotubes, and magnetic carbon nanocomposites as potential catalysts for AOPs. Depending on the catalysts used and their composition, these materials can release $\cdot OH$ radicals from H_2O_2 by various methods, even at neutral pH. Releasing the $\cdot OH$ radical from H_2O_2 has been extensively studied using cerium-based solid solutions as catalysts. In CeO_2-based CWPO reactions, oxygen vacancies frequently serve a dual purpose. The vacancies not only make electron transmission easier, but they also provide adsorption sites. To enhance the catalytic activity of CeO_2, environmental scientists frequently dope the crystal lattice with transition metals having varying valency. A notable reduction at 70 °C in COD and total organic carbon (TOC) was seen when zirconium was doped into CeO_2, which in turn enhanced the catalytic action of nanoceria catalysts concerning water-phase oxidation of 2,4-dichlorophenol, 4-chlorophenol, and 2,4-dichlorophenoxyacetic acid [128]. Surface area lengths and X-ray diffraction proved that the catalysts were constant and reclaimable for up to six contends. The defected Ce^{3+} sites that can be transformed to Ce^{4+} species in H_2O_2 were enhanced in the Zr-doped ceria catalysts, mostly attributable to their increased surface Ce content and specific surface area [115]. The positive effects of vanadium doping on chlorinated pollutants with CWPO were observed when the ceria lattice was combined with the dopant, demonstrating the synergy between the two materials [127]. Scavenging investigations showed that hydroxyl radicals were formed, which activated the process.

The CWPO treatment of fabricated quinoline wastewater by porous CeCu oxide compounds proved extremely useful and long-lasting. With a remarkable TOC removal and reduced Cu^{2+} leaching throughout a broad pH extent, the solid Cu mixed in the CeO_2 lattice showed a 98% oxidation alteration for quinoline [117]. The selectivity toward full phenol mineralization was reduced when CuO to CeO_2 catalysts [175]. The degradation activity of MnO_2–CeO_2 regarding phenol was similarly negatively affected by the presence of potassium as a dopant [226]. The pH and results from the WPO of phenol-like pollutants using Mn/TiO_2–CeO_2 composite showed a free-radical approach and a synergistic effect of the dopant and ceria [157].

To reproduce Fenton-like action, researchers often choose iron as a doping agent with ceria. When chlorinated organics were doped with iron, the reducing power of ceria rose. There was a noticeable drop in COD and TOC between 50 and 70 °C, and the reusable catalysts successfully degraded the target pollutants entirely [133]. At temperatures between 80 and 100 °C, iron samples exhibited substantially higher activity in a different investigation using Fe_2O_3/CeO_2 and WO_3/CeO_2 catalysts. The catalysts exhibited remarkable stability in the acidic reaction media with minimal iron leaching [174]. The oxidation of tetracycline was enhanced in the instance of Fe^0/CeO_2 nanocomposite, which combined nanoscale zero-valent iron with CeO_2 [297]. The importance of surface-bound ·OH in the Fenton reaction was demonstrated by radical scavenging experiments. Fe^0/CeO_2 generated $^•$OH at a far higher rate in the Fenton system than bare nanoscale zero-valent iron. Surface oxygen vacancies enhanced the catalyst's adsorption capabilities and sped up electron transport, which aided in the breakdown of H_2O_2. Fe/TiO_2–CeO_2 composites effectively promoted the oxidation of dye waste products, including H-acid, into biodegradable species [300].

The characteristics of the CeO_2 often change when other metal oxides are added. The impact of doping $Fe_2O_3/\gamma\text{-}Al_2O_3$ with CeO_2 on the composition and catalytic activity of azo dyes catalytic WPO was assessed [156]. Adding CeO_2 increased the catalytic activity of wastewater by the Fe_2O_3 crystals. Almost no iron leaching was observed throughout the leaching, indicating that the catalysts were stable. Furthermore, the surface area of the resulting catalyst was greatly enhanced by the addition of CeO_2 to TiO_2. The CeO_2-doped equivalents of virgin perovskites $LaMO_3$ (Cu, Fe) were also reported to exhibit heterogeneous Fenton-like catalysts for breaking bisphenol. Scavenging studies and EPR analysis showed that surface-bound ·OH was the primary decomposer of the pollutant [102].

In addition to ceria-based catalysts, supporting ceria catalysts are also utilized in WPO processes. Recent work has investigated the possibility of conducting WPO reactions on CeO_2 based on porous materials such as xerogels and activated carbon. One example is the catalytic oxidation of methyl green dye, which was accomplished using CeO_2-carbon xerogel composite catalysts. The explanation for higher efficiency was determined to be a synergistic effect caused by the incorporation of essential oxygen groups and the redox characteristics of CeO_2 at the surface of the frame [217]. Research on quinoline mineralization by a CWPO with granular activated carbon impregnated with ceria revealed that the power law model could accurately depict the oxidation process kinetics [283].

6.3 Catalytic Ozonation Process

Although ozone has a lot of potential applications in water treatment technologies, its selectivity for organic contaminants makes it unsuitable for use in wastewater with significant COD. The process also necessitates complex equipment, which in turn increases operational costs. When applied to ozonation, heterogeneous catalysts increase reactivity with various organic contaminants at gentler settings, reducing cost [75]. The potential of cerium oxide as an ozonation catalyst for eliminating certain organic molecules has been investigated. The catalysts used in ozonating oxalic acid, aniline, and a reactive dye were cerium oxides produced by hydrothermal synthesis or precipitation techniques [197]. Total organic carbon (TOC) reductions were nearly 100% in every instance. Researchers observed that ozone promoted the oxidation of Ce^{3+} to Ce^{4+} in acid media (HNO_3), leading to a higher conversion of phenol than uncatalyzed ozonation. This investigation found that ozone and the cerium redox pair worked synergistically [177]. Using CeO_2 nanorods, nanocubes, and nano octahedra, researchers have lately investigated the impact of ceria geometry on the catalytic ozonation of p-nitrophenol. Because of variations of coordination atoms, the density of defects and oxygen vacancies, and the surface basic sites, ceria nanorods exhibited the highest degrading efficiency regardless of the surface area [6]. Light pre-treatment, which increased the surface oxygen vacancy concentration of ceria, improved its capacity to catalyze the degradation of stearic acid, according to another study. This discovery drives home the point that CeO_2 efficacy in catalytic ozonation is highly dependent on the quantities of oxygen defect vacancies [58]. Investigations exploring the impacts of various cerium sources on phenol ozonation have shown that cerium nitrate, as opposed to ceric ammonium nitrate, is the superior precursor [66]. The efficacy of 4-chlorophenol and 4-phenolsulfonic acid removal by ceria films on glass utilizing the spray pyrolysis procedure in conjunction with catalytic ozonation. According to TOC, the obtained ceria films have a higher catalytic activity than conventional ozonation [100]. Interesting research also focused on the production of bromate by catalytic ozonation of p-chloronitrobenzene over CeO_2. Since both bromide and bromate contain oxygen, they were absorbed on the active sites of CeO_2 when O_3 was broken down in solution, which explains why CeO_2 did not adsorb them [268]. Another study found that to eliminate UV254, the best catalyst for ozonating filtered raw river water was CeO_2. The study also utilized synthetic goethite, MgO, and CeO_2. Concurrently, fulvic and humic acids were effectively oxidized [267].

As ceria supports in the catalytic ozonation process, carbon-based materials such as activated carbon and carbon xerogels have been investigated [10]. The enhanced catalytic activity in the ozonation process is mainly attributed to the synergetic effect of carbon and CeO_2 interface structures [73]. Composites with a higher carbon content exhibited more catalytic activity [200]. In a related study, the same group demonstrated comparable results when sulfanilic acid, aniline, and benzenesulfonic acid were catalytically ozonated using a CeO_2/activated carbon composite [74]. The synergistic impact of activated carbon and CeO_2 was proven by the mineralization of oxalic acid and aniline over a CeO_2/activated carbon composite. The surface concentration of Ce^{3+} species had an effect on the overall catalytic performance [90]. Activated carbon-containing ceria and multi-walled carbon nanotubes promote the degradation of pyruvic and maleic acids, sulfamethoxazole [91], erythromycin [92], and bezafibrate [93], among others. The mineralization of the target compounds was enhanced by both materials as compared to single ozonation. Dimethyl phthalate can be broken down with the help of Ozone, and CeO_2/activated carbon catalyst works well for this. [145]. With CeO_2/activated carbon, the mineralization rate constant was about 2.5 times higher than that achieved with pure activated carbon (> 7.5 times) and with O_3 (> 7.5 times). One area where CeO_2/activated carbon really shines is in the Ozonation of p-toluenesulfonic acid, according to another study. The idea that the catalyst was better was based on the fact that its surface had more hydroxyl radicals [49]. To produce the oxidized byproducts, these radicals may react rapidly with the pollutants of interest and any intermediates. An improved and more stable method for ozonating oxalic acid in water was found using a catalyst made of rod-shaped ceria coated on activated carbon [76]. The active site for ozonation of landfill leachate on the surface of CeO_2/activated carbon was proposed as a viable strategy for improving the catalytic property of CeO_2/activated carbon [213]. Another study discovered that oxalic acid and the textile Reactive Blue 5 dye were catalysts for the ozonation of CeO_2-carbon xerogel composites with different compositions [198].

Table 8 shows a list of organic pollutants that have been catalytically ozonated using catalysts containing cerium and catalysts supported by cerium oxides. Foreign atom modifications to the ceria lattice created oxygen vacancies, largely responsible for the catalysts' outstanding activity.

Table 8 Evaluation of the catalytic activity of CeO_2-based nano catalyst by the ozonation process

Catalyst	Support	Target contaminants	Oxidation time	Removal efficiency (%)	References
Ce–Mn–O_x	MCM-41	Oxalic acid	20 min	90.3	[116]
CeO_2	MCM-48	Atrazine	20 min	67	[151]
CeO_2	Red mud	Bezafibrate	5 min	96	[279]
CeO_2	Oxidized C nanotubes	Phenol	60 min	96	[264]
CeO_2	ZSM5 zeolites	Nitrobenzene	60 min	86.3	[39]
CeO_2	CuO	Oxalate	25 min	97	[299]
$V_{1-x}Ce_xO_2$	Al_2O_3/SiO_2	Chloronitrophenol	4 h	100	[166]
$Fe_3O_4@SiO_2@CeO_2$	–	Acetylsalicylic acid	60 min	81	[50]
CeO_2–TiO_2	–	Tetracycline	60 min	65	[302]
Ru/CeO_2	–	Succinic acid	60 min	100	[53]
Mn–CeO_x	γ-Al_2O_3	Bromaminic acid	120 min	40	[277]
Ce–Ti–Zr–O_x	–	Oxalic acid	120 min	89	[232]
$Zr_{1-x}Ce_xO_2$	Al_2O_3/ SiO_2/TiO_2	Trichlorophenol	4 h	100	[167]
$Ce_{0.1}Fe_{0.9}OOH$	–	Sulfamethazine	30 min	46	[24]
$Zr_{1-x}Ce_xO_2$	SiO_2	Toluene	15 min	80	[247]

6.4 Catalytic Wet Air Oxidation (WAO) Process

WAO process uses molecular oxygen as an oxidizing agent. The process uses high pressure and temperature to fully break down pollutants and produce active oxygen species, such as hydroxyl radicals. It is commonly used to treat effluent resistant to direct biological purification because of their high organic matter content (COD on the 10–100 g/L). Complexes with complicated structures are reduced to simpler complexes to facilitate their degradation by other means. Because certain low molecular weight molecules, such as propionic and acetic acids, acetaldehyde, ethanol, and methanol, resist further transformation to carbon dioxide, the WAO method does not achieve total mineralization of organics. Ammonia, a stable byproduct of organic nitrogen compound transformation, is the primary product under WAO operating conditions. Therefore, wastewater that needs further treatment, usually in a traditional biological treatment facility, undergo the WAO process as a pre-treatment step. To reduce the intensity of the necessary conditions, the WAO process has made extensive use of ceria and ceria-based catalysts [243]. The catalytic performance of noble metal can be enhanced when CeO_2 is utilized as a support, which is why this direction has been extensively researched [257].

Owing to its inherent ability to store oxygen, CeO_2 support facilitated the movement of oxygen from the gas phase to the catalyst, including metal-reinforced ceria catalysts made of Pt and Ru. The acid–base catalyst and redox characteristics also significantly influenced the WAO reactions of aniline, phenol, and acetic acid [138]. N-butanol and acetic acid oxidation were seen when ceria was loaded onto Pt catalysts assisted on CeO_2/Al_2O_3 mixed oxides, improving their reducibility [230]. The CWAO of oleic acid was carried out using a ceria catalyst supported by platinum. Additionally, two oxidation pathways were found by sequential carboxy-decarboxylation, which produced high acetic acid by breaking C–C bonds in the alkyl chain and creating CO_2 [141]. CWAO of aniline, ammonia, phenol, and carboxylic acids was performed in a batch reactor using Pd, Ru, and Pt adopted on CeO_2. Good performance in CWAO depended on ceria's oxygen transfer capacity [27]. Depending on the metal/support contact and the geometry of the support, the behavior of Ru/CeO_2 catalysts in the WAO of acetic, succinic, and acrylic acids changed [196]. Similarly, a study was conducted on the WAO process using Pt and Ru catalysts aided on industrial CeO_2 and mixed oxide $Zr_{0.1}(Ce_{0.75}Pr_{0.25})_{0.9}O_2$. This type of support has no substantial impact on the intrinsic activity of ruthenium, in contrast to platinum. The WAO of methyl tert-butyl ether used Ag nanoparticles to assist ZrO_2/CeO_2. The surface silver atoms' D-electron density increased due to ceria reducing Ag_2O [96]. Additionally, Table 9 includes a list of catalysts reported to use CeO_2-based catalysts. In contrast to the uncatalyzed WAO process, most catalysts function under mild circumstances, which is an intriguing observation.

Table 9 The catalytic activity of CeO_2-based catalysts by WAO process

Catalyst	Support	Target contaminants	Removal (%)	References
CeO_2	MnO_2	Phenol	53	[4]
Mn/Ce	–	Acrylic acid	97	[234]
Mn/CeO_2	–	Aniline	96	[142]
MnO_2/CeO_2	–	Phenol	65	[103]
Pt/$Ce_xZr_{1-x}O_2$	–	Phenol	95	[192]
Pt/CeO_2/$MnCeO_x$	–	Phenol	80	[13]
CeO_2	γ-Al_2O_3	Phenol	100	[41]
Ru/CeO_2	Active carbon	Acrylic acid, phenol	100	[195]
Ce/TiO_2	–	Phenol	77	[285]
Pt/ceria	Al_2O_3	Oxalic acid	74	[221]
CuO/CeO_2	–	Black liquor discharge	77	[85]
CeO_2	γ-Al_2O_3	Phenol	100	[34]
NiO/CeO_2	–	Wastewater of the paper industry	62	[126]

6.5 Photochemical AOPs Process

Photochemical AOPs are simpler, less expensive, more effective, and more ecologically friendly than chemical AOPs. Photochemical oxidation relies on visible light, often known as ultraviolet (UV) radiation. Further improvements to water treatment efficiencies have come from a variety of method integrations, including AOP combinations, AOP integrations with biological processes, and the use of ultrasound in conjunction with AOP. Consequently, catalysts and strong oxidants like O_3 and H_2O_2 have been combined with UV radiations [20]. The efficiency of AOPs is enhanced by doped heterogeneous CeO_2, which increases photocatalytic activity and speeds up the process. CeO_2 has a broad bandgap, but it is cheap, photostable, can store oxygen, and reduces power consumption. With their large band gap, strong redox capacity, high oxygen storage/release capability, eco-friendliness, photostability, and affordability, photocatalysts made from nanoceria hold a lot of potential for environmental sustainability. Researchers have proposed a variety of techniques to increase the photocatalytic ability of CeO_2, additionally, the synthesis conditions affect the surface characteristics of CeO_2. Researchers found that surface oxygen defects were more significant for the photocatalytic oxidation of toluene than total voids in oxygen concentration in a study using multiple nanostructured rods and cubes with various physicochemical characteristics. Greater surface area meant greater bulk and surface defects were concentrated there, as there were more accessible crystal planes compared to nanocubes [158].

Compared to CeO_2 produced by air annealing, those produced by vacuum annealing showed superior photocatalytic activity when degrading methyl orange under both ultraviolet and visible light [46]. Researchers have discovered that a deep eutectic solvent improves the durability and production of CeO_2 NPs for flume quine degradation under UV-C irradiation. Pollutant elimination by ceria nanoparticles mediated by UV-C was attributed to ·OH radicals [114]. As demonstrated in Fig. 3b, the catalyst oxygen vacancies helped the Fe/CeO_2 create $^{\bullet}OH$ during the decomposition of hydrogen peroxide, the TEM image of Fe/CeO_2 presented in Fig. 3a [45]. A qualitative investigation of metal photocatalytic assets was conducted by doping CeO_2 with cobalt, titanium, iron, and manganese [291]. The absorption shifted to visible region due to doping, and the band was expanded to absorb a widespread wavelength as illustrated in Fig. 4. Methylene blue (MB) degraded because of this doping, with the degradation efficiency increasing from Co > Mn > Ti > Fe. The precipitation and impregnation approaches produced an effective degrading agent for methyl orange due to enhancement in its surface area and photocatalytic efficacy, the catalyst displayed higher overall performance. The degrading efficiency outperformed pure ceria by a factor of 3.87. Figure 5 shows the photodegradation pathway [36].

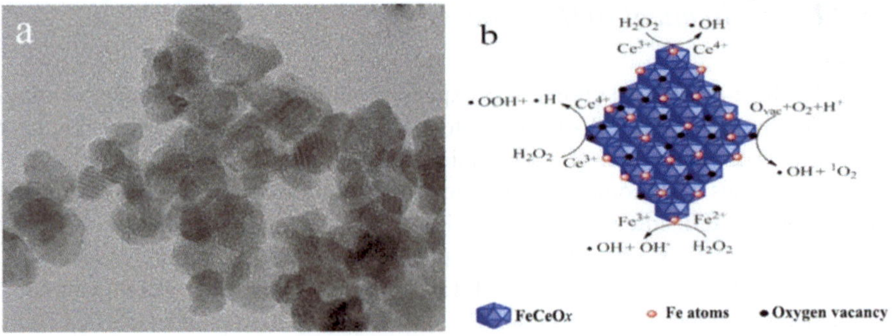

Fig. 3 **a** TEM of Fe/CeO_2, and **b** oxygen vacancies mechanism. Reproduced from [45] with permission, Copyright 2017 Elsevier

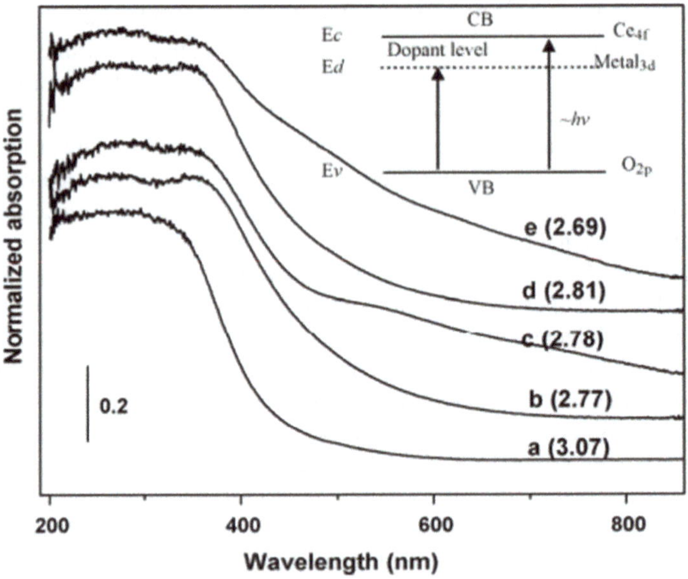

Fig. 4 UV–Vis spectra of Mn, Fe, Co, and Ti-doped CeO_2 NPs. Reproduced from [291] with permission, Copyright 2008 Elsevier

6.6 *Supported CeO_2 Photocatalysts for Water Management*

Doping or supporting CeO_2 with branded photocatalysts like ZnO or TiO_2 can also increase photodegradation effectiveness [99]. The photocatalytic degradation of MB by ZnO/Ce_2O_3 photocatalyst under UV light displayed a promising performance [262]. Under UV light, CeO_2–ZnO binary oxide catalysts loaded with palladium enhanced the photocatalytic degradation of MTBE [229]. TiO_2 and CeO_2-integrated husk of rice silica catalysts photo-degraded MB under UV irradiation in a manner

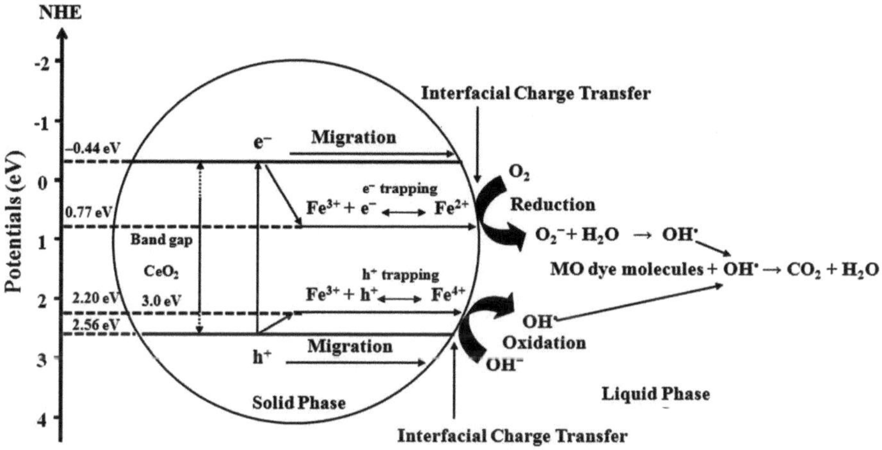

Fig. 5 Photodegradation mechanism of Fe/CeO_2. Reproduced from [36] with permission, Copyright 2014 Springer Nature

similar to that of the pristine anatase TiO_2 [5]. When exposed to UV light, Fe/ TiO_2 coatings with a CeO_2 layer showed heterogeneous photocatalysts that broke down the MB and methyl orange (MO) [211]. It has been reported that CeO_2–TiO_2 nanocomposite displayed a potential photocatalytic degradation toward rhodamine B (RhB) photochemically under UV and visible light [121]. Additionally, this study confirms the remarkable ability of Ce–TiO_2 composite as antibacterial activity against *E. coli*, in addition to their significant oxidation activity and superhydrophilicity. Furthermore, the catalyst activity of TiO_2 can be enhanced by doping with gold and/ or CeO_2. This is because the interaction between the two materials which increases the separation of charge, and enhanced photocatalytic oxidation of 2-propanol over these catalysts [81]. Caffeine can be photocatalytically broken down by CeO_2/TiO_2 catalyst in the absence of oxidants, according to an intriguing study [189].

There have also been reports of other support materials, including polymers, activated carbon, and polyaniline (PANI) [263]. When exposed to ultraviolet light, polyaniline, and zinc oxide nanocomposite showed remarkable photocatalytic efficacy in degrading Rhodamine B in effluent water [224]. Due to a synergistic effect that decreases the photoinduced electron–hole couple recombination degree, the first-order rate constant was eight times higher when compared to cerium oxide nanoparticles when using polyaniline. Under visible light irradiation, PANI/ceria nanocomposites degraded methylene blue dye effectively in 150 min [261]. The photocatalytic degradation of 4-chlorophenol by CeO_2/activated carbon occurred at the CeO_2 surface, whereas the adsorption occurs at activated carbon, a superior adsorbent. The rate constant was increased by 5.6 times and the quantum produce by 2.9 times due to a synergy effect [281]. Compared to pure CeO_2 and Ti_3C_2, the photocatalytic activity of CeO_2/Ti_3C_2 toward RhB under UV irradiation significantly improved [301].

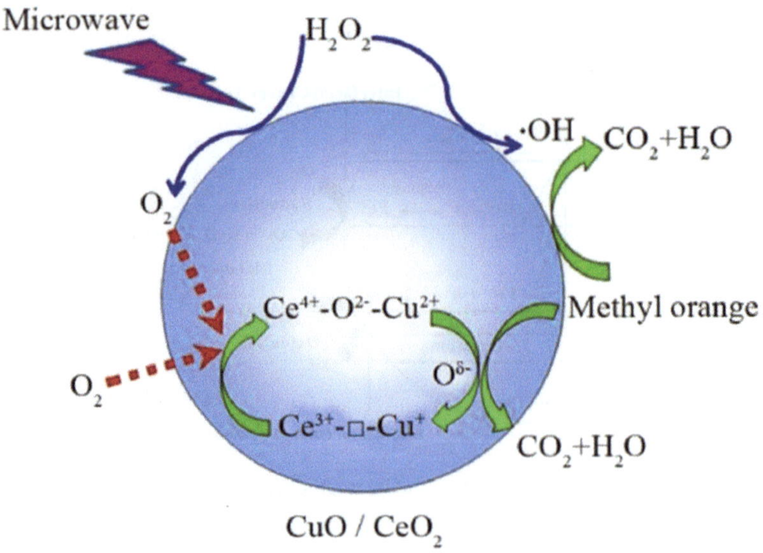

Fig. 6 Process of CuO/CeO_2 degradation. Reproduced from [280] with permission, Copyright 2014 American Chemical Society

The catalyst improved the oxidation of contaminants by raising the surface-absorbed oxygen concentration [42]. When compared to CeO_2, CuO/CeO_2 composite exhibited superior stability and photoactivity. They were utilized for the photodegradation of MO dye and were both inexpensive and economically beneficial [293]. Figure 6 shows the deterioration mechanism [280]. The photocatalytic activity was greatly enhanced when supported ceria catalysts were doped with noble metals. When exposed to visible light, anionic and cationic dyes were destroyed by AgBr/Ag_3PO_4/CeO_2 nanocomposites. The stability of the catalyst was greatly improved, and charge recombination was reduced after doping CeO_2 [67]. Samples of Pd/CeO_2-clinoptilolite were great photocatalysts for the decomposition of p-xylene because their nanoscale structure enabled improved distribution of noble metals. A chemical pathway centered around the adsorption displacement of species was postulated as the explanation of p-xylene oxidation [19]. Adding gold to nanocomposite fibrous membranes made of poly(vinylidene fluoride-co-hexafluoropropylene) and ZnO significantly enhanced their photocatalytic degradation when exposed to visible light [173].

6.7 *Sono Chemically AOPs*

Wastewater can be treated using ultrasound waves in two ways: indirectly, through chemical processes, or directly, through physical processes. The chemical process involves the creation of radical species at high sound frequencies employing homolytic fission of oxygen and water molecules. On the other hand, the physical process is characterized by expanding and contracting bubbles, which generate formidable breaking forces at 2000–5000 K and $\approx 6 \times 10^4$ kPa. When water molecules undergo sonolysis, they release reactive radicals into the surrounding water, which can be combined by organic species or undergo pyrolysis to break down organic contaminants [186, 208]. An ultrasonic system's oxidizing power can be amplified by using heterogeneous catalysts. By supplying more nuclei, heterogeneous catalysts speed up the pyrolysis of H_2O molecules and the production of ·OH, which speeds up cavitation bubbles [52].

Sonocatalytic degradation of water contaminants using cerium oxide is a relatively new application [227]. For the sonocatalytic breakdown of caffeine, CeO_2 with a mixture of rod and cube crystallite forms has been investigated. Operating settings were crucial regarding how efficiently caffeine degraded [193]. Over time, Fe(III) was reduced to Fe(II) and Ce(IV) was oxidized to Ce(IV) by employing $FeCeO_x$ in the sonocatalytic hydrolysis of diclofenac in water [44]. The substantial reduction of oxygen vacancies in $FeCeO_x$ that ensued from the reaction validated its role in the ultrasonic. The macrolide compounds antibiotic tylosin has been successfully catalytically degraded in water using WS_2 nanosheets interspersed ceria [289]. It was fascinating to see that sonocatalysis was hindered by water hardness. Reactive Red 84 was found to be efficiently broken down by a CeO_2/biochar nanocomposite under ultrasonic irradiation [123]. The efficiency of sono-, photo-, and photo-assisted sonocatalytic degradation of 4-nitrophenol utilizing porous $FeVO_4$ coated on CeO_2 nanocubes ($FeVO_4/CeO_2$) was examined. In comparison to $FeVO_4$ and CeO_2, core–shell nanostructured $FeVO_4/CeO_2$ performed better throughout a wide pH scope [65]. The photo-assisted sonocatalytic breakdown of Brilliant Green dye has also been demonstrated using a bimetallic catalyst known as NiO-impregnated CeO_2 [225].

6.8 Electrochemically AOP

One technique to make hydroxyl radicals electrochemically is through direct anodic oxidation, while another is the electro-Fenton method, which involves the indirect electrocatalytic synthesis of Fenton's reagent in situ. Electrochemical AOPs have a stellar reputation for being highly efficient, versatile, compatible with a wide range of systems, and environmentally friendly. One advantage of electrochemical AOP is that it can regulate and produce hydroxyl radicals in situ without adding chemical reagents to the material. Another is that it can immediately discharge corrected effluents into natural waters. Anodic oxidation is an electrochemically direct and chemical-free method of producing ·OH radicals [256]. When anodes with an elevated oxygen formation rate oxidize water, they directly create ·OH radicals at the anode surface [77]. Theoretically, the process might have a catalytic mechanism [185]. By combining organic oxidation with electrolysis of water, heterogeneous radicals of hydroxyl M(·OH) are produced. It is possible to conduct photo assisted or ultrasound-enhanced anodic oxidation with or without a catalyst.

Structures created using CeO_2 have been extensively utilized in electrochemical AOPs. A CeO_2 anode is used to mineralize the model insecticide endosulfan, which shows how the catalytic activities of endosulfan work in tandem with its plasma excitation [169]. As electrocatalysts, composites of Co_3O_4 and CeO_2 degraded tetracycline persulfate with relative ease. The short-lived species generated in this case were confirmed to be oxidizing hydroxyl radicals and sulfate radicals by means of electron spin resonance and radical scavenging [63]. Thin films of deposited by electrolysis CeO_2 and CeO_2–CoO polymeric oxides were sufficient for the electrochemical degradation of phenol [95]. Cyclic voltammetry (CV) and electrochemical impedance spectroscopy (EIS) investigations to examine its efficiency toward urea electro-oxidation. The results revealed that the electrocatalytic activity of $CeNi_2$/NiO was enhanced at high pH at varying urea concentrations. By amplifying structural defects, charge transfer, and surface roughness at Ni sites, CeNi assembly during electrodeposition enhances electroactivity [2]. Ceria composite electrolyte materials doped with Gd^{3+}, Sm^{3+}, and Nd^{3+} were used to successfully electrochemically cure C.I. Reactive Orange 107 [215]. By employing graphite electrodes as anodes and cathodes in a NaCl electrolyte, the possibility of removing dyes and TOC from a simulated C.I. Reactive Blue 198 effluent via samarium doped CeO_2 catalyst. While electro-oxidation was shown to eliminate all hues, complete mineralization could only be achieved with electrocatalytic oxidation [214]. A comparable investigation on a Ce-doped Ti/nano TiO_2/PbO_2 electrode similarly demonstrated that a reduced charge transfer resistance and a big electrochemical active surface area contribute to the electrode's strong electrocatalytic activity in water [282]. Reactive brilliant blue was successfully electrocatalytically oxidized using Ce-modified Ti-PbO_2 electrodes. The reasons included unique surface hydrophobicity and synergistic effects of cerium and PbO_2 [163].

7 CeO_2-Based Adsorbents for Water Treatment

Various contaminants can also be removed from wastewater using adsorption techniques. This procedure is easy and inexpensive to implement when pollutant concentrations are low. The fact that this procedure does not necessitate any pre-treatment makes it appealing to many researchers. Their use is very straightforward, and adsorption produces no hazardous by-products. Chemical and physical adsorption are the two primary categories of adsorption. In chemical adsorption, the adsorbent and adsorbate form a strong connection due to electron transfer. On the other hand, the Vander-Walls force between the adsorbent and adsorbate makes physical adsorption bonds relatively weaker.

Additionally, organic dyes, herbicides, and harmful metal ions can all be adsorbed using CeO_2. The adsorption effectiveness of pure CeO_2 has been enhanced through doped CeO_2. Using the Langmuir isotherm model, Pb, Zn, and Cu were degraded using virgin ceria [178]. Pb was discovered to be the most rapidly deteriorated of the three metals. In this case, the pseudo-second-order kinetics models the reaction. Degradation is time, and temperature dependent, with an ideal pH of 5.6. Arsenic was likewise destroyed by pure CeO_2 [203]. The isotherm model, a first-order reaction, was utilized by Freundlich and Langmuir for the reduction of arsenate and cadmium utilizing activated carbon doped into CeO_2 [228]. No pre-oxidation procedure was required for the adsorption, and the deterioration followed a Freundlich isotherm. Using the catalyst, an efficiency of 90% was achieved in 60 min. CeO_2 is ready to extract cadmium from effluent [47]. A larger quantity was used to remove cadmium during the adsorption, which obeyed a pseudo-second-order reaction. Their effectiveness was impressive, but their deterioration period may be as long as 72 h. Pure CeO_2 was used to extract phosphate [220]. The isotherm investigated the many variables affecting pollutant adsorption according to the equation $R^2 = 0.98$. Phosphate removal efficiency was enhanced by adding biochar to the catalyst [270]. The adsorption obeyed the Freundlich and Langmuir isotherm models, and the photocatalytic efficiency remarkable enhanced. The maximum adsorption efficiency was achieved at 10 min, and at a pH of 3.0.

The sol–gel approach was used to prepare CeO_2, which effectively eliminated acid black 210 [112]. The isotherm study shows that the procedure was effective, with an adsorption capacity of 175.75 mg/g. It has been reported that doping MnO_2 with CeO_2 greatly improved the degraded efficiency of arsenic (V), and the adsorption was obeyed the Freundlich model [97]. Furthermore, doped CeO_2 accelerated the degradation of EBT dye [241]. The removal efficiency was enhanced by doping the CeO_2 with silica xerogel. The adsorption follows the pseudo-second-order reaction and Freundlich and Langmuir isotherm models. Using pure CeO_2, fluoride was able to be broken down, and the adsorption process followed the Langmuir model, and the estimated adsorption capacity was 80.64 mg/g [54]. According to the Freundlich isotherm model, chromium was degraded by means of hexamethylenetetramine doped CeO_2 [219]. The degradation was accelerated by doping, and followed a pseudo-second-order mechanism.

The Langmuir adsorption model was followed in the degradation of As (III) and As (V) utilizing pure CeO_2 [149]. Due to its large surface area, the catalyst demonstrated exceptional adsorption efficiency. By adding CeO_2 to carbon nanotubes, the fluoride was extracted from the effluent [55]. With a working pH ranging from 3.0 to 7.4, the catalyst was synthesized utilizing the heat treatment approach. Carbon nanotubes doped with CeO_2 were used to extract arsenate [206]. This catalyst effectively removes contaminants from wastewater while being environmentally friendly and cost-effective. The incorporation of CeO_2 into graphene allowed the removal of arsenate from wastewater [290]. Its adsorption behavior was consistent with the Langmuir isotherm model. The adsorption capacity was 62.32 mg/g at pH 4.0 and 41.31 mg/g at pH 7.0. Phosphate degradation was facilitated by the utilization of prepared SiO_2/CeO_2 composite [143]. Its adsorption behavior was consistent with the Langmuir isotherm model, and the adsorption efficiency was 99%. The adsorption capacity reached 110 mg/g. The adsorption of contaminants by catalysts based on CeO_2 is summarized in Table 10.

Table 10 Heterogeneously doped CeO_2 for use in water treatment

Catalyst	Support	Isotherm model	Target pollutant	References
CeO_2	–	Langmuir	Pb, Zn, Cu	[178]
CeO_2	–	Freundlich and Langmuir	Arsenic	[203]
CeO_2	Activated carbon	Freundlich	As(III), As(V)	[228]
CeO_2	–	Freundlich	Cd	[47]
CeO_2	–	Lineal relation	Phosphate	[220]
CeO_2	Biochar	Freundlich and Langmuir	Phosphate	[270]
CeO_2	–	Freundlich and Langmuir	Acid black 210	[112]
CeO_2	MnO_2	Freundlich	As(V)	[97]
CeO_2	Silica Xerogel	Freundlich and Langmuir	Erichrome Black T	[241]
CeO_2	–	Langmuir	Fluoride	[54]
CeO_2	Hexamethylenetetramine	Freundlich	Cr(VI)	[219]
CeO_2	–	Langmuir	As(III), As(V)	[149]
CeO_2	Carbon nanotubes	Freundlich and Langmuir	Fluoride	[55]
CeO_2	Carbon nanotubes	Freundlich and Langmuir	Arsenate	[206]
CeO_2	Graphene	Langmuir	Arsenate	[290]
CeO_2	SiO_2	Langmuir	Phosphate	[143]

8 Filtration

One common method for separating contaminants from water in desalination and other industrial operations is membrane filtration. Now, polymers make up most of the membranes. Stability, durability, and selectivity are frequently lacking in these materials. Further, the separation effectiveness is reduced due to the hydrophobic nature of the membrane, which leads to significant biofouling. Some properties, like antibacterial activity, stability, hydrophilicity, and increased smoothness, have led to claims that nanoparticle embedded membranes are better than polymeric membranes [129, 179]. A thin-film membrane with nanoparticles of CeO_2 was reported [139]. Reduced fouling was seen as hydrophobic impurities were removed by the CeO_2-embedded membrane. Its antibacterial efficacy against *Escherichia coli* was also better than that of pure polyamide membrane.

9 Conclusion

Water treatment methods can make good use of CeO_2, a substance with unique redox, transport, and oxygen storage capabilities that make it technologically relevant. The surface characteristics and catalytic activities of CeO_2 are affected by several variables, including the synthesis method, the nature of the precursors, and the reaction temperature. Doping the CeO_2 with additional metal ions that have a lower valence and/or are smaller in size is one way to alter properties of ceria; this allows one to create stable compounds with new physical and chemical characteristics while simultaneously improving the performance of the metal oxide involved. Most of the time, ceria and the dopant or support worked together in a complementary fashion. Consequently, these substances have been put to good use in solving environmental problems, such as treating water. There has been a lot of success with CeO_2-based catalysts and adsorbents, but there are still some issues, like surface deactivation. Additional research on the creation of multicomponent ceria technologies that allow for better control over surface morphology is highlighted in this chapter.

References

1. Abbasi Z, Haghighi M, Fatehifar E, Rahemi N (2012) Comparative synthesis and physicochemical characterization of CeO_2 nanopowder via redox reaction, precipitation and sol–gel methods used for total oxidation of toluene. Asia-Pac J Chem Eng 7(6):868–876. https://doi.org/10.1002/apj.652
2. Abd El-Lateef HM, Almulhim NF, Mohamed IMA (2020) Physicochemical and electrochemical investigations of an electrodeposited $CeNi_2$@NiO nanomaterial as a novel anode electrocatalyst material for urea oxidation in alkaline media. J Mol Liq 297:111737. https://doi.org/10.1016/j.molliq.2019.111737
3. Abdelmonem HA, Hassanein TF, Sharafeldin HE, Gomaa H, Ahmed ASA, Abdel-lateef AM, Allam EM, Cheira MF, Eissa ME, Tilp AH (2024) Cellulose-embedded polyacrylonitrile/amidoxime for the removal of cadmium (II) from wastewater: adsorption performance and proposed mechanism. Colloids Surf, A 684:133081. https://doi.org/10.1016/j.colsurfa.2023.133081
4. Abecassis-Wolfovich M, Jothiramalingam R, Landau MV, Herskowitz M, Viswanathan B, Varadarajan TK (2005) Cerium incorporated ordered manganese oxide OMS-2 materials: improved catalysts for wet oxidation of phenol compounds. Appl Catal B 59(1):91–98. https://doi.org/10.1016/j.apcatb.2005.01.001
5. Adam F, Muniandy L, Thankappan R (2013) Ceria and titania incorporated silica based catalyst prepared from rice husk: Adsorption and photocatalytic studies of methylene blue. J Colloid Interface Sci 406:209–216. https://doi.org/10.1016/j.jcis.2013.05.066
6. Afzal S, Quan X, Lu S (2019) Catalytic performance and an insight into the mechanism of CeO_2 nanocrystals with different exposed facets in catalytic ozonation of p-nitrophenol. Appl Catal B 248:526–537. https://doi.org/10.1016/j.apcatb.2019.02.010
7. Aghamohammadi S, Haghighi M, Maleki M, Rahemi N (2017) Sequential impregnation vs. sol-gel synthesized Ni/Al_2O_3-CeO_2 nanocatalyst for dry reforming of methane: effect of synthesis method and support promotion. Mol Catal 431:39–48. https://doi.org/10.1016/j.mcat.2017.01.012
8. Ahmed ASA, Elshehy EA, Cheira MF (2024) 4—Functionalized hetero 2D material-based optical, electronic, magnetic, and mechanical properties. In: El-Said WA, Ghany N (eds)

Functionalization of two-dimensional materials and their applications. Woodhead Publishing, pp 115–149. https://doi.org/10.1016/B978-0-323-89955-0.00008-X
9. Ahmed HE, Iqbal Y, Aziz MH, Atif M, Batool Z, Hanif A, Yaqub N, Farooq WA, Ahmad S, Fatehmulla A, Ahmad H (2021) Green synthesis of CeO_2 nanoparticles from the abelmoschus esculentus extract: evaluation of antioxidant, anticancer, antibacterial, and wound-healing activities. Molecules 26(15):4659. https://doi.org/10.3390/molecules26154659
10. Akhtar J, Amin NS, Aris A (2011) Combined adsorption and catalytic ozonation for removal of sulfamethoxazole using Fe_2O_3/CeO_2 loaded activated carbon. Chem Eng J 170(1):136–144. https://doi.org/10.1016/j.cej.2011.03.043
11. Al-Abadleh HA, Grassian VH (2003) Oxide surfaces as environmental interfaces. Surf Sci Rep 52(3):63–161. https://doi.org/10.1016/j.surfrep.2003.09.001
12. Al-Mhyawi SR, Bader DMD, Bajaber MA, El Dayem SMA, Ragab AH, Abd El-Rahem KA, Gado MA, Atia BM, Cheira MF (2023) Zirconium oxide with graphene oxide anchoring for improved heavy metal ions adsorption: isotherm and kinetic study. J Market Res 22:3058–3074. https://doi.org/10.1016/j.jmrt.2022.11.121
13. Arena F, Italiano C, Spadaro L (2012) Efficiency and reactivity pattern of ceria-based noble metal and transition metal-oxide catalysts in the wet air oxidation of phenol. Appl Catal B 115–116:336–345. https://doi.org/10.1016/j.apcatb.2011.12.035
14. Arumugam A, Karthikeyan C, Haja Hameed AS, Gopinath K, Gowri S, Karthika V (2015) Synthesis of cerium oxide nanoparticles using *Gloriosa superba* L. leaf extract and their structural, optical and antibacterial properties. Mater Sci Eng C 49:408–415. https://doi.org/10.1016/j.msec.2015.01.042
15. Aruna ST, Kini NS, Rajam KS (2009) Solution combustion synthesis of CeO_2–$CeAlO_3$ nanocomposites by mixture-of-fuels approach. Mater Res Bull 44(4):728–733. https://doi.org/10.1016/j.materresbull.2008.09.034
16. Aruna ST, Mukasyan AS (2008) Combustion synthesis and nanomaterials. Curr Opin Solid State Mater Sci 12(3):44–50. https://doi.org/10.1016/j.cossms.2008.12.002
17. Arunachalam T, Karpagasundaram M, Rajarathinam N (2017) Ultrasound assisted green synthesis of cerium oxide nanoparticles using *Prosopis juliflora* leaf extract and their structural, optical and antibacterial properties. Mater Sci Pol 35(4):791–798. https://doi.org/10.1515/msp-2017-0104
18. Aseyd Nezhad S, Es-haghi A, Tabrizi MH (2020) Green synthesis of cerium oxide nanoparticle using L. leaf extract, its characterization and biological activities. Appl Organomet Chem 34(2):e5314. https://doi.org/10.1002/aoc.5314
19. Asgari N, Haghighi M, Shafiei S (2013) Synthesis and physicochemical characterization of nanostructured Pd/ceria-clinoptilolite catalyst used for p-xylene abatement from waste gas streams at low temperature. J Chem Technol Biotechnol 88(4):690–703. https://doi.org/10.1002/jctb.3887
20. Athanasekou CP, Romanos GE, Katsaros FK, Kordatos K, Likodimos V, Falaras P (2012) Very efficient composite titania membranes in hybrid ultrafiltration/photocatalysis water treatment processes. J Membr Sci 392–393:192–203. https://doi.org/10.1016/j.memsci.2011.12.028
21. Avgouropoulos G, Ioannides T (2003) Selective CO oxidation over CuO-CeO_2 catalysts prepared via the urea–nitrate combustion method. Appl Catal A Gen 244(1):155–167. https://doi.org/10.1016/S0926-860X(02)00558-6
22. Ayub MA, Sohail MI, Umair M, Zia ur Rehman M, Usman M, Sabir M, Rizwan M, Ali S, Ahmad Z (2019) Chapter Eight—Cerium oxide nanoparticles: Advances in synthesis, prospects and application in agro-ecosystem. In: Verma SK, Das AK (eds) Comprehensive analytical chemistry, vol 87. Elsevier, pp 209–250. https://doi.org/10.1016/bs.coac.2019.10.003
23. Babitha KK, Sreedevi A, Priyanka KP, Sabu B, Varghese T (2015) Structural characterization and optical studies of CeO_2 nanoparticles synthesized by chemical precipitation. Indian J Pure Appl Phys 53:596–603
24. Bai Z, Wang J, Yang Q (2017) Advanced treatment of municipal secondary effluent by catalytic ozonation using Fe_3O_4-CeO_2/MWCNTs as efficient catalyst. Environ Sci Pollut Res 24(10):9337–9349. https://doi.org/10.1007/s11356-017-8619-z

25. Baldim V, Bedioui F, Mignet N, Margaill I, Berret JF (2018) The enzyme-like catalytic activity of cerium oxide nanoparticles and its dependency on Ce_{3+} surface area concentration. Nanoscale 10(15):6971–6980. https://doi.org/10.1039/C8NR00325D
26. Balouiri M, Sadiki M, Ibnsouda SK (2016) Methods for in vitro evaluating antimicrobial activity: a review. J Pharma Anal 6(2):71–79. https://doi.org/10.1016/j.jpha.2015.11.005
27. Barbier J, Oliviero L, Renard B, Duprez D (2005) Role of ceria-supported noble metal catalysts (Ru, Pd, Pt) in wet air oxidation of nitrogen and oxygen containing compounds. Top Catal 33(1):77–86. https://doi.org/10.1007/s11244-005-2509-1
28. Bayyappagari B, Shaik K, Nasina MR, Inturu O, Dugasani SR (2018) Effect of Fe substitution on optical and magnetic properties of CeO_2 nanoparticles. Optik 154:821–827. https://doi.org/10.1016/j.ijleo.2017.10.025
29. Bhatt P, Rani A (2013) Textile dyeing and printing industry: an environmental hazard. Asian Dyer 10:51–54
30. Bhosale RR, Kumar A, AlMomani F, Alxneit I (2016) Sol–gel derived CeO_2–Fe_2O_3 nanoparticles: synthesis, characterization and solar thermochemical application. Ceram Int 42(6):6728–6737. https://doi.org/10.1016/j.ceramint.2016.01.042
31. Bulai IM, Venturino E (2016) Biodegradation of organic pollutants in a water body. J Math Chem 54(7):1387–1403. https://doi.org/10.1007/s10910-016-0603-1
32. Chadwick AV, Savin SLP (2009) EXAFS study of nanocrystalline CeO_2 samples prepared by sol–gel and ball-milling routes. J Alloys Compd 488(1):1–4. https://doi.org/10.1016/j.jallcom.2006.10.164
33. Chaisuk C, Wehatoranawee A, Preampiyawat S, Netiphat S, Shotipruk A, Mekasuwandumrong O (2011) Preparation and characterization of CeO_2/TiO_2 nanoparticles by flame spray pyrolysis. Ceram Int 37(5):1459–1463. https://doi.org/10.1016/j.ceramint.2010.11.018
34. Chang L, Chen IP, Lin S-S (2005) An assessment of the suitable operating conditions for the CeO_2/γ-Al_2O_3 catalyzed wet air oxidation of phenol. Chemosphere 58(4):485–492. https://doi.org/10.1016/j.chemosphere.2004.09.011
35. Channei D, Inceesungvorn B, Wetchakun N, Phanichphant S, Nakaruk A, Koshy P, Sorrell CC (2013) Photocatalytic activity under visible light of Fe-doped CeO_2 nanoparticles synthesized by flame spray pyrolysis. Ceram Int 39(3):3129–3134. https://doi.org/10.1016/j.ceramint.2012.09.093
36. Channei D, Inceesungvorn B, Wetchakun N et al (2014) Photocatalytic degradation of methyl orange by CeO_2 and Fe–doped CeO_2 films under visible light irradiation. Sci Rep 4:5757. https://doi.org/10.1038/srep05757
37. Cheira MF (2021) Synthesis of aminophosphonate-functionalised ZnO/polystyrene-butadiene nanocomposite and its characteristics for uranium adsorption from phosphoric acid. Int J Environ Anal Chem 101(12):1710–1734. https://doi.org/10.1080/03067319.2019.1686493
38. Cheira MF, Ahmed ASA, Elshehy EA (2024) 7—Advances of 2D nanostructure-based membranes for water treatment and radioactive pollutants removal. In: El-Said WA, Ghany N (eds) Functionalization of two-dimensional materials and their applications. Woodhead Publishing, pp 209–270. https://doi.org/10.1016/B978-0-323-89955-0.00004-2
39. Chen C, Yan X, Yoza BA, Zhou T, Li Y, Zhan Y, Wang Q, Li QX (2018) Efficiencies and mechanisms of ZSM5 zeolites loaded with cerium, iron, or manganese oxides for catalytic ozonation of nitrobenzene in water. Sci Total Environ 612:1424–1432. https://doi.org/10.1016/j.scitotenv.2017.09.019
40. Chen H-I, Chang H-Y (2005) Synthesis of nanocrystalline cerium oxide particles by the precipitation method. Ceram Int 31(6):795–802. https://doi.org/10.1016/j.ceramint.2004.09.006
41. Chen IP, Lin S-S, Wang C-H, Chang L, Chang J-S (2004) Preparing and characterizing an optimal supported ceria catalyst for the catalytic wet air oxidation of phenol. Appl Catal B 50(1):49–58. https://doi.org/10.1016/j.apcatb.2003.12.019
42. Cheng J, Song L, Wu R, Li S, Sun Y, Zhu H, Qiu W, He H (2020) Promoting effect of microwave irradiation on CeO_2-TiO_2 catalyst for selective catalytic reduction of NO by NH_3. J Rare Earths 38(1):59–69. https://doi.org/10.1016/j.jre.2019.04.014

43. Cheng Y, Feng Q, Ren X, Yin M, Zhou Y, Xue Z (2015) Adsorption and removal of sulfonic dyes from aqueous solution onto a coordination polymeric xerogel with amino groups. Colloids Surf, A 485:125–135. https://doi.org/10.1016/j.colsurfa.2015.09.002
44. Chong S, Zhang G, Wei Z, Zhang N, Huang T, Liu Y (2017) Sonocatalytic degradation of diclofenac with $FeCeO_x$ particles in water. Ultrason Sonochem 34:418–425. https://doi.org/10.1016/j.ultsonch.2016.06.023
45. Chong S, Zhang G, Zhang N, Liu Y, Huang T, Chang H (2017) Diclofenac degradation in water by $FeCeO_x$ catalyzed H_2O_2: influencing factors, mechanism and pathways. J Hazard Mater 334:150–159. https://doi.org/10.1016/j.jhazmat.2017.04.008
46. Choudhury B, Chetri P, Choudhury A (2014) Oxygen defects and formation of Ce_{3+} affecting the photocatalytic performance of CeO_2 nanoparticles. RSC Adv 4(9):4663–4671. https://doi.org/10.1039/C3RA44603D
47. Contreras AR, García A, González E, Casals E, Puntes V, Sánchez A, Font X, Recillas S (2012) Potential use of CeO_2, TiO_2 and Fe_3O_4 nanoparticles for the removal of cadmium from water. Desalin Water Treat 41(1):296–300. https://doi.org/10.1080/19443994.2012.664743
48. Dahle JT, Arai Y (2015) Environmental geochemistry of cerium: applications and toxicology of cerium oxide nanoparticles. Int J Environ Res Public Health 12(2):1253–1278. https://doi.org/10.3390/ijerph120201253
49. Dai Q, Wang J, Chen J, Chen J (2014) Ozonation catalyzed by cerium supported on activated carbon for the degradation of typical pharmaceutical wastewater. Sep Purif Technol 127:112–120. https://doi.org/10.1016/j.seppur.2014.01.032
50. Dai Q, Wang J, Yu J, Chen J, Chen J (2014) Catalytic ozonation for the degradation of acetylsalicylic acid in aqueous solution by magnetic CeO_2 nanometer catalyst particles. Appl Catal B 144:686–693. https://doi.org/10.1016/j.apcatb.2013.05.072
51. Darroudi M, Hoseini SJ, Kazemi Oskuee R, Hosseini HA, Gholami L, Gerayli S (2014) Food-directed synthesis of cerium oxide nanoparticles and their neurotoxicity effects. Ceram Int 40(5):7425–7430. https://doi.org/10.1016/j.ceramint.2013.12.089
52. Darvishi Cheshmeh Soltani R, Jorfi S, Safari M, Rajaei M-S (2016) Enhanced sonocatalysis of textile wastewater using bentonite-supported ZnO nanoparticles: Response surface methodological approach. J Environ Manage 179:47–57. https://doi.org/10.1016/j.jenvman.2016.05.001
53. Delanoë F, Acedo B, Karpel Vel Leitner N, Legube B (2001) Relationship between the structure of Ru/CeO_2 catalysts and their activity in the catalytic ozonation of succinic acid aqueous solutions. Appl Catal B 29(4):315–325. https://doi.org/10.1016/S0926-3373(00)00212-5
54. Dhillon A, Kumar Sharma T, Soni SK, Kumar D (2016) Fluoride adsorption on a cubical ceria nanoadsorbent: function of surface properties. RSC Adv 6(92):89198–89209. https://doi.org/10.1039/C6RA16962G
55. Di ZC, Li YH, Peng XJ, Luan ZK, Liang J (2007) Adsorption of fluoride by aligned carbon nanotubes supported ceria nanoparticles. In: Solid state phenomena, pp 1221–1224. https://doi.org/10.4028/3-908451-30-2.1221
56. Djilani C, Zaghdoudi R, Djazi F, Bouchekima B, Lallam A, Modarressi A, Rogalski M (2015) Adsorption of dyes on activated carbon prepared from apricot stones and commercial activated carbon. J Taiwan Inst Chem Eng 53:112–121. https://doi.org/10.1016/j.jtice.2015.02.025
57. Djilani C, Zaghdoudi R, Modarressi A, Rogalski M, Djazi F, Lallam A (2012) Elimination of organic micropollutants by adsorption on activated carbon prepared from agricultural waste. Chem Eng J 189–190:203–212. https://doi.org/10.1016/j.cej.2012.02.059
58. Dong Y, Chen SJ, Yu JL (2012) A systematic review and meta-analysis of long-term development of early term infants. Neonatology 102(3):212–221. https://doi.org/10.1159/000338099
59. Du H, Wang Y, Arandiyan H, Younis A, Scott J, Qu B, Wan T, Lin X, Chen J, Chu D (2017) Design and synthesis of CeO_2 nanowire/MnO_2 nanosheet heterogeneous structure for enhanced catalytic properties. Mater Today Commun 11:103–111. https://doi.org/10.1016/j.mtcomm.2017.03.002

60. Durán-Jiménez G, Hernández-Montoya V, Montes-Morán MA, Bonilla-Petriciolet A, Rangel-Vázquez NA (2014) Adsorption of dyes with different molecular properties on activated carbons prepared from lignocellulosic wastes by Taguchi method. Microporous Mesoporous Mater 199:99–107. https://doi.org/10.1016/j.micromeso.2014.08.013
61. Dutta D, Mukherjee R, Patra M, Banik M, Dasgupta R, Mukherjee M, Basu T (2016) Green synthesized cerium oxide nanoparticle: a prospective drug against oxidative harm. Colloids Surf B: Biointerfaces 147:45–53. https://doi.org/10.1016/j.colsurfb.2016.07.045
62. Eissa ME, Sakr AK, Hanfi MY, Sayyed MI, Al-Otaibi JS, Abdel-lateef AM, Cheira MF, Abdelmonem HA (2023) Physicochemical investigation of mercury sorption on mesoporous thioacetamide/chitosan from wastewater. Chemosphere 341:140062. https://doi.org/10.1016/j.chemosphere.2023.140062
63. El Hajji M, Hallaoui A, Bazzi L, Jbara O, Benlhachemi A, Hilali M, Tara A, Bakiz B (2015) Electrodegradation of phenol using pure and nanocomposite cerium oxides. J Mater Environ Sci 6(3):673–681
64. Elshehy EA, Cheira MF, Alhindawy IG, Ahmed ASA (2024) 2—Synthesis methodology for size and shape control of two-dimensional materials. In: El-Said WA, Ghany N (eds) Functionalization of two-dimensional materials and their applications. Woodhead Publishing, pp 19–57. https://doi.org/10.1016/B978-0-323-89955-0.00011-X
65. Eshaq G, Wang S, Sun H, Sillanpaa M (2020) Superior performance of $FeVO_4@CeO_2$ uniform core-shell nanostructures in heterogeneous Fenton-sonophotocatalytic degradation of 4-nitrophenol. J Hazard Mater 382:121059. https://doi.org/10.1016/j.jhazmat.2019.121059
66. Esmailpour AA, Moradi S, Yun J, Scott J, Amal R (2019) Promoting surface oxygen vacancies on ceria via light pretreatment to enhance catalytic ozonation. Catal Sci Technol 9(21):5979–5990. https://doi.org/10.1039/C9CY01450K
67. Eswar NK, Katkar VV, Ramamurthy PC, Madras G (2015) Novel $AgBr/Ag_3PO_4$ decorated ceria nanoflake composites for enhanced photocatalytic activity toward dyes and bacteria under visible light. Ind Eng Chem Res 54(33):8031–8042. https://doi.org/10.1021/acs.iecr.5b01993
68. Falcony C, Aguilar-Frutis MA, García-Hipólito M (2018) Spray pyrolysis technique; high-k dielectric films and luminescent materials: a review. Micromachines 9(8):414. https://doi.org/10.3390/mi9080414
69. Fan Z, Meng F, Gong J, Li H, Hu Y, Liu D (2016) Enhanced photocatalytic activity of hierarchical flower-like CeO_2/TiO_2 heterostructures. Mater Lett 175:36–39. https://doi.org/10.1016/j.matlet.2016.03.136
70. Farahmandjou M, Dastpak M (2018) Fe-loaded CeO_2 nanosized prepared by simple co-precipitation route. Phys Chem Res 6(4):713–720. https://doi.org/10.22036/pcr.2018.132220.1486
71. Farahmandjou M, Dastpak M (2020) Synthesis of Fe-doped CeO_2 nanoparticles prepared by solgel method. J Sci Islam Repub Iran 31(1):39–43. https://doi.org/10.22059/jsciences.2020.256813.1007255
72. Farahmandjou M, Zarinkamar M, Firoozabadi T (2016) Synthesis of cerium oxide (CeO_2) nanoparticles using simple co-precipitation method. Rev Mex Fis 62:496–499
73. Faria PCC, Monteiro DCM, Órfão JJM, Pereira MFR (2009) Cerium, manganese and cobalt oxides as catalysts for the ozonation of selected organic compounds. Chemosphere 74(6):818–824. https://doi.org/10.1016/j.chemosphere.2008.10.016
74. Faria PCC, Órfão JJM, Pereira MFR (2009) Activated carbon and ceria catalysts applied to the catalytic ozonation of dyes and textile effluents. Appl Catal B 88(3):341–350. https://doi.org/10.1016/j.apcatb.2008.11.002
75. Faria PCC, Órfão JJM, Pereira MFR (2009) Mineralization of substituted aromatic compounds by ozonation catalyzed by cerium oxide and a cerium oxide-activated carbon composite. Catal Lett 127(1):195–203. https://doi.org/10.1007/s10562-008-9670-7
76. Feng J, Zhang X, Fu J, Chen H (2018) Catalytic ozonation of oxalic acid over rod-like ceria coated on activated carbon. Catal Commun 110:28–32. https://doi.org/10.1016/j.catcom.2018.03.001

77. Feng L, van Hullebusch ED, Rodrigo MA, Esposito G, Oturan MA (2013) Removal of residual anti-inflammatory and analgesic pharmaceuticals from aqueous systems by electrochemical advanced oxidation processes, A review. Chem Eng J 228:944–964. https://doi.org/10.1016/j.cej.2013.05.061
78. Feng Y, Song Q, Lv W, Liu G (2017) Degradation of ketoprofen by sulfate radical-based advanced oxidation processes: kinetics, mechanisms, and effects of natural water matrices. Chemosphere 189:643–651. https://doi.org/10.1016/j.chemosphere.2017.09.109
79. Fernandes M, Rb Singh K, Sarkar T, Singh P, Pratap Singh R (2020) Recent applications of magnesium oxide (MgO) nanoparticles in various domains. Adv Mater Lett 11(8):1–10. https://doi.org/10.5185/amlett.2020.081543
80. Ferreira NS, Angélica RS, Marques VB, de Lima CCO, Silva MS (2016) Cassava-starch-assisted sol–gel synthesis of CeO_2 nanoparticles. Mater Lett 165:139–142. https://doi.org/10.1016/j.matlet.2015.11.107
81. Fiorenza R, Bellardita M, D'Urso L, Compagnini G, Palmisano L, Scirè S (2016) Au/TiO_2-CeO_2 catalysts for photocatalytic water splitting and VOCs oxidation reactions. Catalysts 6(8):121. https://doi.org/10.3390/catal6080121
82. Gado MA, Atia BM, Cheira MF, Abdou AA (2021) Thorium ions adsorption from aqueous solution by amino naphthol sulphonate coupled chitosan. Int J Environ Anal Chem 101(10):1419–1436. https://doi.org/10.1080/03067319.2019.1683552
83. Gadore V, Ahmaruzzaman M (2021) Tailored fly ash materials: a recent progress of their properties and applications for remediation of organic and inorganic contaminants from water. J Water Process Eng 41:101910. https://doi.org/10.1016/j.jwpe.2020.101910
84. Galvín AP, Ayuso J, García I, Jiménez JR, Gutiérrez F (2014) The effect of compaction on the leaching and pollutant emission time of recycled aggregates from construction and demolition waste. J Clean Prod 83:294–304. https://doi.org/10.1016/j.jclepro.2014.07.074
85. Garg A, Mishra IM, Chand S (2008) Catalytic oxidative treatment of diluted black liquor at mild conditions using copper oxide/cerium oxide catalyst. Water Environ Res 80(2):136–141. https://doi.org/10.2175/106143007X220752
86. Gnanam S, Rajendran V (2011) Synthesis of CeO_2 or α–Mn_2O_3 nanoparticles via sol–gel process and their optical properties. J Sol-Gel Sci Technol 58(1):62–69. https://doi.org/10.1007/s10971-010-2356-9
87. Godinho MJ, Gonçalves RF, S Santos LP, Varela JA, Longo E, Leite ER (2007) Room temperature co-precipitation of nanocrystalline CeO_2 and $Ce_{0.8}Gd_{0.2}O_{1.9-\delta}$ powder. Mater Lett 61(8):1904–1907. https://doi.org/10.1016/j.matlet.2006.07.152
88. Gomaa H, Shenashen MA, Cheira MF, Sueki K, Seaf El-Nasr TA, Selim MM, El-Safty SA (2023) Green extraction of uranium (238U) from natural radioactive resources. Chem Eng J 461:142014. https://doi.org/10.1016/j.cej.2023.142014
89. Gomaa H, Shenashen MA, Cheira MF, Sueki K, Seaf El-Nasr TA, Selim MM, El-Safty SA (2023) A novel, spongy mesoporous hybrid bio-adsorbents derived from agricultural waste for highly selective thorium recovery. J Clean Prod 402:136819. https://doi.org/10.1016/j.jclepro.2023.136819
90. Gonçalves A, Silvestre-Albero J, Ramos-Fernández EV, Serrano-Ruiz JC, Órfão JJM, Sepúlveda-Escribano A, Pereira MFR (2012) Highly dispersed ceria on activated carbon for the catalyzed ozonation of organic pollutants. Appl Catal B 113–114:308–317. https://doi.org/10.1016/j.apcatb.2011.11.052
91. Gonçalves AG, Órfão JJM, Pereira MFR (2013) Ceria dispersed on carbon materials for the catalytic ozonation of sulfamethoxazole. J Environ Chem Eng 1(3):260–269. https://doi.org/10.1016/j.jece.2013.05.009
92. Gonçalves AG, Órfão JJM, Pereira MFR (2014) Ozonation of erythromycin over carbon materials and ceria dispersed on carbon materials. Chem Eng J 250:366–376. https://doi.org/10.1016/j.cej.2014.04.012
93. Gonçalves AG, Órfão JJM, Pereira MFR (2015) Ozonation of bezafibrate over ceria and ceria supported on carbon materials. Environ Technol 36(6):776–785. https://doi.org/10.1080/09593330.2014.961563

94. Goulart C, Djurado E (2013) Synthesis and sintering of Gd-doped CeO_2 nanopowders prepared by ultrasonic spray pyrolysis. J Eur Ceram Soc 33(4):769–778. https://doi.org/10.1016/j.jeurceramsoc.2012.09.029
95. Guan R, Yuan X, Wu Z, Jiang L, Zhang J, Li Y, Zeng G, Mo D (2019) Efficient degradation of tetracycline by heterogeneous cobalt oxide/cerium oxide composites mediated with persulfate. Sep Purif Technol 212:223–232. https://doi.org/10.1016/j.seppur.2018.11.019
96. Guerra-Que Z, Torres-Torres G, Pérez-Vidal H, Cuauhtémoc-López I, de Los E, Monteros A, Beltramini JN, Frías-Márquez DM (2017) Silver nanoparticles supported on zirconia-ceria for the catalytic wet air oxidation of methyl tert-butyl ether. RSC Adv 7(6):3599–3610. https://doi.org/10.1039/C6RA25684H
97. Gupta K, Bhattacharya S, Chattopadhyay D, Mukhopadhyay A, Biswas H, Dutta J, Ray NR, Ghosh UC (2011) Ceria associated manganese oxide nanoparticles: synthesis, characterization and arsenic(V) sorption behavior. Chem Eng J 172(1):219–229. https://doi.org/10.1016/j.cej.2011.05.092
98. Gupta V, Rakesh M (2019) Catalytic wet peroxide oxidation (CWPO) of 2-Hydroxybenzoic acid and contaminated industrial effluent using $LnMO_3$(Ln= La or Pr and M= Fe or Fe-Co). J Water Process Eng 27:58–66. https://doi.org/10.1016/j.jwpe.2018.11.011
99. Guzmán C, del Ángel G, Gómez R, Galindo-Hernández F, Ángeles-Chavez C (2011) Degradation of the herbicide 2,4-dichlorophenoxyacetic acid over Au/TiO_2–CeO_2 photocatalysts: effect of the CeO_2 content on the photoactivity. Catal Today 166(1):146–151. https://doi.org/10.1016/j.cattod.2010.09.009
100. Guzmán IC, Rodríguez JL, Poznyak T, Chairez I, Hernández I, Hernández RT (2020) Catalytic ozonation of 4-chlorophenol and 4-phenolsulfonic acid by CeO_2 films. Catal Commun 133:105827. https://doi.org/10.1016/j.catcom.2019.105827
101. Hadi A, Yaacob II (2007) Novel synthesis of nanocrystalline CeO_2 by mechanochemical and water-in-oil microemulsion methods. Mater Lett 61(1):93–96. https://doi.org/10.1016/j.matlet.2006.04.013
102. Hammouda SB, Zhao F, Safaei Z, Babu I, Ramasamy DL, Sillanpää M (2017) Reactivity of novel Ceria-Perovskite composites CeO_2–$LaMO_3$ (MCu, Fe) in the catalytic wet peroxidative oxidation of the new emergent pollutant 'Bisphenol F': characterization, kinetic and mechanism studies. Appl Catal B 218:119–136. https://doi.org/10.1016/j.apcatb.2017.06.047
103. Hamoudi S, Larachi F, Cerrella G, Cassanello M (1998) Wet oxidation of phenol catalyzed by unpromoted and platinum-promoted manganese/cerium oxide. Ind Eng Chem Res 37(9):3561–3566. https://doi.org/10.1021/ie980081w
104. Hassanin MA, El-Gendy HS, Cheira MF, Atia BM (2021) Uranium ions extraction from the waste solution by thiosemicarbazide anchored cellulose acetate. Int J Environ Anal Chem 101(3):351–369. https://doi.org/10.1080/03067319.2019.1667984
105. Hassanin MA, Negm SH, Youssef MA, Sakr AK, Mira HI, Mohammaden TF, Al-Otaibi JS, Hanfi MY, Sayyed MI, Cheira MF (2022) Sustainable remedy waste to generate SiO_2 functionalized on graphene oxide for removal of U(VI) Ions. Sustainability 14(5):2699. https://doi.org/10.3390/su14052699
106. Hassanzadeh-Tabrizi SA, Mazaheri M, Aminzare M, Sadrnezhaad SK (2010) Reverse precipitation synthesis and characterization of CeO_2 nanopowder. J Alloys Compd 491(1):499–502. https://doi.org/10.1016/j.jallcom.2009.10.243
107. Hassanzadeh-Tabrizi SA, Taheri-Nassaj E (2011) Synthesis of high surface area Al_2O_3–CeO_2 composite nanopowder via inverse co-precipitation method. Ceram Int 37(4):1251–1257. https://doi.org/10.1016/j.ceramint.2010.12.004
108. Hazzaa R, Hussein M (2015) Adsorption of cationic dye from aqueous solution onto activated carbon prepared from olive stones. Environ Technol Innov 4:36–51. https://doi.org/10.1016/j.eti.2015.04.002
109. He Q, Shen Y, Xiao K, Xi J, Qiu X (2016) Alcohol electro-oxidation on platinum–ceria/graphene nanosheet in alkaline solutions. Int J Hydrog Energy 41(45):20709–20719. https://doi.org/10.1016/j.ijhydene.2016.07.205

110. He Y, Huang W, Chen R, Zhang W, Lin H (2015) Enhanced electrochemical oxidation of organic pollutants by boron-doped diamond based on porous titanium. Sep Purif Technol 149:124–131. https://doi.org/10.1016/j.seppur.2015.05.008
111. Hendy MA, Kashar TI, Allam EM, Gado MA, Yahia NS, Cheira MF (2024) Synthesis of pyridine dicarboxylic acid functionalized and crosslinked to polyvinyl alcohol/polyamide for thorium capturing from aqueous solution. ChemistrySelect 9(28):e202402329. https://doi.org/10.1002/slct.202402329
112. Hu J, Deng W, Chen D (2017) Ceria hollow spheres as an adsorbent for efficient removal of acid dye. ACS Sustain Chem Eng 5(4):3570–3582. https://doi.org/10.1021/acssuschemeng.7b00396
113. Iqbal A, Cevik E, Mustafa A, Qahtan TF, Zeeshan M, Bozkurt A (2024) Emerging developments in polymeric nanocomposite membrane-based filtration for water purification: a concise overview of toxic metal removal. Chem Eng J 481:148760. https://doi.org/10.1016/j.cej.2024.148760
114. Iqbal J, Shah NS, Sayed M, Muhammad N, Rehman S-u, Khan JA, Haq Khan ZU, Howari FM, Nazzal Y, Xavier C, Arshad S, Hussein A, Polychronopoulou K (2020) Deep eutectic solvent-mediated synthesis of ceria nanoparticles with the enhanced yield for photocatalytic degradation of flumequine under UV-C. J Water Process Eng 33:101012. https://doi.org/10.1016/j.jwpe.2019.101012
115. Issa Hamoud H, Azambre B, Finqueneisel G (2016) Reactivity of ceria–zirconia catalysts for the catalytic wet peroxidative oxidation of azo dyes: reactivity and quantification of surface Ce(IV)-peroxo species. J Chem Technol Biotechnol 91(9):2462–2473. https://doi.org/10.1002/jctb.4836
116. Jeirani Z, Soltan J (2016) Ozonation of oxalic acid with an effective catalyst based on mesoporous MCM-41 supported manganese and cerium oxides. J Water Process Eng 12:127–134. https://doi.org/10.1016/j.jwpe.2016.07.004
117. Jiao Z, Zhou G, Zhang H, Shen Y, Zhang X, Li J, Gao X (2018) Effect of calcination temperature on catalytic performance of CeCu oxide in removal of quinoline by wet hydrogen peroxide oxidation from water. J Braz Chem Soc 29 (11):2233–2243. https://doi.org/10.21577/0103-5053.20180100
118. Kadam AA, Lade HS, Lee DS, Govindwar SP (2015) Zinc chloride as a coagulant for textile dyes and treatment of generated dye sludge under the solid state fermentation: hybrid treatment strategy. Biores Technol 176:38–46. https://doi.org/10.1016/j.biortech.2014.10.137
119. Kandil AE-HT, Atia BM, El-Dars FMSE, Hussein MYM, Cheira MF (2024) Synthesis, characterization, and kinetic approach of a polyvinyl chloride embedded-thiosemicarbazide sequestering agent for efficient uranium adsorption from aqueous solution. J Water Process Eng 67:106144. https://doi.org/10.1016/j.jwpe.2024.106144
120. Kannan SK, Sundrarajan M (2014) A green approach for the synthesis of a cerium oxide nanoparticle: characterization and antibacterial activity. Int J Nanosci 13(03):1450018. https://doi.org/10.1142/S0219581X14500185
121. Kasinathan K, Kennedy J, Elayaperumal M, Henini M, Malik M (2016) Photodegradation of organic pollutants RhB dye using UV simulated sunlight on ceria based TiO_2 nanomaterials for antibacterial applications. Sci Rep 6(1):38064. https://doi.org/10.1038/srep38064
122. Khadar Y, A B, Devarajan VP, Subramanian R (2017) Hydrothermal synthesis of gadolinium (Gd) doped cerium oxide (CeO_2) nanoparticles: characterization and antibacterial activity. Orient J Chem 33:2405–2411. https://doi.org/10.13005/ojc/330533
123. Khataee A, Gholami P, Kalderis D, Pachatouridou E, Konsolakis M (2018) Preparation of novel CeO_2-biochar nanocomposite for sonocatalytic degradation of a textile dye. Ultrason Sonochem 41:503–513. https://doi.org/10.1016/j.ultsonch.2017.10.013
124. Kim JY, Kim U, Cho W-S (2009) Synthesis of ceria nanosphere by ultrasonic spray pyrolysis. J Korean Ceram Soc 46:249–252. https://doi.org/10.4191/KCERS.2009.46.3.249
125. Kumar A, Jayeoye TJ, Mohite P, Singh S, Rajput T, Munde S, Eze FN, Chidrawar VR, Puri A, Prajapati BG, Parihar A (2024) Sustainable and consumer-centric nanotechnology-based materials: an update on the multifaceted applications, risks and tremendous opportunities. Nano Struct Nano Objects 38:101148. https://doi.org/10.1016/j.nanoso.2024.101148

126. Kumar S, Sharma C (2015) NiO–CeO_2 nano-catalysts: synthesis, characterization and application in catalytic wet air oxidation of wastewater. Mater Express 5(5):419–428
127. Kunjachan C, Sreevalsan A, Kurian M (2017) Effect of vanadium on the catalytic activity of ceria towards wet peroxide oxidation of persistent and biorecalcitrant chlorinated organics. J Water Process Eng 19:42–50. https://doi.org/10.1016/j.jwpe.2017.07.007
128. Kunjachan C, Sreevalsan A, Kurian M (2018) On the redox nature of ceria-zirconia mixed nanocatalysts and its effect on catalytic wet peroxide oxidation of chlorinated organic pollutants. J Environ Chem Eng 6(4):4151–4159. https://doi.org/10.1016/j.jece.2018.06.009
129. Kurian M (2020) Cerium oxide based materials for water treatment—a review. J Environ Chem Eng 8(5):104439. https://doi.org/10.1016/j.jece.2020.104439
130. Kurian M, Joy M, Raj D (2012) Hydroxylation of phenol over rare earth exchanged iron pillared montmorillonites. J Porous Mater 19(5):633–640. https://doi.org/10.1007/s10934-011-9514-5
131. Kurian M, Kunjachan C (2014) Investigation of size dependency on lattice strain of nanoceria particles synthesised by wet chemical methods. Int Nano Lett 4(4):73–80. https://doi.org/10.1007/s40089-014-0122-7
132. Kurian M, Kunjachan C (2015) Effect of lattice distortion on physical properties and surface morphology of nanoceria framework with incorporation of iron/zirconium. Nano Struct Nano Objects 1:15–23. https://doi.org/10.1016/j.nanoso.2015.06.002
133. Kurian M, Kunjachan C, Sreevalsan A (2017) Catalytic degradation of chlorinated organic pollutants over $Ce_xFe_{1-x}O_2$ (x: 0, 0.25, 0.5, 0.75, 1) nanocomposites at mild conditions. Chem Eng J 308:67–77. https://doi.org/10.1016/j.cej.2016.09.039
134. Kurian M, Nair DS (2014) On the efficiency of cobalt zinc ferrite nanoparticles for catalytic wet peroxide oxidation of 4-chlorophenol. J Environ Chem Eng 2(1):63–69. https://doi.org/10.1016/j.jece.2013.11.026
135. Kurian M, Nair DS (2015) Manganese zinc ferrite nanoparticles as efficient catalysts for wet peroxide oxidation of organic aqueous wastes. J Chem Sci 127(3):537–546. https://doi.org/10.1007/s12039-015-0806-1
136. Kusmierek E (2020) A CeO_2 semiconductor as a photocatalytic and photoelectrocatalytic material for the remediation of pollutants in industrial wastewater: a review. Catalysts 10(12). https://doi.org/10.3390/catal10121435
137. Kyzas GZ, Matis KA (2015) Nanoadsorbents for pollutants removal: a review. J Mol Liq 203:159–168. https://doi.org/10.1016/j.molliq.2015.01.004
138. Lafaye G, Barbier J, Duprez D (2015) Impact of cerium-based support oxides in catalytic wet air oxidation: conflicting role of redox and acid–base properties. Catal Today 253:89–98. https://doi.org/10.1016/j.cattod.2015.01.037
139. Lakhotia A, Notani V, LeDoux C Malware economics and its implication to anti-malware situational awareness. In: 2018 International conference on cyber situational awareness, data analytics and assessment (Cyber SA), 11–12 June 2018 2018, pp 1–8. https://doi.org/10.1109/CyberSA.2018.8551388
140. Lei W, Mi Y, Feng R, Liu P, Hu S, Yu J, Liu X, Rodriguez JA, Wang J-o, Zheng L, Tang K, Zhu S, Liu G, Liu M (2018) Hybrid 0D–2D black phosphorus quantum dots–graphitic carbon nitride nanosheets for efficient hydrogen evolution. Nano Energy 50:552–561. https://doi.org/10.1016/j.nanoen.2018.06.001
141. Levasseur B, Renard B, Barbier Jr J, Duprez D (2006) Catalytic wet air oxidation of oleic acid on ceria-supported platinum catalyst.effect of pH. React Kinetics Catal Lett 87(2):269–279. https://doi.org/10.1007/s11144-006-0034-2
142. Levi R, Milman M, Landau MV, Brenner A, Herskowitz M (2008) Catalytic wet air oxidation of aniline with nanocasted Mn−Ce-oxide catalyst. Environ Sci Technol 42(14):5165–5170. https://doi.org/10.1021/es703070h
143. Li C-m, Zhang Y-s, Wang X-p, Yin X-b, Luo N-n, Khayambashi A, Wei Y-z (2019) The synthesis and characterization of hydrous cerium oxide nanoparticles loaded on porous silica micro-sphere as novel and efficient adsorbents to remove phosphate radicals from water. Microporous Mesoporous Mater 279:73–81. https://doi.org/10.1016/j.micromeso.2018.11.034

144. Li K, Lyu T, He J, Jang BWL (2020) Selective hydrogenation of acetylene over Pd/CeO_2. Front Chem Sci Eng 14(6):929–936. https://doi.org/10.1007/s11705-019-1912-2
145. Li L, Ye W, Zhang Q, Sun F, Lu P, Li X (2009) Catalytic ozonation of dimethyl phthalate over cerium supported on activated carbon. J Hazard Mater 170(1):411–416. https://doi.org/10.1016/j.jhazmat.2009.04.081
146. Li M, Guha S, Zangmeister R, Tarlov MJ, Zachariah MR (2011) Method for determining the absolute number concentration of nanoparticles from electrospray sources. Langmuir 27(24):14732–14739. https://doi.org/10.1021/la202177s
147. Li Q, Qi Y, Gao C (2015) Chemical regeneration of spent powdered activated carbon used in decolorization of sodium salicylate for the pharmaceutical industry. J Clean Prod 86:424–431. https://doi.org/10.1016/j.jclepro.2014.08.008
148. Li Q, Zhang Q, Cui H, Ding L, Wei Z, Zhai J (2013) Fabrication of cerium-doped lead dioxide anode with improved electrocatalytic activity and its application for removal of Rhodamine B. Chem Eng J 228:806–814. https://doi.org/10.1016/j.cej.2013.05.064
149. Li R, Li Q, Gao S, Shang JK (2012) Exceptional arsenic adsorption performance of hydrous cerium oxide nanoparticles: Part A. Adsorption capacity and mechanism. Chem Eng J 185–186:127–135. https://doi.org/10.1016/j.cej.2012.01.061
150. Li T, Hu X, Liu C, Tang C, Wang X, Luo S (2016) Efficient photocatalytic degradation of organic dyes and reaction mechanism with $Ag_2CO_3/Bi_2O_2CO_3$ photocatalyst under visible light irradiation. J Mol Catal A: Chem 425:124–135. https://doi.org/10.1016/j.molcata.2016.10.001
151. Li W, Lu X, Xu K, Qu J, Qiang Z (2015) Cerium incorporated MCM-48 (Ce-MCM-48) as a catalyst to inhibit bromate formation during ozonation of bromide-containing water: efficacy and mechanism. Water Res 86:2–8. https://doi.org/10.1016/j.watres.2015.05.052
152. Li Y, Shi J-Q, Qu R-J, Feng M-B, Liu F, Wang M, Wang Z-Y (2012) Toxicity assessment on three direct dyes (D-BLL, D-GLN, D-3RNL) using oxidative stress bioassay and quantum parameter calculation. Ecotoxicol Environ Saf 86:132–140. https://doi.org/10.1016/j.ecoenv.2012.09.023
153. Lin M, Fu ZY, Tan HR, Tan JPY, Ng SC, Teo E (2012) Hydrothermal synthesis of CeO_2 nanocrystals: Ostwald ripening or oriented attachment? Cryst Growth Des 12(6):3296–3303. https://doi.org/10.1021/cg300421x
154. Lin Y, Yang Z, Cheng J (2007) Preparation, characterization and antibacterial property of cerium substituted hydroxyapatite nanoparticles. J Rare Earths 25(4):452–456. https://doi.org/10.1016/S1002-0721(07)60455-4
155. Liu IT, Hon M-H, Teoh LG (2013) Structure and optical properties of CeO_2 nanoparticles synthesized by precipitation. J Electron Mater 42(8):2536–2541. https://doi.org/10.1007/s11664-013-2617-9
156. Liu Y, Sun D (2007) Effect of CeO2 doping on catalytic activity of Fe_2O_3/γ-Al2O3 catalyst for catalytic wet peroxide oxidation of azo dyes. J Hazard Mater 143(1):448–454. https://doi.org/10.1016/j.jhazmat.2006.09.043
157. Lopes RJG, Perdigoto MLN, Quinta-Ferreira RM (2012) Tailored investigation and characterization of heterogeneous {Mn, Cu}/TiO_2 catalysts embedded within a ceria-based framework for the wet peroxide oxidation of hazardous pollutants. Appl Catal B 117–118:292–301. https://doi.org/10.1016/j.apcatb.2012.01.025
158. López JM, Gilbank AL, García T, Solsona B, Agouram S, Torrente-Murciano L (2015) The prevalence of surface oxygen vacancies over the mobility of bulk oxygen in nanostructured ceria for the total toluene oxidation. Appl Catal B Environ 174–175:403–412. https://doi.org/10.1016/j.apcatb.2015.03.017
159. López T, Rojas F, Alexander-Katz R, Galindo F, Balankin A, Buljan A (2004) Porosity, structural and fractal study of sol–gel TiO_2–CeO_2 mixed oxides. J Solid State Chem 177(6):1873–1885. https://doi.org/10.1016/j.jssc.2004.01.013
160. Lu P, Li C, Zeng G, He L, Peng D, Cui H, Li S, Zhai Y (2010) Low temperature selective catalytic reduction of NO by activated carbon fiber loading lanthanum oxide and ceria. Appl Catal B 96(1):157–161. https://doi.org/10.1016/j.apcatb.2010.02.014

161. Lu Y, Duan L, Sun Z, Chen J (2021) Flame spray pyrolysis synthesized CuO–CeO_2 composite for catalytic combustion of C_3H_6. Proc Combust Inst 38(4):6513–6520. https://doi.org/10.1016/j.proci.2020.07.122
162. Luo Y, Yang F, Mathieu J, Mao D, Wang Q, Alvarez PJJ (2014) Proliferation of multidrug-resistant new delhi metallo-β-lactamase genes in municipal wastewater treatment plants in Northern China. Environ Sci Technol Lett 1(1):26–30. https://doi.org/10.1021/ez400152e
163. Lyu J, Han H, Wu Q, Ma H, Ma C, Dong X, Fu Y (2019) Enhancement of the electrocatalytic oxidation of dyeing wastewater (reactive brilliant blue KN-R) over the Ce-modified Ti-PbO_2 electrode with surface hydrophobicity. J Solid State Electrochem 23(3):847–859. https://doi.org/10.1007/s10008-018-04170-9
164. Ma R, Yin L, Li L, Zhang S, Wen T, Zhang C, Wang X, Chen Z, Hayat T, Wang X (2018) Comparative investigation of Fe_2O_3 and $Fe_{1-x}S$ nanostructures for uranium decontamination. ACS Appl Nano Mater 1(10):5543–5552. https://doi.org/10.1021/acsanm.8b01059
165. Ma X, Lu P, Wu P (2018) Effect of structure distortion and oxygen vacancy on ferromagnetism in hydrothermally synthesized CeO_2 with isovalent Zr^{4+} doping. Ceram Int 44(13):15989–15994. https://doi.org/10.1016/j.ceramint.2018.06.032
166. Maddila S, Dasireddy VDBC, Jonnalagadda SB (2014) Ce-V loaded metal oxides as catalysts for dechlorination of chloronitrophenol by ozone. Appl Catal B 150–151:305–314. https://doi.org/10.1016/j.apcatb.2013.12.036
167. Maddila S, Dasireddy VDBC, Oseghe EO, Jonnalagadda SB (2013) Ozone initiated dechlorination and degradation of trichlorophenol using Ce–Zr loaded metal oxides as catalysts. Appl Catal B 142–143:129–141. https://doi.org/10.1016/j.apcatb.2013.05.012
168. Malleshappa J, Nagabhushana H, Prasad BD, Sharma SC, Vidya YS, Anantharaju KS (2016) Structural, photoluminescence and thermoluminescence properties of CeO_2 nanoparticles. Optik 127(2):855–861. https://doi.org/10.1016/j.ijleo.2015.10.114
169. Manoj Kumar Reddy P, Mahammadunnisa S, Subrahmanyam C (2014) Catalytic non-thermal plasma reactor for mineralization of endosulfan in aqueous medium: a green approach for the treatment of pesticide contaminated water. Chem Eng J 238:157–163. https://doi.org/10.1016/j.cej.2013.08.087
170. Mansfield E, Tyner KM, Poling CM, Blacklock JL (2014) Determination of nanoparticle surface coatings and nanoparticle purity using microscale thermogravimetric analysis. Anal Chem 86(3):1478–1484. https://doi.org/10.1021/ac402888v
171. Mantilla A, Jácome-Acatitla G, Morales-Mendoza G, Tzompantzi F, Gómez R (2011) Photoassisted degradation of 4-chlorophenol and p-cresol using mgal hydrotalcites. Ind Eng Chem Res 50(5):2762–2767. https://doi.org/10.1021/ie1006883
172. Maqbool Q, Nazar M, Maqbool A, Pervez MT, Jabeen N, Hussain T, Franklin G (2018) CuO and CeO_2 nanostructures green synthesized using olive leaf extract inhibits the growth of highly virulent multidrug resistant bacteria. Front Pharmacol 9. https://doi.org/10.3389/fphar.2018.00987
173. Marselli B, Garcia-Gomez J, Michaud P-A, Rodrigo MA, Comninellis C (2003) Electrogeneration of hydroxyl radicals on boron-doped diamond electrodes. J Electrochem Soc 150(3):D79
174. Massa P, Dafinov A, Cabello FM, Fenoglio R (2008) Catalytic wet peroxide oxidation of phenolic solutions over Fe_2O_3/CeO_2 and WO_3/CeO_2 catalyst systems. Catal Commun 9(7):1533–1538. https://doi.org/10.1016/j.catcom.2007.12.025
175. Massa P, Ivorra F, Haure P, Fenoglio R (2011) Catalytic wet peroxide oxidation of phenol solutions over CuO/CeO_2 systems. J Hazard Mater 190(1):1068–1073. https://doi.org/10.1016/j.jhazmat.2011.03.033
176. Masui T, Fujiwara K, Machida K-i, Adachi G-y, Sakata T, Mori H (1997) Characterization of cerium(IV) oxide ultrafine particles prepared using reversed micelles. Chem Mater 9(10):2197–2204. https://doi.org/10.1021/cm970359v
177. Matheswaran M, Balaji S, Chung SJ, Moon IS (2007) Studies on cerium oxidation in catalytic ozonation process: a novel approach for organic mineralization. Catal Commun 8(10):1497–1501. https://doi.org/10.1016/j.catcom.2006.12.017

178. Meepho M, Sirimongkol W, Ayawanna J (2018) Samaria-doped ceria nanopowders for heavy metal removal from aqueous solution. Mater Chem Phys 214:56–65. https://doi.org/10.1016/j.matchemphys.2018.04.083
179. Mishra SR, Ahmaruzzaman M (2021) Cerium oxide and its nanocomposites: structure, synthesis, and wastewater treatment applications. Mater Today Commun 28:102562. https://doi.org/10.1016/j.mtcomm.2021.102562
180. Mohammadi-Aloucheh R, Habibi-Yangjeh A, Bayrami A, Latifi-Navid S, Asadi A (2018) Green synthesis of ZnO and ZnO/CuO nanocomposites in Mentha longifolia leaf extract: characterization and their application as anti-bacterial agents. J Mater Sci: Mater Electron 29(16):13596–13605. https://doi.org/10.1007/s10854-018-9487-0
181. Mokkelbost T, Kaus I, Grande T, Einarsrud M-A (2004) Combustion synthesis and characterization of nanocrystalline CeO_2-based powders. Chem Mater 16(25):5489–5494. https://doi.org/10.1021/cm048583p
182. Mondal A, Adhikary B, Mukherjee D (2015) Room-temperature synthesis of air stable cobalt nanoparticles and their use as catalyst for methyl orange dye degradation. Colloids Surf, A 482:248–257. https://doi.org/10.1016/j.colsurfa.2015.05.011
183. Mondal DK, Mondal C, Roy S (2016) Catalytic wet air oxidation of aqueous solution of phenol in a fixed bed reactor over Ru catalysts supported on ceria promoted MCM-41. RSC Adv 6(115):114383–114395. https://doi.org/10.1039/C6RA22080K
184. Mooney JB, Radding SB (1982) Spray pyrolysis processing. Annu Rev Mater Sci 12:81–101. https://doi.org/10.1146/annurev.ms.12.080182.000501
185. Morselli D, Campagnolo L, Prato M, Papadopoulou EL, Scarpellini A, Athanassiou A, Fragouli D (2018) Ceria/gold nanoparticles in situ synthesized on polymeric membranes with enhanced photocatalytic and radical scavenging activity. ACS Appl Nano Mater 1(10):5601–5611. https://doi.org/10.1021/acsanm.8b01227
186. Mahamuni NN, Adewuyi YG (2010) Advanced oxidation processes (AOPs) involving ultrasound for waste water treatment: a review with emphasis on cost estimation. Ultrason Sonochem 17:990–1003
187. Nair DS, Kurian M (2017) Catalytic peroxide oxidation of persistent chlorinated organics over nickel-zinc ferrite nanocomposites. J Water Process Eng 16:69–80. https://doi.org/10.1016/j.jwpe.2016.12.010
188. Nair DS, Kurian M (2017) Heterogeneous catalytic oxidation of persistent chlorinated organics over cobalt substituted zinc ferrite nanoparticles at mild conditions: reaction kinetics and catalyst reusability studies. J Environ Chem Eng 5(1):964–974. https://doi.org/10.1016/j.jece.2017.01.021
189. Ndabankulu VO, Maddila S, Jonnalagadda SB (2019) Ceria doped TiO_2 as photocatalyst for water treatment under visible light. IOP Conf Ser: Mater Sci Eng 668(1):012011. https://doi.org/10.1088/1757-899X/668/1/012011
190. Negm SH, Abd El-Magied MO, El Maadawy WM, Abdel Aal MM, Abd El Dayem SM, Taher MA, Abd El-Rahem KA, Rashed MN, Cheira MF (2022) Appreciatively efficient sorption achievement to U(VI) from the El Sela Area by ZrO_2/chitosan. Separations 9(10):311. https://doi.org/10.3390/separations9100311
191. Norzaee S, Taghavi M, Djahed B, Kord Mostafapour F (2018) Degradation of Penicillin G by heat activated persulfate in aqueous solution. J Environ Manage 215:316–323. https://doi.org/10.1016/j.jenvman.2018.03.038
192. Nousir S, Keav S, Barbier J, Bensitel M, Brahmi R, Duprez D (2008) Deactivation phenomena during catalytic wet air oxidation (CWAO) of phenol over platinum catalysts supported on ceria and ceria–zirconia mixed oxides. Appl Catal B 84(3):723–731. https://doi.org/10.1016/j.apcatb.2008.06.010
193. Nur Fadzeelah AK, Abdullah AZ, Zubir NA, Abd Razak AH, Azha NA (2019) Sonocatalytic degradation of caffeine using CeO_2 catalyst: parametric and reusability studies. J Phys: Conf Ser 1349(1):012147. https://doi.org/10.1088/1742-6596/1349/1/012147
194. Oei BC, Ibrahim S, Wang S, Ang HM (2009) Surfactant modified barley straw for removal of acid and reactive dyes from aqueous solution. Biores Technol 100(18):4292–4295. https://doi.org/10.1016/j.biortech.2009.03.063

195. Oliviero L, Barbier J, Duprez D, Guerrero-Ruiz A, Bachiller-Baeza B, Rodríguez-Ramos I (2000) Catalytic wet air oxidation of phenol and acrylic acid over Ru/C and Ru–CeO_2/C catalysts. Appl Catal B: Environ 25(4):267–275. https://doi.org/10.1016/S0926-3373(99)00141-1
196. Oliviero L, Barbier J, Labruquère S, Duprez D (1999) Role of the metal–support interface in the total oxidation of carboxylic acids over Ru/CeO_2 catalysts. Catal Lett 60(1):15–19. https://doi.org/10.1023/A:1019026100843
197. Orge CA, Órfão JJM, Pereira MFR (2011) Catalytic ozonation of organic pollutants in the presence of cerium oxide–carbon composites. Appl Catal B 102(3):539–546. https://doi.org/10.1016/j.apcatb.2010.12.036
198. Orge CA, Órfão JJM, Pereira MFR (2012) Carbon xerogels and ceria–carbon xerogel materials as catalysts in the ozonation of organic pollutants. Appl Catal B 126:22–28. https://doi.org/10.1016/j.apcatb.2012.06.029
199. Orge CA, Órfão JJM, Pereira MFR, Duarte de Farias AM, Fraga MA (2012) Ceria and cerium-based mixed oxides as ozonation catalysts. Chem Eng J 200–202:499–505. https://doi.org/10.1016/j.cej.2012.06.088
200. Orge CA, Órfão JJM, Pereira MFR, Duarte de Farias AM, Neto RCR, Fraga MA (2011) Ozonation of model organic compounds catalysed by nanostructured cerium oxides. Appl Catal B 103(1):190–199. https://doi.org/10.1016/j.apcatb.2011.01.026
201. Oturan MA, Aaron J-J (2014) Advanced oxidation processes in water/wastewater treatment: principles and applications. A review. Crit Rev Environ Sci Technol 44(23):2577–2641. https://doi.org/10.1080/10643389.2013.829765
202. Pandian CJ, Palanivel R, Dhananasekaran S (2015) Green synthesis of nickel nanoparticles using Ocimum sanctum and their application in dye and pollutant adsorption. Chin J Chem Eng 23(8):1307–1315. https://doi.org/10.1016/j.cjche.2015.05.012
203. Pang J-H, Liu Y, Li J, Yang X-J (2019) Solvothermal synthesis of nano-CeO_2 aggregates and its application as a high-efficient arsenic adsorbent. Rare Met 38(1):73–80. https://doi.org/10.1007/s12598-018-1072-4
204. Patil BB, Pawar SH (2011) Structural, morphological and electrical properties of spray deposited nano-crystalline CeO_2 thin films. J Alloys Compd 509(2):414–420. https://doi.org/10.1016/j.jallcom.2010.09.045
205. Pavan Kumar CSS, Pandeeswari R, Jeyaprakash BG (2014) Structural, morphological and optical properties of spray deposited Mn-doped CeO_2 thin films. J Alloys Compd 602:180–186. https://doi.org/10.1016/j.jallcom.2014.02.143
206. Peng X, Luan Z, Ding J, Di Z, Li Y, Tian B (2005) Ceria nanoparticles supported on carbon nanotubes for the removal of arsenate from water. Mater Lett 59(4):399–403. https://doi.org/10.1016/j.matlet.2004.05.090
207. Periyat P, Laffir F, Tofail SAM, Magner E (2011) A facile aqueous sol–gel method for high surface area nanocrystalline CeO_2. RSC Adv 1(9):1794–1798. https://doi.org/10.1039/C1RA00524C
208. Pétrier C, Francony A (1997) Ultrasonic waste-water treatment: incidence of ultrasonic frequency on the rate of phenol and carbon tetrachloride degradation. Ultrason Sonochem 4(4):295–300. https://doi.org/10.1016/S1350-4177(97)00036-9
209. Phokha S, Pinitsoontorn S, Maensiri S (2012) Room-temperature ferromagnetism in Co-doped CeO_2 nanospheres prepared by the polyvinylpyrrolidone-assisted hydrothermal method. J Appl Phys 112(11):113904. https://doi.org/10.1063/1.4766273
210. Primo A, Marino T, Corma A, Molinari R, García H (2011) Efficient visible-light photocatalytic water splitting by minute amounts of gold supported on nanoparticulate CeO_2 obtained by a biopolymer templating method. J Am Chem Soc 133(18):6930–6933. https://doi.org/10.1021/ja2011498
211. Protsenko VS, Vasil'eva EA, Tsurkan AV, Kityk AA, Korniy SA, Danilov FI (2017) Fe/TiO_2 composite coatings modified by ceria layer: electrochemical synthesis using environmentally friendly methanesulfonate electrolytes and application as photocatalysts for organic dyes degradation. J Environ Chem Eng 5(1):136–146. https://doi.org/10.1016/j.jece.2016.11.034

212. Pundir S, Priya R, Singh K, Kaur H, Choudhary P (2023) A systematic study on synthesis of CeO_2 nanoparticles by various routes. IOP Conf Ser: Earth Environ Sci 1110(1):012030. https://doi.org/10.1088/1755-1315/1110/1/012030
213. Qin H, Chen H, Zhang X, Yang G, Feng Y (2014) Efficient degradation of fulvic acids in water by catalytic ozonation with CeO_2/AC. J Chem Technol Biotechnol 89(9):1402–1409. https://doi.org/10.1002/jctb.4222
214. Rajkumar K, Muthukumar M, Mangalaraja RV Electrocatalytic oxidation of reactive blue 198 using Sm^{3+} doped cerium oxide as a catalyst. In: International conference on nanoscience, engineering and technology (ICONSET 2011), 28–30 Nov 2011, pp 14–18. https://doi.org/10.1109/ICONSET.2011.6167900
215. Rajkumar K, Muthukumar M, Mangalaraja RV (2015) Electrochemical degradation of C.I. Reactive Orange 107 using Gadolinium (Gd^{3+}), Neodymium (Nd^{3+}) and Samarium (Sm^{3+}) doped cerium oxide nanoparticles. Int J Industr Chem 6(4):285–295. https://doi.org/10.1007/s40090-015-0051-y
216. Ramachandran M, Subadevi R, Sivakumar M (2019) Role of pH on synthesis and characterization of cerium oxide (CeO_2) nano particles by modified co-precipitation method. Vacuum 161:220–224. https://doi.org/10.1016/j.vacuum.2018.12.002
217. Rashwan WE, Fathy NA, Elkhouly SM (2018) A novel catalyst of ceria-nanorods loaded on carbon xerogel for catalytic wet oxidation of methyl green dye. J Taiwan Inst Chem Eng 88:234–242. https://doi.org/10.1016/j.jtice.2018.04.011
218. Ravishankar TN, Ramakrishnappa T, Nagaraju G, Rajanaika H (2015) Synthesis and characterization of CeO_2 nanoparticles via solution combustion method for photocatalytic and antibacterial activity studies. ChemistryOpen 4(2):146–154. https://doi.org/10.1002/open.201402046
219. Recillas S, Colón J, Casals E, González E, Puntes V, Sánchez A, Font X (2010) Chromium VI adsorption on cerium oxide nanoparticles and morphology changes during the process. J Hazard Mater 184(1):425–431. https://doi.org/10.1016/j.jhazmat.2010.08.052
220. Recillas S, García A, González E, Casals E, Puntes V, Sánchez A, Font X (2012) Preliminary study of phosphate adsorption onto cerium oxide nanoparticles for use in water purification; nanoparticles synthesis and characterization. Water Sci Technol 66(3):503–509. https://doi.org/10.2166/wst.2012.185
221. Roy S, Saroha AK (2014) Ceria promoted γ-Al_2O_3 supported platinum catalyst for catalytic wet air oxidation of oxalic acid: kinetics and catalyst deactivation. RSC Adv 4(100):56838–56847. https://doi.org/10.1039/C4RA06529H
222. Sadykov VA, Mezentseva NV, Bobrova LN, Smorygo OL, Eremeev NF, Fedorova YE, Bespalko YN, Skriabin PI, Krasnov AV, Lukashevich AI, Krieger TA, Sadovskaya EM, Belyaev VD, Shmakov AN, Vinokurov ZS, Bolotov VA, Tanashev YY, Korobeynikov MV, Mikhailenko MA (2019) Chapter 12—Advanced materials for solid oxide fuel cells and membrane catalytic reactors. In: Sadykov VA (ed) Advanced nanomaterials for catalysis and energy. Elsevier, pp 435–514. https://doi.org/10.1016/B978-0-12-814807-5.00012-7
223. Sajeevan AC, Sajith V (2016) Synthesis of stable cerium zirconium oxide nanoparticle—diesel suspension and investigation of its effects on diesel properties and smoke. Fuel 183:155–163. https://doi.org/10.1016/j.fuel.2016.06.048
224. Samai B, Bhattacharya SC (2018) Conducting polymer supported cerium oxide nanoparticle: enhanced photocatalytic activity for waste water treatment. Mater Chem Phys 220:171–181. https://doi.org/10.1016/j.matchemphys.2018.08.050
225. Sancheti SV, Saini C, Ambati R, Gogate PR (2018) Synthesis of ultrasound assisted nanostuctured photocatalyst (NiO supported over CeO_2) and its application for photocatalytic as well as sonocatalytic dye degradation. Catal Today 300:50–57. https://doi.org/10.1016/j.cattod.2017.02.047
226. Santiago AFJ, Sousa JF, Guedes RC, Jerônimo CEM, Benachour M (2006) Kinetic and wet oxidation of phenol catalyzed by non-promoted and potassium-promoted manganese/cerium oxide. J Hazard Mater 138(2):325–330. https://doi.org/10.1016/j.jhazmat.2006.05.118

227. Sathishkumar P, Mangalaraja RV, Pugazhenthiran N, Niño Galeano MA, Gracia-Pinilla MA, Mansilla HD, Anandan S (2017) Structural investigation and sonocatalytic efficiency of $Ce_{0.9}Nd_{0.1}O_{1.95}$ and $Ce_{0.9}Pr_{0.1}O_{1.95}$ nanocatalysts. Mater Chem Phys 200:241–249. https://doi.org/10.1016/j.matchemphys.2017.07.066
228. Sawana R, Somasundar Y, Iyer VS, Baruwati B (2017) Ceria modified activated carbon: an efficient arsenic removal adsorbent for drinking water purification. Appl Water Sci 7(3):1223–1230. https://doi.org/10.1007/s13201-016-0398-z
229. Seddigi ZS, Bumajdad A, Ansari SP, Ahmed SA, Danish EY, Yarkandi NH, Ahmed S (2014) Preparation and characterization of Pd doped ceria–ZnO nanocomposite catalyst for methyl tert-butyl ether (MTBE) photodegradation. J Hazard Mater 264:71–78. https://doi.org/10.1016/j.jhazmat.2013.10.070
230. Sedjame H-J, Fontaine C, Lafaye G, Barbier J Jr (2014) On the promoting effect of the addition of ceria to platinum based alumina catalysts for VOCs oxidation. Appl Catal B 144:233–242. https://doi.org/10.1016/j.apcatb.2013.07.022
231. Selvan RK, Augustin CO, Šepelák V, Berchmans LJ, Sanjeeviraja C, Gedanken A (2008) Synthesis and characterization of $CuFe_2O_4/CeO_2$ nanocomposites. Mater Chem Phys 112(2):373–380. https://doi.org/10.1016/j.matchemphys.2008.05.094
232. Shan C, Xu Y, Hua M, Gu M, Yang Z, Wang P, Lu Z, Zhang W, Pan B (2018) Mesoporous Ce-Ti-Zr ternary oxide millispheres for efficient catalytic ozonation in bubble column. Chem Eng J 338:261–270. https://doi.org/10.1016/j.cej.2018.01.046
233. Shrestha R, Ban S, Devkota S, Sharma S, Joshi R, Tiwari AP, Kim HY, Joshi MK (2021) Technological trends in heavy metals removal from industrial wastewater: a review. J Environ Chem Eng 9(4):105688. https://doi.org/10.1016/j.jece.2021.105688
234. Silva AMT, Marques RRN, Quinta-Ferreira RM (2004) Catalysts based in cerium oxide for wet oxidation of acrylic acid in the prevention of environmental risks. Appl Catal B 47(4):269–279. https://doi.org/10.1016/j.apcatb.2003.09.019
235. Singh J, Dutta T, Kim K-H, Rawat M, Samddar P, Kumar P (2018) 'Green' synthesis of metals and their oxide nanoparticles: applications for environmental remediation. J Nanobiotechnology 16(1):84. https://doi.org/10.1186/s12951-018-0408-4
236. Singh KRB, Nayak V, Sarkar T, Singh RP (2020) Cerium oxide nanoparticles: properties, biosynthesis and biomedical application. RSC Adv 10(45):27194–27214. https://doi.org/10.1039/D0RA04736H
237. Singh RK, Sadhasivam T, Sheeja GI, Singh P, Srivastava ON (2013) Effect of different sized CeO_2 nano particles on decomposition and hydrogen absorption kinetics of magnesium hydride. Int J Hydrogen Energy 38(14):6221–6225. https://doi.org/10.1016/j.ijhydene.2012.12.060
238. Sisubalan N, Ramkumar VS, Pugazhendhi A, Karthikeyan C, Indira K, Gopinath K, Hameed ASH, Basha MHG (2024) Retraction Note: ROS-mediated cytotoxic activity of ZnO and CeO_2 nanoparticles synthesized using the Rubia cordifolia L. leaf extract on MG-63 human osteosarcoma cell lines. Environ Sci Pollut Res 31(54):63426–63426. https://doi.org/10.1007/s11356-024-35507-9
239. Song H, Wu R, Yang J, Dong J, Ji G (2018) Fabrication of CeO_2 nanoparticles decorated three-dimensional flower-like BiOI composites to build p-n heterojunction with highly enhanced visible-light photocatalytic performance. J Colloid Interface Sci 512:325–334. https://doi.org/10.1016/j.jcis.2017.10.080
240. Srinivasan MP, Uthiram C, Ayeshamariam A, Kaviyarasu K, Punithavelan N (2023) Dielectric performance of CeO_2/ZnO core–shell nanocomposite with their structural, optical and morphological properties. J King Saud Uni—Sci 35(2):102508. https://doi.org/10.1016/j.jksus.2022.102508
241. Sriram G, Uthappa UT, Rego RM, Kigga M, Kumeria T, Jung H-Y, Kurkuri MD (2020) Ceria decorated porous diatom-xerogel as an effective adsorbent for the efficient removal of Eriochrome Black T. Chemosphere 238:124692. https://doi.org/10.1016/j.chemosphere.2019.124692

242. Steffen W, Richardson K, Rockström J, Cornell SE, Fetzer I, Bennett EM, Biggs R, Carpenter SR, de Vries W, de Wit CA, Folke C, Gerten D, Heinke J, Mace GM, Persson LM, Ramanathan V, Reyers B, Sörlin S (2015) Planetary boundaries: guiding human development on a changing planet. Science 347(6223):1259855. https://doi.org/10.1126/science.1259855
243. Sushma SAK (2018) Biodegradability enhancement of industrial organic raffinate containing pyridine and its derivatives by CWAO using ceria promoted MnO_x/Al_2O_3 catalyst at atmospheric pressure. Chem Eng J 334:985–994. https://doi.org/10.1016/j.cej.2017.10.100
244. Suwanboon S, Amornpitoksuk P, Bangrak P (2018) The improvement of the band gap energy and antibacterial activities of CeO_2/ZnO nanocomposites prepared by high energy ball milling. Chiang Mai J Sci 45:1129–1137
245. Tbessi I, Benito M, Molins E, Liorca J, Touati A, Sayadi S, Najjar W (2019) Effect of Ce and Mn co-doping on photocatalytic performance of sol-gel TiO_2. Solid State Sci 88:20–28. https://doi.org/10.1016/j.solidstatesciences.2018.12.004
246. Tekbaş M, Yatmaz HC, Bektaş N (2008) Heterogeneous photo-Fenton oxidation of reactive azo dye solutions using iron exchanged zeolite as a catalyst. Microporous Mesoporous Mater 115(3):594–602. https://doi.org/10.1016/j.micromeso.2008.03.001
247. Teramoto Y, Kosuge K, Sugasawa M, Kim H-H, Ogata A, Negishi N (2015) Zirconium/cerium oxide solid solutions with addition of SiO_2 as ozone-assisted catalysts for toluene oxidation. Catal Commun 61:112–116. https://doi.org/10.1016/j.catcom.2014.12.022
248. Thovhogi N, Diallo A, Gurib-Fakim A, Maaza M (2015) Nanoparticles green synthesis by Hibiscus Sabdariffa flower extract: main physical properties. J Alloys Compd 647:392–396. https://doi.org/10.1016/j.jallcom.2015.06.076
249. Tianshu Z, Hing P, Huang H, Kilner J (2001) The effect of Fe doping on the sintering behavior of commercial CeO_2 powder. J Mater Process Technol 113(1):463–468. https://doi.org/10.1016/S0924-0136(01)00600-8
250. Tilp AH, Akl HM, T. Kandil AE-H, Cheira MF, Gado HS, Salah BA (2024) Extraction of La, Ce, and Nd using an ionic liquid and its application to rare earth leachate derived from phosphogypsum. J Rare Earths. https://doi.org/10.1016/j.jre.2024.07.005
251. Tiwari S, Khatun N, Shrivastava T, Kumar S, Liu S-W, Biring S, Sen S (2018) Structural, optical and mechanical properties of sol-gel synthesized Mn-doped CeO_2. Superlattices Microstruct 122:316–323. https://doi.org/10.1016/j.spmi.2018.07.033
252. Tok AIY, Boey FYC, Dong Z, Sun XL (2007) Hydrothermal synthesis of CeO_2 nano-particles. J Mater Process Technol 190(1):217–222. https://doi.org/10.1016/j.jmatprotec.2007.02.042
253. Tok AIY, Luo LH, Boey FYC (2004) Carbonate co-precipitation of Gd_2O_3-doped CeO_2 solid solution nano-particles. Mater Sci Eng A 383(2):229–234. https://doi.org/10.1016/j.msea.2004.05.071
254. Tolan D, El-Sawaf A, Ahmed ASA, Nassar A, Mohamed NM, Alhindawy IG, Elshehy EA, Utgikar V (2024) Enhanced photocatalytic activity of (In–Sr–P) tridoped TiO_2/Bi_2O_3 composite loaded on mesoporous carbon: a facile sol-hydrothermal synthesis approach. Mater Chem Phys 322:129570. https://doi.org/10.1016/j.matchemphys.2024.129570
255. Tomova D, Iliev V, Eliyas A, Rakovsky S (2015) Promoting the oxidative removal rate of oxalic acid on gold-doped CeO_2/TiO_2 photocatalysts under UV and visible light irradiation. Sep Purif Technol 156:715–723. https://doi.org/10.1016/j.seppur.2015.10.070
256. Traid HD, Vera ML, Ares AE, Litter MI (2015) Porous titanium dioxide coatings obtained by anodic oxidation for photocatalytic applications. Procedia Mater Sci 9:619–626. https://doi.org/10.1016/j.mspro.2015.05.038
257. Tran ND, Besson M, Descorme C, Fajerwerg K, Louis C (2011) Influence of the pretreatment conditions on the performances of CeO_2-supported gold catalysts in the catalytic wet air oxidation of carboxylic acids. Catal Commun 16(1):98–102. https://doi.org/10.1016/j.catcom.2011.09.014
258. Tseng CHT, Paul BK, Chang C-H, Engelhard MH (2013) Continuous precipitation of ceria nanoparticles from a continuous flow micromixer. Int J Adv Manuf Technol 64(1):579–586. https://doi.org/10.1007/s00170-012-4428-1

259. Umale SV, Tambat SN, Sontakke SM (2017) Combustion synthesized CeO_2 as an anodic material in dye sensitized solar cells. Mater Res Bull 94:483–488. https://doi.org/10.1016/j.materresbull.2017.07.004
260. Vallet-Regí M, Conde F, Nicolopoulos S, Ragel CV, Gonzalez-Calbet J (1996) Synthesis and characterization of CeO_2 obtained by spray pyrolysis method. Mater Sci Forum 235:291–296. https://doi.org/10.4028/www.scientific.net/MSF.235-238.291
261. Vidya J, Balamurugan P (2019) Photocatalytic degradation of Methylene Blue using PANi/Ceria nanocomposite under visible light irradiation. Desalin Water Treat 156:349–356. https://doi.org/10.5004/dwt.2019.23934
262. Vuppala V, Motappa MG, Venkata SS, Sadashivaiah PH (2012) Photocatalytic degradation of methylene blue using a zinc oxide-cerium oxide catalyst. Eur J Chem 3(2):191–195
263. Wang C, Ao Y, Wang P, Hou J, Qian J (2010) Preparation, characterization and photocatalytic activity of a novel composite photocatalyst: ceria-coated activated carbon. J Hazard Mater 184(1):1–5. https://doi.org/10.1016/j.jhazmat.2010.07.034
264. Wang J, Quan X, Chen S, Yu H, Liu G (2019) Enhanced catalytic ozonation by highly dispersed CeO_2 on carbon nanotubes for mineralization of organic pollutants. J Hazard Mater 368:621–629. https://doi.org/10.1016/j.jhazmat.2019.01.095
265. Wang J, Shi X, Chen L, Li H, Mao M, Zhang G, Yi H, Fu M, Ye D, Wu J (2021) Enhanced performance of low Pt loading amount on Pt-CeO_2 catalysts prepared by adsorption method for catalytic ozonation of toluene. Appl Catal A 625:118342. https://doi.org/10.1016/j.apcata.2021.118342
266. Wang K, Yang Y, Zhang TC, Liang Y, Wang Q (2019) Degradation of methylene blue with magnetic Co-doped Fe_3O_4@FeOOH nanocomposites as heterogeneous catalysts of peroxymonosulfate. RSC Adv 9(31):17664–17673. https://doi.org/10.1039/C9RA01671F
267. Wang Q, Yang Z, Chai B, Cheng S, Lu X, Bai X (2016) Heterogeneous catalytic ozonation of natural organic matter with goethite, cerium oxide and magnesium oxide. RSC Adv 6(18):14730–14740. https://doi.org/10.1039/C5RA21674E
268. Wang Q, Yang Z, Ma J, Wang J, Wang L, Guo M (2016) Study on the mechanism of cerium oxide catalytic ozonation for controlling the formation of bromate in drinking water. Desalin Water Treat 57(33):15533–15546. https://doi.org/10.1080/19443994.2015.1079261
269. Wang S, Wang W, Liu Q, Zhang M, Qian Y (2000) Preparation and characterization of cerium (IV) oxide thin films by spray prolysis method. Solid State Ion 133(3):211–215. https://doi.org/10.1016/S0167-2738(00)00762-1
270. Wang Y, Xie X, Chen X, Huang C, Yang S (2020) Biochar-loaded Ce^{3+}-enriched ultra-fine ceria nanoparticles for phosphate adsorption. J Hazard Mater 396:122626. https://doi.org/10.1016/j.jhazmat.2020.122626
271. Wen J, Xie J, Chen X, Li X (2017) A review on g-C_3N_4-based photocatalysts. Appl Surf Sci 391:72–123. https://doi.org/10.1016/j.apsusc.2016.07.030
272. Wen X-J, Niu C-G, Ruan M, Zhang L, Zeng G-M (2017) AgI nanoparticles-decorated CeO_2 microsheets photocatalyst for the degradation of organic dye and tetracycline under visible-light irradiation. J Colloid Interface Sci 497:368–377. https://doi.org/10.1016/j.jcis.2017.03.029
273. Wen X-J, Niu C-G, Zhang L, Liang C, Zeng G-M (2018) A novel Ag_2O/CeO_2 heterojunction photocatalysts for photocatalytic degradation of enrofloxacin: possible degradation pathways, mineralization activity and an in depth mechanism insight. Appl Catal B 221:701–714. https://doi.org/10.1016/j.apcatb.2017.09.060
274. Wen X-J, Zhang C, Niu C-G, Zhang L, Zeng G-M, Zhang X-G (2017) Highly enhanced visible light photocatalytic activity of CeO_2 through fabricating a novel p–n junction BiOBr/CeO_2. Catal Commun 90:51–55. https://doi.org/10.1016/j.catcom.2016.11.018
275. Witkowska A, Lewandowska AU (2016) Water soluble organic carbon in aerosols (PM1, PM2.5, PM10) and various precipitation forms (rain, snow, mixed) over the southern Baltic Sea station. Sci Total Environ 573:337–346. https://doi.org/10.1016/j.scitotenv.2016.08.123
276. Wu J, Zheng W, Chen Y (2022) Definition of photocatalysis: current understanding and perspectives. Curr Opin Green Sustain Chem 33:100580. https://doi.org/10.1016/j.cogsc.2021.100580

277. Wu Z, Zhang G, Zhang R, Yang F (2018) Insights into mechanism of catalytic ozonation over practicable mesoporous Mn-CeO_x/γ-Al_2O_3 catalysts. Ind Eng Chem Res 57(6):1943–1953. https://doi.org/10.1021/acs.iecr.7b04516
278. Xiao H, Ai Z, Zhang L (2009) Nonaqueous sol−gel synthesized hierarchical CeO_2 nanocrystal microspheres as novel adsorbents for wastewater treatment. J Phys Chem C 113(38):16625–16630. https://doi.org/10.1021/jp9050269
279. Xu B, Qi F, Sun D, Chen Z, Robert D (2016) Cerium doped red mud catalytic ozonation for bezafibrate degradation in wastewater: efficiency, intermediates, and toxicity. Chemosphere 146:22–31. https://doi.org/10.1016/j.chemosphere.2015.12.016
280. Xu D, Cheng F, Lu Q, Dai P (2014) Microwave enhanced catalytic degradation of Methyl Orange in aqueous solution over CuO/CeO_2 catalyst in the absence and presence of H_2O_2. Ind Eng Chem Res 53(7):2625–2632. https://doi.org/10.1021/ie4033022
281. Xu J (2011) Synergy effect on a suspended mixture of ceria and activated carbon for the photocatalytic degradation of phenol. Powder Technol 210(1):1–5. https://doi.org/10.1016/j.powtec.2011.02.001
282. Xu M, Wang Z, Wang F, Hong P, Wang C, Ouyang X, Zhu C, Wei Y, Hun Y, Fang W (2016) Fabrication of cerium doped Ti/nanoTiO_2/PbO_2 electrode with improved electrocatalytic activity and its application in organic degradation. Electrochim Acta 201:240–250. https://doi.org/10.1016/j.electacta.2016.03.168
283. Yadav B, Srivastava VC (2017) Catalytic peroxidation of recalcitrant quinoline by ceria impregnated granular activated carbon. Clean Technol Environ Policy 19(5):1547–1555. https://doi.org/10.1007/s10098-016-1315-8
284. Yadav TP, Srivastava ON (2012) Synthesis of nanocrystalline cerium oxide by high energy ball milling. Ceram Int 38(7):5783–5789. https://doi.org/10.1016/j.ceramint.2012.04.025
285. Yang S, Zhu W, Wang J, Chen Z (2008) Catalytic wet air oxidation of phenol over CeO_2-TiO_2 catalyst in the batch reactor and the packed-bed reactor. J Hazard Mater 153(3):1248–1253. https://doi.org/10.1016/j.jhazmat.2007.09.084
286. Yao P, Huang Y, Jiao Y, Xu H, Wang J, Chen Y (2023) Soot oxidation over Pt-loaded CeO_2-ZrO_2 catalysts under gasoline exhaust conditions: soot-catalyst contact efficiency and Pt chemical state. Fuel 334:126782. https://doi.org/10.1016/j.fuel.2022.126782
287. Younis A, Loucif A (2021) Defects mediated enhanced catalytic and humidity sensing performance in ceria nanorods. Ceram Int 47(11):15500–15507. https://doi.org/10.1016/j.ceramint.2021.02.117
288. Younis A, Shirsath SE, Shabbir B, Li S (2018) Controllable dynamics of oxygen vacancies through extrinsic doping for superior catalytic activities. Nanoscale 10(39):18576–18585. https://doi.org/10.1039/C8NR03801E
289. Yousef Tizhoosh N, Khataee A, Hassandoost R, Darvishi Cheshmeh Soltani R, Doustkhah E (2020) Ultrasound-engineered synthesis of WS_2@CeO_2 heterostructure for sonocatalytic degradation of tylosin. Ultrason Sonochem 67:105114. https://doi.org/10.1016/j.ultsonch.2020.105114
290. Yu L, Ma Y, Ong CN, Xie J, Liu Y (2015) Rapid adsorption removal of arsenate by hydrous cerium oxide–graphene composite. RSC Adv 5(80):64983–64990. https://doi.org/10.1039/C5RA08922K
291. Yue L, Zhang X-M (2009) Structural characterization and photocatalytic behaviors of doped CeO_2 nanoparticles. J Alloy Compd 475(1):702–705. https://doi.org/10.1016/j.jallcom.2008.07.096
292. Yulizar Y, Juliyanto S, Sudirman, Apriandanu DOB, Surya RM (2021) Novel sol-gel synthesis of CeO_2 nanoparticles using *Morinda citrifolia* L. fruit extracts: structural and optical analysis. J Mol Struct 1231:129904. https://doi.org/10.1016/j.molstruc.2021.129904
293. Zabilskiy M, Djinović P, Tchernychova E, Tkachenko OP, Kustov LM, Pintar A (2015) Nanoshaped CuO/CeO_2 materials: effect of the exposed ceria surfaces on catalytic activity in N_2O decomposition reaction. ACS Catal 5(9):5357–5365. https://doi.org/10.1021/acscatal.5b01044

294. Zamani A, Marjani AP, Alimoradlu K (2017) Walnut shell-templated ceria nanoparticles: green synthesis, characterization and catalytic application. Int J Nanosci 17(06):1850008. https://doi.org/10.1142/S0219581X18500084
295. Zhang C, Li S, Wu G, Gong J (2014) Synthesis of stable Ni-CeO_2 catalysts via ball-milling for ethanol steam reforming. Catal Today 233:53–60. https://doi.org/10.1016/j.cattod.2013.08.013
296. Zhang D, Qian Y, Shi L, Mai H, Gao R, Zhang J, Yu W, Cao W (2012) Cu-doped CeO_2 spheres: synthesis, characterization, and catalytic activity. Catal Commun 26:164–168. https://doi.org/10.1016/j.catcom.2012.05.001
297. Zhang N, Chen J, Fang Z, Tsang EP (2019) Ceria accelerated nanoscale zerovalent iron assisted heterogenous Fenton oxidation of tetracycline. Chem Eng J 369:588–599. https://doi.org/10.1016/j.cej.2019.03.112
298. Zhang T, Bonnaud L, Raquez J-M, Poorteman M, Olivier M, Dubois P (2020) Cerium salts: an efficient curing catalyst for benzoxazine based coatings. Polymers 12(2):415. https://doi.org/10.3390/polym12020415
299. Zhang T, Li W, Croué J-P (2012) A non-acid-assisted and non-hydroxyl-radical-related catalytic ozonation with ceria supported copper oxide in efficient oxalate degradation in water. Appl Catal B 121–122:88–94. https://doi.org/10.1016/j.apcatb.2012.03.021
300. Zhao B, Shi B, Zhang X, Cao X, Zhang Y (2011) Catalytic wet hydrogen peroxide oxidation of H-acid in aqueous solution with TiO_2–CeO_2 and Fe/TiO_2–CeO_2 catalysts. Desalination 268(1):55–59. https://doi.org/10.1016/j.desal.2010.09.050
301. Zhou W, Zhu J, Wang F, Cao M, Zhao T (2017) One-step synthesis of Ceria/Ti_3C_2 nanocomposites with enhanced photocatalytic activity. Mater Lett 206:237–240. https://doi.org/10.1016/j.matlet.2017.06.117
302. Zhu Y, Quan X, Chen F, Fan X, Feng Y (2012) CeO_2-TiO_2 coated ceramic membrane with catalytic ozonation capability for treatment of tetracycline in drinking water. Sci Adv Mater 4(12):1191–1199. https://doi.org/10.1166/sam.2012.1413

Study of Nickel Oxide (NiO) in Wastewater Treatment

Mahmoud A. Ahmed, Mohamed A. Ahmed, and Ashraf A. Mohamed

Abstract The growing prevalence of various heavy metals and organic pollutants, such as dyes, pharmaceuticals, and agrochemicals, in wastewater poses significant environmental and public health concerns, necessitating the development of green, innovative, low-cost, and highly effective treatment technologies. Conventional methods often fall short in effectively eliminating these pollutants, prompting the exploration of advanced materials and techniques. This chapter examines recent advances in wastewater treatment using NiO-based nanocomposites, focusing on their enhanced adsorptive and photocatalytic capabilities. The chapter examines the underlying adsorption mechanisms, emphasizing the critical role of operational factors like pH, adsorbent dosage, and contact time in maximizing pollutants' removal efficiencies. Furthermore, the chapter explores the utilization of NiO-based nanocomposites as highly effective photocatalysts, leveraging the NiO's light-harvesting abilities and charge separation dynamics to facilitate organic pollutants' degradation. The synergistic effects observed when NiO is combined with complementary nanomaterials, such as carbon-based structures, semiconductors, and other metal oxides, to generate hybrid nanocomposites have been shown to significantly enhance both the adsorption and photocatalytic activity. The chapter also addresses the challenges associated with the practical implementation of NiO-based nanocomposites, such as recyclability and stability, while discussing emerging strategies and future research directions to overcome these barriers. The comprehensive insights offered in this chapter help to further the ongoing efforts to develop effective, sustainable, and scalable solutions for removing a wide range of pollutants from wastewater, paving the way for better environmental remediation and water resource management.

Keywords Pollutants' removal · Nanotechnology · NiO-based nano-photocatalysts · Adsorption and photocatalytic mechanisms

M. A. Ahmed (✉)
Veolia Water Technologies, Cairo, Egypt
e-mail: mahmoudmahmoud_p@sci.asu.edu.eg

M. A. Ahmed · M. A. Ahmed · A. A. Mohamed
Department of Chemistry, Faculty of Science, Ain Shams University, Cairo 11566, Egypt

A. Ahmad et al. (eds.), *Metal Oxide Based Nanophotocatalyst for Wastewater Purification*, Smart Nanomaterials Technology,
https://doi.org/10.1007/978-981-96-7683-5_6

1 Introduction

Recently, the global landscape has seen rapid economic growth, increased energy needs, heightened industrial activity, and urban expansion [1, 2]. While these changes have led to many positive outcomes, they have also introduced significant challenges for the environment, human health, and ecosystems. A major concern linked to these transformations is the extensive generation and use of organic pollutants, which have raised alarms due to their harmful effects on the environment [3]. These carbon-containing chemicals are notorious for their ability to resist breakdown in the environment and can remain for extended periods, threatening both ecosystems and humans [4, 5]. Organic pollutants are regularly released into the air, water, and soil through human activities. Their origins are varied, stemming from industrial operations, agricultural activities, waste incineration, and certain consumer goods. Specific types of organic pollutants, such as dyes, surfactants, and halogenated organic pesticides, e g., polychlorinated biphenyls (PCBs), and polybrominated diphenyl ethers (PBDEs), are particularly concerning [6–9]. These compounds are utilized across diverse sectors, including agriculture, manufacturing, electrical appliances, and fire retardants, resulting in their widespread presence in the environment [10]. The contamination caused by organic pollutants poses serious threats to ecosystems and biodiversity. These pollutants can accumulate in living organisms, leading to biomagnification within food chains [11]. Consequently, organisms at higher trophic levels, including humans, are exposed to elevated concentrations of these pollutants. This accumulation can have dire effects on wildlife, contributing to reproductive issues, developmental defects, and population declines [12]. In humans, exposure has been linked to a range of health problems, such as immune system impairments, endocrine disruptions, neurological issues, and an increased likelihood of cancer. The enduring nature and harmful properties of organic pollutants highlight the urgent need for effective management and remediation strategies. Efforts to alleviate the environmental consequences of these pollutants have centered on the development and application of sophisticated treatment technologies. A variety of methods—chemical, physical, and biological—are being utilized to tackle the issue of organic pollutants and their environmental impact [13–17]. Among these, adsorption and photocatalysis have emerged as effective, innovative methods that offer distinct advantages over traditional approaches [18, 19].

Adsorption is a process where organic pollutants adhere to the surface of solid materials, often referred to as adsorbents [20]. Common adsorbents include activated carbon, zeolites, and biochar [21–23]. Adsorption is valued for its simplicity, cost-effectiveness, and ability to remove a wide range of pollutants from air, water, and soil [24]. The porous structure of adsorbents provides a large surface area, facilitating the trap of pollutants. However, the effectiveness of adsorption can be influenced by factors such as the nature of the pollutant, the characteristics of the adsorbent and environmental conditions like pH and temperature. Furthermore, while adsorption can effectively reduce pollutants' concentrations, it does not destroy them; instead, it merely transfers them to another medium, necessitating subsequent treatment or disposal of the spent adsorbent and the accumulated adsorbate [25, 26]. In contrast, photocatalysis is a process that employs light, typically ultraviolet (UV) light, to activate a photocatalyst which then generates reactive oxygen species capable of decomposing organic pollutants into less harmful substances, such as carbon dioxide and water [27, 28]. This method is particularly beneficial for the degradation of organic pollutants that resist other forms of degradation. Photocatalysis offers several advantages over traditional methods, such as biological treatment and chemical oxidation. It not only effectively mineralizes organic pollutants but also operates under mild conditions and does not generate secondary pollutants. However, the efficiency of photocatalysis can be limited by factors such as light availability, the presence of competing substances, and the photocatalyst's reusability [29, 30]. In summary, adsorption and photocatalysis offer complementary approaches to the removal and degradation of organic pollutants. While adsorption provides a straightforward method for capturing pollutants, photocatalysis presents a powerful tool for their mineralization.

Metal oxides have garnered significant attention in the fields of adsorption and photocatalysis due to their unique chemical, physical, and structural properties [31–33]. Their versatility and effectiveness in removing a wide array of organic pollutants from various environmental matrices make them valuable materials for environmental remediation. Metal oxides, such as titanium dioxide (TiO_2), zinc oxide (ZnO), iron oxide (Fe_2O_3 and Fe_3O_4), aluminum oxide (Al_2O_3), and other metal oxides exhibit unique properties, including high thermal stability, tunable surface area, and variable oxidation states [33–37]. These characteristics contribute to their effectiveness in adsorption and photocatalytic applications. The surface properties of various metal oxides, including hydroxyl groups, oxygen vacancies, and active sites, significantly influence their interaction with adsorbates and photocatalytic behavior [38, 39]. The presence of surface hydroxyl groups on metal oxides enhances their ability to adsorb organic molecules through hydrogen bonding and dipole interactions. Further, the electronic structure of metal oxides defines their photocatalytic efficiency. Upon illumination, metal oxides generate electron–hole pairs in their core structure. The generated electrons can reduce adsorbed species, while the holes can oxidize organic pollutants or water, producing reactive oxygen species ROS (e.g., hydroxyl, peroxyl, and superoxide radicals) that effectively degrade organic pollutants [40, 41]. Among them, nickel oxide (NiO) is a transition metal oxide that has garnered increasing interest in environmental remediation due to its versatile attributes as both a photocatalyst and an adsorbent [42, 43].

NiO functions as a photocatalyst by utilizing its semiconductor features to facilitate various photoreactions under light irradiation. One of the most significant features of NiO is its band gap energy of approximately 3.6 eV, which allows it to absorb ultraviolet (UV) light and generate electron–hole pairs upon excitation [44, 45]. When exposed to UV light, the excited electrons can effectively reduce adsorbed pollutants, while the corresponding holes participate in oxidation reactions. This dual reactivity leads to the generation of highly ROS. Moreover, the photocatalytic efficiency of NiO is influenced by various factors, including its morphological characteristics. The synthesis methods employed can dictate the particle size, shape, and surface area of NiO, all of which have a direct bearing on its photocatalytic activity. Nanosized NiO particles typically possess a significantly higher surface area, thereby enhancing light absorption and providing more active sites for pollutant degradation [46, 47]. Additionally, ongoing research into the doping of NiO with metals or non-metals aims to narrow the band gap, enabling the material to harness visible light, which would substantially enhance its applicability in real-world settings, particularly in harnessing solar energy for photocatalytic processes [48, 49]. Further, NiO exhibits remarkable potential for removing both organic and inorganic pollutants from aqueous solutions. The surface chemistry of NiO is characterized by the presence of hydroxyl groups and oxygen vacancies, which contribute to its strong affinity for various pollutants. These functionalities facilitate adsorption through electrostatic interactions, surface complexation, and hydrogen bonding, allowing NiO to effectively capture a diverse range of pollutants, including heavy metals and organic dyes.

In recent years, there has been a significant surge in academic publications centered on the application of NiO nanocomposites for pollutant removal through photocatalysis and adsorption, as illustrated in Fig. 1. Initially, the annual output of research in this field was quite low, indicating minimal interest. However, beginning in 2010, both the number of publications and citations have risen rapidly, highlighting an increasing focus on this topic. The vast majority of these publications are journal articles (94.6%), with a minimal percentage comprising review articles (0.096%) and conference proceedings (4.1%), as indicated in Fig. 1. This disparity suggests a notable lack of comprehensive and updated literature review, which is crucial for researchers and the broader scientific community seeking thorough insights into the role of NiO nanocomposites as adsorbents and photocatalysts. Hence, the aim of this chapter is to provide a comprehensive overview of the recent advancements in wastewater treatment technologies utilizing NiO-based nanocomposites. It seeks to elucidate the enhanced adsorptive and photocatalytic properties of these materials, explore the mechanisms underlying their effectiveness in removing organic pollutants, and heavy metals from wastewater, and analyze the synergistic benefits of hybrid nanocomposite structures. Additionally, the chapter aims to identify the key factors influencing pollutant removal efficiencies, address practical challenges associated with the implementation of NiO-based nanocomposites, and propose future research directions to enhance their recyclability and stability, ultimately contributing to the development of sustainable solutions for environmental remediation and water resource management.

2 Organic Pollutants

Organic pollutants are a diverse group of chemical substances that have been identified as significant environmental pollutants, arising predominantly from various industrial, agricultural, and domestic activities [50]. They can be systematically classified into multiple categories based on their sources and chemical properties. For instance, pharmaceuticals encompass an array of bioactive chemicals

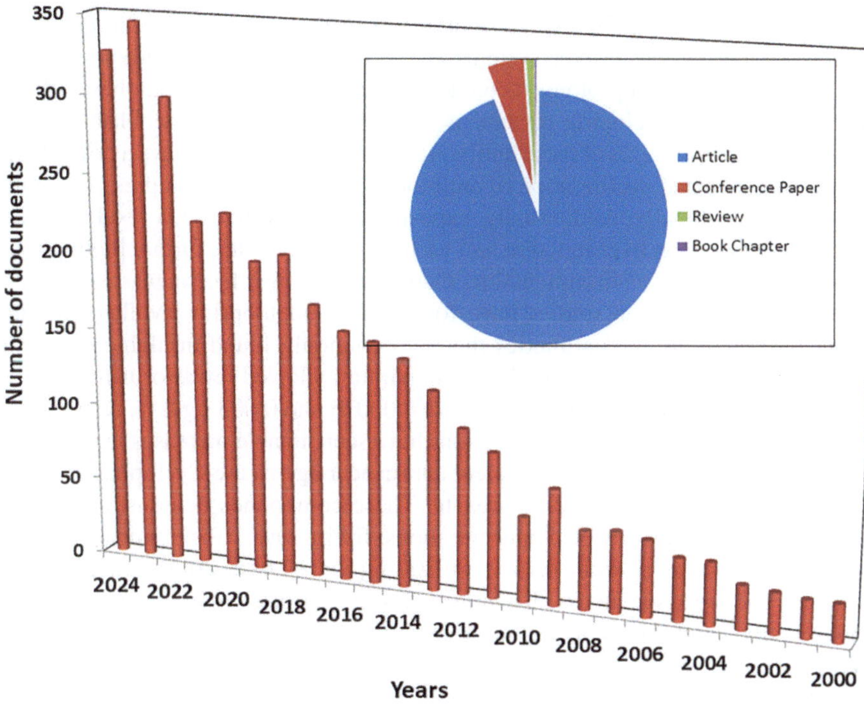

Fig. 1 Number of publications in Scopus database reporting g-NiO-based composites for wastewater treatment: keywords "(NiO) and ((adsorption*) or photocatalysis))"

used for medical purposes, including antibiotics, analgesics, anti-inflammatory and hormones, which are designed to affect biological systems [51–53]. These compounds commonly enter the environment through various channels: they are excreted by humans and animals, improperly disposed of, or pass through wastewater treatment plants that cannot fully remove these substances [54, 55]. The environmental persistence of pharmaceuticals raises concerns about their long-term effects on ecosystems. Over the past decade, the global consumption of pharmaceuticals has witnessed an extraordinary upswing, increasing by more than 2.5 times. One of the major issues is that some drug by-products are not only toxic but can exhibit toxicity levels that equal or exceed those of their original compounds. This poses a daunting challenge for conventional wastewater treatment plants, which often lack the infrastructure and technology needed to effectively eliminate these hazardous substances. As a result, antibiotics, along with other pharmaceuticals, have been detected in various environmental matrices, including soil, surface water, groundwater, sediment, and even atmospheric air [56, 57]. Such widespread distribution raises alarming questions about the long-term impacts on ecosystems and human health. Notably, tetracyclines (TCs), which are among the most widely prescribed antibiotics for both human and veterinary use, exemplify the potential risks associated with antibiotic pollution. Although tetracyclines are effective for treating infections, studies indicate that over 70% of these drugs are excreted unchanged through urine and feces, entering wastewater systems and, consequently, the environment [18]. Research has shown that prolonged contact with tetracycline can impair bone formation, especially in developing organisms, due to its effects on osteoblast activity. Furthermore, the presence of these antibiotics in the environment plays a critical role in the emergence and proliferation of antibiotic-resistant bacteria, a pressing public health concern globally. Additionally, exposure to tetracyclines can result in acute and chronic effects on liver function and gastrointestinal health, further highlighting the need for enhanced environmental regulations and more effective waste treatment solutions [18, 58]. Some studies have found that trimethoprim levofloxacin, and sulfamethoxazole are the antibiotics that are commonly found in high amounts in water sources [59, 60]. For example, the maximum allowed levels of trimethoprim and sulfamethoxazole in drinking water are 0.00025 and 0.003 $\mu g\ L^{-1}$, respectively [61]. Furthermore, anti-cancer medications have seen notable increases in usage and development. Among the most prevalent anti-cancer agents are Cyclophosphamide, Ifosfamide, Tamoxifen, and Methotrexate. These compounds demonstrate significant variability in their detected concentrations; for instance, Ifosfamide levels have been recorded to range from 0.14 to 86,200 ng L^{-1}. These drugs generally lack specificity and are associated with a range of adverse effects, including genotoxicity, teratogenicity, cytotoxicity, as well as potential nephrotoxic and hepatotoxic effects [61].

On the other hand, organic dyes are extensively utilized in the textile, food, and cosmetic industries to impart color to materials. These dyes are classified according to factors such as their color features, origin, intended uses, and molecular structure, following a color index created by the American Association of Textile Chemists in conjunction with an international network of color experts, as demonstrated in Fig. 2a [62, 63]. Natural dyes are obtained from substances found in nature, such as minerals, clays, and materials obtained from plants and animals [64]. However, these dyes have limitations in terms of durability and color range, as they tend to fade when exposed to sunlight or washed. In contrast, synthetic dyes are produced in huge quantities to meet global demand for these substances [65]. A report by the Federation of Indian Chambers of Commerce reflects that dye production in developed countries has been declining since 1980, with Asian countries now taking the lead in production. China currently provides about 40% of the world's textile dyes, while India contributes about 9%. The improper disposal of dye wastewater often results in significant environmental pollution, as many synthetic dyes are resistant to degradation and can persist in natural water bodies for long periods. The primary environmental hazard posed by organic dyes is their impact on aquatic ecosystems; they can obstruct light penetration, hindering photosynthesis in aquatic plants, and lead to oxygen depletion due to the increased load of organic matter associated with dye dispersion [66]. Long-term exposure to these pollutants can degrade water quality, resulting in detrimental effects on fish and other aquatic organisms, potentially causing behavioral changes, reproductive issues, and even mortality. Furthermore, the accumulation of these dyes in food chains poses risks to human health, particularly for communities that rely on these water bodies for fishing and recreation [67]. Some organic dyes have been identified as carcinogenic, raising alarms about the potential exposure of local populations [68]. For instance, extensive usage of hair dyes by individuals has been associated with an elevated susceptibility to the development of cancers impacting the blood and bone marrow. This correlation can be attributed to the existence of specific ingredients, namely p-toluenediamine (PTD), p-aminophenol (PAP), and p-phenylenediamine. Furthermore, professionals in the hairdressing and barbering sector exhibit a raised risk of bladder cancer [69]. Moreover, the application of acidic dyes poses a robust hazard to health, manifesting in toxic impacts on the ocular organs, respiratory system, and skin [70]. The aesthetic impact of dye pollution can deter tourism and recreational activities, impacting local economies that are dependent on clean water sources. The presence of even a small amount of dyestuff in water (less than 1 ppm) is easily noticeable and affects the visual appeal and transparency of water bodies. Figure 2b depicts the ecotoxicological ramifications of textile effluent containing dyes on living microorganisms and the ecosystem.

Phenols and biphenyls are industrial chemicals that are used in a wide variety of applications ranging from disinfectants to resins, leather, petrochemicals, and plastics [71]. Their introduction into the environment primarily occurs through industrial discharges, accidental spills, or the degradation of products containing these compounds [72]. According to a study conducted by in 2017 [73], it is estimated that approximately 10 million tons of phenolic waste is released into the ecosystem annually without undergoing any form of treatment [72, 73]. To address these concerns, regulatory bodies such as the United States Environmental Protection Agency (EPA) have established guidelines for acceptable phenol levels in drinking water and water bodies [74]. The EPA sets a maximum permissible level of 1 ppm for phenol in water bodies, however, for drinking water, the permissible level of phenol is set at 1 ppb [72, 74]. The long-term exposure to phenol can lead to various health issues, including vertigo, sudden weight loss, diarrhea, and loss of appetite (anorexia) [75]. The stringent regulations regarding phenol discharge are primarily driven by its high reactivity, which makes it prone to interact and react with other compounds present in water. Phenol exhibits a strong affinity for reacting with various organic and inorganic substances, leading to the formation of potentially harmful byproducts [76]. For instance, Phenol can react with inorganic substances, especially in the presence of UV radiation and nitrite ions, resulting in the formation of 2-nitrophenol and 4-nitrophenol in water bodies [77]. The potent electron-withdrawing ability of $-NO_2$ groups makes their elimination from the natural environment challenging due to the complex biodegradation process involved [77]. Another notable reaction occurs during the bacterial degradation of pentachlorophenol, resulting in the formation of chlorinated catechol and tetrachlorophenol compounds.

On the other hand, Pesticides are widely used to manage agricultural pests and protect crop yields, yet their extensive application raises significant public health and environmental concerns [78]. Comprising various chemicals-including herbicides, insecticides, and fungicides-pesticides can enter the environment through direct application, runoff, and atmospheric deposition [79]. Many pesticides are

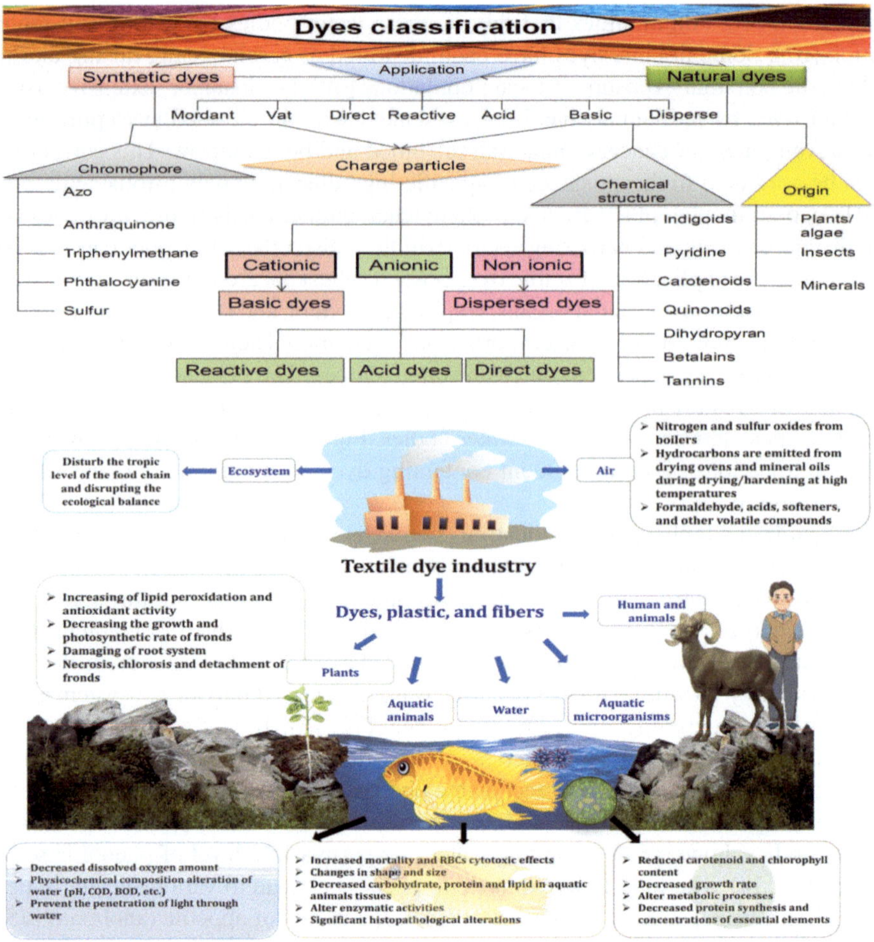

Fig. 2 **a** Classification of dyes based on structure, and applications, **b** ecotoxicological ramifications of textile effluent containing dyes on living microorganisms and the ecosystem

designed to be biologically active at low concentrations, raising considerable toxicity concerns. Long-term exposure to pesticides has been linked to a myriad of health issues in humans, including neurological disorders, reproductive problems, and increased cancer risk. Notably, certain pesticides are known to induce oxidative stress in mammals, generating reactive oxygen species that can damage cellular structures, leading to chronic health conditions [80]. Environmentally, pesticides threaten biodiversity; they can reduce populations of non-target species, such as pollinators and beneficial insects, disrupting ecosystems, and food chains. Additionally, pesticide resistance in target pests can lead to a cycle of increased chemical use which exacerbates the problem [81]. Persistent organic pollutants among pesticides can accumulate in soil and sediment, representing a long-term environmental hazard.

3 Effluent Treatment

Effluent remediation involves a series of three distinct stages (Fig. 3a, b) [82]. The initial stage, referred to as primary treatment, focuses on aeration and filtration to the separation of undissolved particles and solids from the wastewater. Following this, the secondary treatment stage employs biological concept to further purify the effluent by decreasing its biological oxygen demand (BOD) and converting organically bound C, H, N, and P into its inorganic counterparts [83]. Lastly, in the tertiary stage, advanced techniques like membrane filtration, and advanced oxidation phenomena are utilized to promote the quality of the water while eliminating the produced inorganics. The primary objective of effluent treatment is to ensure that the water meets the prescribed standards set by regulatory bodies such as the World Health Organization (WHO) and environmental agencies. Table 1 Displays a comparison of different remediation methods.

3.1 Coagulation/Flocculation

The utilization of flocculants and coagulation has emerged as a comprehensive and effective solution for the removal of a diverse array of pollutants from wastewater streams [84]. Fine or colloidal particles present in water often possess electrical charges on their surfaces or are coated with hydrophilic substances, leading to a stable dispersion owing to repulsive forces. Coagulants, such as aluminum sulfate, ferric chloride, and organic polymers, play a fundamental role in addressing this challenge by adsorbing or reacting with these colloidal particles, neutralizing their electrical charges [85]. This charge neutralization weakens the repulsive forces, allowing the particles to aggregate and form larger flocs. The process of flocculation then follows, where flocculants, including synthetic polymers and natural materials like sodium alginate (SA), and chitosan, collect the coagulated particles and facilitate the formation of larger, easily removable flocs [86]. The charge neutralization and subsequent flocculation processes can effectively destabilize and aggregate a wide range of organic and inorganic pollutants, and decrease color, COD, heavy metals, dyes, organisms, and oil and grease, making coagulants and flocculants suitable for comprehensive wastewater treatment [87]. The simplicity of operation, cost-effectiveness, and reliable performance of flocculants and coagulants has made them an attractive option for effluent treatment. For example, the researchers reported that $FeCl_3$ was effective in removing 79–82% of reactive dyes, while $FeCl_2$ was able to remove only 58–64% [88]. In another investigation, at pH 6.0, $FeCl_3$ was able to reduce COD levels by 63% and remove 44% of reactive yellow 84 [89]. In contrast, at pH 7.0, the elimination rates dropped to 23% for COD and 41% for reactive blue 49. In a separate study, the performance of polyaluminum chloride and aluminum sulfate was evaluated across different pH levels and dosages to determine the optimal conditions for treating samples with different levels of suspended solids [90]. The results of the

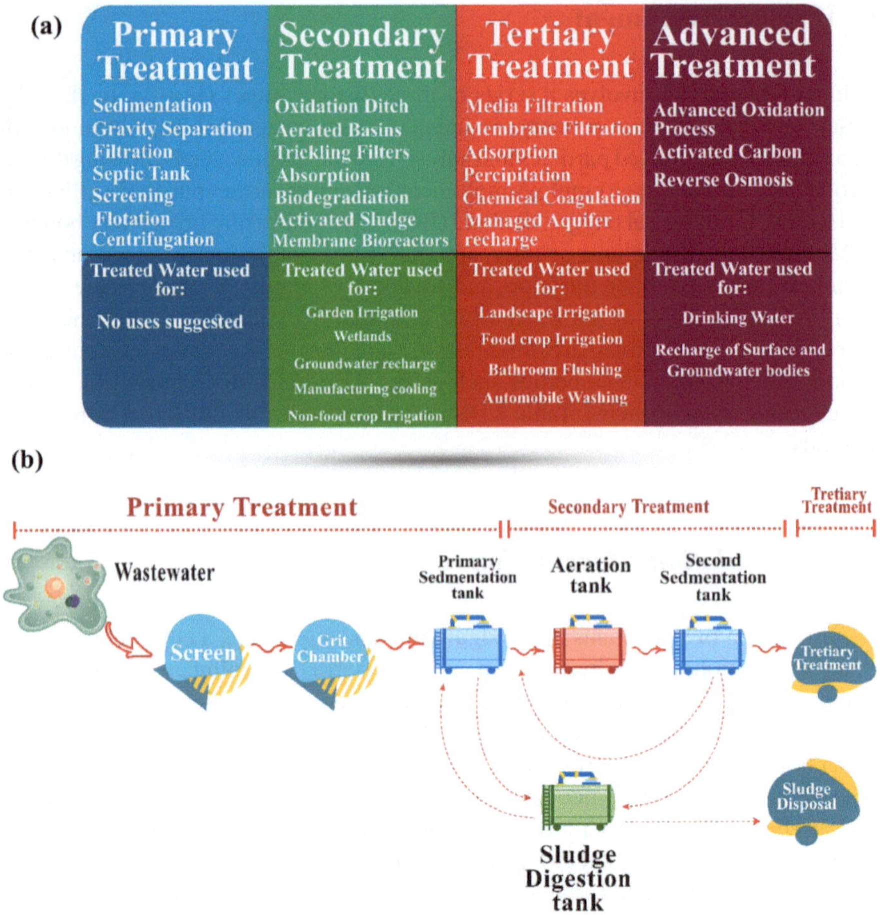

Fig. 3 **a** Steps in wastewater treatment process in industries, **b** flow chart for wastewater treatment including primary, secondary, and tertiary treatment

study reflected that coagulation was efficient in significantly minimizing turbidity in samples with low to moderate levels, using a relatively low level of 10–20 mg/L of either coagulant [90].

3.2 *Ion Exchange Method*

Ion exchange is a pivotal process in advanced wastewater treatment, particularly for the removal of specific ionic pollutants such as heavy metals, hazardous anionics, and radionuclides [91]. However, it can also play a role in the removal of certain organic pollutants, particularly ionizable species such as certain dyes, phenols, and organic

Table 1 Comparison of different remediation methods

Mechanical methods	Advantages	Disadvantages
Sedimentation	1. No energy requirement 2. Excellent reproducibility 3. Low operational and maintenance costs due to gravity dependence 4. Capable of treating large volumes of wastewater efficiently	1. Ineffective for fine particles (< 0.01 mm) and colloids due to low settling rates 2. Space requirements: requires significant area for large-scale operations 3. Continuous removal of sludge is necessary to prevent quality issues
Dissolved air flotation (DAF)	1. Effluent quality: provides superior removal of oils, fats, and suspended solids, yielding high-quality effluent 2. Rapid process: efficient treatment for varying influent qualities and flow rates 3. Space efficiency: smaller footprint compared to conventional technologies like sedimentation alone	1. Significant energy and maintenance costs associated with air generation and system upkeep 2. Pretreatment requirement: often necessitates pretreatment for optimal performance due to the presence of specific contaminants 3. Requires careful engineering to ensure effective separation
Filtration	1. Effective removal: capable of removing a wide range of particulates and some pathogens 2.Flexibility: modular design allows scalability and integration into various systems 3. Can enhance downstream biological and chemical treatments by eliminating solids	1. Clogging: filters can quickly become clogged, necessitating regular maintenance and backwashing 2. Energy use: depending on pressure-driven systems, filtration can be more energy-intensive than passive methods
Screening	1. Protects downstream equipment from large solids and debris which can cause failures 2. Ease of operation: generally easy to operate and manage minimal technical skills required 3. Cost-effective: significantly reduces the load on subsequent treatment stages	1. Only removes larger solids (typically > 1 mm); finer particles may require additional processes 2. Regular cleaning: screens must be regularly cleaned to ensure continuous operation (mechanical or manual)

(continued)

Table 1 (continued)

Mechanical methods	Advantages	Disadvantages
Centrifugation	1. Very effective at separating various contaminants, including fine solids; can achieve low solid concentrations in effluent 2. Capable of rapid processing of sludge and other suspended materials 3. Smaller equipment footprint compared to traditional settling tanks	1. Requires considerable energy and maintenance, making it expensive to operate 2. More complex systems may require skilled operators and regular maintenance to prevent breakdowns 3. Can generate significant noise and vibration during operation
Physicochemical methods		
Membrane filtration	1. Can remove pathogens, bacteria, and solids down to the nanometer range, producing high-quality effluent 2. Generates less sludge compared to conventional methods 3.Consume small area	1. High capital and operating costs 2. Membranes are prone to fouling
Coagulation and flocculation	1. High efficiency for fine particles 2. Removes metals, color, and turbidity 3. Increases the overall effectiveness of sedimentation and filtration processes	1. Chemical use 2. Needs careful management of chemical dosing to avoid residual chemicals in treated water 3. Sludge production 4. Multiple process step
Adsorption	1. Low cost 2. Relatively high performance 3. Design simplicity 4. Non-toxic 5. Regeneration of adsorbents is often possible	1. Low selectivity of adsorbent 2. Disposal problems 3. Not suitable for high concentration pollutants
Ozonation	1. Highly effective in disinfection and mineralization of organic pollutants 2. Produces no harmful residues 3. No need to alter pH and temperature	1. Ozone is unstable and must be generated on-site 2. Requires careful handling and safety considerations 3. In ozone generation, toxicity issues, and fire hazards may occur

(continued)

Table 1 (continued)

Mechanical methods	Advantages	Disadvantages
Advanced oxidation processes (AOPs)	1. Effective for a wide range of organic contaminants, including those that are resistant to conventional treatment methods 2. Can achieve high degradation rates	1. Higher operational costs due to the need for chemicals and specialized equipment 2. Potential for toxicity of intermediates
Ion exchange	1. Possible to regenerate resin 2. Small area requirement	1. Fouling issue
Biological methods		
Aerobic treatment	1. Simplicity of activity 2. Enhances nitrogen and phosphorus removal	1. Foaming problems may arise 2. Cost expensive
Anaerobic treatment	1. Producing biogas 2. Less environmental pollution 3. Reduces sludge volume 4. Can be applied for large capacities	1. Longer retention times required 2. Potential for odor emission

acids [92, 93]. This capability is due to the presence of functional groups on the ion exchange resins that can interact with charged organic and/or inorganic species. The resins are usually composed of cross-linked organic polymers with a high surface area, embedded with functional groups that can selectively bind ions of opposite charge. The ion exchange process can be categorized into two main types: cation exchange and anion exchange. Cation exchange resins contain negatively charged functional groups, such as sulfonic, phosphonic, or carboxylic acid groups, which attract and hold positively charged ions (cations) like calcium, magnesium, heavy metals or other positively charged inorganic and organic species [94]. Conversely, anion exchange resins possess positively charged functional groups, such as quaternary ammonium groups, which attract negatively charged ions (anions) like nitrates, nitrites, chromates, arsenates, sulfates, or negatively charged inorganic and organic moieties [95, 96]. The process begins with the wastewater being passed through a column filled with the ion exchange resin. As the water flows through, ions in the solution are exchanged with ions on the resin. This exchange is driven by the relative affinity of the resin for different ions, which is influenced by factors such as ionic charge, size, and concentration. The selectivity of the resin can be fine-tuned by modifying the functional groups or by using mixed-bed resins that contain both cation and anion exchange sites [97]. One of the key advantages of ion exchange is its ability to achieve high levels of purification and selectivity, making it ideal for applications where specific ion removal is critical [98, 99]. However, the process is not without challenges. The presence of competing ions can reduce the efficiency of the exchange process, and the resins can become fouled by organic matter or precipitated solids, necessitating periodic regeneration [100, 101]. Regeneration involves flushing the resin with a concentrated solution of the exchange ions, which displaces the accumulated contaminants and restores the resin's capacity. Advanced applications of ion exchange in wastewater treatment often involve integration with other treatment processes, such as reverse osmosis or advanced oxidation, to enhance overall treatment efficacy. Additionally, research into novel resin materials, such as those incorporating nanotechnology or bio-based polymers, is ongoing to improve the efficiency, selectivity, and sustainability of the ion exchange process. For example, cation exchange resins were found to achieve a removal level of 99.9% for the Basic Blue 3 dye after just one hour of exposure. However, the performance of this removal was adversely impacted by anionic surfactants, which disrupted the process [102]. Furthermore, when the dye level was at 10^{-2} M, the anion exchange resin Amberlite IRA-400 successfully removed 97% of the anionic dye Acid Orange 10 from the wastewater [103].

3.3 *Biological Methods*

Biological processes are a cornerstone of advanced wastewater treatment, leveraging the metabolic capabilities of microorganisms to degrade organic pollutants and transform them into less harmful substances [104]. These processes are primarily divided into aerobic and anaerobic treatments, each with distinct mechanisms and applications [105]. In aerobic treatment, microorganisms require oxygen to metabolize organic matter. This process is typically carried out in activated sludge systems, trickling filters, or biofilm reactors. In activated sludge systems, wastewater is aerated in large tanks, promoting the growth of aerobic bacteria that consume organic pollutants. The biomass, or sludge, is then separated from the treated water, with a portion being recycled to maintain microbial populations. Aerobic processes are highly effective at reducing biochemical oxygen demand (BOD) and removing nutrients such as nitrogen and phosphorus through nitrification and denitrification pathways [106, 107]. Anaerobic treatment, on the other hand, occurs in the absence of oxygen and is particularly suited for high-strength industrial wastewaters. In anaerobic digesters, microorganisms break down organic matter through a series of biochemical reactions, ultimately producing biogas, a mixture of methane and carbon dioxide, as a valuable byproduct [108, 109]. This process not only reduces the organic load but also generates renewable energy, making it an attractive option for sustainable wastewater management. Anaerobic treatment is often used in conjunction with aerobic processes to achieve comprehensive treatment and energy recovery [105]. Advanced biological processes also include the use of specialized microbial consortia for the degradation of specific pollutants, such as xenobiotics and emerging pollutants. Techniques such as bioaugmentation, where specific strains of microorganisms are introduced to enhance degradation, and bio-stimulation, which involves optimizing environmental conditions to promote microbial activity, are employed to tackle recalcitrant compounds [110]. Moreover, recent advancements in molecular biology and genetic engineering have opened new avenues for enhancing biological treatment processes. The development of genetically modified microorganisms with enhanced degradation capabilities and the application of metagenomics to understand microbial community dynamics are paving the way for more efficient and targeted wastewater treatment solutions. For instance, under conditions that have been optimized, peroxidases extracted from the bio-waste of Luffa acutangula (commonly known as luffa) and soybean were utilized to enzymatically degrade MO dye. These enzymes achieved remarkable performances of 75.3% and 81.4%, respectively [111]. Furthermore, research has explored the ability of different microbial strains to break down methyl orange (MO) [112]. Notably, the strain Pseudomonas putida SKG-1 exhibited a remarkable removal efficiency of 92.8% for the dye Orange II (OII) after a contact period of 96 h [113].

3.4 Membrane-Based Filtration Methods

Membrane technology presents a sophisticated method for efficiently removing salts and organic pollutants such as COD, BOD, TOC, from wastewater. Additionally, this technique effectively segregates proteins, large chemical entities, colloids, and microbial contaminants [16]. Membrane separation involves directing effluent through membranes with extremely small pores that reject particles exceeding the membrane pore size, thereby effectively isolating pollutants from purified water [14, 114]. To maintain optimal filtration performance, it is crucial to routinely clean the fouling layers that result from residual materials [114]. The performance of membrane approach is generally assessed by evaluating both pollutant rejection and permeates flow rates. As illustrated in Fig. 4, this approach can be classified according to membrane pore size [115]. Microfiltration (MF) employs membranes with pore diameters ranging from 0.1 to 10 μm to eliminate larger colloids, microorganisms, and suspended particles from wastewater [115]. Ultrafiltration (UF) membranes are characterized by pore sizes between 10 and 100 nm that selectively remove macromolecules while allowing smaller molecules and dissolved ions to trans-pass. Furthermore, nanofiltration (NF) operates with pore sizes ranging from 1 to 10 nm, facilitating the removal of dyes and the reduction of effluent salinity. Meanwhile, reverse osmosis (RO) membranes, with pore sizes ranging from 0.1 to 1.0 nm, demonstrate exceptional efficacy in eliminating nearly all organic pollutants and in reducing salinity by removing mono-, di-, and tri-valent ions [15, 16].

3.5 Ozonation Methods

Ozonation is an advanced oxidation operation that plays a critical role in wastewater treatment, particularly for the degradation of recalcitrant organic substances

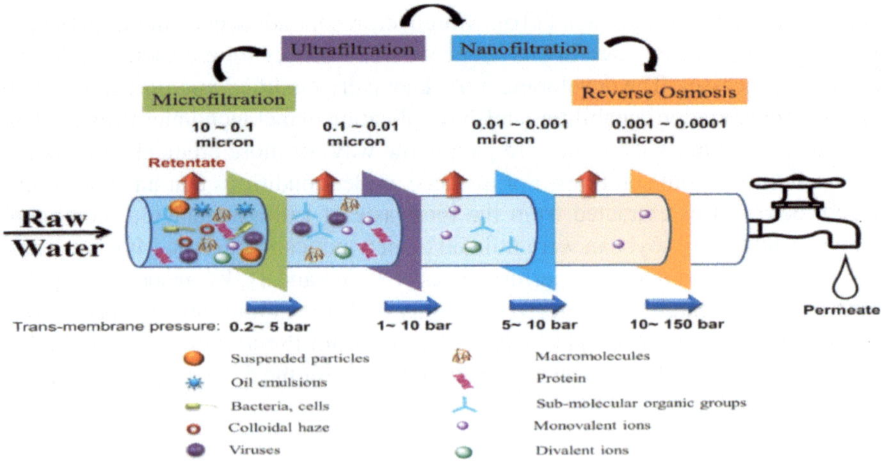

Fig. 4 Pollutants that can be eliminated by membrane processes

and the disinfection of pathogenic microorganisms [116]. This process involves the generation and application of ozone (O_3), a powerful oxidizing agent, which reacts with a wide range of pollutants [116]. Ozone is produced on-site using ozone generators that typically employ corona discharge or ultraviolet radiation to convert oxygen (O_2) into ozone. The ozonation process is highly effective in breaking down complex organic molecules, including agrochemicals, pharmaceuticals, personal care products, and other emerging pollutants that are often resistant to conventional treatment methods [117]. Ozone reacts with these compounds through direct and indirect oxidation processes. Ozone molecules can directly attack the pollutant, or decompose in water to form hydroxyl radicals ($^{\cdot}OH$) that are even more reactive [118]. This dual mechanism allows for the comprehensive degradation of a wide range of pollutants into smaller, more biodegradable molecules, or complete mineralization into carbon dioxide and water. One of the significant advantages of ozonation is its ability to achieve high levels of disinfection without the formation of harmful chlorinated byproducts, which are a concern with traditional chlorination methods. Ozone effectively inactivates bacteria, viruses, and protozoa by damaging their cell walls and disrupting their metabolic functions. This makes ozonation particularly valuable in applications where pathogen removal is critical, such as in the treatment of drinking water and wastewater reuse. However, the implementation of ozonation in wastewater treatment is not without challenges [119]. The high reactivity of ozone necessitates careful control of dosing and contact time to optimize treatment efficiency while minimizing operational costs. Additionally, ozone is unstable and must be generated on-site, requiring significant energy input and specialized equipment. The formation of certain byproducts, such as bromate in bromide-rich waters, also necessitates careful monitoring and management. For example, the use of ozonation in the decolorization of reactive dyes lead to the generation of toxic byproducts, such as short-chain carboxylic acids and harmful mineralized substances [120]. In contrast, incorporating $CuFe_2O_4$ nanoparticles (NPs) with photocatalytic ozonation results in byproducts that are relatively harmless, including nitrate (NO_3^-) and sulfate (SO_4^{2-}).

3.6 *Adsorption*

Adsorption is a sophisticated and highly effective process used in wastewater treatment to remove a diverse array of pollutants, including heavy metals, and organic compounds, such as dyes, pharmaceuticals, and pesticides [121]. This process involves the adhesion of molecules from the liquid phase onto the surface of a solid adsorbent [122]. The efficiency of adsorption is largely determined by the characteristics of the adsorbent, such as its surface area, porosity, and the nature of its surface functional groups [122] (Fig. 5a). There are two main types of adsorptions, namely, physical adsorption (physisorption) and chemical adsorption (chemisorption) [123]. Physisorption is governed by weak van der Waals forces and is typically reversible, making it suitable for the removal of volatile organic compounds [124]. Chemisorption, on the other hand, involves the formation of stronger covalent bonds between the adsorbate and the adsorbent surface, often resulting in irreversible interactions [125]. This type is ideal for pollutants that require more robust binding. The adsorption process involves several key steps. Initially, bulk diffusion, film diffusion occurs, where adsorbate molecules move from the bulk solution to the boundary layer surrounding the adsorbent (Fig. 5b). This is followed by intraparticle diffusion, where the adsorbate pollutant molecules penetrate the pores of the adsorbent and diffuse into its interior. The rate of adsorption is influenced by these diffusion steps, as well as factors such as adsorbent dose, pollutant's concentration, temperature, medium pH, and contact time. The adsorption capacity of an adsorbent is often described by isotherm models, such as the Langmuir, Freundlich, and Sips isotherms, which offer insights into the interactions between the targeted pollutants and the adsorbent surface. Kinetic models describe the rate of adsorption and help optimize process conditions. Recent advancements in adsorption technology focus on developing novel adsorbent materials with enhanced properties. These include the use of nanomaterials, bio-based adsorbents, modified clays, and their smart composites which offer improved adsorption capacities and selectivity for specific pollutants. Additionally, the regeneration and reuse of adsorbents are critical considerations for the economic and environmental sustainability of the adsorption process.

3.7 *Photocatalysis*

Photocatalysis, particularly involving metal oxides, is a cutting-edge approach in wastewater treatment, offering the potential to degrade a wide array of organic

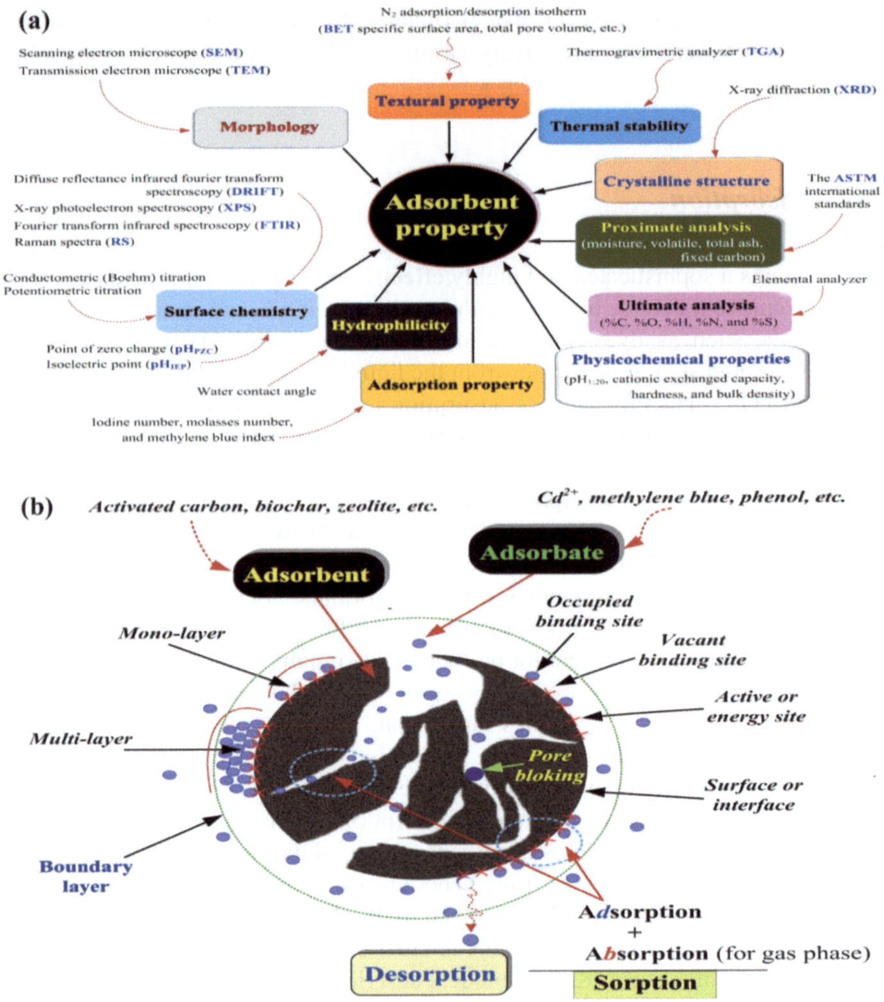

Fig. 5 **a** Features of an adsorbent estimated by various analysis, **b** a basic terms employed in adsorption technology

pollutants. Metal oxides, such as titanium dioxide (TiO_2), zinc oxide (ZnO), tungsten oxide (WO_3), and others are commonly used as photocatalysts due to their stability, non-toxicity, and strong oxidative capabilities. These materials are activated by light, typically in the ultraviolet (UV) range, to generate electron–hole pairs [126]. These pairs initiate redox reactions that produce reactive oxygen species, which are highly effective in breaking down complex organic molecules into simpler, non-toxic compounds [127]. The efficiency of metal oxide photocatalysts is influenced by several factors, including their bandgap energy, surface area, and crystallinity [128, 129]. The bandgap determines the wavelength of light required for activation; for instance, TiO_2 has a bandgap of about 3.2 eV, making it suitable for UV light activation [130]. To enhance the photocatalytic activity and extend light absorption into the visible spectrum, various strategies have been employed, such as doping with metals (e.g., with silver, iron, and others) or non-metals (e.g., nitrogen, sulfur, and others), and creating composite materials that frequently combine different metal oxides [31, 131–134]. Nano structuring of metal oxides is another advanced approach to improve photocatalytic performance. By reducing the particle size to the nanoscale, the surface area is significantly increased, providing more active sites for photocatalytic reactions [135, 136]. Additionally, nanostructures can facilitate better charge separation and reduce the recombination rate of electron–hole pairs, thereby enhancing the overall efficiency of the photocatalytic process. Despite the promising capabilities of metal oxide photocatalysts, challenges remain, such as the recombination of electron–hole pairs and the need for efficient recovery and reuse of the photocatalyst. Research is ongoing to develop more robust and efficient photocatalysts, including the exploration of novel metal oxides and hybrid and smart materials that can operate under visible light, thus utilizing solar energy more effectively [137–139].

4 NiO-Based Photocatalysts

4.1 Optical Features and Electronic Structure of NiO

NiO displays intriguing optical features and possesses a distinctive electronic structure, rendering it a subject of great interest in numerous scientific disciplines [140]. Optically, NiO is renowned for its characteristic deep green color, which arises from its electronic transitions within the visible range [141]. The absorption spectrum of NiO exhibits pronounced absorption in the ultraviolet (UV) region, extending into the visible range [142, 143]. This absorption phenomenon is attributed to charge transfer transitions between the oxygen $2p$ and nickel $3d$ orbitals. The energy bandgap of NiO typically resides around 3.4–4.0 eV, classifying it as a wide-bandgap material [144–146]. This bandgap dictates the energy of photons that NiO can absorb, making it suitable for applications in the UV region. The electronic structure of NiO is intricately linked to its cubic crystal structure, which typically adopts a rock salt-like configuration (NaCl structure). In this arrangement, each nickel ion is encompassed by six oxygen ions in an octahedral coordination. The electronic configuration of nickel in NiO is $[Ar]3d^8$, with two unpaired electrons occupying the $3d$ orbitals [147]. The electronic band structure of NiO unveils a fundamental bandgap that separates the valence and conduction bands. The valence band primarily comprises oxygen $2p$ orbitals, while the conduction band predominantly originates from nickel $3d$ orbitals. The energy disparity between these bands governs the bandgap of NiO.

NiO is also recognized as a p-type semiconductor [148]. This means that it exhibits a predominance of positive charge carriers, known as "holes," in its electronic structure [149]. The p-type behavior of NiO arises from the presence of oxygen vacancies and nickel interstitials, which create acceptor states within the bandgap. The presence of oxygen vacancies or other defects can perturb the band structure, thereby influencing the optical and electronic properties of NiO. Many studies have examined the effect of oxygen vacancies on the reactivity of NiO in applications like catalysis, photocatalysis, sensing, and electronic devices [150–153]. These studies used computational and experimental approaches to investigate the impact of oxygen vacancies on the performance of NiO. The primary limitation of NiO semiconductor in photocatalytic processes is its quick charge carriers recombination phenomena and wide band gap, which restricts its performance in the visible light region. This poses a challenge for the degradation of complex organic molecules, as it necessitates efficient photon generation and charge carrier separation to initiate effective radical-driven redox reactions.

4.2 Enhancements and Optimization of the Performance of NiO-Based Photocatalysts

To address this limitation, extensive research has been conducted to enhance the photocatalytic performance of NiO. Strategies such as constructing NiO heterojunctions, doping NiO with other elements, as well as hybridizing NiO with carbon-based materials, have been explored [144, 154].

4.2.1 Formation of Heterojunctions Between NiO and Other Semiconductors

The efficiency of photocatalytic processes of NiO can be significantly enhanced through the construction of heterojunctions with other semiconductors, which can be classified into various types, including Z-scheme, Type II, p–n junctions, and S-scheme (Fig. 6) [8, 45, 155, 156]. Each of these heterojunction designs plays a pivotal role in improving charge separation and transfer efficiency, thereby facilitating effective redox reactions under visible light illumination [157]. Additionally, in Type II, S-scheme, and Z-scheme heterojunctions, a p–n junction can be created when p-type and n-type semiconductors are combined [158]. For instance, the p–n heterojunction mechanism is a pivotal route for enhancing the photocatalytic performance of NiO, a p-type semiconductor (rich in holes), by pairing it with an n-type semiconductor (rich in electrons). In this process, charge carriers diffuse across the junction: electrons from the n-type side move into the p-type region, where they recombine with holes, while holes from the p-type diffuse into the n-type side. This recombination creates a depletion region, a zone that is largely void of free charge carriers, resulting in the establishment of an internal electric field across the junction. When the heterojunction is exposed to light (irradiation), exciting electrons from p-type into the n-type [156, 159, 160]. At this point, the internal electric field becomes increasingly vital. The electric field at the junction actively drives the newly generated photogenerated electrons from the p-type conduction band toward the n-type semiconductor [160]. Simultaneously, the holes created in the valence band are drawn toward the p-type side. This directional movement occurs because the energy levels of the CB and VB in the p-type are generally higher than those in the n-type material, allowing for efficient electron transfer. Thus, the researchers synthesized a $NiO/Zn_3In_2S_6$ employing ultrasonic-assisted deposition approach [156]. Unlike bare NiO, the pristine $Zn_3In_2S_6$ (denoted as ZIS_6) and the 25% NiO/ZIS_6 reflected the ability to absorb light across both visible and ultraviolet ranges. The estimated bandgap energies for NiO, ZIS_6, and the 25% NiO/ZIS_6 are 3.49 eV, 2.87 eV, and 2.85 eV, respectively. Additionally, electrochemical impedance spectroscopic (EIS) analysis and Mott-Schottky plots revealed that the formation of a p–n heterojunction between NiO and ZIS_6 significantly promotes the separation and migration of photoinduced charge carriers. In another investigation, the interface of the p–n junction in the NiO/TiO_2, combined with the existence of transient Ni^{3+} states, promotes the generation of photo-induced electrons by minimizing the recombination rate, and thus promoting the photocatalytic performance [161]. Utilizing the p–n NiO/TiO_2, a remarkable 98% efficiency in mineralization of MO dye was achieved within 60 min. The researchers reported that the performance of p–n NiO–ZnO junction for of crystal violet (CV) dye mineralization was more robust than pristine ZnO and NiO [162]. The activities of NiO, ZnO and NiO–ZnO were reported in the ratio of 1: 3.8: 7.8, respectively. Other researchers reported that the electric field between the p–n junction in the synthesized material has a fundamental role in enhancing the catalytic activities, and the analysis confirmed the successful p-n creation in the $S/N_{0.5}/Z_{0.5}$ [163].

In the Z-scheme, the unique configuration allows for an effective pathway for charge carrier transfer, enhancing the separation and utilization of photogenerated electrons (e^-) and holes (h^+). The Z-shaped transfer pathway highlights how the spatial arrangement of the semiconductors facilitates the movement of charge carriers in a manner that maximizes their availability for subsequent chemical reactions [155]. In the direct Z-scheme configuration, two semiconductors are directly coupled, allowing for a seamless transfer of charge carriers. Thus, the all-solid-state Z-scheme employs an electron mediator or conductor, which plays a crucial role in linking the two semiconductors while enhancing the efficiency of electron transfer. This mediator can effectively shuttle electrons, further inhibiting recombination and improving overall photocatalytic performance. For example, the researchers prepared $NiO/g–C_3N_4$ Z-pattern using a direct ultrasonic pathway [45]. Studies of the scavengers effects reflected that the Z-scheme helped in separating the electron–hole pairs created in the presence of solar irritation. Based on experimental characterization and DFT calculation, other researchers fabricated NiO/Bi_2MoO_6 heterostructure and confirmed that the formation of Z-scheme could greatly enhance the charge separation operation [164]. Further, the researchers fabricated $NiO/CDs/BiVO_4$ as all-solid-state Z-scheme composite that exhibits excellent performance, in which carbon dots (CDs) serve as electron mediator [165]. Furthermore, the Z-scheme $NiO/NiFe_2O_4/Fe_2O_3$ composite was fabricated through the partial chemical

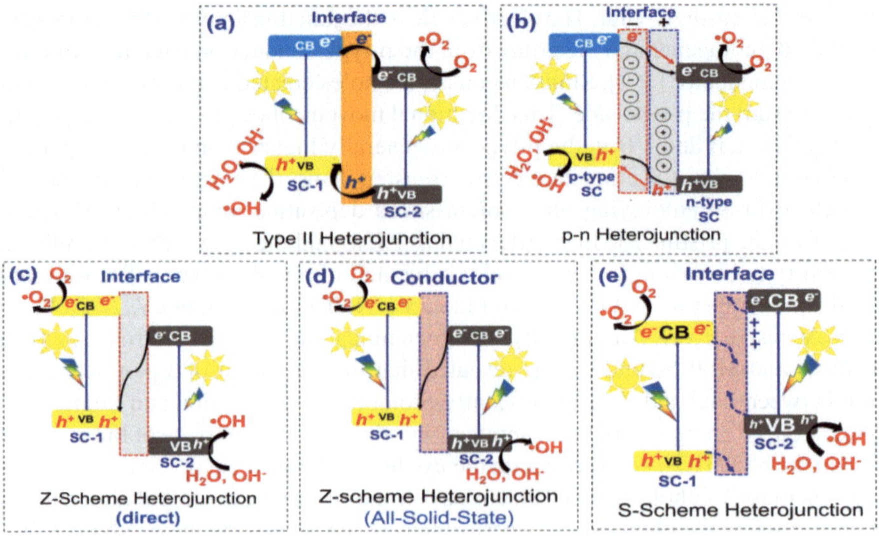

Fig. 6 Diagram illustrating different photocatalyst heterojunction interfaces: **a** Type-II junction, **b** p–n junction, **c** Direct Z-scheme junction, **d** All-solid-state Z-scheme junction, and **e** S-scheme junction

combustion of $Ni(OH)_2$ and $Fe(OH)_3$ [166]. This process resulted in promising interfaces between $NiFe_2O_4$ and NiO, as well as between Fe_2O_3 and $NiFe_2O_4$, which enhanced the migration of photo-induced electrons. Compared to pristine Fe_2O_3 or NiO, the prepared $NiO/NiFe_2O_4/Fe_2O_3$ composite exhibited a smaller arc radius in EIS Nyquist plots, reduced PL intensities, and increased photocurrent density. These findings indicate that the Z-scheme composite demonstrated superior performance relative to its individual components [166]. In another study, the researchers developed a defected NiO/TiO_2 composite with oxygen vacancies through a two-step hydrothermal process followed by a single calcination step (Fig. 7a, b). This approach resulted in the in-situ creation of oxygen vacancies and Ti^{3+} ions, which contributed to forming a Z-scheme band structure [167]. These intrinsic defects served as mid-gap states within TiO_2, enabling direct Z-scheme charge migration between the VB of NiO and the Ti^{3+}/ oxygen vacancy states. This mechanism effectively minimized the recombination of photogenerated electrons and holes. Meanwhile, the arrangement of band structure resulting from constructing the NiO/TiO_2 heterojunction promoted the photogenerated carries activity in the photocatalytic reduction and oxidation reactions (Fig. 7c).

Further, the S-scheme mechanism is critical in facilitating efficient charge migration and separation [168]. Upon forming the junction, the energy bands of the two materials bend, creating a profile that resembles the letter "S". The electrons in the CB of the n-type semiconductor are attracted toward the VB p-type side due to the built-in electric field [167]. This movement occurs across the junction, helping to separate the electron from its associated hole. For instance, the charge transfer dynamics in NiO/

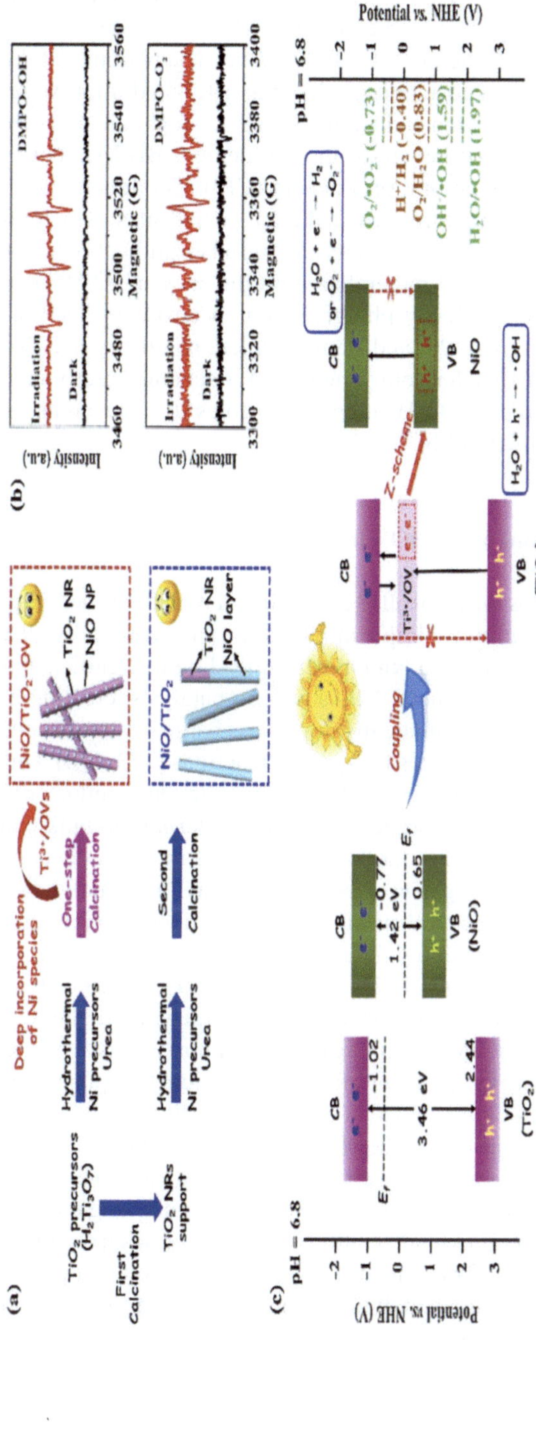

Fig. 7 **a** Schematic illustration for the generation of surface Ti^{3+} and oxygen defects on NiO/TiO_2-OV catalyst, **b** ESR spectra of hydroxyl and superoxide radicals over the NiO/TiO_2-OV catalyst under irradiation, and **c** the proposed mechanism for its enhanced performance in the photocatalytic reactions

BiOCl-7 heterostructures was altered by shifting from a type-II configuration to an S-scheme arrangement [169]. This transition was achieved through pH modulation, which optimized the energy band structure of BiOCl and created numerous oxygen vacancies. Notably, the S-scheme NiO/BiOCl-7 heterojunctions exhibited significantly enhanced photocatalytic activity, achieving a 1.7-fold improvement over the type-II NiO/BiOCl configuration and a remarkable 4.3-fold enhancement compared to pure BiOCl samples. In another study, the S-scheme S-Bi_2WO_6/NiO was created for the efficient mineralization of ciprofloxacin (CIP). The functional groups of S-Bi_2WO_6/NiO and unique structure provided many active sites, which promoted the separation and migration of charge carriers and the adsorption of reactants. The 0.35 S-Bi_2WO_6/NiO achieved a CIP mineralization efficiency of 92.5% within 90 min of exposure to visible light. X-ray photoelectron spectroscopy (XPS), Ultraviolet photoelectron spectroscopy (UPS) and Density functional theory (DFT) calculations reflected that the carrier transfer conformed to an S-scheme heterojunction mechanism. Both hydroxyl radicals ($^{\cdot}OH$) and positive holes (h^+) played an essential part in degrading CIP, as shown in Fig. 8 [170].

4.2.2 Metal Deposition onto NiO Photocatalysts

Metal deposition involves the controlled anchoring of metal NPs onto the surface of a NiO [171, 172]. This seemingly simple modification can profoundly alter the photocatalytic properties of the pristine NiO material, leading to significant performance enhancements. The deposited metal NPs act as co-catalysts, creating a synergistic interaction with the NiO that dramatically improves charge separation, enhances light absorption, and provides active sites for specific reactions [173, 174]. Unlike doping, which involves incorporating impurities directly into the semiconductor lattice, metal deposition modifies the surface properties without altering the bulk characteristics of the semiconductor. This surface modification creates interfaces between the metal and semiconductor, introducing new electronic pathways and catalytic sites [8, 175]. The choice of metal, its size and distribution, and the deposition method employed all play crucial roles in determining the effectiveness of this strategy.

The deposition of metals onto NiO can significantly enhance its performance through various mechanisms. For instance, when a metal with a higher work function, such as gold (Au), platinum (Pt), silver (Ag), or palladium (Pd), is deposited onto NiO, a Schottky junction is formed at the interface [8]. This junction plays a crucial role in photocatalytic processes by acting as an electron sink. It effectively attracts photogenerated electrons from the conduction band of NiO, facilitating their transfer to the metal. This process creates a spatial separation between electrons and holes, which is essential for reducing their recombination rate. By minimizing recombination, the lifetime of charge carriers is extended, allowing them to participate more effectively in photocatalytic reactions. This enhanced charge separation and prolonged carrier lifespan lead to improved photocatalytic efficiency, making the system more effective for applications such as solar energy conversion and environmental remediation [176, 177].

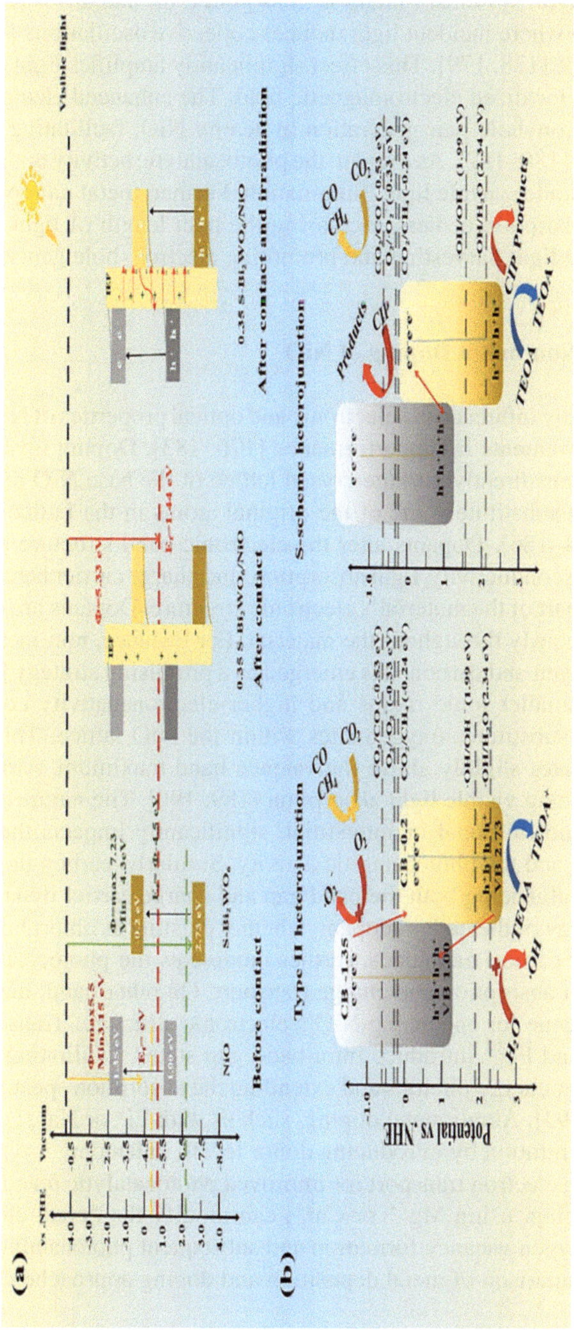

Fig. 8 The diagram of S-scheme mechanism for the 0.35 S-Bi_2WO_6/NiO heterojunction (**a**); Illustration of band structure of S-scheme 0.35 S-Bi_2WO_6/NiO heterojunction and surface redox reactions (**b**)

Surface Plasmon Resonance (SPR) is another critical factor that enhances performance in photocatalytic systems. Metals such as gold (Au) and silver (Ag) exhibit SPR, a phenomenon where incident light induces collective oscillations of electrons on the metal's surface [178, 179]. This effect significantly amplifies light absorption and creates a strong localized electromagnetic field. The enhanced electromagnetic field increases electron–hole pair generation in nearby NiO, facilitating more efficient charge transfer [180–182]. As a result, the photocatalytic activity is significantly boosted, especially under visible light illumination. Further, metal nanoparticles act as scattering and absorption centers, increasing the path length of light within the NiO, thus improving light harvesting and promoting electron–hole generation.

4.2.3 Metal and Non-metal Doping of NiO

Doping can profoundly influence the electronic and optical properties of NiO, leading to substantial improvements in its performance [176, 183]. Doping involves introducing impurity atoms directly into the crystal lattice of the base NiO material [8]. These dopant atoms substitute some of the original atoms in the lattice or occupy interstitial sites [184–186]. Dopants alter the electronic band structure of the NiO material, influencing conductivity, light absorption, and charge carrier behavior. They become an integral part of the material's electronic structure. Dopants are distributed relatively homogeneously throughout the material. For instance, non-metal doping, exemplified by nitrogen and carbon, has emerged as a promising strategy [187, 188]. Nitrogen, with its smaller ionic radius and higher electronegativity compared to oxygen, effectively substitutes oxygen sites within the NiO lattice. This substitution creates N 2*p* states slightly above the valence band maximum, narrowing the band gap and enhancing visible light absorption [189, 190]. The nature of N incorporation, whether substitutional or interstitial, significantly impacts the extent of band gap narrowing and the photocatalytic activity. Similarly, carbon doping introduces C 2*p* states, influencing both the band gap and charge carrier dynamics [191, 192]. The morphology of the carbon dopant, whether existing as amorphous carbon, graphitic carbon, or carbon nanotubes, further modulates the photocatalytic properties, affecting light absorption and charge transport. On other hand, metal doping presents another avenue for enhancing NiO's electronic structure. Transition metal dopants like Cu^{2+} and Fe^{3+} introduce intra-band gap states, facilitating electronic transitions with lower energy photons and extending the absorption spectrum toward the visible region [193]. Alkali metal doping, such as with Li^{+} or Na^{+}, can increase charge carrier concentration by introducing donor levels, enhancing n-type conductivity and facilitating electron transport for improved photocatalytic reactions. Alkaline earth metal doping, using Mg^{2+} or Ca^{2+}, can modify the defect chemistry of NiO, influencing oxygen vacancy formation and subsequent photocatalytic activity. Table 2 shows a comparison of metal deposition and doping approaches.

Table 2 Comparison of metal deposition and doping approaches

Feature	Metal deposition	Doping
Mechanism of enhancement	Schottky barrier formation, SPR effect (for specific metals), enhanced light absorption, co-catalytic activity, increased surface area	Modified band structure (band gap narrowing, intra-band gap states), enhanced charge separation (trapping sites), increased charge carrier mobility, modified surface properties
Location of modification	Primarily on the surface of NiO	Within the NiO lattice
Nature of modification	Formation of a metal–semiconductor interface; physical presence of metal nanoparticles	Introduction of foreign atoms into the NiO crystal structure
Light absorption enhancement	Primarily through plasmon resonance (for specific metals) and light scattering; can also indirectly enhance absorption in NiO through improved charge separation	Directly modifies the band gap or introduces intra-band gap states, leading to increased absorption in the visible region
Charge separation enhancement	Schottky barrier promotes electron transfer from NiO to metal	Dopants act as electron or hole traps, preventing recombination
Charge carrier mobility impact	Can indirectly improve mobility in NiO by reducing recombination	Can directly enhance mobility by altering the electronic structure and conductivity of NiO
Material stability	Potential for metal leaching or sintering over time, especially under harsh conditions	Generally more stable as dopants are incorporated into the lattice
Examples	Au, Pt, Ag, Pd, Fe deposition on NiO	N, C, Cu, Fe, Al, Li doping of NiO
Disadvantages	Potential for metal leaching, light shielding at high loadings, cost of noble metals	Can be challenging to control dopant concentration and distribution, optimization can be complex

4.2.4 Carbon-Based NiO Nanocomposites

The integration of carbon-based substances with NiO presents a compelling strategy for significantly enhancing its photocatalytic performance. Carbon materials, such as graphene (G), graphene oxide (GO), reduced graphene oxide (rGO), graphitic carbon nitride (g-C_3N_4), carbon quantum dots (CQDs), and chitosan (CS), offer a diverse range of advantageous properties, including high surface area, excellent electrical conductivity, tunable electronic properties, and good chemical stability, making them ideal substances for boosting the NiO photocatalytic activity [194–197]. The primary mechanism responsible for enhanced photocatalysis in NiO-carbon composites is the improved charge separation. The carbon material acts as an electron acceptor, creating a pathway for photogenerated electrons in NiO to migrate to the carbon material, effectively separating them from positively charged holes. This separation minimizes electron–hole recombination, allowing the charge carriers to effectively participate in redox reactions with adsorbed species on the catalyst surface. Furthermore, some carbon materials can enhance light absorption, broadening the spectral response of NiO and improving its utilization of solar energy. The high surface area of many carbon materials also contributes to increased reactant adsorption and active site availability, further boosting photocatalytic efficiency. For instance, graphene, a set of two-dimensional sheets of sp^2 hybridized carbon atoms, exhibits remarkable electrical conductivity and a large surface area. Integrating graphene with NiO facilitates efficient charge separation by acting as an electron acceptor, drawing photogenerated electrons away from the NiO and hindering their recombination with holes [126, 198]. This extended lifespan of charge carriers enables them to participate in redox reactions at the catalyst surface [126]. Furthermore, graphene's high surface area provides ample sites for reactant adsorption, enhancing the overall photocatalytic efficiency. The strong π-π interactions between graphene and organic pollutants also contribute to improved adsorption and degradation. Graphene is characterized by its zero-bandgap energy, massless fermions, exceptional electron mobility (200,000 $cm^2 V^{-1} s^{-1}$), huge specific surface area (2630 $m^2 g^{-1}$), high thermal conductivity (approximately 5000 W $m^{-1} K^{-1}$), and significant transparency (optical transmittance of 97.7%) [20, 126]. Graphene oxide (GO) and reduced graphene oxide (rGO) are commonly used variants of graphene [19, 20, 24, 66, 199]. Table 3 presents a comparison of the features of graphene-based materials when used as supports for photocatalysts or adsorbents. For instance, rGO/NiO was produced using a microwave-assisted in-situ co-precipitation and hydrothermal technique. The experimental findings revealed that this composite significantly enhanced the photocatalytic activity for methylene blue (MB) dye photo-mineralization, achieving a 93% efficiency compared to the negligible activity of rGO alone and the 65% activity of bare NiO NPs. Within these nanocomposites, the transfer of electronic charges occurs from NiO nanoparticles to rGO, effectively mitigating the recombination of electron–hole pairs [200]. In another study, GO/NiO–TiO_2 nanocomposite was developed for dual applications in photocatalysis and electrochemical oxygen evolution reactions (OER) [201]. The researchers found that GO enhanced the electronic conductivity, while Ni^{2+} and Ti^{4+} ions offer additional active sites. The composite's robust activity is attributed to its efficient charge separation, and low band gap energy [201].

Moreover, carbon quantum dots (CQDs), zero-dimensional carbon nanomaterials, exhibit unique optical properties, including tunable photoluminescence and upconversion capabilities. Incorporating CQDs into NiO photocatalytic systems can enhance visible light absorption and promote charge separation [202, 203]. CQDs can act as photosensitizers, absorbing light in the visible region and transferring energy to NiO, thereby extending its photocatalytic activity into the visible spectrum. Moreover, CQDs can also act as electron reservoirs, facilitating charge transfer and suppressing the e^-—hole recombination. Furthermore, CQD/NiO/g-C_3N_4 ternary hybrid was fabricated and exhibited a robust mineralization rate of MO, attributed to the numerous active sites and conductive pathways provided by CQDs [204].

Further, chitosan (CS), a naturally derived biopolymer, has emerged as promising supporting materials for enhancing the photocatalytic activity of NiO or other metal oxides [205]. For instance, CS mitigates this issue through its electron scavenging

Table 3 Comparison of between graphene and its counterparts

Features	Graphene	GO	rGO
Method of synthesis	Mechanical exfoliation, CVD, etc.	Chemical oxidation of graphite	Chemical or thermal reduction of GO
Electrical conductivity	Extremely high	Low	Moderate to high
Surface area	High (2630 m^2/g) theoretical	Moderate, varies with oxidation levels	High (but lower than pure graphene)
Surface functionalization	Minimal (pure form)	High (various functional groups)	Moderate (some groups remain)
Photocatalytic activity	High, due to excellent carrier mobility and reduced recombination rates	Lower than pure graphene, requires activation	Improved compared to GO
Adsorption efficiency	Very high, favored for organic pollutants and heavy metals	High, especially for polar and ionic substances	Better than GO due to improved surface area and conductivity
Chemical stability	Very stable	Stable but may degrade under extreme conditions	Stable, but may retain some sensitivity to oxidation
Practical challenges	Expensive and complex synthesis	Solubility can complicate processing, lower conductivity	Quality can vary based on reduction method

capabilities. The $-NH_2$ and –OH groups present in chitosan chains act as traps for photogenerated electrons from NiO, effectively separating them from the holes [205]. This prolonged lifespan of charge carriers increases their likelihood of participating in redox reactions, ultimately boosting photocatalytic activity [205]. Additionally, chitosan significantly enhances the NiO's adsorption capacity through its functional groups which can interact directly with pollutants [206–208]. These interactions can involve hydrogen bonding, electrostatic forces, and chelation, leading to improved adsorption capacity and selectivity for specific pollutants. Some studies reported that the interaction between chitosan and NiO extended the light absorption range of the composite into the visible region. This enhanced visible light absorption translates to better utilization of solar energy, a critical factor for sustainable photocatalysis. Moreover, the interaction between CS and NiO can subtly modify NiO's electronic structure, potentially leading to a reduced band gap. This narrowing of the band gap further enhances visible light absorption and hence the overall photocatalytic efficiency. For instance, the photocatalytic mineralization of MB dye employing chitosan/NiO and chitosan achieved 96.1 and 21%, respectively [209]. In another study, diffuse reflectance spectrometric (DRS) data reflected the reduction in band gap energy of pristine NiO/ZnO from 2.70 to 2.32 eV upon hybridization with CS in CS-NiO/ZnO nanohybrid [210].

5 NiO-Based Adsorbents

Metal oxides have emerged as promising adsorbents for wastewater treatment due to their high surface area, diverse functionalities, and tunable properties. Their effectiveness stems from their ability to interact with various pollutants, including heavy metals, and organic molecules including dyes, pharmaceuticals and agrochemicals through mechanisms like electrostatic attraction, surface complexation, ion exchange, and H-bonding interactions. Among the various metal oxides, nickel oxide (NiO) has garnered considerable attention due to its relatively low cost, ease of synthesis, and good adsorption capacity for a wide range of pollutants. The porous structure of NiO, often achieved through controlled synthesis methods, provides a large surface area for pollutants' removal [211]. Furthermore, the surface hydroxyl groups present on NiO can facilitate the adsorption process through hydrogen bonding and complex formation with pollutant molecules. For instance, the researchers reported NiO (111) nanosheets exhibiting adsorption capacity (q_{max}) of 220, 35.15 and 30.4 mg g^{-1} for fuchsin acid, Congo red and reactive brilliant red dyes, respectively [212]. In another study, NiO was synthesized, through a straightforward thermal decomposition of nickel ascorbate, and was capable of removing 97.2% of ponceau xylidine dye from aqueous solutions [213]. FESEM images revealed clusters of pores in the NiO NPs, likely due to gas evolution during its calcination process. Other researchers prepared NiO NPs using a microwave-assisted rout, for the adsorptive removal of Bromophenol blue (BB) dye [214]. The prepared NiO NPs exhibited a pore volume of 0.068 cm^3/g, surface area of 78.37 m^2/g, pore size of 34.92 Å, and an isoelectric point (IEP) is of 8.1. Adsorption isotherms data fitted the Langmuir model. The calculated q_{max} values were 58.48 79.37 and 93.46 mg/g at 30, 45 and 60 °C, respectively.

Therefore, NiO-based adsorbents can be further enhanced by modifying their structure and composition; e g., by doping with other metal ions or incorporating NiO into nanocomposite materials that can significantly improve its adsorption performance. For example, surface functionalization of NiO with organic moieties or polymers can tailor its selectivity toward specific target pollutants.

6 Conclusion and Future Directions

This chapter has explored a variety of wastewater treatment methods, highlighting the limitations and advantages of conventional approaches in addressing relatively persistent pollutants like dyes, pharmaceuticals and agrochemicals. A particular focus has been placed on the potential of NiO-based nanocomposites for advanced wastewater treatment due to their adsorptive and photocatalytic properties. While pure NiO suffers from limited photoactivity due to its wide band gap, restricting its effective utilization of the solar spectrum, significant advancements have been made in enhancing its performance. Strategies such as tailored structural modifications and hybridization with other nanomaterials, including carbon-based structures, semiconductors, and other metal oxides, have demonstrated synergistic effects leading to improved adsorption and photocatalytic performances for the degradation of pollutants. While metal doping of NiO has yielded inconsistent results, the construction of p–n junctions with carbonaceous materials like graphene, CQDs, chitosan and g-C_3N_4 presents a promising avenue for optimizing charge carrier dynamics and enhancing overall photocatalytic efficiency.

Looking ahead, several avenues of research and development can be pursued to optimize the use of NiO in wastewater treatment:

1. Future studies should focus on developing more effective and stable NiO-based nanocomposites that can withstand harsh environmental conditions while retaining their favorable adsorptive and photocatalytic properties over extended periods.
2. Research is expected to concentrate on creating ternary or quaternary NiO hybrids incorporating carbon-based materials. These materials possess outstanding conductivity and electron-accepting properties, which can boost charge carrier dynamics and enhance photocatalytic efficiency.
3. Conduct pilot-scale studies to evaluate the performance and long-term stability of NiO-based systems in real-world applications, such as wastewater treatment plants or air purification systems.
4. Develop scalable and cost-effective methods for synthesizing high-quality NiO nanomaterials and fabricating functional devices. Explore sustainable and environmentally friendly synthesis routes.
5. Investigate strategies for integrating NiO-based photocatalysts and adsorbents with existing water and air purification technologies to enhance their efficiency and reduce costs.

By pursuing these advanced research directions, we can push the boundaries of NiO-based nanomaterials and pave the way for their widespread implementation in addressing critical environmental and energy challenges. A key emphasis should be placed on bridging the gap between fundamental research and practical applications.

References

1. Cohen B (2004) Urban growth in developing countries: a review of current trends and a caution regarding existing forecasts. World Dev 32(1):23–51
2. Ahmed MA, Mahmoud SA, Mohamed AA (2024) Unveiling the complexities of microbiologically induced corrosion: mechanisms, detection techniques, and mitigation strategies. Front Environ Sci Eng 18(10):120
3. Adel M, Ahmed MA, Mohamed AA (2021) Effective removal of indigo carmine dye from wastewaters by adsorption onto mesoporous magnesium ferrite nanoparticles. Environ Nanotechnol Monit Manage 16:100550
4. Thamaraiselvi C et al (2024) Effective reduction of carbon-containing pollutants in coffee cherry pulping wastewater using natural polysaccharide from *Tamarindus indica L.* seeds. Environ Res 259:119527

5. Khan I et al (2024) Biopolymer-carbonaceous composites, progress, and adsorptive mitigation of water pollutants. Int J Biol Macromol 133379
6. Kim M, Han J (2024) Treatment techniques for removal of polybrominated diphenyl ethers (PBDEs) from real wastewater: limitations, challenges, and future research directions. J Water Process Eng 63:105463
7. Wang S et al (2024) The uptake and degradation of polychlorinated biphenyls in constructed wetlands planted with Myriophyllum aquaticum. Environ Sci Pollut Res 31(11):17115–17123
8. Makhado KP et al (2024) Current updates on p-type nickel oxide (NiO) based photocatalysts towards decontamination of organic pollutants from wastewater. Mater Today Sustain 100664
9. Al-Wasidi AS, Hegazey RM, Abdelrahman EA (2024) Efficient removal of methylene blue dye from aqueous media using facilely synthesized magnesium borate/magnesium oxide nanostructures. Molecules 29(14):3392
10. Helali S et al (2024) Structural analysis and adsorption studies of (PbO, MgO) metal oxide nanocomposites for efficient methylene blue dye removal from water. Materials 17(12):2890
11. Tison L et al (2024) Transfer and bioaccumulation of pesticides in terrestrial arthropods and food webs: state of knowledge and perspectives for future research. Chemosphere 142036
12. Iyiola AO et al (2024) Organic chemical pollutants within water systems and sustainable management strategies. Water crises and sustainable management in the global south. Springer, pp 211–251
13. Song Q et al (2024) Biochar-based composites for removing chlorinated organic pollutants: applications, mechanisms, and perspectives. Environ Sci Ecotechnol 100420
14. Ahmed MA, Mahmoud SA, Mohamed AA (2024) Nanomaterials-modified reverse osmosis membranes: a comprehensive review. RSC Adv 14(27):18879–18906
15. Ahmed M, Amin S, Mohamed A (2023) Fouling in reverse osmosis membranes: monitoring, characterization, mitigation strategies and future directions. Heliyon 9:e14908
16. Ahmed MA, Mohamed AA (2024) Evaluation and optimization of antiscalant substances for enhanced reverse osmosis performance. J Saudi Chem Soc 101923
17. Deng Y et al (2024) Prospects, advances and biological applications of MOF-based platform for the treatment of lung cancer. Biomater Sci 12(15):3725–3744
18. Ahmed MA, Ahmed MA, Mohamed AA (2023) Removal of 4-nitrophenol and indigo carmine dye from wastewaters by magnetic copper ferrite nanoparticles: kinetic, thermodynamic and mechanistic insights. J Saudi Chem Soc 27(6):101748
19. Adel M, Ahmed MA, Mohamed AA (2020) Effective removal of cationic dyes from aqueous solutions using reduced graphene oxide functionalized with manganese ferrite nanoparticles. Compos Commun 22:100450
20. Adel M, Ahmed MA, Mohamed AA (2021) Synthesis and characterization of magnetically separable and recyclable crumbled $MgFe_2O_4$/reduced graphene oxide nanoparticles for removal of methylene blue dye from aqueous solutions. J Phys Chem Solids 149:109760
21. Yu Q et al (2024) Silica-alumina zeolite adsorbents for oxygen generation via pressure swing adsorption: mechanisms and challenge. Chem Eng J 148788
22. Tan J, Zhou L (2024) Two birds with one stone: one-pot hydrothermal treatment of dragon fruit peel to simultaneously prepare efficient biochar adsorbent and fluorescent carbon quantum dots. Ind Crops Prod 210:118086
23. Mushtaq S et al (2024) Utilizing sludge-based activated carbon for targeted leachate mitigation in wastewater treatment. Environ Res 118326
24. Adel M, Ahmed MA, Mohamed AA (2021) A facile and rapid removal of cationic dyes using hierarchically porous reduced graphene oxide decorated with manganese ferrite. FlatChem 26:100233
25. Akhtar MS, Ali S, Zaman W (2024) Innovative adsorbents for pollutant removal: exploring the latest research and applications. Molecules 29(18):4317
26. Liaquat I et al (2024) Exploring zeolite-based composites in adsorption and photocatalysis for toxic wastewater treatment: preparation, mechanisms, and future perspectives. Environ Pollut 123922

27. Ahmed, M.A. and A.A. Mohamed, *Advances in ultrasound-assisted synthesis of photocatalysts and sonophotocatalytic processes: A review.* Iscience, 2023.
28. Shakil M et al (2023) NiO and Ag–Cd co-doped NiO nanoparticles: study of photocatalytic degradation of rhodamine B dye for wastewater treatment. Int J Environ Sci Technol 20(2):2021–2036
29. Mohtaram MS et al (2024) Photocatalytic degradation of tetracycline using a novel WO3–ZnO/AC under visible light irradiation: optimization of effective factors by RSM-CCD. Environ Pollut 347:123746
30. Samuel O et al (2023) Photocatalytic degradation of recalcitrant aromatic hydrocarbon compounds in oilfield-produced water: a critical review. J Clean Prod 415:137567
31. Raub AAM et al (2024) Photocatalytic activity enhancement of nanostructured metal-oxides photocatalyst: a review. Nanotechnology
32. Sundaram GA et al (2024) Introduction and advancements in room-temperature ferromagnetic metal oxide semiconductors for enhanced photocatalytic performance. ChemEngineering 8(2):36
33. Shanaah HH et al (2023) Photocatalytic degradation and adsorptive removal of emerging organic pesticides using metal oxide and their composites: recent trends and future perspectives. Sustainability 15(9):7336
34. Pascariu P, Gherasim C, Airinei A (2023) Metal oxide nanostructures (MONs) as photocatalysts for ciprofloxacin degradation. Int J Mol Sci 24(11):9564
35. Chowdhury PR et al (2023) Photocatalysis: TiO_2, ZnO, and species of iron oxides. Nanoremediation. Elsevier, pp 101–126
36. Davoodi Y, Shafeeyan MS (2024) Tailoring photocatalytic performance through Fe-doped TiO_2/ZnO for effective remediation of organic contaminants. Water Resour Ind 31:100239
37. Christan CS et al (2024) Amorphous vs. crystalline Al_2O_3 nanoparticles: a comparative study in photocatalytic dye degradation. Nanomater Energy 1–19
38. Hu Q et al (2024) Enhanced removal of phosphate from aqueous solutions by oxygen vacancy-rich MgO microspheres: performance and mechanism. Chemosphere 355:141776
39. Singh PP et al (2024) The pivotal role of oxygen vacancy and surface hydroxyl in the adsorption and activation of CO_2 on ceria-zirconia mixed oxide. Mol Catal 555:113855
40. Jain N et al (2024) Heterostructured core-shell metal oxide-based nanobrushes for ultrafast UV photodetectors. Mater Sci Eng R Rep 160:100826
41. Clavero C (2014) Plasmon-induced hot-electron generation at nanoparticle/metal-oxide interfaces for photovoltaic and photocatalytic devices. Nat Photonics 8(2):95–103
42. Gomaa I et al (2024) Acoustic-assisted fabrication, characterization, and photocatalytic application of Ni_2O_3/NiO/rGO nanocomposites. Curr Anal Chem
43. Agalya S, Nehru L, Sagadevan S (2024) Green-synthesized nickel oxide nanoparticles for sustainable wastewater treatment and enhanced bacterial control. Ceram Int
44. Shifu C et al (2008) Preparation and activity evaluation of p–n junction photocatalyst NiO/TiO_2. J Hazard Mater 155(1–2):320–326
45. Ahmed MA, Ahmed MA, Mohamed AA (2024) Fabrication of NiO/g-C_3N_4 Z-scheme heterojunction for enhanced photocatalytic degradation of methylene blue dye. Opt Mater 151:115339
46. Akbari A et al (2020) Effect of nickel oxide nanoparticles as a photocatalyst in dyes degradation and evaluation of effective parameters in their removal from aqueous environments. Inorg Chem Commun 115:107867
47. Yu J, Wang W, Cheng B (2010) Synthesis and enhanced photocatalytic activity of a hierarchical porous flowerlike p–n junction NiO/TiO_2 photocatalyst. Chem Asian J 5(12):2499–2506
48. Li J-S et al (2023) Effect of drift layer doping and NiO parameters in achieving 8.9 kV breakdown in 100 μm diameter and 4 kV/4 A in 1 mm diameter NiO/β-Ga_2O_3 rectifiers. J Vac Sci Technol A 41(4)
49. Aejitha S et al (2023) Effect of La-doping on NiO photocatalyst for enhancing photocatalytic degradation performance under visible light irradiation: DFT calculations and degradation mechanism. Inorg Chem Commun 156:111172

50. Kordbacheh F, Heidari G (2023) Water pollutants and approaches for their removal. Mater Chem Horiz 2(2):139–153
51. Salleh KM, Abd Rashid NF (2024) Potential seaweed-derived bioactive compounds for pharmaceutical applications. Mar Biomass: Biorefinery Bioprod Environ Bioremed 297
52. Khan HK, Rehman MYA, Malik RN (2020) Fate and toxicity of pharmaceuticals in water environment: an insight on their occurrence in South Asia. J Environ Manage 271:111030
53. Chaturvedi P et al (2021) Prevalence and hazardous impact of pharmaceutical and personal care products and antibiotics in environment: a review on emerging contaminants. Environ Res 194:110664
54. Ghosh S et al (2023) A review on ciprofloxacin removal from wastewater as a pharmaceutical contaminant: covering adsorption to advanced oxidation processes to computational studies. Mater Today Commun 107500
55. Mosharaf MK et al (2024) Wastewater reuse and pharmaceutical pollution in agriculture: uptake, transport, accumulation and metabolism of pharmaceutical pollutants within plants. Chemosphere 143055
56. Sanusi IO et al (2023) Occurrence, environmental impact and fate of pharmaceuticals in groundwater and surface water: a critical review. Environ Sci Pollut Res 30(39):90595–90614
57. Mary Ealias A, Meda G, Tanzil K (2024) Recent progress in sustainable treatment technologies for the removal of emerging contaminants from wastewater: a review on occurrence, global status and impact on biota. Rev Environ Contam Toxicol 262(1):16
58. Antos J et al (2024) Tetracyclines contamination in European aquatic environments: a comprehensive review of occurrence, fate, and removal techniques. Chemosphere 141519
59. Mohammadzadeh M, Leiviskä T (2023) Iron-modified peat and magnetite-pine bark biosorbents for levofloxacin and trimethoprim removal from synthetic water and various pharmaceuticals from real wastewater. Ind Crops Prod 195:116491
60. Gracia-Marín E et al (2024) Comprehensive study on the potential environmental risk of temporal antibiotic usage through wastewater discharges. Chemosphere 346:140587
61. Rahimi F et al (2024) Advances in three-dimensional electrochemical degradation: a comprehensive review on pharmaceutical pollutants removal from aqueous solution. Chemosphere 142620
62. Vaiano V, De Marco I (2023) Removal of azo dyes from wastewater through heterogeneous photocatalysis and supercritical water oxidation. Separations 10(4):230
63. Mishra S, Cheng L, Maiti A (2021) The utilization of agro-biomass/byproducts for effective bio-removal of dyes from dyeing wastewater: a comprehensive review. J Environ Chem Eng 9(1):104901
64. Bishal A et al (2023) Natural dyes: its origin, categories and application on textile fabrics in brief. Eur Chem Bull 12(8):9780–9802
65. Islam T et al (2023) Impact of textile dyes on health and ecosystem: a review of structure, causes, and potential solutions. Environ Sci Pollut Res 30(4):9207–9242
66. Ahmed MA, Ahmed MA, Mohamed AA (2022) Facile adsorptive removal of dyes and heavy metals from wastewaters using magnetic nanocomposite of zinc ferrite@ reduced graphene oxide. Inorg Chem Commun 144:109912
67. Al-Tohamy R et al (2022) A critical review on the treatment of dye-containing wastewater: ecotoxicological and health concerns of textile dyes and possible remediation approaches for environmental safety. Ecotoxicol Environ Saf 231:113160
68. Keshava C et al (2023) Application of systematic evidence mapping to identify available data on the potential human health hazards of selected market-relevant azo dyes. Environ Int 176:107952
69. Babić Ž et al (2022) Association of hairdressing with cancer and reproductive diseases: a systematic review. J Occup Health 64(1):e12351
70. Ramamurthy K et al (2024) Hues of risk: investigating genotoxicity and environmental impacts of azo textile dyes. Environ Sci Pollut Res 1–22
71. Mukherjee M et al (2024) Bio-degradation of phenol and phenolic compounds. Advancements in bio-systems and technologies for wastewater treatment. Springer, pp 207–224

72. Pavithra KG et al (2023) A review on recent advancements in extraction, removal and recovery of phenols from phenolic wastewater: challenges and future outlook. Environ Res 237:117005
73. Li S-W et al (2017) Preparation of mesoporous Cs-POM@ MOF-199@ MCM-41 under two different synthetic methods for a highly oxidesulfurization of dibenzothiophene. J Hazard Mater 337:208–216
74. Said KAM et al (2021) A review of technologies for the phenolic compounds recovery and phenol removal from wastewater. Process Saf Environ Prot 151:257–289
75. Aloo S-O et al (2023) Insights on dietary polyphenols as agents against metabolic disorders: obesity as a target disease. Antioxidants 12(2):416
76. Patil P et al (2023) A review on interaction of phenolic pollutant with other pollutants in the binary adsorption system. Desalin Water Treat 285:213–241
77. Haleem A et al (2023) A comprehensive review on adsorption, photocatalytic and chemical degradation of dyes and nitro-compounds over different kinds of porous and composite materials. Molecules 28(3):1081
78. Zhou W, Li M, Achal V (2024) A comprehensive review on environmental and human health impacts of chemical pesticide usage. Emerg Contam 100410
79. Abaineh A et al (2024) Risks of pesticides on aquatic ecosystems and pesticide management effectiveness in Ethiopia. Int J Environ Sci Technol 1–16
80. Dmitrieva VA, Tyutereva EV, Voitsekhovskaja OV (2024) What can reactive oxygen species (ROS) tell us about the action mechanism of herbicides and other phytotoxins? Free Radical Biol Med
81. Shehzad M et al (2023) Recent insights into pesticide resistance mechanisms in Plutella xylostella and possible management strategies. Environ Sci Pollut Res 30(42):95296–95311
82. Micheal A, Salam HA, Ghanem G (2022) Implementation of secondary and tertiary treated wastewater in concrete manufacturing. In: IOP conference series: earth and environmental science. IOP Publishing
83. Azizi N et al (2022) Evaluation of conventional wastewater treatment plants efficiency to remove microplastics in terms of abundance, size, shape, and type: a systematic review and meta-analysis. Mar Pollut Bull 177:113462
84. Koul B et al (2022) Application of natural coagulants in water treatment: a sustainable alternative to chemicals. Water 14(22):3751
85. Akinnawo SO, Ayadi PO, Oluwalope MT (2023) Chemical coagulation and biological techniques for wastewater treatment. Ovidius Univ Ann Chem 34(1):14–21
86. Kurniawan SB et al (2023) Coagulation–flocculation of aquaculture effluent using biobased flocculant: from artificial to real wastewater optimization by response surface methodology. J Water Process Eng 53:103869
87. Husen AK et al (2024) COD, color, and turbidity reduction from surface water using natural coagulants: investigation and optimization. Progr Eng Sci 1(2–3):100007
88. Butt M et al (2005) Spectrophotometric estimation of colour in textile dyeing wastewater. J Chem Soc Pak 27(6):627
89. Kim T-H et al (2004) Comparison of disperse and reactive dye removals by chemical coagulation and Fenton oxidation. J Hazard Mater 112(1–2):95–103
90. Daryabeigi Zand A, Hoveidi H (2015) Comparing aluminium sulfate and poly-aluminium chloride (PAC) performance in turbidity removal from synthetic water. J Appl Biotechnol Rep 2(3):287–292
91. Hafeez MA et al (2023) Recent advances in Fenton-like treatment of radioactive ion exchange resins. Chem Eng J Adv 14:100461
92. Yanardağ D, Edebali S (2024) Adsorptive removal of malachite green dye from aqueous solution by ion exchange resins. Biomass Conv Bioref 14(4):5699–5710
93. Handle KF, Bakry R, Bonn GK (2023) Phosphonium based anion exchange resin for enrichment of phenolic acids. J Chromatogr A 1700:464049
94. Shahid MK et al (2023) Ion exchange process for removal of microconstituents from water and wastewater. In: Microconstituents in the environment: occurrence, fate, removal and management. p 303–320

95. Li L, Liu Y (2024) Adsorption of perchlorate on different quaternary ammonium-modified resins: influencing factors and mechanisms. Sep Purif Technol 130364
96. Tang Y, Wen Q, Chen Z (2023) Simultaneous removal of nitrogen and phosphorus nutrients from secondary effluent by magnetic resin containing two types of quaternary ammonium adsorption sites: preparation, characterization, and application. Chem Eng J 477:147137
97. Palazhchenko OY, Ferguson JP, Cook WG (2023) Iron hydrolysis and lithium uptake on mixed-bed ion exchange resin at alkaline pH. Nucl Eng Technol 55(10):3665–3676
98. Tekinalp Ö et al (2023) Cation exchange membranes and process optimizations in electrodialysis for selective metal separation: a review. Membranes 13(6):566
99. Eshbobaev J et al (2023) Field trial of solar-powered ion-exchange resin for the industrial wastewater treatment process. Eng Proc 37(1):47
100. Wiercik P, Frączek B, Chrobot P (2020) Fouling of anion exchanger by image and FTIR analyses. J Environ Chem Eng 8(3):103761
101. Wiercik P et al (2021) Organic fouling of strongly basic anion exchange resin by image and FTIR analyses. In: Book of abstracts
102. Wawrzkiewicz M (2013) Removal of CI basic blue 3 dye by sorption onto cation exchange resin, functionalized and non-functionalized polymeric sorbents from aqueous solutions and wastewaters. Chem Eng J 217:414–425
103. Marin NM et al (2019) Removal of the acid orange 10 by ion exchange and microbiological methods. Int J Environ Sci Technol 16:6357–6366
104. Thirumalaivasan N et al (2024) Utilization of fungal and bacterial bioremediation techniques for the treatment of toxic waste and biowaste. Front Mater 11:1416445
105. Chan YJ et al (2009) A review on anaerobic–aerobic treatment of industrial and municipal wastewater. Chem Eng J 155(1–2):1–18
106. McCarty PL (2018) What is the best biological process for nitrogen removal: when and why? ACS Publications
107. Siddiqui MI et al (2022) Aeration control strategy design based on dissolved oxygen and redox potential profiles for nitrogen and phosphorus removal from sewage in a sequencing batch reactor. J Water Process Eng 50:103259
108. Kong Z et al (2019) Challenges and prospects for the anaerobic treatment of chemical-industrial organic wastewater: a review. J Clean Prod 231:913–927
109. Atelge M et al (2020) A critical review of pretreatment technologies to enhance anaerobic digestion and energy recovery. Fuel 270:117494
110. Muter O (2023) Current trends in bioaugmentation tools for bioremediation: a critical review of advances and knowledge gaps. Microorganisms 11(3):710
111. Chiong T et al (2016) Enzymatic treatment of methyl orange dye in synthetic wastewater by plant-based peroxidase enzymes. J Environ Chem Eng 4(2):2500–2509
112. Carolin CF, Kumar PS, Joshiba GJ (2021) Sustainable approach to decolourize methyl orange dye from aqueous solution using novel bacterial strain and its metabolites characterization. Clean Technol Environ Policy 23:173–181
113. Garg SK et al (2012) Biodecolorization of textile dye effluent by Pseudomonas putida SKG-1 (MTCC 10510) under the conditions optimized for monoazo dye orange II color removal in simulated minimal salt medium. Int Biodeterior Biodegradation 74:24–35
114. Adel M et al (2022) Characterization of fouling for a full-scale seawater reverse osmosis plant on the Mediterranean sea: membrane autopsy and chemical cleaning efficiency. Groundw Sustain Dev 16:100704
115. Liao Y et al (2018) Progress in electrospun polymeric nanofibrous membranes for water treatment: fabrication, modification and applications. Prog Polym Sci 77:69–94
116. Rekhate CV, Srivastava J (2020) Recent advances in ozone-based advanced oxidation processes for treatment of wastewater—a review. Chem Eng J Adv 3:100031
117. Wang J, Chen H (2020) Catalytic ozonation for water and wastewater treatment: recent advances and perspective. Sci Total Environ 704:135249
118. Asghar A et al (2022) Influence of water matrix on the degradation of organic micropollutants by ozone based processes: a review on oxidant scavenging mechanism. J Hazard Mater 429:128189

119. Lim S et al (2022) Ozonation of organic compounds in water and wastewater: a critical review. Water Res 213:118053
120. Constapel M et al (2009) Degradation of reactive dyes in wastewater from the textile industry by ozone: analysis of the products by accurate masses. Water Res 43(3):733–743
121. Ahmed MA, Ahmed MA, Mohamed AA (2023) Adsorptive removal of tetracycline antibiotic onto magnetic graphene oxide nanocomposite modified with polyvinylpyrroilidone. React Funct Polym 191:105701
122. Ahmed MA, Mohamed AA (2023) A systematic review of layered double hydroxide-based materials for environmental remediation of heavy metals and dye pollutants. Inorg Chem Commun 148:110325
123. Bai Y et al (2024) FCF-LDH/BiVO4 with synergistic effect of physical enrichment and chemical adsorption for efficient reduction of nitrate. Green Energy Environ 9(7):1112–1121
124. Ahmad A et al (2023) Carbon-integrated semiconductor photocatalysts for removal of volatile organic compounds in indoor environments. Chem Eng J 452:139436
125. Gil A (2023) Classical and new insights into the methodology for characterizing adsorbents and metal catalysts by chemical adsorption. Catal Today 423:114016
126. Ahmed MA, Mohamed AA (2023) Recent progress in semiconductor/graphene photocatalysts: synthesis, photocatalytic applications, and challenges. RSC Adv 13(1):421–439
127. Pattnaik A et al (2023) Current perspective of nano-engineered metal oxide based photocatalysts in advanced oxidation processes for degradation of organic pollutants in wastewater. Chem Eng Res Des 190:667–686
128. Rahman MZ et al (2023) Key strategies for enhancing H_2 production in transition metal oxide based photocatalysts. Angew Chem 135(48):e202305385
129. Peng D, Zhang Y (2023) Engineering of mixed metal oxides photocatalysts derived from transition-metal-based layered double hydroxide towards selective oxidation of cyclohexane under visible light. Appl Catal A 653:119067
130. Ahmed TY et al (2024) Band structure study of pure and doped anatase titanium dioxide (TiO_2) using first-principle-calculations: role of atomic mass of transition metal elements (TME) on band gap reduction. Opt Quant Electron 56(7):1249
131. Malumi SO et al (2023) Enhance and performance evolution of silver-doped titanium dioxide dye-sensitized solar cells using different dyes. J Eng Ind Res 4(4):189–200
132. Chai X, Gao R-T, Wang L (2024) Electron transport kinetics via ZnO with ultralow Fe dopant for stable oxygen reduction to H_2O_2. Chem Eng J 484:149527
133. Zhang T et al (2024) Boosting electrochemical ammonia synthesis via NO_x reduction over sulfur-doped copper oxide nanoneedle arrays. Adv Energy Mater 14(30):2400790
134. Luo J et al (2024) In-situ electronic modulation of ultra-high-capacity S-modified Cu/Cu_2O electrodes for energy storage applications. Chem Eng J 485:149945
135. Uma K, Senthilkumar G, Balamurugan K (2023) Nano metal oxide (CuO/Co_3O_4) based feed stock filament for FDM. J Mech Sci Technol 37(9):4817–4823
136. Kumar R (2023) Metal oxides-based nano/microstructures for photodegradation of microplastics. Adv Sustain Syst 7(6):2300033
137. Rajan ST, Senthilnathan J, Arockiarajan A (2023) Sputter-coated N-enriched mixed metal oxides (Ta_2O_5-Nb_2O_5-N) composite: a resilient solar driven photocatalyst for water purification. J Hazard Mater 452:131283
138. Geldasa FT et al (2023) Experimental and computational study of metal oxide nanoparticles for the photocatalytic degradation of organic pollutants: a review. RSC Adv 13(27):18404–18442
139. Bhuyan A, Ahmaruzzaman M (2024) Ultrasonic-assisted synthesis of highly efficient and robust metal oxide QDs immobilized-MOF-5/Ni-Co-LDH photocatalyst for sunlight-mediated degradation of multiple toxic dyes. J Alloy Compd 972:172781
140. Zhou X et al (2025) Unveiling the mechanistic synergy in Mn-doped NiO catalysts with atomic-burry structure: enhanced CO oxidation via Ni–OH and Mn bifunctionality. Sep Purif Technol 354:129330
141. Ma C-Y et al (2024) Deep insights into novel hexa-coordinated phenoxy-bridged trinuclear Ni (II) asymmetric salamo-like complexes affected by different types of combined solvents. J Mol Struct 1298:137071

142. Fernández HM et al (2024) Phase transitions in NiO during the oxygen evolution reaction assessed via electrochromic phenomena through operando UV–Vis spectroscopy. Electrochim Acta 498:144626
143. Ali A et al (2024) Synthesis, characterization, optical and dielectric properties of NiO and Zn-doped NiO nanostructures: toward advanced applications. Adv Condens Matter Phys 2024(1):6645827
144. Xiao Y et al (2024) Defect engineering of W^{6+}-doped NiO for high-performance black smart windows. Nano Res 17(4):3043–3052
145. Nabi G et al (2023) Bandgap tuning by controlled growth of Mo doped NiO nanoparticles and their functional role as excellent photocatalytic degradation agent. Inorg Chem Commun 157:111448
146. Phogat P et al (2024) Phase transition of thermally treated polyhedral nano nickel oxide with reduced band gap. MATEC Web Conf
147. Joyner NA et al (2024) The electronic structure of diatomic nickel oxide. Phys Chem Chem Phys 26(29):19646–19657
148. Tsay C-Y et al (2023) Photoresponse of solution-processed transparent heterojunction ultraviolet photodetectors composed of n-type ZTO and p-type NiO-based semiconductor thin films. Mater Chem Phys 295:127143
149. Da Y et al (2023) Single-atom Pt doping induced p-type to n-type transition in NiO nanosheets toward self-gating modulated electrocatalytic hydrogen evolution reaction. ACS Nano 17(18):18539–18547
150. Bhattu M, Acevedo R, Alhadrawi M (2024) NiO as a promising photocatalyst for wastewater purification: a review. E3S Web Conf
151. Soni V et al (2022) Photocatalytic transition-metal-oxides-based p–n heterojunction materials: synthesis, sustainable energy and environmental applications, and perspectives. J Nanostruct Chem 1–38
152. Ananthi S et al (2023) Synthesis, analysis and characterization of camellia sinensis mediated synthesis of NiO nanoparticles for ethanol gas sensor applications. Sens Actuators, B Chem 387:133742
153. Nemati SS et al (2024) An innovative sensor for sensitive determination of phenobarbital in real samples using CeO_2/CuO/MnO and CeO_2/CuO/NiO ion-sensitive field effect transistors. IEEE Sens J
154. Nie S et al (2023) Insights into selective mechanism of NiO-TiO_2 heterojunction to H_2 and CO. ACS Sens 8(11):4121–4131
155. Dharmalingam K et al (2024) Direct Z-scheme heterojunction impregnated MoS_2–NiO–CuO nanohybrid for efficient photocatalyst and dye-sensitized solar cell. Sci Rep 14(1):14518
156. Li Y et al (2024) Boosting the photocatalytic hydrogen evolution performance by fabricating the NiO/$Zn_3In_2S_6$ pn heterojunction. Appl Surf Sci 642:158622
157. Latifian P et al (2023) Fabrication of Ti-doped Bi_2S_3/NiO pn heterojunction with enhanced visible-light–driven photocatalytic activity. J Mol Liq 376:121445
158. Li J et al (2022) Facile synthesis of Z-scheme NiO/α-MoO_3 pn heterojunction for improved photocatalytic activity towards degradation of methylene blue. Arab J Chem 15(1):103513
159. Gul T et al (2023) Photodegradation of orange II dye using pn junction NiO/TiO_2 composite, and assessment of its biological activities. J Saudi Chem Soc 27(4):101654
160. Li Y et al (2024) In situ topotactic formation of the pn-type inorganic intergrowth bulk heterojunction NiO (Al)/CuO (Ni, Al) for photocatalytic CO_2 methanation. Appl Catal B 341:123282
161. Rajendran S et al (2020) Nanosized titania-nickel mixed oxide for visible light photocatalytic activity. J Mol Liq 311:113328
162. Saeed M et al (2023) Synthesis of pn NiO-ZnO heterojunction for photodegradation of crystal violet dye. Alex Eng J 65:561–574
163. Abdullah H, Gultom NS, Kuo D-H (2020) Depletion-zone size control of p-type NiO/n-type Zn (O, S) nanodiodes on high-surface-area SiO_2 nanoparticles as a strategy to significantly enhance hydrogen evolution rate. Appl Catal B 261:118223

164. Gong W et al (2023) In-situ construction of direct Z-scheme NiO/Bi2MoO6 heterostructure arrays with enhanced room temperature ether sensing properties under visible light irradiation. J Hazard Mater 458:131936
165. Fu Y et al (2019) All-solid-state Z-scheme system of NiO/CDs/BiVO4 for visible light-driven efficient overall water splitting. Chem Eng J 358:134–142
166. Ma X et al (2021) Construction of dual Z-scheme NiO/NiFe_2O_4/Fe_2O_3 photocatalyst via incomplete solid state chemical combustion reactions for organic pollutant degradation with simultaneous hydrogen production. Int J Hydrogen Energy 46(62):31659–31673
167. Wang Z et al (2022) BiOBr/NiO S-scheme heterojunction photocatalyst for CO_2 photoreduction. Solar RRL 6(1):2100587
168. Reddy NR et al (2022) Construction of various morphological ZnO-NiO S-scheme nanocomposites for photocatalytic dye degradation. Inorg Chem Commun 146:110107
169. Wang K et al (2024) Energy band modulating of NiO/BiOCl heterojunction with transition from type-II to S-scheme for enhancing photocatalytic CO_2 reduction. Chem Eng J 497:154711
170. Li Z et al (2023) Bi-functional S-scheme S-Bi_2WO_6/NiO heterojunction for photocatalytic ciprofloxacin degradation and CO_2 reduction: mechanisms and pathways. Sep Purif Technol 310:123197
171. Karimi-Maleh H et al (2020) Tuning of metal oxides photocatalytic performance using Ag nanoparticles integration. J Mol Liq 314:113588
172. Benseghier SA et al (2024) Low-cost synthesis of Ag-doped NiO thin films: their structural, optical, photocatalytic and antimicrobial properties. Inorg Chem Commun 159:111787
173. Khiar H et al (2024) (Ag–NiO)/g-C_3N_4 with enhanced photocatalytic activity for hydrogen evolution under simulated sunlight. Int J Hydrogen Energy 91:39–49
174. Liang Y et al (2019) Active {1 1 1}-faceted ultra-thin NiO single-crystalline porous nanosheets supported highly dispersed Pt nanoparticles for synergetic enhancement of gas sensing and photocatalytic performance. Appl Surf Sci 471:124–133
175. Al Boukhari J et al (2019) Synthesis, characterization, optical and magnetic properties of pure and Mn, Fe and Zn doped NiO nanoparticles. Chem Phys 516:116–124
176. Sankar S et al (2016) Photocatalytic properties of Mn-doped NiO spherical nanoparticles synthesized from sol-gel method. Optik 127(22):10727–10734
177. Sheena P et al (2019) Microstructural characterization and modified spectral response of cobalt doped NiO nanoparticles. Mater Chem Phys 229:412–420
178. Leong KH et al (2014) Synthesis of surface plasmon resonance (SPR) triggered Ag/TiO_2 photocatalyst for degradation of endocrine disturbing compounds. Appl Surf Sci 319:128–135
179. Wajid N et al (2023) Dual role of Au-NPs for Schottky effect and SPR electron injection over CdS surfaces for photocatalytic applications. Mater Chem Phys 306:128062
180. Zheng J et al (2018) Solution-processed, silver-doped NiO_x as hole transporting layer for high-efficiency inverted perovskite solar cells. ACS Appl Energy Mater 1(2):561–570
181. Hwang J-D, Lee Y-H (2024) Studies of NiO/Ag/NiO transparent conducting electrodes on NiO and ZnO Schottky diodes. Phys Chem Chem Phys 26(31):20807–20813
182. Ding Y et al (2010) Preparation and characterization of NiO–Ag nanofibers, NiO nanofibers, and porous Ag: towards the development of a highly sensitive and selective non-enzymatic glucose sensor. J Mater Chem 20(44):9918–9926
183. Chen W et al (2018) Understanding the doping effect on NiO: toward high-performance inverted perovskite solar cells. Adv Energy Mater 8(19):1703519
184. Zunger A, Malyi OI (2021) Understanding doping of quantum materials. Chem Rev 121(5):3031–3060
185. Dozzi MV, Selli E (2013) Doping TiO_2 with p-block elements: effects on photocatalytic activity. J Photochem Photobiol, C 14:13–28
186. Bie C et al (2021) Design, fabrication, and mechanism of nitrogen-doped graphene-based photocatalyst. Adv Mater 33(9):2003521
187. Marschall R, Wang L (2014) Non-metal doping of transition metal oxides for visible-light photocatalysis. Catal Today 225:111–135

188. Nasirian M et al (2018) Enhancement of photocatalytic activity of titanium dioxide using non-metal doping methods under visible light: a review. Int J Environ Sci Technol 15:2009–2032
189. Wang Xh et al (2019) Nitrogen-doped NiO nanosheet array for boosted electrocatalytic N_2 reduction. ChemCatChem 11(18):4529–4536
190. Wang C et al (2021) N-doped NiO nanosheet arrays as efficient electrocatalysts for hydrogen evolution reaction. J Electron Mater 50(9):5072–5080
191. Kou T et al (2020) Carbon doping switching on the hydrogen adsorption activity of NiO for hydrogen evolution reaction. Nat Commun 11(1):590
192. Shkir M et al (2019) A facile one-step flash combustion synthesis and characterization on C doped NiO nanostructures. Mater Sci Semicond Process 100:106–112
193. Ponnusamy P et al (2016) Structural, optical and magnetic properties of undoped NiO and Fe-doped NiO nanoparticles synthesized by wet-chemical process. Mater Charact 114:166–171
194. Soofivand F, Salavati-Niasari M (2017) Step synthesis and photocatalytic activity of NiO/graphene nanocomposite under UV and visible light as an effective photocatalyst. J Photochem Photobiol, A 337:44–53
195. Mateo D, Albero J, García H (2018) Graphene supported NiO/Ni nanoparticles as efficient photocatalyst for gas phase CO_2 reduction with hydrogen. Appl Catal B 224:563–571
196. Kang S-Z et al (2012) Composite photocatalyst containing Eosin Y and multiwalled carbon nanotubes loaded with CuO/NiO: mixed metal oxide as an active center of H_2 evolution from water. Appl Surf Sci 258(16):6029–6033
197. Nemera DJ et al (2022) Hybridization of nickel oxide nanoparticles with carbon dots and its application for antibacterial activities. Luminescence 37(6):965–970
198. Adel M et al (2022) Removal of heavy metals and dyes from wastewater using graphene oxide-based nanomaterials: a critical review. Environ Nanotechnol Monit Manage 18:100719
199. Ahmed MA, Ahmed MA, Mohamed AA (2023) Synthesis, characterization and application of chitosan/graphene oxide/copper ferrite nanocomposite for the adsorptive removal of anionic and cationic dyes from wastewater. RSC Adv 13(8):5337–5352
200. Sadhukhan S et al (2020) Synthesis of RGO/NiO nanocomposites adopting a green approach and its photocatalytic and antibacterial properties. Mater Chem Phys 247:122906
201. Noor S et al (2020) Energy harvesting for electrochemical OER and solar photocatalysis via dual functional GO/TiO_2-NiO nanocomposite. J Clean Prod 277:123280
202. Tao F, Dong Y, Yang L (2023) NiO/g-C_3N_4 quantum dots for photocatalytic CO_2 reduction. Appl Surf Sci 638:158044
203. Pebdeni AB et al (2021) Green synthesis of carbon quantum dots doped on nickel oxide nanoparticles as recyclable visible light photocatalysts for enhanced degradation of malachite green. ChemistrySelect 6(20):5034–5042
204. Noor S et al (2022) Role of conductive channels via CQDs on NiO/g-C_3N_4 Z-scheme composite as a bi-functional photocatalyst. Int J Hydrogen Energy 47(86):36517–36529
205. Ahmed MA, Mohamed AA (2023) The use of chitosan-based composites for environmental remediation: a review. Int J Biol Macromol 242:124787
206. Marand SA, Almasi H, Marand NA (2021) Chitosan-based nanocomposite films incorporated with NiO nanoparticles: physicochemical, photocatalytic and antimicrobial properties. Int J Biol Macromol 190:667–678
207. Choo CK et al (2016) Synthesis and characterization of NiO nano-spheres by templating on chitosan as a green precursor. J Am Ceram Soc 99(12):3874–3882
208. El-Shafai NM et al (2022) Advanced applications of the nanohybrid membrane of chitosan/nickel oxide for photocatalytic, electro-biosensor, energy storage, and supercapacitors. J Energy Storage 50:104626
209. El-Newehy MH et al (2024) Photocatalytic degradation of methylene blue using chitosan/NiO nanocomposite: a comprehensive exploration. Opt Mater 152:115531
210. Yasin A et al (2023) Photocatalytic and antibacterial potential of chitosan supported nickel oxide/zinc oxide composite synthesized by alcohothermal method. Water Air Soil Pollut 234(9):592

211. Wang H-J et al (2012) Film-form dye adsorbents of NiO: synthesis, biological activity and application on dye sorption. Chem Eng J 209:442–450
212. Song Z et al (2009) NiO (111) nanosheets as efficient and recyclable adsorbents for dye pollutant removal from wastewater. Nanotechnology 20(27):275707
213. Hosny NM et al (2020) Synthesis, magnetic and adsorption of dye onto the surface of NiO nanoparticles. J Mater Sci: Mater Electron 31:8413–8422
214. Al-Aoh HA (2018) Adsorption performances of nickel oxide nanoparticles (NiO NPs) towards bromophenol blue dye (BB). Desalin Water Treat 110:229–238

Study of Silver Oxide (Ag_2O) in Wastewater Treatment

Gaurav Uniyal, Chetan Kumar, Shivam Pandey, and Ritu Painuli

Abstract Nanotechnology holds great potential for improving water quality and controlling water pollution. The distinguishing features of nanosized materials lead to enhanced performances in photocatalytic activity making nanomaterials play a major role in a broad range of applications in water treatment. Nanomaterials can effectively remove noxious and organic pollutants from waste water and can efficiently kill bacteria. As an effective and efficient oxidation process, photocatalysis finds extensive use in the treatment of wastewater. The unique physical and chemical properties of metal oxide nanoparticles make them most versatile materials in the environmental remediation. Nevertheless, there is an urge for developing cleaner, green, and more sustainable methods for the synthesis of nanomaterials by employing cheaper and more ecofriendly methods as compared to the traditional synthesis methods. Therefore, the present chapter explores the studies via plant and microbial-mediated greener synthesis on the preparation of silver oxide nanoparticles. The photocatalytic mechanism as well as the varied photocatalytic applications of silver oxide nanoparticles for their potential use in waste water treatment is also highlighted.

Keywords Nanomaterials · Silver oxide nanoparticles · Green synthesis · Photocatalysis · Waste water treatment

---

G. Uniyal · S. Pandey · R. Painuli (✉)
Department of Chemistry, School of Applied and Life Sciences, Uttaranchal University, Dehradun, Uttarakhand, India
e-mail: ritsjune8.h@gmail.com

C. Kumar
School of Pharmaceutical and Population Health Informatics, Faculty of Pharmacy, DIT University, Dehradun, Uttarakhand, India

A. Ahmad et al. (eds.), *Metal Oxide Based Nanophotocatalyst for Wastewater Purification*, Smart Nanomaterials Technology,
https://doi.org/10.1007/978-981-96-7683-5_7

1 Introduction

Water scarcity has emerged as a significant challenge worldwide [1]. A major contributor to this issue is the deterioration of water quality, resulting from the contamination of water bodies by undesirable pollutants. The main classes of pollutants include nutrients, heavy metals and dyes. According to a World Bank evaluation, effluents released by the textile industry contribute to 17–20% of global water contamination [2, 3]. Textile effluents consist of colored materials known as dyes. The presence of dyes in water bodies leads to odor and discoloration (changing from colorless to colored), a decrease in photosynthesis, and reduced oxygen productivity in the aquatic ecosystem [4–7]. When water bodies reach a hypoxic state due to diminished photosynthesis, the mortality rates of macrophytes, phytoplankton, and aquatic animals increase, resulting in higher biochemical oxygen demand (BOD) and lower dissolved oxygen (DO) levels [8–10]. Furthermore, polluted water can severely impact agricultural productivity, as contaminated irrigation sources can introduce toxins into crops, posing health risks to consumers. The complications of water pollution are extensive, impacting not only human health but also ecosystems and biodiversity. Contaminated water sources can lead to the spread of waterborne disease. Additionally, heavy metals and toxic substances in water can bioaccumulate in aquatic organisms, leading to harmful effects on wildlife and disrupting food chains. The Water Quality Index (WQI) serves as a numerical measure that clearly indicates the levels of water contamination. This index is influenced by a variety of physico-chemical, biological, and toxicological factors that directly affect water quality. Key parameters defining water quality include pH, dissolved oxygen (DO), chemical oxygen demand (COD), biological oxygen demand (BOD), nutrients (particularly nitrite, nitrate, and phosphate), suspended and dissolved solids, and various metals [11, 12]. There are numerous conventional methods for wastewater treatment, for instance, reverse osmosis, solvent extraction, precipitation, membrane filtration, distillation, oxidation, adsorption, electrodialysis, etc. [13–19]. These methods often use enormous energy and can get more complicated when pollutants are transferred across different fluids, wastes, and byproducts produced during wastewater treatment.

However, nanomaterials are primarily used to overcome major water and wastewater treatment complications. A nanometer, or trillionth of a meter, is called a "nanomaterial." Nanoscience and nanotechnology (nanoparticles) indicate that owing to their surface area effect and quantum confinement effects, they may find use in wastewater treatment, environmental sensing, apparel manufacturing, and cosmetics applications. The surface effect is the ratio of surface area to volume, which is inversely related to the nanoparticles' radius. Their nanoscale size promotes higher adsorption affinity, higher reactivity, and improved catalytic efficiency owing to the large surface area and high surface energy. As the nanostructures exhibit quantum confinement effects, their optical, magnetic, and electrical conductivity characteristics differ from those of their respective bulk materials. A nanostructure's effects also enhanced its photocatalytic degradation activity [20–22]. According to Usmani et al.

[23] and Yang et al. [24], water purification process uses metal, oxide nanoparticles, carbon nanotubes, zerovalent nanoparticles, and nano-composites.

2 Metal Oxide Nanostructures

Metal oxide nanoparticles consist exclusively of metal precursors. These nanoparticles have a significant impact on various fields such as chemistry, material sciences. Thermal components can create a vast array of oxide compounds. These can take on a wide range of structural shapes with an electronic configuration that may display, semiconductor, metallic, or insulator properties. The distinct physical and chemical properties of metal oxide nanoparticles have drawn special interest due to their enormous potential for environmental remediation. The applications of metal oxide nanoparticles range from textiles, biomedical applications, heavy metals, removal, toxic gas detection, and the photocatalytic breakdown of organic pollutants [25–27]. The narrow band gap (energy difference between the valence and conduction bands) of the metal oxide nanoparticles is what makes them selective, as their band gap profile is similar to that of semiconductors. Metal oxides such Ag_2O, TiO_2, $WOSnO_2$, and $SrTiO_3$ as well as some chalcogenide meals like ZnS, CdS have been extensively studied as photocatalysts. Nano-metal oxides are extensively explored for their significant roles in photocatalysis, focusing on eliminating colors and breaking down persistent organic pollutants, particularly dyes which present a major threat to the environment. Materials like Ag_2O and CuO, which have small band gaps, are able to efficiently absorb the entire light spectrum and demonstrate excellent photocatalytic applications [4, 5, 28, 29]. These met al., oxide nanoparticles are employed in a number of wastewater treatment processes, including adsorption, photocatalysis, and antibacterial and antifungal activities. They are chemically stable, significant and doesn't have any negative effects [30]. Silver oxide nanoparticles have great potential for visible light-driven photocatalytic removal of toxic pollutants due to their narrow bandgap, which allows them to absorb the entire solar spectrum [31].

Metal oxide nanoparticles has emerged as the materials with significant photocatalytic properties under light sources. Among various physical and chemical methods, green synthesis has emerged as a cheaper and environmentally friendly method for synthesizing nanostructured materials. The present chapter mainly explains the working photocatalytic mechanism and greener synthesis routes for the silver oxide nanoparticles. It also focused on the photocatalytic degradation applications of silver oxide nanoparticles in wastewater treatment. This will help the researchers to synthesize low-cost, greener, and effective photocatalyst for the treatment of wastewater.

3 Green-Synthesis of Silver and Silver Oxide (Ag_2O) Nanoparticles

According to Shume et al. [32], Ag_2O nanoparticles can be produced by numerous methods including chemical, physical (vapor condensation, ball milling), and biosynthetic (using microorganisms and plants as a source of capping and reducing agent instead of expensive and hazardous organic solvents, chemical reducing and stabilizers). Because of their affordability and ecologically friendly nature, the green synthesis method for nanoparticles is regarded as the best substitute method when considering environmental concerns. Therefore, this section mainly focused on the green synthesis of Ag_2O NPs.

3.1 Plant-Mediated Green Synthesis

Due to the presence of certain phytochemicals in the plant extracts that functioned as both stabilizers and reducing agents, Ag-NPs formation generally produces a chemically stable biosynthesis product from $AgNO_3$ Percussor. In comparison with microbe-mediated synthesis, plant-mediated synthesis offers the advantages of speed and scalability. According to Sutan et al. [33], certain plant extracts contain phytochemicals that are unable to demonstrate their dual function as capping and reducing agents; therefore, Ag-NPs oxidize into Ag_2O-NPs. The chemically unstable product AgOH (silver hydroxide) is produced when Ag^+ (from $AgNO_3$ aqua solution) reacts with the (–OH) hydroxyl group of phytocomponents (in the plant extract). AgOH is then further oxidized or hydrolyzed into Ag_2O-NPs [34, 35]. Li et al. [36], using a medicinal plant *(lippia citriodora)* powder added with silver nitrate in a specific stoichiometric ratio, synthesized high purity, thermal stability silver oxide nanoparticles with 20 nm average particle size. Muthukumar et al. [37], used an aqueous extract of *Amaranth* to produce the husk-shaped silver oxide nanoparticles. The extract was gradually added to the silver nitrate solution while stirring constantly with a magnetic stirrer for 20 min at 350 °C. Sharma and Srivastava [35], utilized aqueous *Artocarpus hetrophyllus* plant extract for producing spherical silver oxide nanoparticles. Through gradually adding leaf extract to silver nitrate in a 1:1 ratio, silver oxide nanoparticles were hydrothermally synthesized by heating the mixture continuously for 60 min at 60 to 80 °C. Fayyadh and Jaduaa Alzubaidy [38] used similar method to synthesized antimicrobial active silver oxide nanoparticles but in this method aqueous extract of *Lawsonia inermis (Heena plant)* was used instead of *Artocarpus hetrophyllus.*

Suresh et al. [39] designed the anti-inflammatory active distorated cubic-shaped nanoparticles through mixing silver nitrate solution with *Pavetta indica linn* leaf extracts and stirring constantly for 24 h at 100–120 °C. Aiswariya and Jose [40] reported successfully synthesized silver oxide nanoparticles for photocatalytic degradation of green dye (malachite) by mixing aqueous rhizome extract of *Curcuma zanthorrhiza* (containing eleven bioactive molecules acting as reducing and capping agents) with silver nitrate under sunlight for thirty minutes with constant stirring.

With the aid of *Thunbergia mysorensis* stem and flower extract, Kokika et al. [41], demonstrated the preparation of silver oxide nanoparticles. Because flavonoids and polyphenols were present, extracts from the flowers and stems were utilized separately in this synthesis as a mediator and stabilizing agent. Aqueous stem and flower extracts were blended with silver nitrate in a dark, 60 °C environment while being constantly stirred for two to three hours to develop silver oxide nanoparticles.

3.2 Microbes-Mediated Biosynthesis Method

As microbes have many proteins and intercellular enzymes, so Algae, Fungi, Bacteria, Viruses, and Yeast were used in synthesis methods Yildirim et al. [42]. El-Sheekh et al. [43] used a microbe-mediated green method for the fabrication of quasi-spherical silver oxide nanoparticles by incubating a silver nitrate mixture with *Oscillatoria* sp. microalgal biomass for 24 h. Yildirim et al. [42] produced biological hydrogen using silver oxide nanoparticles. Green-colored microalgae Chlorella vulgaris biomass aqueous extracts were mixed with $AgNO_3$ solution, and the mixture was heated to 60 °C in the dark to produce nanoparticles with an 85 nm diameter.

Using the *endophytic fungus Fusarium oxysporum*, Islam et al. [44] successfully biofabricated water-dispersible mycogenic Ag/Ag_2O-NPs for antibacterial, antifungal, and photocatalytic properties. The fungal mycelia and Ag_2O micro powder were mixed under shaking conditions with a neutral pH at 25 °C to create the spherical-shaped nanoparticles, which ranged in size from 5 to 10 nm. Using a biosynthesis method, Dharmaraj et al. [45] synthesized silver oxide nanoparticles for cytotoxic and antibacterial functions. The supernatant culture of *Bacillus paramycoides,* which was isolated from mangrove sediment, was mixed with silver nitrate solution at a reaction temperature of 35 °C and produced spherical-shaped nanoparticles with a diameter between 25 and 70 nm. Table 1 depicts various biosynthesized silver oxide nanoparticles with different shapes and sizes.

4 Photocatalytic Activity of Ag_2O NPs and Their Composites

Ag_2O-NPs are well-known for its ability to photocatalytically degrade organic pollutants, particularly dyes, when exposed to visible light due to its selectivity, high catalytic activity, and narrow 1.46 eV band gap between the valence and conduction band [27] (Fig. 1).

Table 1 Biosynthesized Ag_2ONPs with their different size and shape

Plant/microbe source	Particle size of Ag_2ONPs with shape	References
Lippa citriodora (leaf extracts)	20 nm/spherical	[36]
Amaranath (leaf extracts)	81 nm/monodispersed husk	[37]
Artocarpus hetrophyllus (Rind extracts)	17 nm/spherical	[35]
Lawsonia inermis (leaf extracts)	39.1 nm/cubic	[38]
Pavetta indica linn (leaf extracts)	49.8 nm width of 469.2 nm/distorted square	[39]
Thunbergia mysorensis (stem & flower extracts)	15 nm,120 nm/spherical	[41]
Oscillatoria sp.	14.42–48.97 nm/Quasi-spherical	[43]
Micro algae Chlorella vulgaris	85 nm/–	[42]
Bacillus paramycoides	25–70 nm/spherical	[45]

4.1 Photocatalysis Mechanism of Ag_2ONPs in Wastewater Treatment

The high photocatalytic activity of silver oxide nanoparticles is attributed to their enhanced stability (Fig. 1). As stated by Wang et al. [46], during the photocatalysis of organic dyes, some metallic Ag partially forms on the surface of Ag_2O nanoparticles, contributing to the improved stability of the photocatalyst. Additionally, the metallic silver is involved in the photocatalytic breakdown of organic pollutants, enhancing the efficiency of the degradation process. When Ag_2O NPs absorb a suitable wavelength (energy of absorbed photon is equivalent to the energy band gap) under visible light irradiation, electrons in Ag_2O are excited from the valence band to their conduction band, developing electron–hole pairs (Eqs. 1 and 2). Due to the more negative potential of silver oxide as that of Ag^+, the photogenerated electrons are captured from the conduction band of silver oxide nanoparticles via Ag^+ lattice (Eq. 3). This results in the reduction of Ag^+ to metallic silver over the surface of silver oxide. The metallic Ag clusters that appear on the surface of the photocatalyst act as a reservoir for electrons in the process of multistep photocatalysis. When a certain quantity of metallic silver has built up, the photogenerated electrons are likely to move to the metallic silver sites and be captured by oxygen dissolved in the solution in a multistep electron transfer process (Eqs. 4 and 5). The removal of the organic pollutant, such as organic dyes, happens when the on the surface of the nano photocatalyst, excited electrons (in conduction band) then reduce to adsorbed dissolved oxygen of water bodies, forming hydrogen peroxide (H_2O_2), which is then further converted into a hydroxyl radical with a hydroxyl ion which is responsible for photo-degradation of aquatic organic pollutants Equations (6–9). Furthermore,

$Ag_2O + hv \rightarrow Ag_2O(e^-_{CB} + h^+_{VB})$ .... (i)

$Ag_2O + hv \rightarrow Ag_2O(e^-_{CB} + h^+_{VB})$ .... (ii)

$Ag_2O + e^-_{CB} \rightarrow Ag^+ + O_2^-$ (electron transfer) ...... (iii)

$Ag^+ + e^-_{CB} \rightarrow Ag$ (electron trapping) .... (iv)

$O_2^- + e^-_{CB} + 2H^+ \rightarrow H_2O_2$ .... (v)

$H_2O_2 + e^-_{CB} \rightarrow OH^- + \cdot OH$ .... (vi)

$H_2O + h^+_{VB} \rightarrow H^+ + \cdot OH$ .... (vii)

Dye + $\cdot OH \rightarrow$ degradation products .... (viii)

Dye + $h^+_{VB} \rightarrow$ degradation products .... (ix)

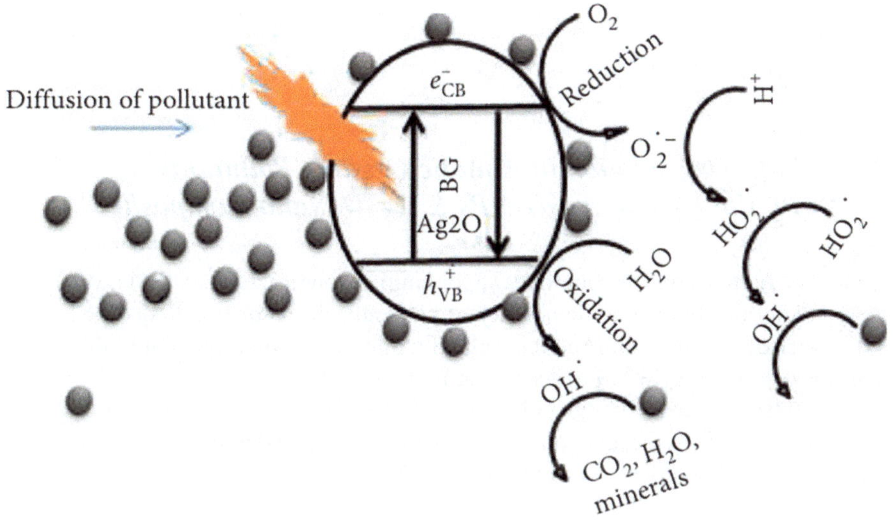

Fig. 1 Schematic representation of photocatalytic degradation (reproduced with consent from Shume et al. [32] CCBY 4.0)

the holes that form in the valence band function as strong oxidizers, oxidizing to adsorbed H_2O and OH^- into a reactive hydroxyl radical that is responsible for the photo-degradation of aquatic organic pollutants [32, 46–48].

$$Ag_2O + hv \rightarrow Ag_2O\left(e^-_{CB} + h^+_{VB}\right) \tag{1}$$

$$Ag_2O + hv \rightarrow Ag_2O\left(e^-_{CB} + h^+_{VB}\right) \tag{2}$$

$$Ag_2O + e^-_{CB} \rightarrow Ag^+ + O_2(\text{electron transfer}) \quad (3)$$

$$Ag^+ + e^-_{CB} \rightarrow Ag(\text{electron trapping}) \quad (4)$$

$$O_2^- + e^-_{CB} + 2H^+ \rightarrow H_2O_2 \quad (5)$$

$$H_2O_2 + e^-_{CB} \rightarrow OH^- + \cdot OH \quad (6)$$

$$H_2O_2 + h^+_{VB} \rightarrow H^+ + \cdot OH \quad (7)$$

$$\text{Dye} + \cdot OH \rightarrow \text{degradation products} \quad (8)$$

$$\text{Dye} + h^+_{VB} \rightarrow \text{degradation products} \quad (9)$$

4.2 *Photo-Degradation of Aquatic Organic Pollutants with the Help of Ag_2O-NPs & Ag_2O-Nano-Composite*

Under UV light irradiation (mercury vapor lamp as source), Li et al. [36] investigated the potential of photocatalytic activity of biosynthesized spherical Ag_2O-NPs with a diameter of 20 nm against acid orange 8 (AO8) as a target dye. Only 40% of the dye was degraded in 150 min, according to their analysis.

AO8 (also known as orange G/Naphthol orange) is a water-soluble dye and toxic due to their carcinogenic and mutagenic nature. Using powdered *Tridax* and *Centella asiatica* plants, Rashmi et al. produced Ag_2O-NPs. One noteworthy finding under UV irradiation was that the *Tridax*-prepared nanoparticles showed greater photocatalytic activity (70% degradation within 140 min) against AO8 than the *Centella asiatica*-prepared NPs, which showed 40% degradation in the same time frame. Ag_2O-NPs produced by Aiswariya and Jose [40] demonstrated their capacity to photo-degrade the hazardous green dye (malachite). The maximum absorption wavelength for Malachite green dye is 618 nm. In the course of analyzing photocatalytic activity, green dye treated with Ag_2O-NPs decolorized and degraded more quickly in the presence of sunlight (94.7% degradation in 4 h) than when exposed to UV irradiation. Only 23.83% degradation was observed in 4 h under UV irrigation. Shah et al. [49] synthesized silver oxide nanoparticles to photo-degrade Methylthioninium chloride (commonly known by the name of methyl blue dye). The average size of the prepared nanoparticles was 38.29 nm and degraded 97.78% of the methyl blue in 180 min when exposed to radiation.

Similarly, in the presence of the reducing agent $NaBH_4$, El-Ghmari et al. [50] also employed Ag_2O-NPs to degrade methyl blue dye as a target aquatic organic pollutant. Ag_2O-NPs were produced using the plant-mediated green method with extracts from the wild plant *Herniaria hirsute (H. hirsute).* Within 35 min, 88.6% photo-degradation was seen at room temperature under white light irradiation.

Laouini et al. [51] used the heterogeneous Ag/Ag2O-NPs for the degradation of methylene blue dye solution similarly in the presence of $NaBH_4$. According to their analysis, 84.5% degradation of dye was observed after 50 min.

Using a green *amaranth* leaf extract, Muthukumar et al. also synthesized Ag_2O-NPs and investigated their potential for photocatalytic degradation activity against caffeine (as a water pollutant). After being exposed to compact fluorescent lamp illumination for 15 h, 99% of the caffeine was found to have been degraded.

Jiang et al. [52], synthesized silver oxide nanoclusters with an average particle size of 8.33 nm and a surface area of 0.4726 m^2/g. The prepared nanoclusters completely degraded methylene orange dye in two minutes when exposed to UV or visible light.

Haq et al. [53] evaluated the photocatalytic activity of Ag_2O nanoparticles (made using 2 dimethyl amino ethanol as a reducing agent) under solar radiation using an aqueous solution of rhodamine 6G dye ($\lambda_{max} = 525$ nm). Within 30 min of the photocatalytic investigation, 98.56 percent of the rhodamine 6G dye was degraded.

The two naphthalene rings that make up Cango dye (CR) are joined by two phenyl rings, and their terminals are joined by sulfonic and amino groups. Malignant bladder growth is caused by CR dye, which is naturally carcinogenic. The photocatalytic dye degradation activity of heterogeneous Ag/Ag_2O NPs, which were made using aqueous extracts of *Phoenix dactylifera* L. leaves, was assessed by Laouini et al. [51] using CR dye. Characteristics showed that the generated heterogeneous Ag/Ag_2O NPs had an oval or spherical form and ranged in size from 28 to 39 nm. After 60 min, 80% photo-degradation of the CR dye was seen in the presence of $NaBH_4$. Warsi et al. [54] synthesized Ag_2O NPs and their nano-composites using WO_3, and analyzed their photocatalytic dye degradation activity using methylene blue and rhodamine B dye. Ag_2O NPs and Ag_2O/WO_3 nano-composites reported degradation values of 86.1% and 90.8% respectively. During their analysis, the Rhodamine B dye degradation values for Ag_2O NPs and Ag_2O/WO_3 nano-composites were 80.7% and 87.25%, respectively.

Durán-Álvarez et al. [55] synthesized the Ag_2O/TiO_2 hetero nano-composite for removal of atenolol (ATL) and carbamazepine (CBZ) from tap water and other secondary effluents under the sunlight. The Ag_2O/TiO_2 hetero nano-composite removes both pharmaceuticals from tap water within 180 min. By using a simultaneous adsorption and photocatalytic degradation mechanism, Al-Jubouri et al. [56] developed an untried efficient nano-composite as a photocatalyst to remove ranitidine (RANI) from aqueous solutions. The silver oxide-coated zeolite (Ag_2O@YZeo) exhibited the highest photocatalytic activity to eliminate RANI from aqueous solution with a degradation of 97.51% in 75 min.

Zia and Riaz [57] produced a polythiophene-sensitized Ag-Ag_2O heterogeneous nano photocatalyst to extract the pharmaceutical compound paracetamol (PCM) from an aqueous solution. Microwave-assisted green synthesis of Ag-Ag_2O nanoparticles was used to synthesize polythiophene-sensitized Ag-Ag_2O heterogeneous nanoparticles using *Osmium sanctum* as a plant source for leaf extracts. The degradation rate of 80% within 20 min of being exposed to microwave radiation was observed for PCM. Table 2 demonstrates various works using silver oxide nanoparticles for the photocatalytic degradation of dyes.

5 Conclusion and Future Perspectives

Metal oxide nanoparticles have demonstrated encouraging outcomes in a variety of applications and are favoured for the absorption of organic contaminants and heavy metals. Therefore, the present chapter highlights the various photocatalytic degradation activities of nanosilver oxide towards pharmaceutical compounds and dyes, which makes them a promising material for wastewater treatment. It is classified as a p-type semiconductor due to its narrow bandgap (1.1 ~ 1.4 eV) and good optical properties. Researchers are more concerned about the environment in recent times, so they prefer green synthesis or biosynthesis over other synthesis approaches. Many green syntheses of silver oxide nanostructures using microbes (*Oscillatoria* sp., Micro algae *Chlorella vulgaris* & *Bacillus paramycoides*) and plants (*Lippa citriodora*, *Artocarpus hetrophyllus*, *Thunbergia mysorensis*, and *Lawsonia inermis*, to name a few) are covered in the literature. All of the photocatalytic degradation of dyes and pharmaceutical compounds, including Acid Orange 8, Methyl Blue, Cango Red, Methyl Orange, Rhodamine B, Green dye (malachite), Azo dye, Ranitidine (RANI), Atenolol (ATL), Carbamazepine (CBZ), and Paracetamol (PCM), has been covered above for wastewater purification using nano-silver oxide semiconductor photocatalysts. Metal oxide nanoparticles, like other nanoparticles, have several applications, but because of their unregulated use and discharge into the environment, there are still certain health hazards. To increase the effectiveness and environmental friendliness of nanoparticle utilization, these issues need to be resolved. Further, to improve the photocatalytic performance of silver oxide, future studies should concentrate on scalability, toxicity evaluation, and the investigation of composite materials.

Table 2 Photocatalytic degradation of organic pollutants (dyes) by silver oxide nanoparticles for wastewater treatment

Nanostructures	Synthesis method	Degraded organic pollutants	Photocatalytic degradation efficiency	References
Ag_2O nanoparticles	Plant-mediated biosynthesis method (leaf extract source-*Lippa citriodora)*	AO8 (Acid Orange 8)	40% degrade within 150 min	[36]
Ag_2O nanoparticles	Plant-mediated biosynthesis method Involutes synthesized same nanoparticles from two different plant source-*Tridax* and *Centella asiatica* plants	Naphthol orange/Orange G/AO8 (Acid Orange 8)	In the same time frame, Tridax-prepared Ag_2O nanoparticles display 30% more efficiency than Centella asiatica prepared nanoparticles that was 70% degradation within 140 min	[48]
Ag_2O nanoparticles	Plant-mediated biosynthesis method Phytochemicals source-*Curcuma zanthorrhiza roxb.* (*Rhizome* extracts)	Water-soluble green dye (Malachite)	In the same time interval 4 h, Ag_2O NPs shows more photocatalytic degradation efficiency (94.7%) in the presence of sunlight irradiation than UV-irradiation (23.83%)	[40]
Ag_2O nanoparticles	Plant-mediated biosynthesis method (Plant source-*Paeonia emodi*)	Methylene (methyl) blue dye/Swiss blue	97.8% degradation was recorded within 180 min (3 h)	[49]

(continued)

Table 2 (continued)

Nanostructures	Synthesis method	Degraded organic pollutants	Photocatalytic degradation efficiency	References
Ag_2O nanoparticles	Plant-mediated biosynthesis method (wild plant *Herniaria hirsute*) (*H. hirsute* used as extract source)	Methyl blue dye	In the presence of reducing agent $NaBH_4$, 97.8% degradation of dye within 35 min was recorded at the room temperature under white irradiation	[50]
Heterogeneous Ag/ Ag_2O-NPs	Green synthesis via Aq. Leaf extracts of *phoenix dactylifera*	Azo dye (methyl blue dye)	In the presence of $NaBH_4$, 84.5% degradation of dye was observed after 50 min	[51]
Silver oxide-coated zeolite (Ag_2O@Y Zeo) composite	Thermal method	Ranitidine (RANI)	Degradation of RANI 97.51% in 75 min	[56]
Ag_2O NPs	Ion exchange method	Methyl orange dye	100% degradation recorded within 2 min under the UV-visible irradiation	[52]
Ag_2O NPs	Chemical reduction method (by using 2-dimethyl amino ethanol as a reducing agent to reduced silver nitrate)	Rhodamine 6G dye	98.56% degradation of dye recorded within 30 min	[53]
Heterogeneous Ag/ Ag_2O NPs	Green synthesis via Aq. Leaf extracts of *phoenix dactylifera*	Cango Red (CR) dye	In the presence of $NaBH_4$, 80% degradation of CR recorded after 1 h under irradiation	[51]

(continued)

Table 2 (continued)

Nanostructures	Synthesis method	Degraded organic pollutants	Photocatalytic degradation efficiency	References
Ag_2O/WO_3 nano-composites	–	Methylene Blue dye, Rhodamine B dye	90.8% and 87.25% degradation recorded for methyl blue and Rhodamine B	[54]
Ag_2O NPs	–	Methylene Blue dye, Rhodamine B dye	86.1% and 80.7% degradation recorded for methyl blue and Rhodamine B	[54]
Ag_2O/TiO_2 Hetero nano-composite	–	Atenolol (ATL) and Carbamazepine (CBZ)	Both pharmaceuticals completely remove from tap water within 180 min via photo-catalytic degradation	Durán-Álvarez et al. [55]

References

1. Tzanakakis VA, Paranychianakis NV, Angelakis AN (2020) Water supply and water scarcity. Water 12:1–16
2. Saini RD (2017) Textile organic dyes: polluting effects and elimination methods from textile waste water. Int J Chem Eng Res 9:121–136
3. Bina R, Maheshwari RY, Yadav RK, Pareek D, Sharma AS (2013) Resolution to provide safe drinking water for sustainability of future perspectives. Res J Chem Environ Sci 1:50–54
4. Yadav M, Lawrence RS, Jalees M, Lawrence K, Tiwari A, Shan M (2018) Biological synthesis and antibacterial activity of silver oxide nanoparticle prepared from Carica papaya root extract. Int J Pure Appl Biosci 6:1632–1639
5. Kim M-J, Kim S-H, Park H-Y, Huh Y-D (2011) Morphological evolution of Ag_2O microstructures from cubes to octapods and their antibacterial activities. Bull Korean Chem Soc 32:3793–3795
6. Manikandan V, Velmurugan P, Park JH, Chang WS, Park YJ, Jayanthi P, Oh BT (2017) Green synthesis of silver oxide nanoparticles and its antibacterial activity against dental pathogens. Biotech 7:1–9
7. Painuli R, Joshi P, Kumar D (2020) Aluminon functionalized silver nanoparticles for the colorimetric detection of aqueous Al(III). Plasmonics 239:1–12
8. Ahmed AI (1995) Reactive dyes development: a review. Textile Dyer Printer 28:19–24
9. Ahmaruzzaman M, Gupta VK (2011) Rice husk and its ash as low-cost adsorbents in water and wastewater treatment. Ind Eng Chem Res 50:13589–13613
10. Annuar MSM, Adnan S, Vikineswary S, Chisti Y (2009) Kinetics and energetics of azo dye decolorization by Pycnoporus sanguineus. Water Air Soil Pollut 202:179–188

11. DeCicco BT (1979) Removal of eutrophic nutrients from wastewater and their bioconversion to bacterial single cell protein for animal feed supplements, Phase II. University of District of Columbia, Water Resources Research Center
12. Larsdotter K (2006) Microalgae for phosphorus removal from wastewater in a Nordic climate. Doctoral dissertation, KTH
13. Mousset E, Doudrick K (2020) A review of electrochemical reduction processes to treat oxidized contaminants in water. Curr Opin Electrochem 22:221–227
14. Saleh TA, Gupta VK (2012) Column with CNT/magnesium oxide composite for lead (II) removal from water. Environ Sci Pollut Res 19:1224–1228
15. Zazouli MA, Kalankesh LR (2017) Removal of precursors and disinfection by-products (DBPs) by membrane filtration from water: a review. J Environ Health Sci Eng 15:1–25
16. Arar Ö, Yüksel Ü, Kabay N, Yüksel M (2014) Various applications of electrode ionization (EDI) method for water treatment-a short review. Desalination 342:16–22
17. Azimi A, Azari A, Rezakazemi M, Ansarpour M (2017) Removal of heavy metals from industrial wastewaters: a review. Chem Bio Eng Rev 4:37–59
18. Yagub MT, Sen TK, Afroze S, Ang HM (2014) Dye and its removal from aqueous solution by adsorption: a review. Adv Coll Interface Sci 209:172–184
19. Zularisam AW, Ismail AF, Salim R (2006) Behaviors of natural organic matter in membrane filtration for surface water treatment-a review. Desalination 194:211–231
20. Bavasso I, Vilardi G, Stoller M, Chianese A, Di Palma L (2016) Perspectives in nanotechnology based innovative applications for the environment. Chem Eng Trans 47:55–60
21. Khan I, Saeed K, Khan I (2019) Nanoparticles: properties, applications and toxicities. Arab J Chem 12:908–931
22. Vilardi G, Stoller M, Verdone N, Di Palma L (2017) Production of nano Zero Valent Iron particles by means of a spinning disk reactor. Chem Eng Trans 57:751–756
23. Amil Usmani M, Khan IH, Bhat AS, Pillai R, Ahmad NK, Mohamad Haafiz M, Oves M (2017) Current trend in the application of nanoparticles for waste water treatment and purification: a review. Curr Org Synth 14:206–226
24. Yang L, Meng F, Qu X, Xia L, Huang F, Qin S, Liu H (2019) Multiple-twinned silver nanoparticles supported on mesoporous graphene with enhanced antibacterial activity. Carbon 155:397–402
25. Chaudhary M, Kumar C, Raghav PS, Painuli R (2024) Sunlight-driven photocatalytic degradation of industrial dyes using *Withania somnifera* decorated MnO_2 nanoparticles. Discover Nano 19:206–211
26. Nair GM, Sajini T, Mathew B (2022) Advanced green approaches for metal and metal oxide nanoparticles synthesis and their environmental applications. Talanta 5:1–11
27. Danish MSS, Estrella LL, Alemaida IMA, Lisin A, Moiseev N, Ahmadi M, Nazari M, Wali M, Zaheb H, Senjyu TP (2021) Photocatalytic applications of metal oxides for sustainable environmental remediation. Metals 11:1–25
28. Manikandan V (2017) Green synthesis of silver oxide nanoparticles and its antibacterial activity against dental pathogens. Biotech 7:1–9
29. Rokade AA, Patil MP, Yoo SI, Lee WK, Park SS (2016) Pure green chemical approach for synthesis of Ag_2O nanoparticles. Green Chem Lett Rev 9:216–222
30. Naseem T, Durrani T (2021) The role of some important metal oxide nanoparticles for wastewater and antibacterial applications, A review. Environ Chem Ecotoxicol 3:59–75
31. Danish MSS, Estrella-Pajulas LL, Alemaida IM, Grilli ML, Mikhaylov A, Senjyu T (2022) Green synthesis of silver oxide nanoparticles for photocatalytic environmental remediation and biomedical applications. Metals 12:1–20
32. Shume WM, Murthy HA, Zereffa EA (2020) A review on synthesis and characterization of Ag_2O nanoparticles for photocatalytic applications. J Chem 1:1–15
33. Şuţan NA, Fierăscu I, Şuţan C, Soare LC, Neblea AM, Somoghi R, Fierăscu RC (2021) In vitro Mito depressive activity of Phyto fabricated silver oxide nanoparticles (Ag_2O-NPs) by leaves extract of Helleborus odrous Waldst. & Kit. ex Willd. Mater Lett 286:1–29

34. Ravichandran S, PaluriV KG, Loganathan K, Kokati Venkata BR (2016) A novel approach for the biosynthesis of silver oxide nanoparticles using aqueous leaf extract of Callistemon lanceolatus (Myrtaceae) and their therapeutic potential. J Exp Nanosci 11:445–458
35. Archana, Sharma SN, Srivastava R (2020) Silver oxide nanoparticles synthesized by green method from Artocarpus Hetrophyllus for antibacterial and antimicrobial applications. Mater Today Proc 28: 332–336
36. Li R, Chen Z, Ren N, Wang Y, Wang Y, Yu F (2019) Biosynthesis of silver oxide nanoparticles and their photocatalytic and antimicrobial activity evaluation for wound healing applications in nursing care. J Photochem Photobiol, B 199:1–19
37. Muthukumar H, Palanirajan SK, Shanmugam MK, Arivalagan P, Gummadi SN (2022) Photocatalytic degradation of caffeine and E. coli inactivation using silver oxide nanoparticles obtained by a facile green co-reduction method. Clean Technol Environ Policy 24:1087–1098
38. Fayyadh AA, Jaduaa Alzubaidy MH (2021) Green-synthesis of Ag_2O nanoparticles for antimicrobial assays. J Mech Behav Mater 30:228–236
39. Suresh S, Pradheesh G, Ramani VA (2018) Biosynthesis and characterization of CuO, MgO and Ag_2O nanoparticles, anti-inflammatory activity and phytochemical screening of the ethanolic extract of the medicinal plant Pavetta indica Linn. J Pharmacogn Phytochem 7:1984–1990
40. Aiswariya KS, Jose V (2022) Bioactive molecules coated silver oxide nanoparticle synthesis from *Curcuma zanthorrhiza* and HR-LCMS monitored validation of its photocatalytic potency towards malachite green degradation. J Cluster Sci 33:1685–1696
41. Kokila NR, Mahesh B, Roopa KP, Prasad BD, Raj K, Manjula SN, Ramu R (2022) Thunbergia mysorensis mediated nano silver oxide for enhanced antibacterial, antioxidant, anticancer potential and in vitro hemolysis evaluation. J Mol Struct 1255:1–32
42. Yildirim O, Tunay D, Ozkaya B, Demir A (2022) Effect of Green synthesized silver oxide nanoparticle on biological hydrogen production. Int J Hydrogen Energy 47:19517–19525
43. El-Sheekh MM, Hassan LH, Morsi HH (2021) Assessment of the in vitro anticancer activities of cyanobacteria mediated silver oxide and gold nanoparticles in human colon CaCo-2 and cervical HeLa cells. Environ Nanotechnol Monit Manage 16:1–55
44. Islam SN, Naqvi SMA, Parveen S, Ahmad A (2021) Application of mycogenic silver/silver oxide nanoparticles in electrochemical glucose sensing; alongside their catalytic and antimicrobial activity. Biotechnology 11:3–42
45. Dharmaraj D, Krishnamoorthy M, Rajendran K, Karuppiah K, Annamalai J, Durairaj K, Ethiraj K (2021) Antibacterial and cytotoxicity activities of biosynthesized silver oxide (Ag_2O) nanoparticles using Bacillus paramycoides. J Drug Deliv Sci Technol 61:1–36
46. Wang X, Li S, Yu H, Yu J, Li S (2011) Ag_2O as a new visible-light photocatalyst: self-stability and high photocatalytic activity. Chem Eur J 17:7777–7780
47. Huang K, Li C, Zheng Y, Wang L, Wang W, Meng X (2020) Recent advances on silver-based photocatalysis: photo corrosion inhibition, visible-light responsivity enhancement, and charges separation acceleration. Sep Purif Technol 283:1–20
48. Rashmi BN, Harlapur SF, Avinash B, Ravikumar CR, Nagaswarupa HP, Anil MR, Gurushantha K, Santosh MS (2020) Facile green synthesis of silver oxide nanoparticles and their electrochemical, photocatalytic and biological studies. Inorg Chem Commun 111:1–15
49. Shah A, Haq S, Rehman W, Waseem M, Shoukat S, Rehman MU (2019) Photocatalytic and antibacterial activities of Paeonia emodi mediated silver oxide nanoparticles. Mater Res Exp 6:1–18
50. El-Ghmari B, Farah H, Ech-Chahad A (2021) A new approach for the green biosynthesis of Silver Oxide nanoparticles Ag_2O, characterization and catalytic application. Bull Chem Reaction Eng Catalysis 16:651–660
51. Laouini SE, Bouafia A, Soldatov AV, Algarni H, Tedjani ML, Ali GA, Barhoum A (2021) Green synthesized of Ag/Ag_2O nanoparticles using aqueous leaves extracts of Phoenix dactylifera L. and their azo dye photodegradation. Membranes 11:1–14
52. Jiang W, Wang X, Wu Z, Yue X, Yuan S, Lu H, Liang B (2015) Silver oxide as superb and stable photocatalyst under visible and near-infrared light irradiation and its photocatalytic mechanism. Ind Eng Chem Res 54:832–841

53. Haq S, Rehman W, Waseem M, Meynen V, Awan SU, Saeed S, Iqbal N (2018) Fabrication of pure and moxifloxacin functionalized silver oxide nanoparticles for photocatalytic and antimicrobial activity. J Photochem Photobiol, B 186:116–124
54. Warsi AZ, Ahmad T, Aziz F, Warsi MF, Munir S, Agboola PO, Shakir I (2024) Synthesis and characterisation of WO_3 and Ag_2O nanoparticles and their nanocomposite for photocatalytic degradation of dyes. Int J Environ Anal Chem 104:2755–2775
55. Durán-Álvarez JC, Cortés-Lagunes S, Mahjoub O, Serrano-Lázaro A, Garduño-Jiménez A, Zanella R (2024) Tapping the Tunisian sunlight's potential to remove pharmaceuticals in tap water and secondary effluents: a comparison of Ag_2O/TiO_2 and BiOI photocatalysts and toxicological insights. Sep Purif Technol 335:1–26
56. Al-Jubouri SM, Sabbar HA, Khudhair EM, Ammar SH, Al Batty S, Khudhair SY, Mahdi AS (2023) Silver oxide-zeolite for removal of an emerging contaminant by simultaneous adsorption-photocatalytic degradation under simulated sunlight irradiation. J Photochem Photobiol, A 442:1–14
57. Zia J, Riaz U (2020) Microwave-assisted degradation of paracetamol drug using polythiophene-sensitized Ag–Ag_2O heterogeneous photocatalyst derived from plant extract. ACS Omega 5:16386–16394

Magnetic Hematite (α-Fe_2O_3) Photocatalyst in Wastewater Treatment

Rozita M. Moattari and Mostafa Mahinroosta

Abstract Hematite (α-Fe_2O_3) has gained significant attention as a photocatalyst for wastewater treatment, particularly in the degradation of pharmaceuticals and organic pollutants. Its unique properties, including a suitable band gap of approximately 2.0–2.2 eV, allow for effective utilization of visible light, making it an attractive candidate for sustainable environmental applications. Photocatalytic processes involving hematite generate reactive oxygen species that can efficiently oxidize organic contaminants, converting them into less harmful substances. Recent studies have demonstrated hematite's effectiveness in degrading various pharmaceutical compounds, achieving notable removal efficiencies under visible light irradiation. However, challenges, such as low charge carrier mobility and high recombination rates of photogenerated electron–hole pairs, limit its overall performance. To enhance its photocatalytic activity, researchers are exploring strategies, such as doping, creating heterojunctions, and optimizing particle morphology to improve charge separation and light absorption. This chapter highlights the current advancements in hematite-based photocatalysts for wastewater treatment and discusses future directions for research aimed at overcoming existing limitations, ultimately contributing to the development of efficient and eco-friendly solutions for water purification.

Keywords Hematite · Fe_2O_3 · Photocatalyst · Water treatment · Wastewater treatment

R. M. Moattari
Center of Excellence for Membrane Science and Technology, Department of Chemical, Petroleum and Gas Engineering, Iran University of Science and Technology, Tehran, Iran

M. Mahinroosta (✉)
Department of Energy, Materials and Energy Research Center, Karaj, Iran
e-mail: mahinroosta2010@gmail.com

A. Ahmad et al. (eds.), *Metal Oxide Based Nanophotocatalyst for Wastewater Purification*, Smart Nanomaterials Technology,
https://doi.org/10.1007/978-981-96-7683-5_8

1 Introduction

Wastewater treatment is a critical process that aims to remove pollutants from sewage and industrial effluents to protect human health and the environment. With increasing urbanization and increasing industrial activity, the amount of wastewater generated has increased, requiring effective treatment solutions to prevent water pollution and reduce the health risks associated with untreated wastewater.

The treatment of wastewater is essential for maintaining ecological balance and public health. Untreated wastewater can lead to serious environmental problems, including algal blooms and eutrophication, which degrade water quality and harm aquatic life. In addition, it poses significant health risks as it facilitates the spread of water-borne diseases, such as cholera, typhoid, and hepatitis [12, 50]. Therefore, effective wastewater treatment is crucial for sustainable water resource management. Wastewater treatment typically involves several sequential processes divided into three main stages: primary, secondary, and tertiary treatment.

Primary treatment utilizes physical methods like sedimentation and flotation to separate and eliminate large solid particles from wastewater. The aim is to reduce the burden on the subsequent treatment stages by eliminating settleable organic and inorganic solids [14, 17].

Secondary treatment uses biological processes to further break down organic matter. Common techniques involve activated sludge systems, trickling filters, and biofilm reactors. Microorganisms play a crucial role in breaking down organic pollutants into simpler compounds [17, 50]. The activated sludge process is particularly popular due to its efficiency in reducing biochemical oxygen demand (BOD) and suspended solids [50].

Tertiary treatment is an advanced stage aimed at removing remaining impurities that were not eliminated during primary and secondary treatment. Techniques may include chemical coagulation, filtration, disinfection (with chlorine or ultraviolet (UV) light), photocatalysis, and advanced oxidation processes. Tertiary treatment improves water quality and makes it suitable for discharge into natural water bodies or for reuse in irrigation and industrial applications [12].

Photocatalysis has emerged as a transformative technology in wastewater treatment, offering innovative solutions to address the growing challenges of water pollution. This process uses light energy to activate photocatalysts that facilitate the breakdown of various pollutants, including stubborn organic pollutants and pathogens. The importance of photocatalysis in wastewater treatment can be understood through its unique advantages, mechanisms, and applications [5, 10, 21].

One of the main reasons for using photocatalysis in wastewater treatment is its effectiveness against persistent organic pollutants, which are difficult to break down using conventional methods. These pollutants, which include pharmaceuticals, dyes, and pesticides, pose significant risks to human health and aquatic ecosystems. Photocatalysis enables the complete mineralization of these compounds into harmless byproducts, such as carbon dioxide and water, providing a comprehensive solution to water pollution problems [5, 33].

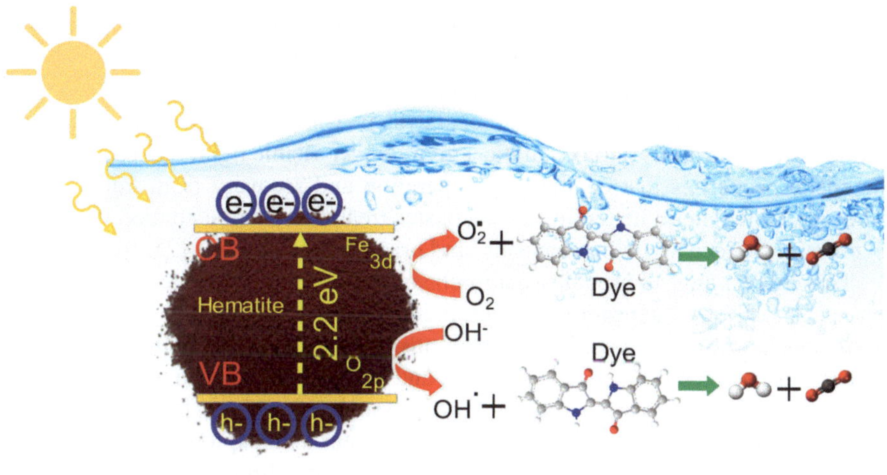

Fig. 1 Photocatalytic degradation of pollutants

The photocatalytic process typically involves semiconductors, such as titanium dioxide (TiO_2) or zinc oxide (ZnO), which generate electron–hole pairs upon light exposure. These charge carriers take part in redox reactions, leading to the production of reactive oxygen species (ROS), such as superoxide anions (O_2^-) and hydroxyl radicals ($^{\cdot}OH$). These ROS are highly reactive and effectively oxidize organic pollutants and break them down into simpler, non-toxic substances [21, 42] (see Fig. 1).

This mechanism not only increases degradation efficiency but also minimizes the formation of secondary pollutants, a common disadvantage of conventional chemical treatments. Photocatalysis is inherently an environmentally friendly technology. It operates under mild conditions–typically ambient temperature and pressure—and can use solar energy as a light source. This decreases reliance on fossil fuels and lowers greenhouse gas emissions linked to wastewater treatment operations. Furthermore, photocatalytic systems can be designed to minimize waste generation, thereby meeting sustainable development goals [10, 33].

Integrating photocatalysis into existing wastewater treatment systems can significantly increase overall efficiency while promoting ecological balance. Photocatalysis is versatile and can be used to treat a wide range of pollutants, including not only organic compounds but also microorganisms. The ability to disinfect pathogens without the use of harmful chemicals makes photocatalytic disinfection an attractive alternative to traditional methods, such as chlorination. Research has demonstrated that photocatalytic processes can effectively inactivate various pathogens, including bacteria and viruses, thereby improving water safety and quality [5, 33].

The cost-effectiveness of photocatalysis is another important aspect that contributes to its relevance in wastewater treatment. While the initial investment in photocatalytic materials may be higher than traditional treatments, long-term operating costs may be lower due to lower chemical consumption and maintenance requirements. Furthermore, advances in photocatalyst synthesis—such as the development of biodegradable or low-cost materials—are making this technology increasingly accessible for widespread application [5, 33].

Magnetic hematite (α-Fe_2O_3) has gained considerable attention due to its unique properties and capabilities as a photocatalyst in wastewater treatment. This iron oxide not only exhibits photocatalytic activity under visible light, but also possesses magnetic properties that enable easy separation and recovery from treated water. The role of magnetic hematite in wastewater treatment can be studied through its synthesis, photocatalytic mechanisms, and practical applications [25, 61].

This chapter aims to provide a comprehensive overview of magnetic α-Fe_2O_3 as a photocatalyst for wastewater treatment, focusing on its unique properties, photocatalytic mechanisms, and practical applications. It seeks to elucidate the effectiveness of hematite in degrading pharmaceuticals and organic pollutants under visible light irradiation while addressing the challenges associated with its low charge carrier mobility and high recombination rates. The chapter also explores recent advancements in enhancing hematite's photocatalytic activity through various strategies, such as doping and heterojunction formation, ultimately contributing to the development of efficient and sustainable solutions for water purification. In addition, this chapter presents novel insights into the dual functionality of magnetic hematite as both a photocatalyst and a magnetic material, facilitating easy recovery post-treatment. It highlights cutting-edge research that demonstrates significant removal efficiencies of persistent organic pollutants using hematite under visible light, a critical advancement in photocatalytic wastewater treatment. By discussing innovative approaches to improve charge separation and light absorption, this work contributes to the growing body of knowledge on eco-friendly wastewater treatment technologies, paving the way for future research directions that can address existing limitations in photocatalytic processes.

2 Properties of Magnetic Hematite

2.1 *Chemical Composition and Structure*

Magnetic hematite is a widely studied iron oxide known for its unique chemical composition, crystalline structure, and significant applications in photocatalysis, particularly in wastewater treatment. Understanding its chemical composition and structural characteristics is essential for optimizing its performance in various environmental applications.

The chemical formula of magnetic hematite is α-Fe_2O_3, which indicates that it is composed of iron (Fe) and oxygen (O) in a certain stoichiometric ratio. In this compound, iron is predominantly in the +3 oxidation state (Fe^{3+}), which is crucial for its photocatalytic properties. The hematite structure can be described as a close-packed arrangement of oxygen anions, with iron cations occupying octahedral interstitials. Each iron ion is coordinated by six oxygen ions and forms distorted

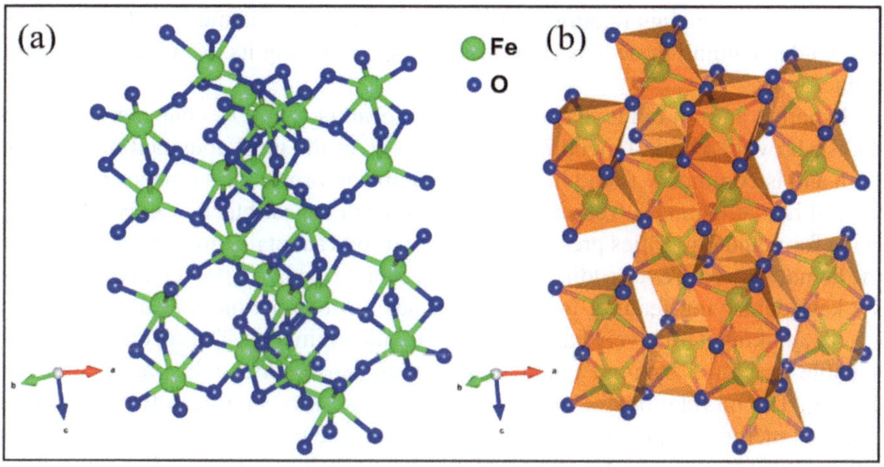

Fig. 2 **a** Ball-and-stick model of α-Fe_2O_3 crystal structure and **b** polyhedral model of α-Fe_2O_3 crystal structure [43]

octahedra, which contribute to the material's unique electronic and optical properties [43, 47].

The crystal structure of magnetic hematite is characterized by a rhombohedral lattice system with the space group R-3c. The lattice parameters are typically given as $a = b = 5.035$ and $c = 13.748$, with angles $\alpha = \beta = 90°$ and $\gamma = 120°$. This arrangement results in a three-dimensional network in which Fe^{3+} ions are coordinated by oxygen ions in a distorted octahedral geometry. The arrangement of these ions results in specific crystallographic planes, such as (012), (104), (110), and others, which can be identified by X-ray diffraction (XRD) analysis [22, 55]. Figure 2a, b illustrates the ball-and-stick model and the polyhedral representation of the rhombohedral crystal structure of α-Fe_2O_3, highlighting the arrangement of Fe^{3+} and O^{2-} ions within the compound. The structure is characterized as corundum and crystallizes in the trigonal R-3c space group. In this configuration, each Fe^{3+} ion is coordinated with six equivalent O^{2-} ions, resulting in a distorted mixture of face-sharing, edge-sharing, and corner-sharing FeO_6 octahedra. The tilting angles of the corner-sharing octahedra vary between 48° and 60°. Among the Fe–O bonds, three exhibit shorter lengths of 1.98 Å, while the other three are longer at 2.13 Å. Additionally, each O^{2-} ion is coordinated to four equivalent Fe^{3+} ions, forming a trigonal pyramid with distorted edges and corner-sharing OFe_4 units [58].

2.2 *Physical Properties*

Magnetic hematite (α-Fe_2O_3) is a fascinating material with unique physical properties, morphology, size characteristics, and magnetic behavior that make it particularly suitable for various applications, including wastewater treatment and photocatalysis. Understanding these aspects is crucial for optimizing its use in environmental technologies.

Hematite, an iron oxide, has the chemical formula α-Fe_2O_3, characterized by its reddish-brown color and metallic sheen. It is one of the most abundant minerals on Earth and is commonly found in sedimentary, metamorphic, and igneous rocks. The color of hematite ranges from red to brown and can appear black or gray depending on the form and impurities present. Its luster exhibits a metallic to submetallic luster. The hematite produces a reddish streak when rubbed against a porcelain plate, which is a key identifying feature. In addition, it has a Mohs hardness of approximately 5.5–6.5, making it relatively hard compared to other minerals. The specific gravity of hematite is about 5.26 g/cm^3, indicating its high density due to the presence of iron [19, 49]. The morphology of magnetic hematite can vary significantly based on the synthesis method employed. Nanoscale hematite (see Fig. 3) particles often exhibit high surface area-to-volume ratios, which enhance their reactivity and photocatalytic performance. These particles can be synthesized through methods, such as sol–gel processes, precipitation, or hydrothermal synthesis, yielding various shapes including spherical, rod-like, or irregular forms [19, 37].

In bulk, hematite typically appears as massive aggregates or crystalline grains. Particle size can range from tens of nanometers for nanoparticles to several micrometers for bulk materials. The size of the hematite particles significantly influences their physical and magnetic properties. Nanoscale particles, due to their small size, can exhibit superparamagnetism, allowing thermal fluctuations to overcome magnetic anisotropy energy barriers. This leads to increased magnetic susceptibility compared to larger particles that exhibit weak antiferromagnetism or paramagnetism [9].

Hematite is primarily classified as an antiferromagnetic material at room temperature and below its Néel temperature (approximately 956 K or 683 °C). In this state, the magnetic moments of adjacent iron ions align in opposite directions, effectively canceling each other out and resulting in no net macroscopic magnetization. However, under certain conditions—such as the presence of impurities or defects in the crystal structure—hematite can exhibit weak ferromagnetic properties due to spin canting. The magnetic properties of hematite are highly temperature-dependent [9]. Below the Morin transition temperature (around – 10 °C), hematite exhibits antiferromagnetic behavior. At room temperature, it may show weak ferromagnetism due to slight deviations from perfect antiparallel alignment of spins. Above the Néel temperature, hematite transitions into a paramagnetic state where thermal energy disrupts the ordered magnetic alignment [13, 35, 61].

The weakly magnetic nature of α-Fe_2O_3 makes it suitable for various applications where magnetic separation is advantageous. For instance, in wastewater treatment processes utilizing photocatalysis, the ability to easily recover and reuse magnetic hematite enhances operational efficiency and reduces costs associated with catalyst loss.

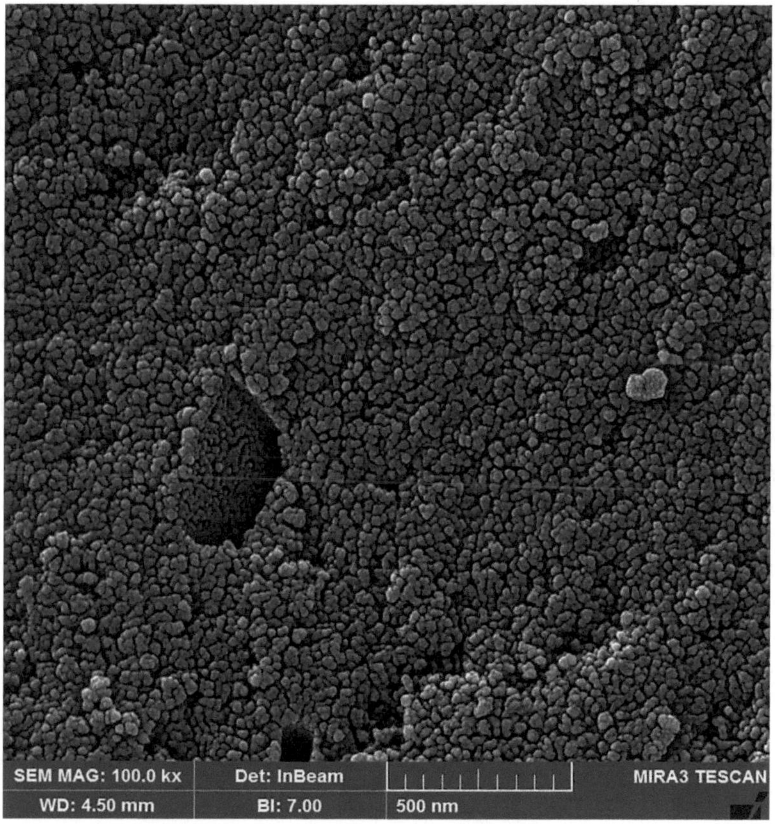

Fig. 3 Hematite particles synthesized by precipitation method

2.3 *Optical Properties*

Magnetic hematite (α-Fe_2O_3) has attracted great interest in various fields, especially photocatalysis and environmental remediation, due to its unique band gap energy and light absorption properties. Understanding these properties is critical to optimizing its performance in applications, such as wastewater treatment and solar energy conversion.

The band gap energy of α-Fe_2O_3 is a critical parameter that influences its electronic properties and photocatalytic efficiency. Hematite is classified as an n-type semiconductor with a band gap energy typically between 2.0 and 2.2 eV. This relatively narrow band gap allows hematite to absorb visible light up to about 600 nm, making it a promising candidate for harnessing solar energy. The band gap can be estimated using the Tauc plot method, where the square of the absorption coefficient is plotted against the photon energy to determine the intercept corresponding to the band gap energy [6, 15].

Several factors can influence the band gap energy of α-Fe_2O_3, including particle size, morphology, and synthesis conditions. Nanoscale hematite particles often exhibit quantum confinement effects, which result in fluctuations in bandgap energy compared to bulk materials. Smaller particles can exhibit an increase in the band gap due to the spatial confinement of the charge carriers, which changes their electronic properties. Furthermore, impurities or defects introduced during synthesis can also change the electronic structure and thus affect the band gap [41].

The optical absorption spectrum of magnetic hematite extends from the UV region into the visible light spectrum, with significant absorption features observed around 400 nm (blue region) and 550 nm (green region). Hematite's intense red color is primarily due to its strong absorption of blue and green light while transmitting red light. This property makes hematite suitable for applications that require activation by visible light, such as photocatalysis [41]. Light absorption in α-Fe_2O_3 is attributed to electronic transitions involving Fe^{3+} ions within octahedral coordination sites. When light interacts with the material, electrons are excited from the valence band into the conduction band, generating electron–hole pairs that are essential for photocatalytic reactions. The specific absorbed wavelengths correspond to ligand-to-metal charge transfer transitions and pair excitations of neighboring Fe^{3+} cations occupying neighboring sites. The morphology and size of hematite nanoparticles significantly influence their light absorption properties. For example, studies have shown that different synthesis methods produce hematite with different shapes (spherical, rod-shaped, etc.) and sizes, which can affect its surface area and therefore its ability to effectively absorb light. Smaller nanoparticles generally exhibit higher light absorption due to larger surface area and increased interaction with incident photons [6, 15, 41].

3 Synthesis Methods

The effectiveness of magnetic hematite as a photocatalyst depends largely on its synthesis method. Various techniques, including sol–gel processes, hydrothermal synthesis, and chemical precipitation, have been used to prepare α-Fe_2O_3 nanoparticles with tailored properties. For example, the use of ionic liquids in the synthesis process has been shown to improve the structural and optical properties of hematite, resulting in improved photocatalytic performance due to increased surface area and better charge separation capabilities [22, 62]. Control over particle size and morphology during synthesis is crucial as these factors have a direct influence on the photocatalytic activity and efficiency of the material [15].

3.1 Hydrothermal Synthesis

Hydrothermal synthesis is a widely used method for producing magnetic hematite nanoparticles (α-Fe_2O_3) that utilizes high temperature and pressure conditions to facilitate the crystallization of materials from aqueous solutions. This technique is particularly advantageous for the synthesis of nanostructured materials with controlled morphology and size, which are critical for applications in photocatalysis, environmental remediation, and energy storage. During hydrothermal synthesis, iron precursors, such as ferric chloride ($FeCl_3$) or iron nitrate ($Fe(NO_3)_3$), are dissolved in water together with stabilizers and pH regulators. The resulting solution is then transferred to a Teflon-lined autoclave, which is sealed and heated to elevated temperatures (typically between 150 and 200 °C) for several hours. The high pressure environment allows the solubility of the reactants to be increased and promotes the nucleation and growth of hematite crystals. The hydrothermal process involves several key steps, including dissolution, nucleation, and growth. In dissolution, the iron precursor dissolves in water, forming iron ions in solution. During nucleation, as the temperature rises, supersaturation occurs, leading to the nucleation of hematite particles. In the growth step, continuous heating allows for the growth of these nuclei into larger crystals. The morphology of the resulting hematite can be influenced by factors, such as temperature, pressure, reaction time, and precursor concentration [32, 38].

One of the key advantages of hydrothermal synthesis is its ability to control the morphology of α-Fe_2O_3 nanoparticles. By varying synthesis parameters like temperature and reaction time, researchers can produce a range of morphologies including spherical nanoparticles, nanorods, microcubes, and flower-like structures. For instance, increasing the reaction time generally leads to larger particle sizes due to continued growth processes [32, 38].

3.2 Solvothermal Synthesis

Solvothermal synthesis is a widely used method for producing magnetic hematite (α-Fe_2O_3) nanoparticles, leveraging the advantages of high temperature and pressure in a solvent medium to facilitate the crystallization of materials. This technique is particularly effective for synthesizing nanostructured materials with controlled morphologies and sizes, which are essential for various applications, including photocatalysis, sensors, and energy storage [59].

3.3 Chemical Precipitation Techniques

Chemical precipitation is a widely used method for synthesizing magnetic hematite nanoparticles (α-Fe_2O_3). In this technique, solid particles are formed from a homogeneous solution through chemical reactions, often resulting in highly pure and well-defined nanostructures. The process is relatively simple, is inexpensive, and allows control of particle size and morphology by adjusting various synthesis parameters [36].

In chemical precipitation, iron salts are dissolved in a solvent to create a precursor solution. A precipitating agent, such as sodium hydroxide (NaOH) or ammonium hydroxide (NH_4OH), is then added to this solution to induce the formation of hematite nanoparticles. The reaction typically occurs under controlled conditions, including temperature and pH, which significantly influence the properties of the resulting hematite [31].

The general reaction mechanism for synthesizing α-Fe_2O_3 via chemical precipitation can include dissolution, precipitation, and formation of hematite [40, 46]. During dissolution, iron(III) salts, such as ferric chloride ($FeCl_3 \cdot 6H_2O$) or iron(III) nitrate ($Fe(NO_3)_3$), are dissolved in deionized water to form a clear solution. In the precipitation step, a precipitating agent is added to the solution while maintaining a specific pH (usually between 10 and 12). For example, adding NaOH leads to the formation of iron hydroxides according to the following reaction:

$$Fe^{3+} + 3OH^{-} \rightarrow Fe(OH)_3 \downarrow \quad (1)$$

The precipitate formed is typically ferric hydroxide ($Fe(OH)_3$), which can further dehydrate to form hematite upon heating. The precipitate is collected through filtration or centrifugation, washed to remove impurities, and then dried. The dried precursor is subsequently calcined at elevated temperatures (typically between 400 and 700 °C) to convert it into α-Fe_2O_3 [53]:

$$2Fe(OH)_3 \rightarrow \alpha - Fe_2O_3 + 3H_2O \quad (2)$$

The morphology and size of α-Fe_2O_3 nanoparticles synthesized by chemical precipitation can be affected by several factors, including the concentration of the precursors, the pH value, and the temperature. Higher concentrations can result in larger particle sizes due to increased nucleation rates. Adjusting pH during synthesis influences the solubility of iron ions and the rate of precipitate formation, thereby affecting particle morphology. The reaction temperature can also influence the crystallization rate and particle growth. For instance, studies have shown that varying the concentration of NaOH during precipitation can yield hematite nanoparticles with different shapes and sizes, such as spherical or irregular aggregates [53].

A key advantage of chemical precipitation is that it utilizes readily available precursors and does not require complex equipment. In addition, the closed system minimizes contamination from external sources, leading to high-purity products.

This method can be easily scaled up for industrial applications without significant modifications [53].

3.4 Sol–Gel Method

The sol–gel synthesis method is a widely adopted technique for producing magnetic hematite (α-Fe_2O_3) nanoparticles. This approach allows for the precise control of particle size, morphology, and chemical composition, making it particularly suitable for applications in photocatalysis, magnetic materials, and biomedical fields.

The sol–gel process entails the transformation from a sol (a colloidal solution) to a gel phase through hydrolysis and condensation reactions. This technique typically starts with iron precursors, such as ferric nitrate [$Fe(NO_3)_3 \cdot 9H_2O$] or iron(III) acetylacetonate, which are dissolved in a solvent (usually water or alcohol). A chelating agent, commonly citric acid or ethylene glycol, is added to stabilize the solution and facilitate the formation of iron oxide nanoparticles during subsequent heating [39, 45].

The sol–gel synthesis of α-Fe_2O_3 includes precursor solution preparation, hydrolysis and condensation, gelation, and heat treatment. During these steps, the iron precursor is dissolved in a solvent and a chelating agent is added to form a homogeneous solution. For example, mixing iron nitrate with citric acid creates a stable precursor solution. When the solution is heated (typically to temperatures around 70–90 °C), hydrolysis occurs, leading to the formation of metal hydroxides. Condensation reactions then occur, which lead to the formation of a gel network with iron oxide particles.

$$Fe(NO_3)_3 + 3H_2O \rightarrow 2Fe(OH)_3 + 3HNO_3 \qquad (3)$$

As water evaporates, the gel becomes more concentrated, and iron hydroxides begin to precipitate out of the solution. The gel is dried at elevated temperatures (usually between 100 and 200 °C) to remove residual solvents and water. The dried gel is then calcined at higher temperatures (typically between 400 and 600 °C) to convert the iron hydroxides into crystalline α-Fe_2O_3.

One of the key advantages of sol–gel synthesis is its ability to control the morphology and size of α-Fe_2O_3 nanoparticles by adjusting various parameters, such as concentration of iron precursor, type of chelating agent, and calcination temperature. Higher concentrations can lead to larger particle sizes due to increased nucleation. Different agents can influence particle growth rates and final morphology. For example, citric acid tends to yield smaller particles compared to other agents. Higher temperatures generally lead to increased crystallinity and larger particle sizes.

Research has shown that sol–gel synthesized hematite can exhibit a range of morphologies, including spherical nanoparticles, nanorods, or aggregated structures depending on these parameters. For instance, one study reported that hematite synthesized using citric acid as a chelating agent produced nanoparticles with an average size around 10–20 nm, which exhibited enhanced magnetic properties.

The sol–gel method minimizes contamination from external sources due to its closed system. This method uses various precursors and solvents, so that individual adaptation to the desired properties is possible. The method also allows precise control of particle size and morphology through careful adjustment of synthesis parameters.

4 Characterization of Hematite

The characterization of hematite (α-Fe_2O_3) nanoparticles is essential for understanding their structural, optical, magnetic, and photocatalytic properties. Various techniques are employed to analyze different aspects of hematite, including its crystallinity, morphology, surface area, and chemical composition. This text provides a comprehensive overview of the key characterization methods used to study hematite nanoparticles.

4.1 XRD Analysis

XRD is one of the most fundamental techniques for characterizing the crystalline structure of hematite. XRD provides information about the phase composition, crystallinity, and lattice parameters. The typical XRD pattern of hematite (Fig. 4) shows distinct peaks corresponding to the rhombohedral crystal structure with space group R-3c. The most prominent peaks are observed at 24.0°, 33.2°, 35.5°, and 49.5° (2θ), which can be indexed to the (104), (110), (113), and (024) planes, respectively. The intensity and position of these peaks can indicate the degree of crystallinity and any potential impurities present in the sample. For example, higher annealing temperatures during synthesis often lead to increased peak intensity in XRD patterns, indicating improved crystallinity of hematite nanoparticles. Additionally, Scherrer equation can be applied to calculate the average crystallite size based on the full width at half maximum of the XRD peaks [18, 54, 56].

4.2 Scanning Electron Microscopy (SEM)

SEM is used to analyze the surface morphology and particle size distribution of hematite nanoparticles. SEM provides high-resolution images that reveal the shape and aggregation behavior of particles (Fig. 5). For instance, SEM images can show

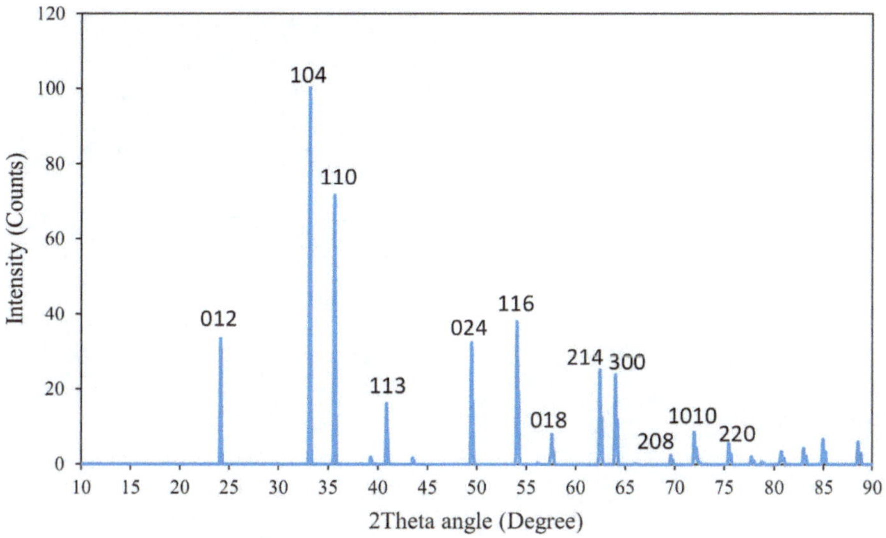

Fig. 4 XRD pattern of hematite

whether hematite particles are spherical, pseudocubes, ellipsoids, peanuts, or irregularly shaped [28]. The particle size can vary significantly depending on the synthesis method, for example, sol–gel synthesis may yield nanoparticles with diameters ranging from 10 nm to several hundred nanometers.

Fig. 5 SEM images of **a** spherical hematite particles and **b** aggregation of hematite particles

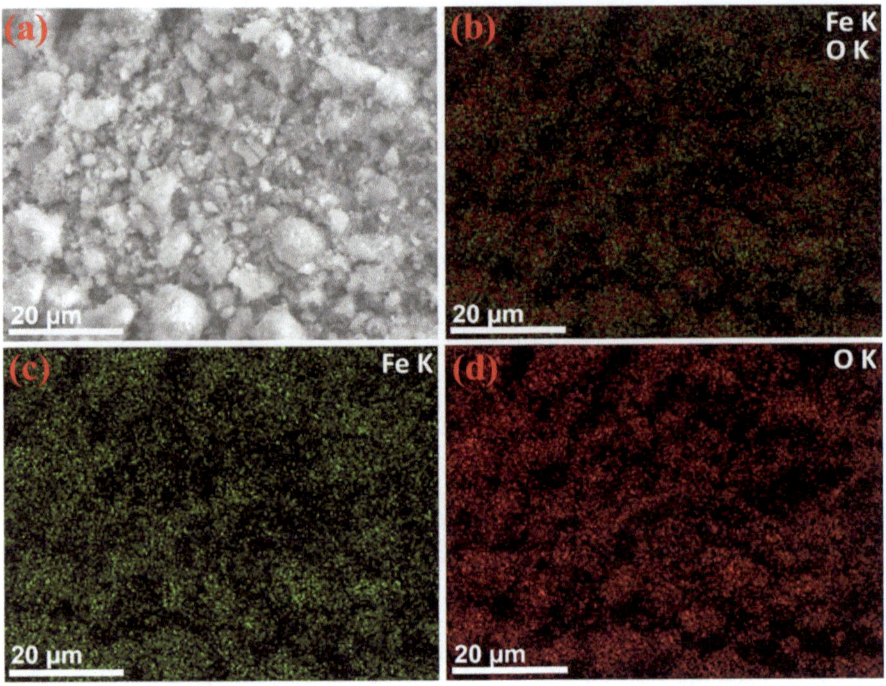

Fig. 6 **a** SEM micrograph and (b–d) EDS mapping of the synthesized α-Fe_2O_3 particles, **b** distribution of all elements, **c** Fe distribution, and **d** O distribution [43]

Energy dispersive X-ray spectroscopy (EDS) can be integrated with SEM to assess the elemental composition of hematite samples. EDS maps (Fig. 6) provide information about the distribution of iron and oxygen within the particles and confirm the presence of hematite without significant contamination from other phases.

4.3 Transmission Electron Microscopy (TEM)

Transmission electron microscopy offers insights into the internal structure and crystallinity of individual hematite nanoparticles at a much higher resolution than SEM. TEM can reveal details about particle morphology, size, and crystal defects. As seen in Fig. 7, the size of hematite nanoparticles typically falls within the nanoscale range, often smaller than 70 nm. The distribution of these particles can vary based on the synthesis method and conditions. High-resolution TEM (HRTEM) allows for direct observation of lattice fringes, providing information on crystallographic orientation and phase purity. For example, HRTEM images can confirm that synthesized hematite consists of single or polycrystalline structures.

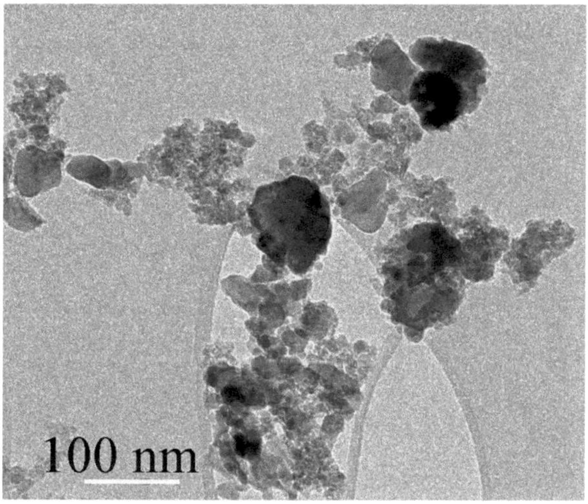

Fig. 7 TEM image of hematite

4.4 *Fourier Transform Infrared (FTIR) Spectroscopy*

FTIR spectroscopy is commonly used to analyze the functional groups present in α-Fe_2O_3 (see Fig. 8). Characteristic peaks associated with Fe–O stretching modes are typically observed at around 469 and 555 cm^{-1}. Furthermore, bands corresponding to hydroxyl groups (−OH) can appear at higher wavenumbers (e.g., about 3412 cm^{-1}), indicating the presence of OH groups from H_2O molecules on the surface of the nanoparticles. These functional groups can influence photocatalytic efficiency by participating in surface reactions during pollutant degradation [18, 54, 55].

4.5 *Surface Area Analysis*

The Brunauer–Emmett–Teller (BET) isotherm is applied to determine the specific surface area and porosity of hematite nanoparticles. By measuring N_2 sorption isotherms at 77 K, researchers can calculate surface area using the BET equation. A higher specific surface area is often correlated with enhanced photocatalytic activity due to increased active sites available for reactions. For instance, a study indicated that the specific surface area of hematite synthesized from waste iron sulfate showed a range between 4 and 52 m^2/g, with variations based on the synthesis conditions, such as the use of different oxidizing agents [51]. Another comprehensive review mentioned that the specific surface area of hematite nanoparticles generally ranges from 10 to 90 m^2/g. This variability is influenced by factors, such as particle size and synthesis techniques [4].

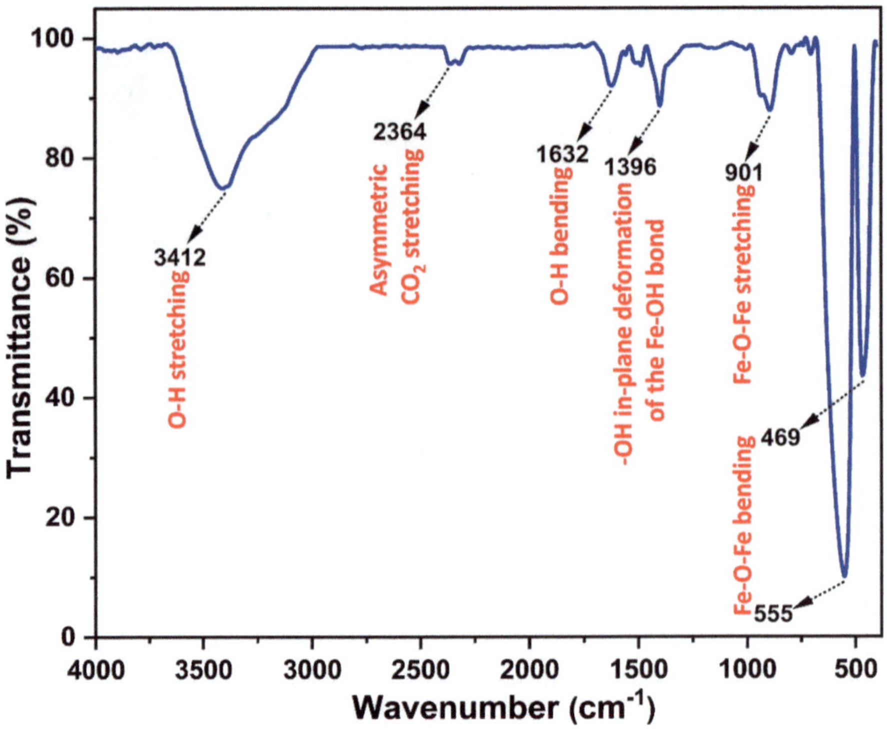

Fig. 8 FTIR spectrum of hematite [43]

4.6 Magnetic Property

The magnetic properties of hematite are typically characterized using vibrating sample magnetometry (VSM). Hematite is generally considered an antiferromagnetic material at room temperature; however, it may exhibit weak ferromagnetism due to spin canting effects in nanoscale particles. VSM measurements provide information about saturation magnetization, coercivity, and remanence, which are critical for applications in magnetic separation or as magnetic carriers in drug delivery systems [29]. Figure 9 shows magnetic behaviors of some commercial hematite samples. Fisher shows the highest coercivity, whereas Aldrich presents a linear, narrow loop. On the other hand, the Alfa sample demonstrates a wasp-waisted loop, marked by a narrowing in the center. This distinctive shape could result from either multiple magnetic phases with varying coercivities or a diverse range of particle size distributions [2].

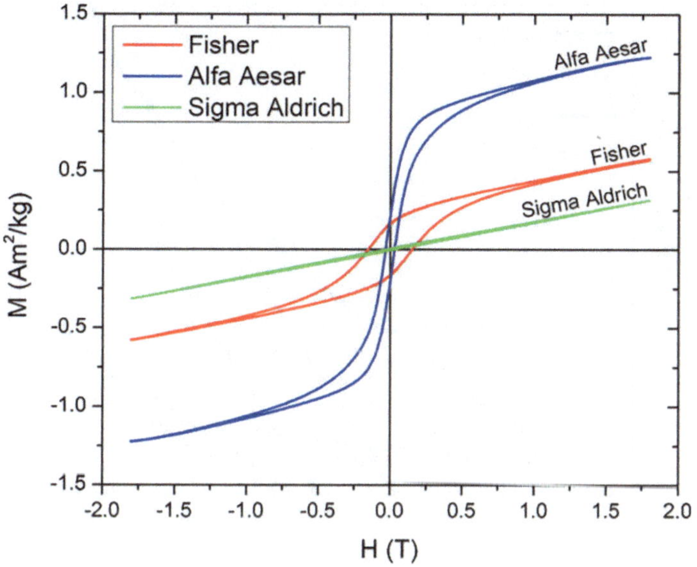

Fig. 9 Hysteresis magnetic loops of commercial hematite samples [2]

4.7 UV–Vis Diffuse Reflectance Spectroscopy (DRS)

UV–Vis DRS is used to investigate the optical properties of hematite nanoparticles. This technique helps determine band gap energy by analyzing light absorption in the UV–visible region. Hematite has a band gap energy typically ranging from 2.0 to 2.2 eV, allowing it to absorb light up to approximately 600 nm. The optical band gap can be calculated using the linear portion of the Tauc plots derived from reflectance data [1]. A study reported that the band gap energy calculated from Tauc plots for pure hematite was found to be around 2.66 eV (see Fig. 10), which decreased with doping (e.g., to 2.31 eV with 4% Zn doping) before increasing again at higher concentrations. This indicates how doping can modify the electronic properties of hematite nanoparticles [52].

5 Mechanisms of Photocatalytic Activity

The photocatalytic reaction mechanism of hematite (α-Fe_2O_3) is a vital research focus because of its applications in environmental remediation, especially in the breakdown of organic pollutants and water purification. Hematite, as a semiconductor, exhibits unique properties that facilitate photocatalytic processes under visible light

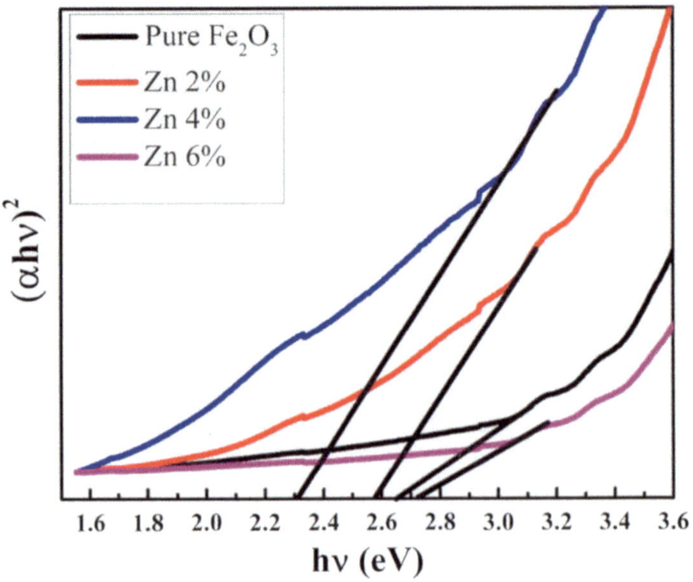

Fig. 10 Tauc plot of Zn-doped hematite samples [52]

irradiation. This text explores the fundamental mechanisms involved in the photocatalytic activity of hematite, including electron–hole pair generation, charge carrier dynamics, and the subsequent chemical reactions leading to pollutant degradation.

5.1 Band Gap and Light Absorption

Hematite has a band gap energy of approximately 2.0–2.2 eV, allowing it to absorb visible light up to about 600 nm. This characteristic makes hematite an attractive photocatalyst for solar energy applications. When hematite is exposed to light, photons with energy greater than the band gap can excite electrons from the valence band to the conduction band, generating electron–hole pairs:

$$Fe_2O_3 + h\nu \rightarrow e^- + h^+ \tag{4}$$

Here, e^- represents the photogenerated electron, and h^+ represents the photogenerated hole. The effectiveness of this process is affected by factors like particle size, morphology, and the presence of surface defects [11].

5.2 *Charge Carrier Dynamics*

Once generated, the electron–hole pairs must migrate to the surface of the catalyst to participate in redox reactions. However, one of the challenges in photocatalysis is the recombination of these charge carriers before they can react with adsorbed species. The recombination rate can be influenced by several factors, such as particle size and surface defects. Smaller nanoparticles typically exhibit higher surface areas, which can enhance charge separation and reduce recombination rates. The presence of surface defects or dopants can act as traps for charge carriers, prolonging their lifetime and enhancing photocatalytic efficiency [18, 22].

5.3 *Surface Reactions*

The photogenerated holes (h^+) and electrons (e^-) participate in various surface reactions, such as oxidation and reduction reactions. These reactions lead to pollutant degradation. The holes can oxidize water molecules or hydroxyl groups adsorbed on the surface of hematite to produce reactive hydroxyl radicals ($^{\cdot}OH$), which are highly effective at degrading organic pollutants:

$$H_2O + h^+ \rightarrow {}^{\cdot}OH + H^+ \tag{5}$$

During reduction reactions, the electrons can reduce oxygen molecules adsorbed on the surface to form superoxide radicals ($O_2{}^-$), which also contribute to the degradation process:

$$O_2 + e^- \rightarrow O_2^- \tag{6}$$

These reactive species ($^{\cdot}OH$ and $O_2{}^-$) are crucial for breaking down complex organic molecules into simpler, non-toxic substances [8, 54].

5.4 *Pollutant Degradation*

The overall photocatalytic degradation process can include adsorption, photolysis, redox, and degradation. Organic pollutants in aqueous solutions are adsorbed onto the surface of hematite nanoparticles. Upon irradiation with visible light, electron–hole pairs are generated. The holes oxidize water or hydroxyl groups to produce reactive radicals, while electrons reduce oxygen to form superoxide radicals. The reactive radicals attack and degrade the adsorbed organic pollutants into smaller molecules or fully mineralize them into CO_2 and H_2O [8, 18].

6 Factors Affecting Photocatalytic Efficiency

The photocatalytic efficiency of hematite (α-Fe_2O_3) is influenced by several key factors, including particle size, morphology, synthesis method, surface area, light absorption characteristics, and reaction conditions. Understanding these parameters is crucial for optimizing hematite as a photocatalyst in applications, such as wastewater treatment and pollutant degradation. This section explores these factors.

6.1 *Particle Size*

The size of hematite nanoparticles significantly impacts their photocatalytic activity. Smaller particles generally exhibit higher surface area-to-volume ratios, which enhances the number of active sites available for reactions. For instance, studies have shown that reducing the particle size to below 10 nm can lead to improved photocatalytic performance due to increased light absorption and reduced recombination rates of electron–hole pairs. Quantitatively, a study reported that hematite nanoparticles with sizes around 20 nm achieved a degradation efficiency of 90% for methylene blue (MB) dye within 120 min under visible light irradiation, while larger particles (approximately 100 nm) exhibited only 60% degradation under the same conditions. This indicates a clear correlation between smaller particle sizes and enhanced photocatalytic activity [11, 26, 54].

6.2 *Morphology*

The morphology of hematite nanoparticles also plays a crucial role in determining their photocatalytic efficiency. Different shapes (e.g., spherical, rod-like, or flower-like structures) can affect light scattering and absorption properties. For example, flower-like structures have been shown to provide more surface area and active sites compared to spherical particles. A study demonstrated that hematite synthesized with a flower-like morphology exhibited a photocatalytic degradation rate constant of $0.068\ min^{-1}$, significantly higher than that of spherical nanoparticles, which had a rate constant of $0.052\ min^{-1}$. This highlights the importance of optimizing morphology for improved photocatalytic performance [11, 26].

6.3 *Synthesis Method*

The method used to synthesize hematite nanoparticles can influence their structural and optical properties, thereby affecting photocatalytic efficiency. Common synthesis methods include sol–gel, hydrothermal, and chemical precipitation techniques. For instance, hematite synthesized via the sol–gel method often exhibits higher crystallinity and better-defined morphologies compared to those produced by chemical precipitation. In one comparative study, hematite nanoparticles synthesized using a sol–gel method showed a band gap energy of 2.25 eV, which is suitable for visible light absorption, whereas those produced via chemical precipitation had a band gap of approximately 2.0 eV, leading to lower photocatalytic activity under visible light. The differences in band gap energy directly correlate with the efficiency of charge carrier generation and subsequent redox reactions [22].

6.4 Surface Area

The specific surface area of hematite nanoparticles is another critical factor influencing their photocatalytic performance. A higher surface area allows for greater adsorption of pollutants and facilitates more efficient interaction with photogenerated reactive species. BET surface area analysis has shown that increasing the surface area from 30 m^2/g to over 100 m^2/g can enhance photocatalytic activity significantly. For example, in experiments involving the degradation of rhodamine B dye, it was observed that samples with a BET surface area of 80 m^2/g achieved a degradation efficiency of 95% after 240 min under UV light irradiation, compared to only 70% for samples with a surface area of 30 m^2/g. This illustrates the importance of maximizing surface area in enhancing photocatalytic reactions [22, 54].

6.5 Light Absorption Characteristics

Hematite's ability to absorb light effectively is crucial for its photocatalytic activity. The band gap energy determines the wavelengths of light that can be absorbed; thus, optimizing this parameter is essential for enhancing photocatalysis under visible light conditions. Hematite typically has a band gap ranging from 2.0 to 2.2 eV, allowing it to absorb visible light efficiently. Studies have shown that modifying the synthesis conditions can tune the band gap energy; for instance, doping hematite with other metals (e.g., Ti or Zn) can lower the band gap further, resulting in improved light absorption and enhanced photocatalytic activity. A recent study indicated that Ti-doped hematite exhibited an increase in degradation efficiency from 75 to 90% for organic dyes under visible light due to improved light absorption characteristics [8, 54].

6.6 Reaction Conditions

Finally, reaction conditions, such as temperature and pH, also significantly affect the photocatalytic efficiency of hematite. Higher temperatures generally increase reaction rates by enhancing the kinetic energy of molecules involved in the degradation process. For example, it was observed that increasing the temperature from room temperature (approximately 25 °C) to 70 °C resulted in an increase in degradation efficiency from 72% to over 95% after 120 min under visible light irradiation. Similarly, pH levels can influence the charge on both the catalyst surface and the pollutants being degraded; optimal pH levels can enhance adsorption and reaction kinetics. Research has shown that maintaining a neutral pH (around 7) often leads to optimal performance for dye degradation processes involving hematite [26].

The photocatalytic efficiency of hematite (α-Fe_2O_3) is significantly influenced by reaction time, which affects the extent of pollutant degradation and the overall kinetics of the photocatalytic process. The duration of the photocatalytic reaction is crucial in determining the efficiency of hematite as a photocatalyst. Longer reaction times generally allow for more extensive interaction between the photogenerated reactive species and the pollutants, leading to higher degradation rates. For example, in a study investigating the degradation of MB dye using hematite nanoparticles, it was noted that extending the reaction time from 30 to 120 min resulted in an increase in degradation efficiency from 50 to 90%. This indicates that longer exposure times facilitate greater pollutant breakdown [22, 47].

The crystallite size of hematite nanoparticles can also change with extended reaction times during synthesis, which subsequently affects their photocatalytic efficiency. Research has indicated that increasing the solvothermal reaction time from 4 to 8 h leads to an increase in crystallite size from approximately 19.7–24.4 nm at a temperature of 180 °C. Larger crystallite sizes can enhance light absorption and reduce recombination rates of charge carriers, thereby improving photocatalytic performance [26].

Optimizing reaction time is essential for maximizing photocatalytic efficiency. A study highlighted that for optimal degradation rates, maintaining a specific reaction time is crucial; for example, under visible light irradiation, hematite nanoparticles exhibited maximum degradation efficiency when the reaction time was optimized to around 120 min, achieving nearly complete degradation of organic dyes. Conversely, shorter reaction times resulted in lower efficiencies due to insufficient interaction between reactive species and pollutants [8].

Longer reaction times can result in higher production of ROS, including $^{\cdot}OH$ and O_2^{-}, which play a vital role in degrading organic pollutants. Studies have shown that extending the reaction time enhances the production of these radicals, correlating with improved photocatalytic activity. For instance, a comparative analysis demonstrated that after 60 min, the concentration of $^{\cdot}OH$ radicals increased significantly with extended exposure time, leading to enhanced degradation rates [8, 54].

7 Applications in Wastewater Treatment

Hematite (α-Fe_2O_3) has emerged as a promising photocatalyst for wastewater treatment due to its stability, low toxicity, and ability to absorb visible light. The photocatalytic applications of hematite in degrading organic pollutants in wastewater are influenced by several quantitative factors, including particle size, morphology, synthesis method, reaction conditions, and the nature of the pollutants. This section provides a quantitative analysis of the photocatalytic efficiency of hematite in wastewater treatment.

A study focused on the use of hematite nanoparticles as a photocatalyst to effectively degrade the toxic Rose Bengal (RB) dye under visible light without any additives (Fig. 11). The research demonstrated that by optimizing parameters like nanoparticle quantity, particle size, and reaction temperature, the degradation of RB dye can be enhanced to 95.33%, with a half-life of 26 min.

The Langmuir–Hinshelwood kinetic equation was utilized to understand the photodegradation kinetics, indicating that the reaction rate is influenced by the size, quantity of nanoparticles, and reaction temperature. Scavenger tests reveal that hydroxyl radicals ($^{\cdot}OH$) are primarily responsible for the photodegradation process, as shown in Fig. 12 [26].

An optimized reaction condition led to a faster reaction rate constant of 0.027 min^{-1}, which was significantly higher compared to previous reports. The study found that the degradation efficiency of reused hematite nanoparticles remains nearly unchanged compared to fresh ones, indicating good stability and reusability [26].

Another study explored an eco-friendly method for synthesizing trimetallic-doped hematite nanoparticles (TD α-Fe_2O_3) using aqueous extracts from *Azadirachta indica*

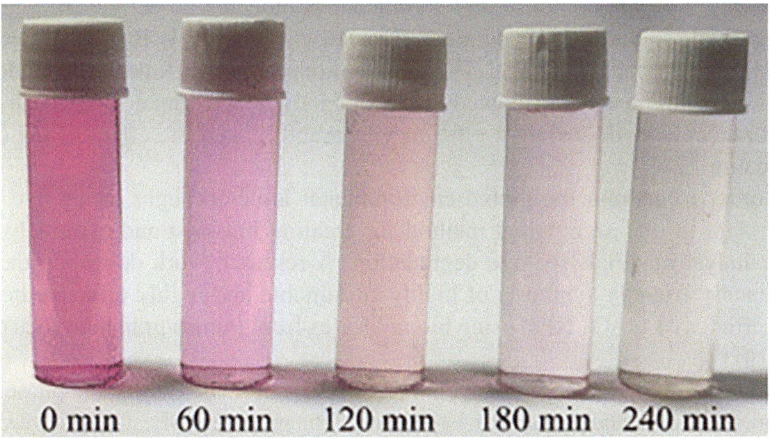

Fig. 11 Photocatalytic removal of RB dye by hematite nanoparticles [26]

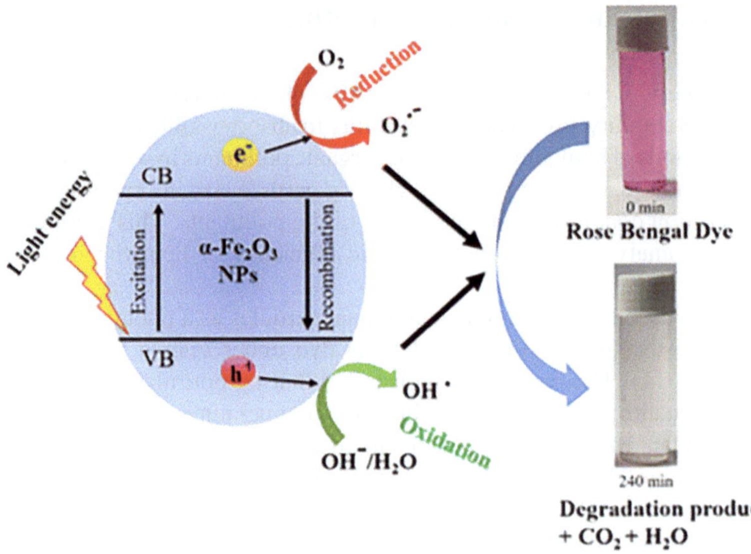

Fig. 12 Mechanism of RB dye degradation [26]

(neem) leaves [24]. The study presents a simple and sustainable approach to synthesizing TD α-Fe_2O_3, utilizing plant-derived biomolecules as capping and stabilizing agents. This method enhances the eco-friendliness and cost-effectiveness of nanoparticle production. The XRD revealed a rhombohedral crystal structure with a crystalline size of 16.2 nm. The SEM showed an average particle size of 15.4 nm with a spheroidal morphology. The UV–Vis spectroscopy indicated a band gap of 2.44 eV, suggesting suitable light absorption properties for photocatalytic applications.

The TD α-Fe_2O_3 demonstrated significant photocatalytic activity in degrading various dyes, including Rhodamine Blue (RhB), Bismarck Brown, Fluorescein Sodium, and Malachite Green. The degradation processes followed pseudo-first-order kinetics. The study evaluated the photocatalytic performance based on degradation percentage, degradation efficiency, reusability, and the effect of pH on dye degradation [24].

In order to address the global environmental issues brought on by dye abuse, it is crucial to find an efficient method for creating low-cost and extremely stable photocatalyst materials for dye degradation. A research work described the environmentally friendly synthesis of highly sustainable and highly effective hematite nanoparticles (α-Fe_2O_3 NPs) using biomolecules from Punica granatum seed extract (Fig. 13) [3].

A number of characterization techniques were used to determine the phase purity and morphology of the prepared α-Fe_2O_3 NPs. The prepared α-Fe_2O_3 NPs were irregularly shaped, with a size of 26–53 nm, according to FESEM and TEM analysis. The BET surface area of the nanoparticles was calculated to range between 31 and 52 m^2/g, and their average pore diameter was 5–54 nm. At room temperature, α-Fe_2O_3 NPs

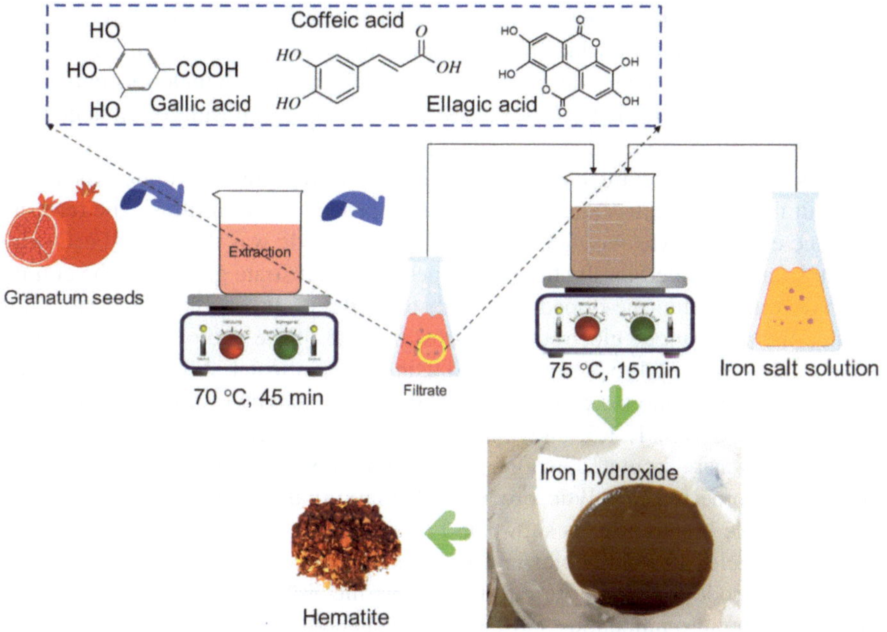

Fig. 13 Synthesis of hematite nanoparticles using biomolecules from Punica granatum seed extract

showed typical ferromagnetic behavior. X-ray photoelectron spectroscopy (XPS) analysis verified the oxidation states of the elements in α-Fe_2O_3 NPs, and the charge on α-Fe_2O_3 NPs was determined to be − 10.6 mV. Using the as-prepared α-Fe_2O_3 NPs, anionic Congo red (CR) and bromophenol blue (BPB) dyes were photocatalytically degraded by solar light. According to experimental results, α-Fe_2O_3 NPs exhibited remarkable photocatalytic activity, degrading roughly 89–42% of CR and 87–96% of BPB in 240 min when exposed to ambient sunlight. Additionally, various reaction conditions that adhered to pseudo-first-order kinetics were used for the photocatalytic degradation experiments. Through radical quenching experiments, the existence of reactive species produced in the reaction system responsible for the degradation of CR and BPB was verified. Outstanding photocatalytic potential of α-Fe_2O_3 NPs indicated that the produced nanoparticles could be used to effectively remove organic dyes from wastewater [3].

In order to remove RhB from textile wastewater, hematite graphene oxide (α-Fe_2O_3-GO) thin-film and α-Fe_2O_3-GO powder nanocomposites were created in a study. The fluorine-doped SnO_2 substrate was coated with α-Fe_2O_3-GO nanomaterials to create a thin layer of nanocomposites. In order to examine the morphology, structure, and characteristics of the produced composites as well as the chemical interactions between α-Fe_2O_3 and GO, various analyses were conducted using XRD, FTIR, Raman spectra, XPS, and FESEM. Using varying concentrations of α-Fe_2O_3-GO, the photocatalytic performance of two synthetic composites was compared. As seen in Fig. 14, the findings demonstrated that powder nanocomposites are superior to thin-film composites in terms of their ability to remove RhB dye. In contrast to thin-film nanocomposites, which had a removal rate of just under 47%, α-Fe_2O_3-GO-5% powder nanocomposites degrade over 64% of the dye. Both materials underwent a reusability test, and the α-Fe_2O_3-GO-5% powder nanocomposites eliminated more dye (up to 63%) in more cycles (6 cycles) [23].

A reactor with a working electrode, counter electrode, and electrolyte was used to carry out the photoelectrocatalytic degradation process for thin films. The reactor's reference electrode was Ag/AgCl, the working electrode was the GO/α-Fe_2O_3 thin film that was deposited on the fluorine-doped SnO_2 substrate glass, and the counter electrode was the fluorine-doped SnO_2 substrate glass. The visible light source illuminated the photoanode electrode. To boost the rate of the photocatalytic reaction, an external voltage of 1–5 V was applied using a potentiostat. After injecting a dye solution (25 mg/L) into the electrolyte, photoelectrocatalytic testing was conducted at intervals of 0, 0.5, 1, 1.5, and 2 h. To assess the reusability of thin layers, the dye solution was recirculated using a peristaltic pump following each testing procedure. The photoelectrocatalytic process schematic diagram in the presence of an external visible light source is shown in Fig. 15. O_2^- and $^{\cdot}OH$ are radicals created when excited electrons and holes in thin films are exposed to xenon light. Near the working electrode's surface, these radicals have the ability to break down dyes and transform them into innocuous substances. The generated voltage improves the dye removal efficiency by accelerating the photocatalytic process's reaction rate [23].

A hydrothermally prepared hematite-containing activated carbon (AC/-Fe_2O_3) nanocomposite is utilized as a photocatalyst to remove tartrazine dye. Iron sand was used to extract hematite, and coffee waste was used to activate activated carbon using hydrochloric acid. The XRD, nitrogen adsorption–desorption, and SEM–EDS were used to characterize the structure and morphology of the photocatalyst. According to these methods, the AC/α-Fe_2O_3 nanocomposite has a specific surface area of 41.52 m^2/g and an average crystallite size of 35.47 nm. The assessment of photocatalytic activity revealed that a number of variables, such as the pH of the dye solution, the dosage of the photocatalyst, the dye concentration, and the duration of irradiation, affected the removal of tartrazine dye. With a pH of 2, a photocatalyst dosage of 250 mg, and an initial tartrazine concentration of 15 ppm, the results demonstrated that the AC/α-Fe_2O_3 nanocomposite can eliminate tartrazine up to 93.69% under solar light and 91.01% under UV light. The irradiation times were 210 min under UV light and 180 min under solar light. When exposed to UV light for 2.5 h and solar light for 2 h, respectively, the AC/Fe_2O_3 nanocomposite can also eliminate tartrazine dye from noodle waste water up to 94–67 and 97–86% [30].

Sn-doped hematite nanoparticles were found to have a higher photocatalytic activity than undoped hematite nanoparticles. According to the results of the recyclability tests, 3 mol% Sn-doped hematite nanoparticles may be employed as a viable reusable photocatalyst for the degradation of dyes [16].

A study examined the synthesis of iron oxide nanoparticles (IONPs) through a green methodology, characterized their properties using various physicochemical techniques, and evaluated their effectiveness in degrading crystal violet dye under

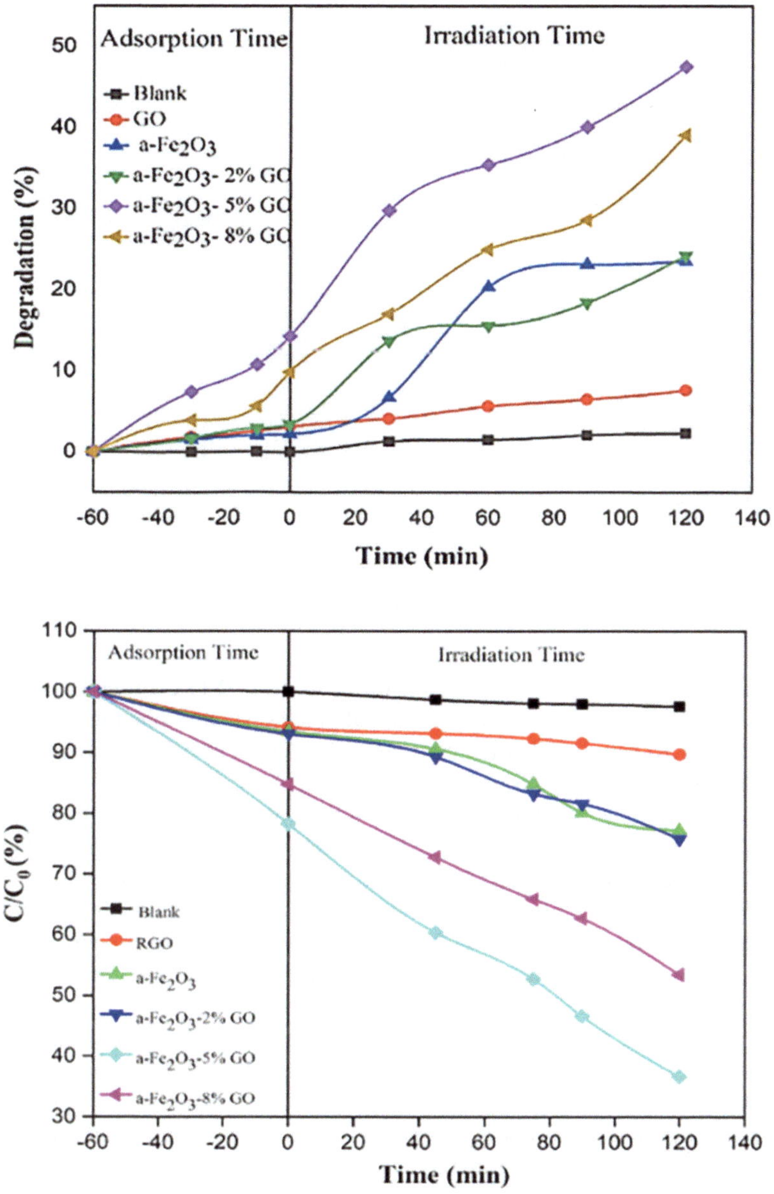

Fig. 14 Photocatalytic results of (Top) thin-film layers and (Bottom) Fe_2O_3-GO powders [23]

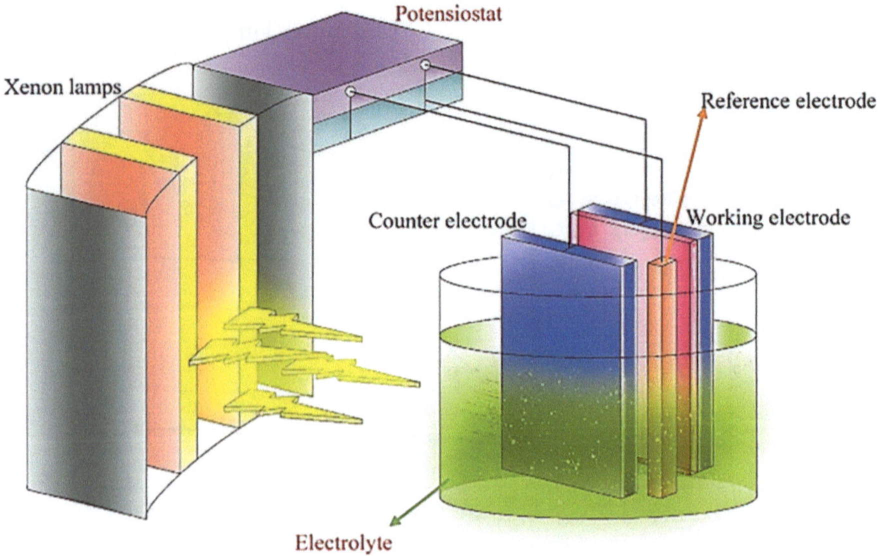

Fig. 15 Schematic of photoelectrochemical reactor [23]

sunlight. The FTIR analyses revealed that the biogenic IONPs exhibited a UV absorption peak at 241 nm and contained functional groups, such as phenols, alkynes, and alkenes. The XRD studies verified their amorphous structure. The nanoparticles had an average diameter of 49.63 ± 9.23 nm and a surface charge of − 5.69 mV. The dye removal efficiency of the synthesized IONPs was tested under both dark and sunlight conditions. Results in Fig. 16 showed that the removal efficiency depended on the dye concentration and exposure time, achieving a peak efficiency of 99.23% after 3.5 h of sunlight exposure. Furthermore, the biogenic IONPs demonstrated antioxidant activity. These findings highlighted the potential of biogenic IONPs as an eco-friendly and efficient solution for removing industrial dyes and remediating contaminated environments and aquatic ecosystems [60].

The removal of pharmaceutical pollutants from aquatic systems had historically posed a significant challenge, demanding innovative and sustainable methods. Semiconductor photocatalysis emerged as a promising eco-friendly approach that harnessed renewable solar energy to address these issues. In a study, researchers partially transformed naturally abundant hematite ores into heterophase compositions containing both α- and γ-phases of Fe_2O_3 through a straightforward and cost-effective single-stage heat treatment method. This synthesis process eliminated the need for solvents or concentrated acids, making it both economical and environmentally sustainable. The XRD analysis verified the successful formation of the α/γ-phases in the transformed Fe_2O_3, and alterations in the local atomic structure of Fe were observed using X-ray absorption spectroscopy.

The hematite was further integrated with graphitic carbon nitride ($g\text{-}C_3N_4$) to form $g\text{-}C_3N_4/Fe_2O_3$ (CNFO) composites, which displayed significantly enhanced

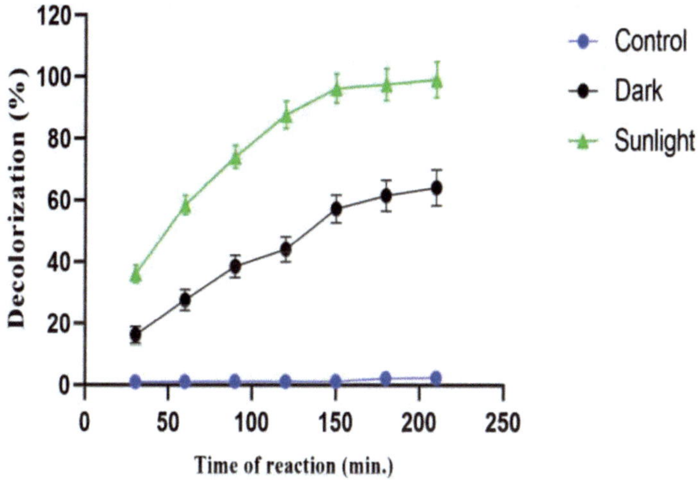

Fig. 16 Decolorization efficiency of biogenic IONPs [60]

photocatalytic properties. Optical analysis revealed a notable improvement in the visible light absorption capabilities of g-C_3N_4 upon combination with hematite. The SEM demonstrated that CN sheets were uniformly grown on the hematite's surface. These CNFO composites were assessed for their effectiveness in degrading a 10 mg/L solution of ofloxacin (OFX) under visible light. As shown in Fig. 17, among the composites, CNFO-5% exhibited the highest photocatalytic activity, achieving a performance 4.8 times greater than pure g-C_3N_4 [48].

The initial rate constant for the photocatalytic breakdown of OFX using CNFO-5% was recorded as 0.1271 min^{-1}. Additionally, the composite maintained its stability and photocatalytic efficiency over five successive degradation cycles. Further analyses, including electrochemical measurements and photoluminescence spectra, validated the efficient separation and transfer of photogenerated charges at the CNFO interface. Scavenger studies identified superoxide anion radicals and photogenerated holes as the primary contributors to OFX degradation. The researchers proposed a Z-scheme heterojunction mechanism to explain the charge transfer dynamics. This work highlighted the potential of cost-effective, abundant, and recyclable natural hematite ores as advanced photocatalysts for the efficient and sustainable removal of pharmaceutical pollutants from wastewater [48].

In a study aimed at developing efficient photocatalysts for environmental remediation, α-Fe_2O_3 fibers were successfully synthesized using the Solution Blow Spinning method, followed by calcination at 800 °C for 2 h. The precursor solution for the fibers was prepared by dissolving 0.8 M iron nitrate and 12 w/v% polyvinylpyrrolidone in a 2:1 mixture of acetic acid and water. The resulting nanostructures were subjected to comprehensive characterization techniques, including SEM, BET isothermal analysis, UV–Vis DRS, XRD, magnetometry, and Mössbauer spectroscopy [7]. The characterization results revealed that the α-Fe_2O_3 fibers had a trigonal crystalline

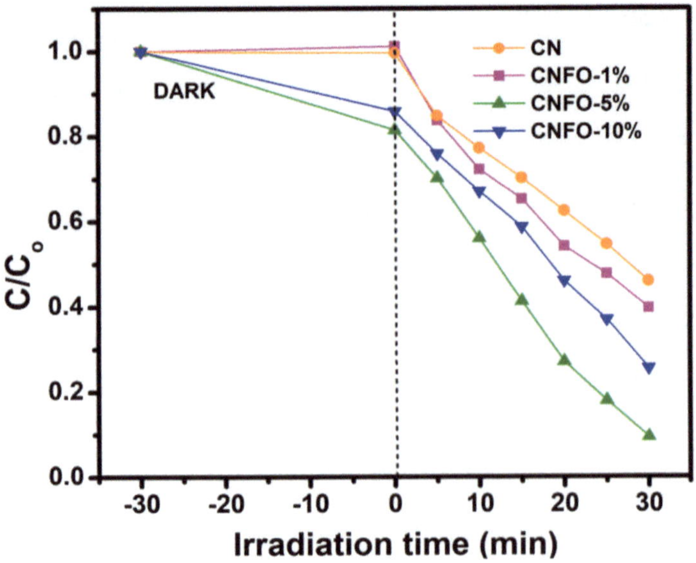

Fig. 17 Photocatalytic performance of CNFO samples in degradation of ofloxacin [48]

structure, with an average fiber diameter of 359 nm and a BET surface area of 2.54 m^2/g. These properties made the fibers suitable for photocatalytic applications. Their effectiveness was tested in the degradation of two organic dyes—MB and crystal violet (CV)—under both visible and UV light irradiations. The fibers demonstrated photodegradation efficiencies of 66% and 48% for MB under visible and UV light, respectively, while achieving higher efficiencies of 90% and 92% for CV under the same conditions after 5 h of irradiation [7]. Additionally, recycling studies highlighted the stability and reusability of the α-Fe_2O_3 fibers, as they maintained their efficiency for CV dye degradation over three consecutive cycles. These results underscored the potential of α-Fe_2O_3 fibers as robust and effective photocatalysts for the treatment of organic pollutants in wastewater, with promising implications for sustainable environmental applications [7].

In a study investigating environmentally friendly methods for synthesizing nanoparticles, researchers developed a green synthesis technique to produce hematite (α-Fe_2O_3) or magnetite (Fe_3O_4) nanoparticles by varying the concentration of *Coffea arabica* L. extract combined with a single iron salt precursor. The resulting nanoparticles were characterized through advanced techniques, including XRD, TEM, Mössbauer spectroscopy, and DC magnetometry. These analyses confirmed that the nanoparticles formed with an average particle size of approximately 5 nm. The optical properties of the synthesized nanoparticles were evaluated using UV–Vis spectroscopy, which revealed band gap energies of 3.09 eV for hematite and 3.23 eV for magnetite. These band gap values suggested their suitability for photocatalytic applications. The photocatalytic activity of the nanoparticles was subsequently tested by degrading methyl orange dye under UV radiation. The results demonstrated that the nanoparticles effectively degraded the dye, highlighting their potential utility in wastewater treatment processes. This green synthesis method provided a sustainable and efficient approach for producing iron oxide nanoparticles with significant potential for environmental remediation applications [27].

The research investigated the photocatalytic properties of hematite (α-Fe_2O_3), a cost-effective metal oxide semiconductor, with a specific focus on the effects of tin (Sn) doping. Hematite had already been identified as a promising material for photoelectrochemical water splitting and photocatalytic applications due to its favorable bandgap and stability. In this study, the researchers sought to enhance the photocatalytic efficiency of hematite nanoparticles by introducing Sn as a dopant, hypothesizing that it would improve charge separation and reduce recombination rates of electron–hole pairs during photocatalytic reactions. The structural and morphological characteristics of the Sn-doped hematite were analyzed using advanced techniques. The XRD confirmed the phase purity and crystallinity of the samples, while TEM provided detailed insights into the size and shape of the nanoparticles. UV–visible spectroscopy was employed to assess the optical properties, demonstrating that Sn doping enhanced light absorption in the visible spectrum compared to undoped samples. This improvement in light absorption was deemed critical for boosting photocatalytic activity under solar irradiation. The photocatalytic performance of the prepared nanoparticles was evaluated using RhB dye as a model organic pollutant. The results showed that Sn-doped hematite NPs exhibited significantly higher photocatalytic activity than their undoped counterparts, confirming that Sn doping effectively improved degradation rates of RhB dye. Furthermore, recyclability tests revealed that nanoparticles with 3 mol% Sn doping maintained their photocatalytic efficiency across multiple cycles, indicating their potential as a reusable catalyst for dye degradation. This study highlighted the dual advantages of Sn doping in hematite nanoparticles: enhancing photocatalytic performance and contributing to sustainable wastewater treatment by providing a reusable catalyst. The findings underscored the importance of developing cost-effective and efficient photocatalysts for mitigating environmental challenges related to dye pollution, which poses serious risks to aquatic ecosystems and human health. Overall, this research offered valuable insights into photocatalysis and paved the way for further advancements in materials for environmental remediation [16].

In another study, hematite nanoparticles (α-Fe_2O_3) were synthesized using a simple hydrothermal method, and their structural, optical, and photocatalytic properties were studied. The XRD analysis confirmed that the nanoparticles exhibited a rhombohedral crystal phase. The FTIR was employed to analyze the stretching vibrations of functional groups present on the surface of the nanoparticles. The optical bandgap of the nanoparticles was estimated to be 2.25 eV through UV–Vis DRS. The SEM revealed that the nanoparticles displayed a spherical morphology. Additionally, their oxidation and reduction properties were studied using cyclic voltammetry. The photocatalytic performance of the α-Fe_2O_3 nanoparticles was evaluated through the degradation of MB dye under natural sunlight. The results demonstrated that the degradation efficiency reached 33% after 2 h, with a pseudo-first-order rate constant calculated at 0.0033 min^{-1}. These findings highlighted the potential of α-Fe_2O_3

nanoparticles as effective photocatalysts for environmental remediation, particularly under natural light conditions [43].

8 Challenges and Limitations

Figure 18 presents challenges and limitations of hematite photocatalysts. Hematite exhibits a very low electron and hole mobility, typically less than 1 $cm^2\ V^{-1}\ s^{-1}$, short hole diffusion length, typically between 2 and 4 nm. This limitation restricts the effective transport of charge carriers to the reaction sites, resulting in reduced photocatalytic efficiency. The short diffusion length means that generated holes may recombine before reaching the surface where reactions occur, leading to poor overall performance in photocatalytic applications [20, 34, 44].

The high recombination rate of charge carriers is another significant limitation for hematite as a photocatalyst. When electrons and holes recombine quickly, it diminishes the availability of charge carriers for the desired photocatalytic reactions. This issue is exacerbated by the intrinsic properties of hematite, which can lead to inefficient utilization of absorbed light energy [34, 44].

Hematite has relatively low electrical conductivity compared to other semiconductor materials. This characteristic hampers the efficient transport of charge

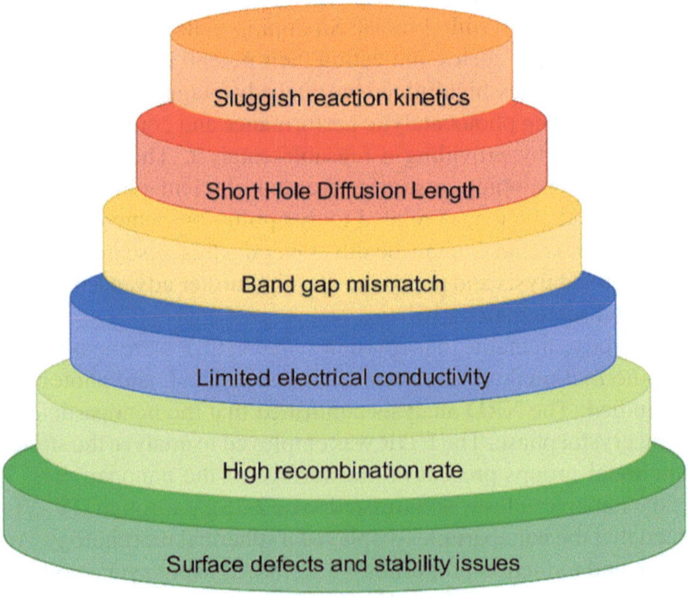

Fig. 18 Challenges and limitations of hematite photocatalysts

carriers within the material, further contributing to the challenges in achieving high photocatalytic performance [34].

While hematite has a band gap energy ranging from 2.0 to 2.2 eV, which allows for visible light absorption, this band gap can also lead to inefficiencies in specific applications due to mismatches with the energy levels required for certain reactions (e.g., water oxidation). This mismatch can limit its effectiveness in photoelectrochemical systems [34].

The presence of oxygen vacancies can have dual effects on hematite's photocatalytic performance. While they can enhance conductivity and improve photocatalytic activity, they can also act as traps for charge carriers, leading to increased recombination rates and reduced efficiency. Moreover, stability under operational conditions (thermal, hydrothermal, and chemical) poses challenges that affect the longevity and reliability of hematite as a photocatalyst [44].

The kinetics of surface reactions, particularly those involved in water oxidation, are often slow when using hematite as a photocatalyst. This sluggishness limits the overall reaction rates and efficiency of processes, such as hydrogen production through water splitting [34].

In addition, scaling up the production of high-quality hematite nanostructures suitable for commercial applications poses significant challenges. Techniques, such as doping or creating nanostructures (e.g., nanopillars or nanorods), are being explored to enhance performance, but these methods can be complex and costly [57].

9 Future Perspectives

The future perspectives of hematite (α-Fe_2O_3) as a photocatalyst are promising, given its abundance, eco-friendliness, and potential for various applications in sustainable energy and environmental remediation. Future research is likely to focus on the design and synthesis of advanced hematite nanostructures, such as heterojunctions (type II, Z-scheme, and S-scheme), which can improve charge carrier generation, separation, and utilization efficiency. These innovations aim to enhance the photocatalytic activity of hematite for applications like hydrogen production, carbon dioxide reduction, and organic pollutant degradation.

Combining hematite with other semiconductor materials can create composite photocatalysts that leverage the strengths of each component. This approach can help overcome limitations, such as low charge mobility and stability issues, while enhancing light absorption across a broader spectrum.

Research will continue to explore methods for band gap engineering to optimize hematite's light absorption capabilities. Strategies may include doping with other elements or creating surface modifications that allow for effective utilization of visible light, which is crucial for practical solar energy applications.

Future advancements may focus on developing environmentally friendly synthesis methods for hematite nanoparticles. Utilizing biological sources or green chemistry principles can reduce reliance on hazardous chemicals and enhance the sustainability of photocatalyst production.

Hematite photocatalysts hold potential for applications in the degradation of pollutants and the recycling of materials within a circular economy framework. This includes addressing plastic waste and other environmental contaminants through efficient photocatalytic processes.

Future research will also need to address the scalability of hematite photocatalysts for industrial applications. This involves developing efficient reactor designs that maximize light exposure and optimize reaction conditions to enhance overall performance in practical settings.

The future of hematite photocatalysis may benefit from interdisciplinary collaboration that combines insights from various fields, such as materials science, chemistry, and engineering. This holistic approach can lead to innovative solutions that address complex challenges in photocatalytic processes.

10 Conclusion

Hematite (α-Fe_2O_3) has emerged as a promising photocatalyst for wastewater treatment, particularly in the degradation of organic pollutants, including pharmaceuticals and dyes. Its unique properties, such as non-toxicity, environmental abundance, and a suitable band gap of approximately 2.0–2.2 eV, enable effective utilization of visible light for photocatalytic processes. The photocatalytic mechanism involves the generation of reactive oxygen species upon light irradiation, which oxidize organic contaminants into less harmful substances. Recent studies have highlighted hematite's effectiveness in degrading various pharmaceutical compounds, achieving significant removal efficiencies. However, challenges, such as low charge carrier mobility and high recombination rates, limit its overall performance. To enhance its photocatalytic activity, researchers are exploring strategies like doping with metals or creating composite materials to improve charge separation and light absorption. Despite these challenges, hematite's potential for sustainable wastewater treatment remains significant. Ongoing research focuses on optimizing its performance through material modifications and reactor designs to facilitate practical applications. As the demand for efficient and eco-friendly wastewater treatment solutions grows, hematite photocatalysts could play a crucial role in addressing environmental pollution and promoting water sustainability.

References

1. Aceto M, Calà E, Gulino F, Gullo F, Labate M, Agostino A, Picollo M (2022) The use of UV-visible diffuse reflectance spectrophotometry for a fast, preliminary authentication of gemstones. Molecules 27:4716
2. Ahmadzadeh M, Romero C, McCloy J (2018) Magnetic analysis of commercial hematite, magnetite, and their mixtures. Aip Adv 8
3. Ahmed A, Usman M, Yu B, Shen Y, Cong H (2021) Sustainable fabrication of hematite (α-Fe_2O_3) nanoparticles using biomolecules of Punica granatum seed extract for unconventional solar-light-driven photocatalytic remediation of organic dyes. J Mol Liq 339:116729
4. Al-Hakkani MF, Gouda GA, Hassan SHA (2021) A review of green methods for phyto-fabrication of hematite (α-Fe_2O_3) nanoparticles and their characterization, properties, and applications. Heliyon 7
5. Al–Ekabi H, Safarzadeh–Amiri A, Sifton W, Story J (1991) Advanced technology for water purification by heterogeneous photocatalysis. Int J Environ Pollut 1:125–136
6. Alpochoritis G, Ntoukas AK, Georgakilas VI (2024) Hematene nanoplatelets with enhanced visible light absorption; the role of aromatic molecules. Molecules 29:3115
7. Araujo RN, Nascimento EP, Firmino HCT, Macedo DA, Neves GA, Morales MA, Menezes RR (2021) α-Fe_2O_3 fibers: an efficient photocatalyst for dye degradation under visible light. J Alloy Compd 882:160683
8. Benhammada A, Trache D, Kesraoui M, Chelouche S (2020) Hydrothermal synthesis of hematite nanoparticles decorated on carbon mesospheres and their synergetic action on the thermal decomposition of nitrocellulose. Nanomaterials 10:968
9. Bødker F, Hansen MF, Koch CB, Lefmann K, Mørup S (2000) Magnetic properties of hematite nanoparticles. Phys Rev B 61:6826
10. Bodzek M, Rajca M (2012) Photocatalysis in the treatment and disinfection of water. Part I. Theoretical backgrounds/Fotokataliza w oczyszczaniu i dezynfekcji wody część i. podstawy teoretyczne. Ecol Chem Eng S 19:489–512
11. Chen Y-H, Tu K-J (2012) Thickness dependent on photocatalytic activity of hematite thin films. Int J Photoenergy 2012:980595
12. Crini G, Lichtfouse E (2018) Wastewater treatment: an overview. In: Green adsorbents for pollutant removal. Springer Nature, Cham
13. De Boer CB, Mullender TAT, Dekkers MJ (2001) Low-temperature behaviour of haematite: susceptibility and magnetization increase on cycling through the Morin transition. Geophys J Int 146:201–216
14. Dhote J, Ingole S, Chavhan A (2012) Review on wastewater treatment technologies. Int J Eng Res Technol 1:1–10
15. Domingo E, Beltrán A, Sanchis R, García T, Solsona B, Galindo F (2018) Photocatalytic activity of mesoporous α-Fe_2O_3 synthesized via soft chemistry and hard template methods for degradation of azo dye orange II. Catal Lett 148:1289–1295
16. Em S, Yedigenov M, Khamkhash L, Molkenova A, Atabaev TS (2020) Sn-doped hematite nanoparticles for potential photocatalytic dye degradation. IOP Conf Ser: Mater Sci Eng 012042
17. Englande Jr AJ, Krenkel P, Shamas J (2015) Wastewater treatment and water reclamation. In: Reference module in earth systems and environmental sciences
18. Faustino LM, McFadzean B, Junior JTG, de Salles L, Filho L (2024) Bulk and surface characterization of distinct hematite morphology: implications for wettability and flotation response. Minerals 14:609
19. Hiralal P, Saremi-Yarahmadi S, Bayer BC, Wang H, Stephan Hofmann KG, Wijayantha U, Amaratunga GAJ (2011) Nanostructured hematite photoelectrochemical electrodes prepared by the low temperature thermal oxidation of iron. Sol Energy Mater Sol Cells 95:1819–1825

20. Jha BK, Chaule S, Jang J-H (2024) Enhancing photocatalytic efficiency with hematite photoanodes: principles, properties, and strategies for surface, bulk, and interface charge transfer improvement. Mater Chem Front 8:2197–2226
21. Jimenez-Relinque E, Lee SF, Plaza L, Castellote M (2022) Synergetic adsorption–photocatalysis process for water treatment using TiO_2 supported on waste stainless steel slag. Environ Sci Pollut Res 29:39712–39722
22. Khalaji AD, Machek P, Jarosova M (2021) α-Fe_2O_3 nanoparticles: synthesis, characterization, magnetic properties and photocatalytic degradation of methyl orange. Adv J Chem-Sect A 4:317–326
23. Khoshnam M, Farahbakhsh J, Zargar M, Mohammad AW, Benamor A, Ang WL, Mahmoudi E (2021) α-Fe_2O_3/graphene oxide powder and thin film nanocomposites as peculiar photocatalysts for dye removal from wastewater. Sci Rep 11:20378
24. Kumar P, Pathak D, Thakur N (2024) Trimetallic doped hematite (α-Fe_2O_3) nanoparticles using biomolecules of Azadirachta indica leaf extract for photocatalytic dye removal: insights into catalyst stability and reusability. Emerg Mater 1–17
25. Kumari H, Sonia S, Ranga R, Chahal S, Devi S, Sharma S, Kumar S, Kumar P, Kumar S (2023) A review on photocatalysis used for wastewater treatment: dye degradation. Water Air Soil Pollut 234:349
26. Kundu S, Sarkar T, Al-Ahmadi AA, Ali E, Bhattacharjee A (2024) Thermally synthesized hematite (α-Fe_2O_3) nanoparticles as efficient photocatalyst for visible light dye degradation. RSC Adv 14:28944–28955
27. León-Flores J, Pérez-Mazariego JL, Marquina M, Gómez R, Escamilla R, Tehuacanero-Cuapa S, Reyes-Damián C, Arenas-Alatorre J (2023) Controlled formation of hematite—magnetite nanoparticles by a biosynthesis method and its photocatalytic removal potential against methyl orange dye. J Cluster Sci 34:2381–2395
28. Li M, Li X, Qi X, Luo F, He G (2015) Shape-controlled synthesis of magnetic iron oxide@ SiO_2–Au@ C particles with core–shell nanostructures. Langmuir 31:5190–5197
29. Liu Q, Barrón V, Torrent J, Qin H, Yu Y (2010) The magnetism of micro-sized hematite explained. Phys Earth Planet Inter 183:387–397
30. Lubis S, Ramli M, Ermanda Y (2020) Hydrothermal synthesis of activated carbon/α-Fe_2O_3 nanocomposite and its application for removing tartrazine dye. IOP Conf Ser: Mater Sci Eng 012061
31. Mansour H, Letifi H, Bargougui R, De Almeida-Didry S, Negulescu B, Autret-Lambert C, Gadri A, Ammar S (2017) Structural, optical, magnetic and electrical properties of hematite (α-Fe_2O_3) nanoparticles synthesized by two methods: polyol and precipitation. Appl Phys A 123:1–10
32. Muhajir M, Puspitasari P, Razak JA (2019) Synthesis and applications of Hematite α-Fe_2O_3: a review. J. Mech. Eng. Sci. Technol 3:51–58
33. Ngulube KF, Abdelhaleem A, Osman AI, Peng L, Nasr M (2024) Advancing sustainable water treatment strategies: harnessing magnetite-based photocatalysts and techno-economic analysis for enhanced wastewater management in the context of SDGs. Environ Sci Pollut Res 1–37
34. Park J, Kang J, Chaule S, Jang J-H (2023) Recent progress and perspectives on heteroatom doping of hematite photoanodes for photoelectrochemical water splitting. J Mater Chem A
35. Pattanayak N, Bhattacharyya A, Chakravarty S, Bajpai A (2019) Weak ferromagnetism and time-stable remanence in hematite: effect of shape, size and morphology. J Phys: Condens Matter 31:365802
36. Powell CD, Lounsbury AW, Fishman ZS, Coonrod CL, Gallagher MJ, Villagran D, Zimmerman JB, Pfefferle LD, Wong MS (2021) Nano-structural effects on Hematite (α-Fe_2O_3) nanoparticle radiofrequency heating. Nano Convergence 8:1–9
37. Raming TP, Winnubst AJA, van Kats CM, Philipse AP (2002) The synthesis and magnetic properties of nanosized hematite (α-Fe_2O_3) particles. J Colloid Interface Sci 249:346–350
38. Rapadas NJ, Balela MDL (2017) Hydrothermal synthesis of hierarchical hematite (α-Fe_2O_3) microstructures for photocatalytic degradation of methyl orange. Philipp J Sci 146:396

39. Rasheed RT, Al-Algawi SD, Kareem HH, Mansoor HS (2018) Preparation and characterization of hematite iron oxide (α-Fe_2O_3) by sol-gel method. Chem Sci J 9
40. Rincón Joya M, Ortega JB, Malafatti JOD, Paris EC (2019) Evaluation of photocatalytic activity in water pollutants and cytotoxic response of α-Fe_2O_3 nanoparticles. ACS Omega 4:17477–17486
41. Rossman GR, Dyar M, McCammon C, Schaefer M (1996) Why hematite is red: correlation of optical absorption intensities and magnetic moments of Fe^{3+} minerals. In: Mineral spectroscopy: A tribute to Roger G. Burns, vol 5. Special Publication, pp 23–27
42. Roy JS, Messaddeq Y (2024) The role of solar concentrators in photocatalytic wastewater treatment. Energies 17:4001
43. Sagadevan S, Sivasankaran RP, Anita Lett J, Fatimah I, Weldegebrieal GK, Léonard E, Le M-V, Soga T (2023) Evaluation of photocatalytic activity and electrochemical properties of hematite nanoparticles. Symmetry 15:1139
44. Sakamoto M, Fujita R, Nishikawa M, Hirazawa H, Ueno Y, Yamamoto M, Takaoka S (2024) Hematite (α-Fe_2O_3) with oxygen defects: the effect of heating rate for photocatalytic performance. Materials 17:395
45. Scopel E, Conti PP, Stroppa DG, Dalmaschio CJ (2019) Synthesis of functionalized magnetite nanoparticles using only oleic acid and iron (III) acetylacetonate. SN Appl Sci 1:147
46. Shariatzadeh SMR, Salimi M, Fathinejad H, Joshaghani AH (2023) Synthesis, characterization, and photocatalytic application of α-Fe_2O_3/α-$AgVO_3$ nanocomposite for removal of methylene blue from aqueous media. Iran J Chem Chem Eng Res 42
47. Shariatzadeh SMR, Salimi M, Fathinejad H, Hassani Joshaghani A (2022) Nanostructured α-Fe_2O_3: solvothermal synthesis, characterization, and effect of synthesis parameters on structural properties. Int J Eng 35:1186–1192
48. Shenoy S, Farahat MM, Chuaicham C, Sekar K, Ramasamy B, Sasaki K (2023) Mixed-phase Fe_2O_3 derived from natural hematite ores/C_3N_4 Z-scheme photocatalyst for ofloxacin removal. Catalysts 13:792
49. Shinde SS, Bansode RA, Bhosale CH, Rajpure KY (2011) Physical properties of hematite α-Fe_2O_3 thin films: application to photoelectrochemical solar cells. J Semicond 32:013001
50. Singh RK, Tirth V, Singh M (2022) Wastewater treatment processes with special reference to activated sludge process in Indian conditions for water use sustainability. In: Wastewater assessment, treatment, reuse and development in India. Springer
51. Splinter K, Möckel R, Hlawacek G, Lendzion-Bieluń Z (2024) Studies of the morphology of hematite synthesized from waste iron sulfate. Molecules 29:3527
52. Suman, Chahal S, Kumar A, Kumar P (2020) Zn doped α-Fe_2O_3: an efficient material for UV driven photocatalysis and electrical conductivity. Crystals 10:273
53. Supattarasakda K, Petcharoen K, Permpool T, Sirivat A, Lerdwijitjarud W (2013) Control of hematite nanoparticle size and shape by the chemical precipitation method. Powder Technol 249:353–359
54. Tadic M, Panjan M, Tadic BV, Lazovic J, Damnjanovic V, Kopani M, Kopanja L (2019) Magnetic properties of hematite (−FeO) nanoparticles synthesized by sol-gel synthesis method: the influence of particle size and particle size distribution. J Electr Eng 70:71–76
55. Valášková M, Tokarský J, Pavlovský J, Prostějovský T, Kočí K (2019) α-Fe_2O_3 nanoparticles/vermiculite clay material: structural, optical and photocatalytic properties. Materials 12:1880
56. Wahab R, Khan F, Al-Khedhairy AA (2018) Hematite iron oxide nanoparticles: apoptosis of myoblast cancer cells and their arithmetical assessment. RSC Adv 8:24750–24759
57. Waite JL, Hunt J, Ji H (2021) Improving photocatalytic performance using nanopillars and micropillars. Materials 14:299
58. Wu W, Wu Z, Yu T, Jiang C, Kim W-S (2015) Recent progress on magnetic iron oxide nanoparticles: synthesis, surface functional strategies and biomedical applications. Sci Technol Adv Mater 16:023501
59. Xue Y, Wang Y (2020) A review of the α-Fe_2O_3 (hematite) nanotube structure: recent advances in synthesis, characterization, and applications. Nanoscale 12:10912–10932

60. Yassin MT, Al-Otibi FO, Al-Askar AA (2023) Photocatalytic removal of crystal violet dye utilizing greenly synthesized iron oxide nanoparticles. Separations 10:513
61. Zheng Z, He J, Zhang Z, Kumar A, Khan M, Lung CW, Lo IMC (2024) Magnetically recyclable nanophotocatalysts in photocatalysis-involving processes for organic pollutant removal from wastewater: current status and perspectives. Environ Sci Nano 11:1784–1816
62. Zhou Z, Yin H, Zhao Y, Zhang J, Li Y, Yuan J, Tang J, Wang F (2021) Synthesis of magnetic α-Fe_2O_3/rutile TiO_2 hollow spheres for visible-light photocatalytic activity. Catalysts 11:396

CuO-Based Photocatalysts in Wastewater Purification

Mahmoud A. Ahmed and Ashraf A. Mohamed

Abstract Heterogeneous photocatalysis has emerged as a promising tool for the remediation of various pollutants in effluent, with nanoscale metal oxide materials, such as cupric oxide (CuO), garnering robust attention owing to their distinctive size-dependent physicochemical features, as well as their potential applications. CuO has been extensively examined as a promising material for photocatalytic degradation, owing to its non-toxicity, low price, and powerful absorption of a substantial portion of the solar spectrum. However, the performance of CuO has been primarily hindered by the quick recombination phenomena as well as the slow mobility of photogenerated charge carriers. This chapter aims to comprehensively summarize the preparation strategies utilized in the fabrication, modification, and application of CuO-based photocatalysts in wastewater treatment. It also discusses viable strategies to promote their photocatalytic role, such as the generation of ternary and binary heterojunctions by, e.g., incorporating carbonaceous materials and metal ions as dopants, and the integration of other semiconductor materials. Furthermore, it delves into the mechanistic insights underlying the photogenerated charge separation and migration processes, as well as the identification of the functional reactive radical species involved in this process. Finally, the chapter concludes with remarks on the current state of research and the unresolved challenges related to CuO photocatalytic processes, highlighting the future perspectives and research directions in this field.

Keyword Photocatalytic pollutants' degradation · Nanomaterials · CuO-based heterostructures · Photocatalytic mechanisms

M. A. Ahmed (✉) · A. A. Mohamed
Chemistry Department, Faculty of Science, Ain Shams University, Cairo, Egypt
e-mail: mahmoudmahmoud_p@sci.asu.edu.eg

A. Ahmad et al. (eds.), *Metal Oxide Based Nanophotocatalyst for Wastewater Purification*, Smart Nanomaterials Technology,
https://doi.org/10.1007/978-981-96-7683-5_9

1 Introduction

The discharge of polluted wastewater presents a formidable challenge to global water security and environmental sustainability [1–5]. Wastewater streams constitute a complex matrix of pollutants, encompassing a diverse range of organic and inorganic pollutants. Persistent organic pollutants (POPs), notorious for their environmental persistence, bioaccumulation, and long-range transport, pose a significant threat to human and ecosystem health [6, 7]. Examples of POPs include polychlorinated biphenyls (PCBs), dioxins, organochlorine pesticides (e.g., DDT), and polycyclic aromatic hydrocarbons (PAHs) [8–11]. These compounds can exert a range of toxic effects, including endocrine disruption, carcinogenicity, and neurotoxicity. Beyond POPs, wastewater also contains other hazardous organic pollutants such as pharmaceuticals and personal care products (PPCPs), including antibiotics, hormones, analgesics, and antimicrobial agents [12, 13]. These emerging pollutants, even at low concentrations, can disrupt endocrine function, promote antibiotic resistance, and pose risks to aquatic life. Industrial effluents often contribute a significant load of organic pollutants, including dyes, solvents, and heavy metals [14]. Dyes, widely used in textile, leather, and paper industries, not only impart undesirable color to water bodies but also reduce light penetration, affecting aquatic ecosystems [15–18]. Heavy metals, such as lead, mercury, cadmium, and chromium, are toxic and can bioaccumulate in the food chain, posing serious health risks. Furthermore, wastewater contains pathogens, nutrients (nitrogen and phosphorus), and suspended solids, which contribute to eutrophication, oxygen depletion, and the spread of waterborne diseases [19–22]. Figure 1 summarizes effluent treatment approaches. Conventional wastewater treatment methods, such as biological processes (activated sludge, trickling filters), coagulation-flocculation, sedimentation, and disinfection, often prove insufficient for the complete removal of these diverse and recalcitrant pollutants [22–27].

Physicochemical methods like ion exchange and membrane filtration are also employed, but they too possess limitations. Ion exchange resins can selectively remove charged pollutants, but they require periodic regeneration and can be susceptible to fouling. Membrane technologies, such as microfiltration, ultrafiltration, nanofiltration, and reverse osmosis, can effectively remove a wide range of pollutants, but they are susceptible to membrane fouling, require high operating pressures, and generate concentrate streams that require further treatment [28]. These conventional methods may simply transfer pollutants from one phase to another rather than achieving complete degradation and mineralization, and they often struggle with low concentrations of emerging pollutants and highly persistent compounds [17]. This necessitates the development of advanced and efficient treatment methodologies capable of addressing the complex pollution challenges posed by wastewater discharge.

Advanced oxidation processes (AOPs) have emerged as a promising class of wastewater treatment approaches capable of addressing the limitations of conventional methods [29]. AOPs are characterized by the in situ generation of highly

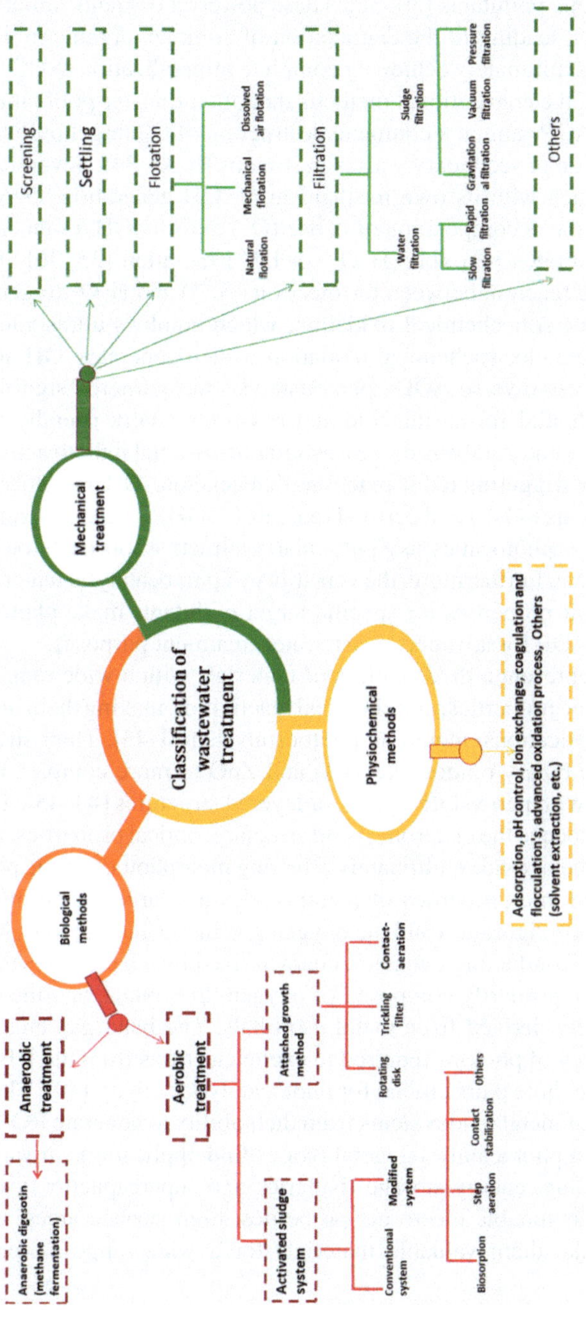

Fig. 1 Various approaches in wastewater treatment

reactive oxygen species (ROS), primarily hydroxyl radicals ($^{\cdot}OH$), which possess a high oxidation potential ($E^{\circ} = 2.8$ V) and can non-selectively attack and mineralize a wide range of organic pollutants [30–32]. These powerful oxidants initiate a cascade of radical reactions, leading to the degradation of complex organic molecules into simpler byproducts, ultimately achieving complete mineralization to CO_2, H_2O, and inorganic salts. Unlike conventional methods that often transfer pollutants from one phase to another, AOPs aim at a complete destruction of the target pollutants, minimizing the formation of secondary wastes in streams [6, 33, 34]. Several AOPs have been developed, each with its own mechanism of $^{\cdot}OH$ generation. Ozonation, for instance, relies on the decomposition of ozone (O_3) to form $^{\cdot}OH$, often catalyzed by the addition of hydrogen peroxide (H_2O_2) or UV irradiation [35, 36]. The Fenton process utilizes the reaction between ferrous iron (Fe^{2+}) and H_2O_2 to generate $^{\cdot}OH$. Other AOPs include sonochemical oxidation, which employs ultrasonic cavitation to generate $^{\cdot}OH$, and electrochemical oxidation, which generates $^{\cdot}OH$ at the anode surface. Among these diverse AOPs, photocatalysis has garnered significant attention due to its potential for sustainable and environmentally friendly wastewater treatment [37–39]. Photocatalysis harnesses solar or artificial light to activate a semiconductor material, triggering redox reactions that generate $^{\cdot}OH$ and other ROS, such as superoxide radicals ($^{\cdot}O_2^-$) and peroxyl radicals ($^{\cdot}OOH$). The use of sunlight as an energy source makes photocatalysis a particularly attractive approach for sustainable wastewater treatment. Furthermore, the versatility of photocatalytic materials and the ability to tailor their properties for specific target pollutants make photocatalysis a highly promising AOP for advanced wastewater treatment purposes.

Metal oxides represent a diverse class of materials with a wide range of crystal structures, electronic properties, and chemical reactivities, making them highly versatile for various applications, including photocatalysis [40–43]. Their structures can range from simple binary oxides like TiO_2 and ZnO to more complex ternary and quaternary oxides with perovskite, spinel, or layered structures [44, 45]. These structural variations influence the electronic band structure, optical properties, and surface characteristics of metal oxides, ultimately affecting their photocatalytic performance [46–48]. The electronic properties of metal oxides are largely determined by the metal's d-electron configuration and the oxygen p-orbital's interactions. Many metal oxides exhibit semiconducting behavior, characterized by a band gap separating the valence band (VB), primarily composed of oxygen $2p$ orbitals, and the conduction band (CB), typically derived from metal d-orbitals. The band gap energy dictates the minimum energy of photons required to excite electrons from the VB to the CB, generating electron–hole pairs crucial for photocatalytic activity [49]. The photocatalytic importance of metal oxides stems from their ability to generate ROS upon light irradiation. Beyond photocatalysis, metal oxides find applications in various fields, including gas sensing, energy storage (batteries and supercapacitors), electronics, and catalysis. Their tunable electronic properties, high surface areas, and diverse functionalities make them valuable materials for a wide range of technological applications.

Among them, cupric oxide (CuO), an earth-abundant and cost-effective material, shows promise in photocatalysis due to its unique structure and optoelectronic properties [50, 51]. Unlike typical transition metal monoxides, CuO's monoclinic crystal structure features distorted square planar CuO_4 and tetrahedral OCu_4 units. This asymmetry enhances surface reactivity, adsorption, and provides a robust surface area for catalytic activity. The monoclinic structure also influences CuO's electronic band structure and charge transport [52]. Further, CuO's p-type semiconductivity and narrow band gap energy (1.2–2.1 eV [53, 54]) enable significant visible light absorption, crucial for solar-driven photocatalysis. Its valence band comprises mainly O 2*p*-orbitals, while the conduction band consists of Cu 3*d*-orbitals. This configuration allows photogenerated holes (VB) and electrons (CB) to drive redox reactions and generate ROS for pollutant degradation. However, CuO suffers from inherent limitations that impede its practical implementation [55, 56]. Specifically, quick recombination, arising from short charge carrier diffusion lengths and lifetimes within the CuO lattice, severely restricts its photocatalytic efficiency [57, 58]. Although possessing a narrow band gap, CuO's light absorption is predominantly confined to the shorter wavelengths of the visible spectrum, leaving a significant portion of solar irradiation unutilized [58–60]. Moreover, photocorrosion, particularly in aqueous media, leads to oxidative degradation of the CuO structure, further diminishing its long-term activity. These drawbacks are compounded by inherently low quantum efficiency and restricted charge carrier mobility, hindering efficient interfacial charge transfer for redox reactions. However, enhancing the performance of CuO as a photocatalyst can be achieved through several strategies, as shown in Fig. 2.

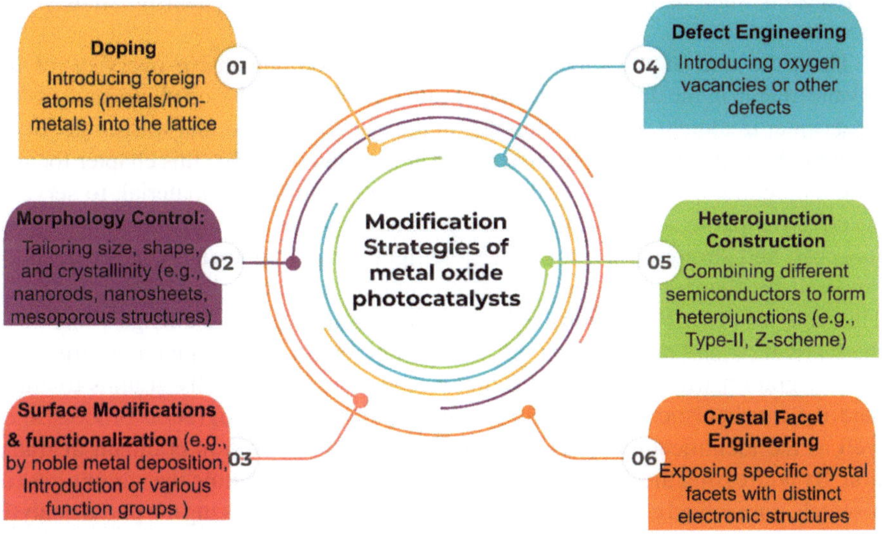

Fig. 2 Summary of modification techniques of copper-based photocatalysts

Analysis using Scopus data mapped the evolving landscape of copper oxide nanocomposite photocatalysis for wastewater treatment. Employing keywords: "CuO oxide" photocatalysis, 740 documents were retrieved as shown in Fig. 3a. These encompass 720 research articles, 20 reviews, and other publication types, as shown in Fig. 3b. Thus, Fig. 3a visualizes the exponential growth in publications, particularly over the past decade, underscoring the burgeoning interest in this field. Furthermore, network visualization mapping of copper oxide-based photocatalysis research output, indexed in PubMed, reveals several key themes and trends, as shown in Fig. 3c. "Water purification" and "water pollutants" occupy a central position, highlighting the dominant application of CuO oxide photocatalysis in environmental remediation. "Adsorption" is another prominent node, indicating the importance of this process alongside photocatalysis. Clusters of keywords related to specific cocatalyst materials (e.g., "iron," "titanium", "metal nanoparticles") appear, showcasing the diversity of research within the field. Terms like "oxidation–reduction," "hydroxyl radical," "hydrogen peroxide," "catalysis," and "light" indicate a focus on the underlying photocatalytic mechanisms. The presence of "photolysis" and "photochemistry" further reinforces this. The inclusion of terms like "surface properties," "particle size," "microscopy," and "materials testing" suggests research efforts aimed at understanding and optimizing the properties of these photocatalysts. The dense network of links illustrates the strong interrelationships between these concepts. The thickness of the lines likely represents the co-occurrence frequency, indicating stronger associations between certain keywords. For instance, "water purification" is strongly linked to "water pollutants" and various specific pollutants. The visualization appears to identify four main clusters, potentially representing different facets of research, such as material synthesis, characterization, mechanistic studies, and specific applications.

Thus, the aim of this chapter is to provide a comprehensive overview of the advancements in the development and optimization of CuO-based photocatalysts for wastewater treatment. Recognizing the limitations of conventional wastewater treatment methods in addressing persistent and emerging pollutants, this chapter focuses on the potential of CuO, an earth-abundant and cost-effective material, to serve as an effective photocatalyst. Despite its promising properties, such as non-toxicity, low cost, and strong absorption of visible light, CuO's photocatalytic efficiency is hindered by the rapid charge carrier recombination and limited charge mobility. This chapter describes various strategies to overcome these challenges, including the fabrication and modification of CuO through doping, heterojunction formation, and integration with other semiconductor materials. Additionally, it aims to elucidate the mechanistic insights into charge separation and migration processes, as well as the identification of reactive radical species involved in pollutant degradation. By addressing these aspects, the chapter aims at highlighting the current state of research, unresolved challenges, and future directions in enhancing the photocatalytic performance of CuO for sustainable wastewater treatment.

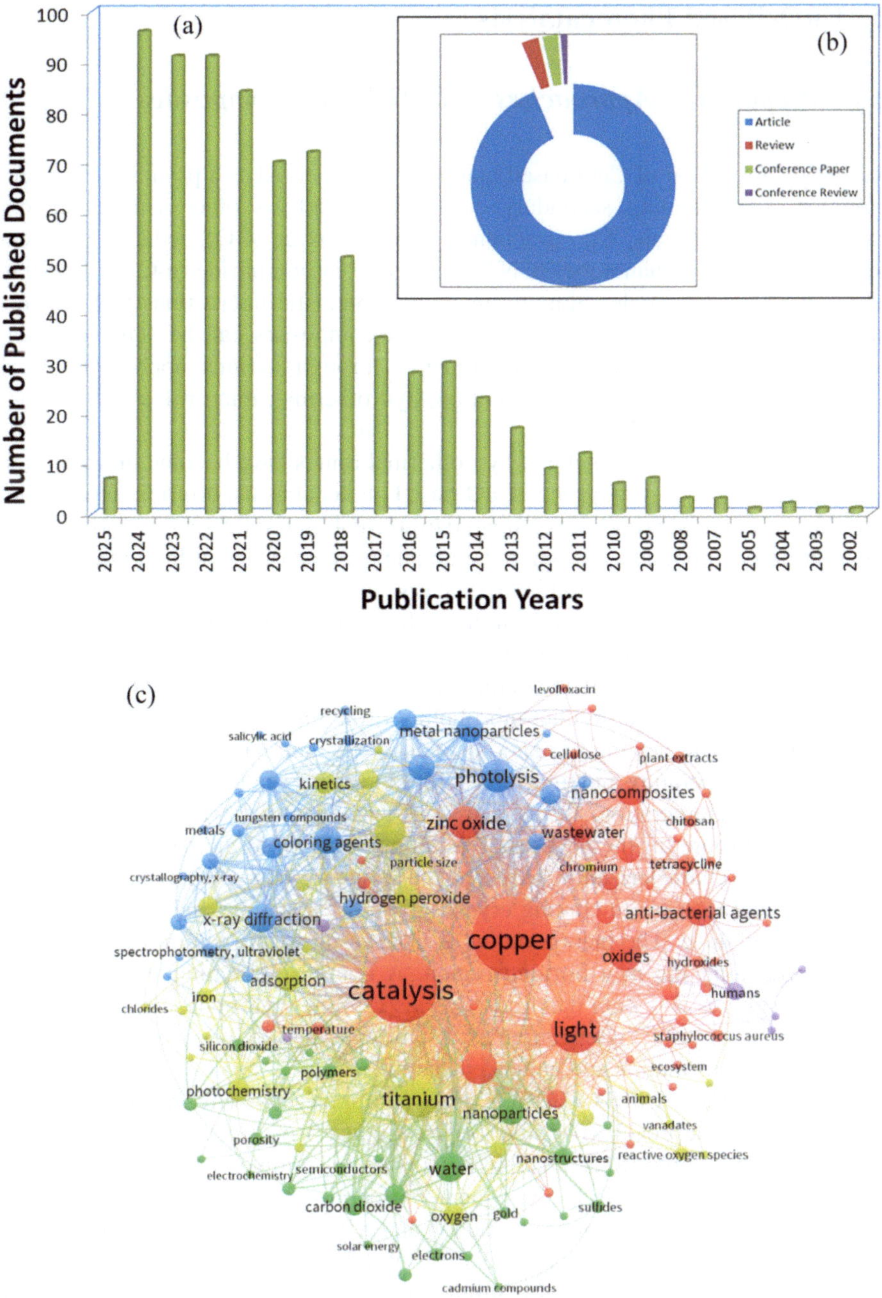

Fig. 3 **a** Number of publications in Scopus database reporting photocatalysis by CuO-based composites for wastewater treatment: keywords "CuO and photocatalysis", **b** type of documents, and **c** bibliometric network mapping of search keywords

2 CuO-Based Photocatalysts

2.1 Preparation Approaches of CuO-Based Composites

Recently, fabrication of CuO-based materials has emerged as a prominent research direction. Incorporating secondary components such as other semiconductors, carbonaceous materials, e.g., graphene, graphene oxide (GO), reduced graphene oxide (rGO), carbon nanotubes (CNTs), graphitic carbon nitride (g-C_3N_4), or plasmonic metal nanoparticles aims to engineer synergistic interactions that enhance photocatalytic performance. The synthesis of composites can be achieved using in situ or ex situ crystallization strategies through various approaches, such as sol–gel, solvothermal, co-precipitation, hydrothermal, microwave-assisted, and ultrasound-assisted methods.

For instance, Solvothermal and hydrothermal routes have become indispensable tools in the synthesis of advanced CuO-based photocatalysts, offering unparalleled control over material properties compared to conventional solid-state methods. These solution-based approaches leverage the unique characteristics of high-temperature, high-pressure environments to manipulate crystal growth, morphology, and composition, ultimately dictating photocatalytic performance [61]. Hydrothermal synthesis, utilizing water as the solvent, provides a facile route to crystalline CuO nanostructures, with solution pH, temperature, and reaction duration acting as critical determinants of the final product morphology. Alkaline conditions generally favor anisotropic growth, leading to nanorods, nanowires, and nanosheets, while precise control over temperature and time allows for tailoring crystallite size and preventing detrimental agglomeration. Thus, ZnO/CuO nanomaterial was fabricated via a hydrothermal route for the photocatalytic mineralization of the pesticide Triclopyr [62]. Material characterization reflected the influence of synthesis parameters on morphological features, including crystal growth, nucleation, and particle size, as shown in Fig. 4a–e. The incorporation of CuO resulted in a minimized ZnO band gap, approaching the near-infrared region. A composite architecture comprising CuO nanospheres and ZnO nanorods was also synthesized hydrothermally, where the utilization of polyethyleneimine as a stabilizing and reducing agent facilitated nanocomposite formation and enhanced stability [63].

In another work, CuO–CoO/TiO_2 composite materials were synthesized using a hydrothermal approach [64]. The TiO_2, $Cu(acac)_2$/$Co(acac)_3$, and $Cu(NO_3)_2$/$Co(NO_3)_2$ were employed as metallic precursors for samples 1 and 2, respectively. The anatase phase of the TiO_2 component was confirmed through X-ray diffraction and Raman spectroscopic analyses. Raman and XPS data provided evidence for the existence of CuO and CoO phases. Distinct morphological features were observed through TEM analysis; sample 1 exhibited nanotube/nanosheet structures, whereas sample 2 displayed nanoparticle morphology. Furthermore, solvothermal synthesis expands this versatility by incorporating organic solvents, opening new avenues for manipulating reaction kinetics and surface properties. The choice of solvent profoundly influenced the final product, where solvents with high boiling points and

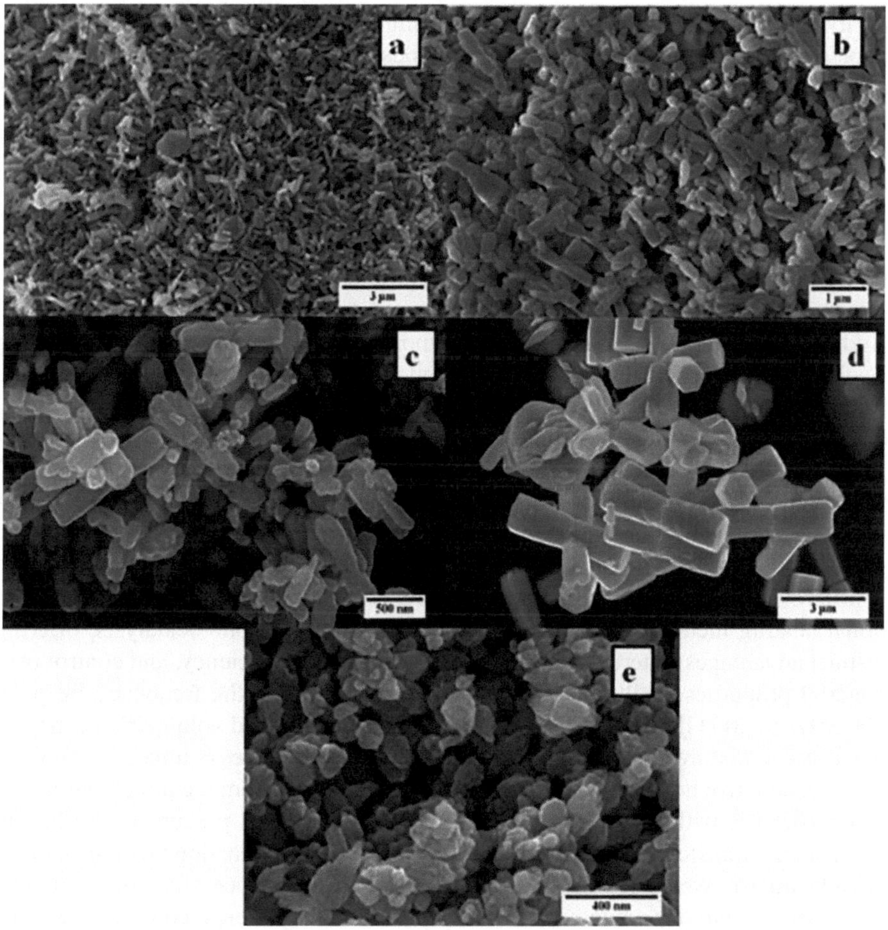

Fig. 4 FESEM image of ZnO/CuO composite with molar compositions (8:1), Zn8Cu, synthesized at various synthesis conditions: **a** $T = 160$ °C, pH = 9; **b** $T = 180$ °C, pH = 9; **c** $T = 200$ °C, pH = 9; **d** $T = 180$ °C, pH = 7; and **e** $T = 180$ °C, pH = 11

low viscosities promoted the growth of well-defined nanostructures. Coordinating solvents can further direct crystal growth, yielding anisotropic morphologies with enhanced surface area and active site exposure. This ability to adjust the solvent environment provides delicate control over the physicochemical properties of the resulting CuO nanomaterials, influencing light absorption, charge carrier dynamics, and ultimately, photocatalytic efficiency. Moreover, the advantages of these solution-based methods are manifold. The lower temperature and pressure requirements compared to solid-state synthesis translate to reduced energy consumption and simplified experimental setups. Precise control over reaction parameters enables fine-tuning of morphology, crystallinity, and particle size distribution, crucial for optimizing photocatalytic performance. The closed reaction environment ensures high purity and homogeneity, minimizing contamination and promoting uniform nucleation and growth. Furthermore, these methods are inherently amenable to doping and composite formation, allowing for the incorporation of secondary components like graphene or plasmonic nanoparticles directly during synthesis, leading to synergistic enhancements in photocatalytic activity. For instance, a one-pot solvothermal technique was utilized to produce TiO_2-rGO-CuO/Fe_2O_3 ternary composites [65]. In a separate study, dendritic CeO_2–CuO was synthesized through a straightforward solvothermal process employing ethylene glycol [66]. The impact of reaction temperature, duration, surfactant type and amount, and precursor stoichiometry on the morphology of the resulting nanomaterials was thoroughly examined. Additionally, CuO/g-C_3N_4 (CuOCN) composite materials were synthesized using a solvothermal approach to assess the ciprofloxacin (CIP) degradation performance [67]. TEM imaging revealed a layered architecture for both the CuO and g-C_3N_4 constituents, indicating close interfacial interaction. Diffuse reflectance spectroscopic (DRS) measurements established a relationship between the quantity of CuO integrated and the optical properties of the composite. Specifically, higher CuO loadings corresponded to an extended visible light absorption range and reduced band gap energy.

Microwave-assisted synthesis has emerged as a compelling alternative to conventional heating methods for the preparation of CuO-based photocatalysts, offering distinct advantages in terms of reaction kinetics, energy efficiency, and control over material properties. Utilizing electromagnetic radiation in the frequency range of 300 MHz to 300 GHz, microwave heating provides rapid and volumetric heating of the reaction mixture, enabling significantly faster reaction rates compared to conventional conductive heating. This accelerated heating arises from the direct interaction of microwaves with the polar molecules in the solvent and precursors, leading to rapid energy transfer and localized superheating within the reaction medium. Consequently, microwave-assisted synthesis drastically reduces processing times, often by orders of magnitude, while simultaneously minimizing energy consumption. The unique heating mechanism of microwave irradiation also offers enhanced control over the nucleation and growth processes during material synthesis. The rapid and uniform heating promotes homogeneous nucleation, leading to a narrower particle size distribution and improved control over morphology. This precise control over crystal growth kinetics allows for the tailoring of specific material properties, such as surface area, pore size distribution, and crystallinity, which are crucial for optimizing adsorption and photocatalytic performance. Moreover, microwave heating can facilitate the formation of metastable phases and unique morphologies that are inaccessible via conventional heating routes, opening avenues for the discovery of novel CuO-based photocatalytic materials. Furthermore, microwave-assisted synthesis offers inherent advantages for the fabrication of CuO composites with other semiconductors, carbon materials, or noble metals. The rapid heating and localized superheating can promote intimate mixing and interfacial bonding between the different components, leading to the formation of highly dispersed and well-integrated composite structures. This enhanced interfacial contact is critical for efficient charge separation and transfer in photocatalytic applications. For instance, in graphene-CuO composites, microwave irradiation can facilitate the uniform anchoring of CuO nanoparticles onto the graphene sheets, maximizing interfacial area and promoting synergistic interactions for enhanced photocatalytic activity. Furthermore, a ZnO/CuO nanocomposite with a 2.0 molar ratio, synthesized via a rapid microwave-assisted hydrothermal method at 70 °C for 10 min, demonstrated a notable photocurrent density exceeding 1.0 mA cm^{-2} [68]. Moreover, silver-doped CuO anchored on graphene oxide (Ag-CuO@GO) was fabricated using an energy-efficient microwave heating technique (Fig. 5a) [69]. The resulting ternary nanocomposite exhibited a reduced optical band gap (Eg) of 1.35 eV that is lower than both bare CuO and Ag-CuO, suggesting enhanced potential for solar energy harvesting and photocatalysis. Furthermore, the suppressed PL observed in the Ag-CuO@GO nanomaterial indicated effective charge carrier separation facilitated by the incorporation of silver and graphene oxide [69].

On the other hand, ultrasound irradiation has emerged as a compelling tool in the synthesis of advanced materials, offering unique advantages for preparing CuO-based photocatalysts with enhanced performance. Distinct from conventional methods, the application of ultrasound introduces acoustic cavitation, a phenomenon characterized by the formation, growth, and implosive collapse of microbubbles within the liquid medium. These implosions generate localized hotspots with transient high temperatures and pressures, along with intense microstreaming and shockwaves. These extreme conditions drastically alter the reaction kinetics, nucleation, and growth dynamics, ultimately influencing the morphology, crystallinity, and surface properties of the resulting CuO nanostructures. The enhanced reactivity under sonochemical conditions stems from several key mechanisms. Firstly, the rapid heating and cooling rates within the cavitation bubbles facilitate the formation of highly reactive species, promoting faster reaction rates and shorter synthesis times. Secondly, the intense microstreaming and shockwaves generated during bubble collapse enhance mass transport and mixing, leading to a more homogeneous distribution of reactants and improved control over particle size distribution. Thirdly, the localized extreme conditions can induce unique phase transformations and promote the formation of metastable phases, offering access to novel CuO nanostructures with distinct properties. Furthermore, the application of ultrasound can facilitate the deagglomeration of particles, preventing aggregation and increasing the active surface area of the catalyst. The versatility of ultrasound-assisted synthesis allows for its integration with various established synthetic routes, including sol–gel, hydrothermal, and solvothermal methods. In the context of CuO synthesis, sonochemical treatment can significantly influence the morphology and structure of the resulting nanomaterials. Studies have demonstrated the formation of unique morphologies such as nanowires, nanorods, nanosheets, and hierarchical structures under ultrasonic irradiation, offering a wider range of controllable morphologies compared to conventional methods. Furthermore, ultrasound can facilitate the incorporation of dopants or secondary components into the CuO lattice, enabling the fabrication of complex composite materials with enhanced photocatalytic activity. For instance, the formation of CuO-carbon-based nanocomposites, where carbon materials like

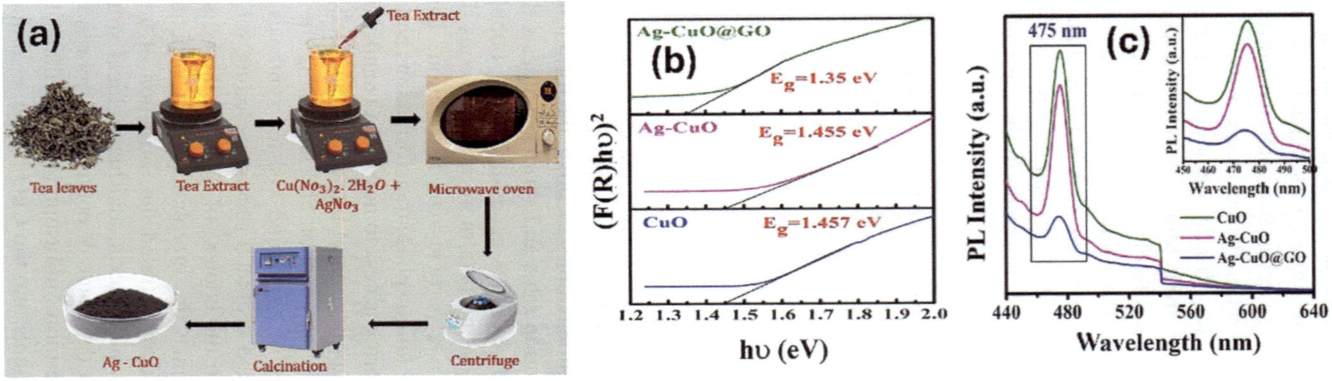

Fig. 5 **a** Step by step synthesis procedure of Ag-CuO NPs, **b** Tauc's plot, and **c** photoluminescence spectra of bare CuO, binary Ag-CuO, and ternary Ag-CuO@GO nanocomposite

graphene or CNTs are intimately integrated with CuO nanoparticles, can be significantly improved through sonochemical treatment, leading to enhanced charge separation and improved photocatalytic performance. The benefits of ultrasound-assisted synthesis extend beyond morphological control, enhancing reactivity and improving mass transport, leading to higher yields and reducing reaction times compared to conventional methods. Moreover, the sonochemical approach can often be carried out at lower temperatures and pressures, reducing energy consumption and simplifying the experimental setup. However, careful optimization of sonication parameters, including frequency, power, and duration, is crucial to achieve the desired results. Excessive sonication can lead to particle fragmentation and degradation, while insufficient sonication may not provide the necessary activation energy for the desired reactions. For instance, S-scheme CuO/CeO_2 heterojunctions were fabricated using a sonochemical method, with optimized CeO_2 and CuO nanoparticle ratios to enhance the overall photocatalytic activity [70]. SEM and TEM analyses confirmed the uniform dispersion of CuO nanoparticles on the CeO_2 surface. Another heterojunction containing different levels of ZnO and CuO was fabricated for the mineralization of RhB, where spherical CuO was deposited onto ZnO by ultrasonic approach [71]. The homogeneous growth of CuO on ZnO enhanced visible light absorption and promoted charge transportation and separation phenomena [71].

The sol–gel method stands as a versatile and widely employed bottom-up approach for synthesizing CuO-based photocatalysts, offering a high degree of control over the final material's physicochemical properties [72]. This technique involves the formation of a stable sol, followed by its transformation into a gel through hydrolysis and condensation reactions [73]. Subsequent thermal treatment converts the gel into the desired CuO phase. The inherent flexibility of the sol–gel process allows for the incorporation of dopants and the formation of composite materials, making it particularly attractive for tailoring photocatalytic performance [74]. A typical sol–gel synthesis of CuO begins with the dissolution of a suitable copper precursor, such as copper nitrate or copper acetate, in a solvent, often an alcohol [75, 76]. The addition of a chelating agent, like citric acid or ethylenediaminetetraacetic acid (EDTA), can enhance the stability and homogeneity of the sol by controlling the rate of hydrolysis and preventing premature precipitation. The temperature, pH, and the water-to-precursor ratio play a crucial role in influencing the hydrolysis and condensation reactions, thus affecting the final morphology and texture of the gel network. Subsequent aging of the sol allows for further polymerization and cross-linking of the metal oxide network, strengthening the gel structure. The transformation of the sol into a gel is achieved through hydrolysis and condensation reactions, which are influenced by factors like temperature, pH, and the water-to-precursor ratio. Hydrolysis involves the reaction of the metal precursor with water, forming hydroxyl groups. Condensation reactions then link these hydroxyl groups, leading to the formation of metal–oxygen-metal bonds and the development of a three-dimensional gel network. The resulting gel, typically a viscous liquid or a solid with interconnected pores, is then subjected to drying and calcination. Drying removes the solvent and residual organic compounds, leading to, e.g., a xerogel or aerogel, depending on the drying method employed. Calcination at elevated temperatures converts the dried gel into the desired CuO phase [77]. The calcination temperature and duration significantly influence the crystallite size, surface area, and pore structure of the final product. Higher calcination temperatures generally lead to larger crystallite sizes and decreased surface area. Careful control of the calcination process is therefore essential for optimizing the photocatalytic performance of the resulting CuO material. Therefore, the sol–gel method offers several advantages for the synthesis of CuO photocatalysts [78]. The low processing temperatures minimize energy consumption and allow for the incorporation of thermally sensitive materials. The homogeneous mixing of precursors at the molecular level ensures uniform distribution of dopants and promotes the formation of highly dispersed composite materials. Furthermore, the sol–gel process offers excellent control over the pore structure and surface area, enabling the design of materials with enhanced light absorption and efficient charge transport. For instance, Fe_3O_4/CuO/ZnO ternary substance was fabricated by sol–gel approach [72]. Moreover, CuO thin films were fabricated by sol–gel assisted by spin coating method [79]. In another study involving the synthesis of CuO using the sol–gel method, XRD analysis of the morphology and structure of the generated metal oxide, confirmed the robust integration of the monoclinic Bravais lattice system [80]. The NPs of CuO were examined, revealing sizes ranging from 52 to 36 nm, which varied depending on the annealing temperatures [80]. In a similar study, SEM analysis revealed micron-sized agglomerated spheres for the sample created using the sol–gel technique. These spheres result from the mutual encapsulation of Al nanoparticles and CuO nanoparticles, with the particles being evenly distributed throughout the agglomerate [81].

In summary, the synthesis of these CuO-based composites employs a variety of techniques, each offering specific advantages and trade-offs. Sol–gel methods offer precise control over stoichiometry and porosity but often require lengthy processing times and high calcination temperatures. Hydrothermal and solvothermal synthesis enable the formation of crystalline materials with tailored morphologies at milder conditions, but scalability can be a challenge. Co-precipitation, is simple and cost-effective, nevertheless it provides limited control over particle size and morphology. Microwave-assisted synthesis offers rapid heating and reduced reaction times but can lead to non-homogeneous particle size distribution. Ultrasound-assisted methods enhance mixing and promote faster reaction kinetics but can complicate morphology control. The optimal choice of fabrication route depends on the specific requirements of the target application and the desired properties of the resulting composite. Further advancements in CuO-based photocatalysis hinge on a deeper understanding of interfacial phenomena within these composite architectures and the development of sophisticated synthetic strategies to precisely tailor morphology, composition, and electronic properties for enhanced photocatalytic performance in diverse applications such as water treatment, CO_2 reduction, and solar energy conversion.

2.2 Characterization Techniques of Composites

Optimizing the photocatalytic performance of CuO-based composites necessitates a profound understanding of the intricate interplay between their structural, morphological, optical, and electronic properties. This demands a sophisticated, multifaceted characterization strategy leveraging advanced techniques. Structural elucidation commences with XRD analysis, employing Rietveld refinement to precisely determine lattice parameters, crystallite size, and phase purity, providing crucial insights into the crystalline nature of the composite [82, 83]. Advanced microscopic techniques like HRTEM, coupled with SAED and EDS mapping, offer atomic-level insights into lattice fringes, defects, interfacial characteristics, and elemental distribution, elucidating the nanoscale architecture and composition [84–86]. SEM provides a comprehensive visualization of surface morphology, particle size distribution, and interparticle connectivity, while electron tomography enables 3D morphological reconstruction, unveiling the spatial arrangement of components within the composite [84, 87–89]. Crucially, nitrogen adsorption–desorption isotherms, analyzed via BET and BJH methods, quantify surface area, pore volume, and pore size distribution, parameters directly influencing reactant adsorption and active site accessibility [90–92]. Surface and interfacial characterization is paramount for understanding the chemical environment and electronic structure. XPS, augmented with depth profiling, provides a detailed analysis of surface elemental composition, chemical states, and defect concentrations, offering valuable insights into surface reactivity [90, 93]. The integration of Auger electron spectroscopy (AES) further enhances surface sensitivity and chemical specificity, enabling the identification of distinct chemical species. FTIR spectroscopy identifies surface functional groups and adsorbed species, complementing XPS data and providing a molecular-level understanding of surface chemistry [32, 94, 95]. Correlating FTIR and Raman spectroscopy yields a comprehensive picture of vibrational modes and chemical bonding within the composite, elucidating the structural integrity and interfacial interactions.

The optical and electronic properties govern the light-harvesting and charge transfer processes crucial for photocatalysis. UV–Vis DRS, coupled with advanced analysis using Tauc plots considering both direct and indirect band gap transitions, accurately determines the optical band gap, a critical parameter dictating the range of absorbed light [96–98]. PL spectroscopy, encompassing both steady-state and time-resolved measurements, probes the radiative and non-radiative recombination dynamics of photogenerated charge carriers, providing insights into charge separation efficiency, exciton lifetimes, and defect-mediated recombination pathways [99–102]. Furthermore, ultrafast techniques like transient absorption spectroscopy (TAS) provide unprecedented temporal resolution, unraveling the dynamics of excited states and charge transfer processes on femtosecond to picosecond timescales. Electrochemical characterization, utilizing Electrochemical impedance spectroscopy (EIS) and Mott-Schottky analysis, quantifies charge carrier density, flat band potential, and interfacial charge transfer resistance, providing crucial information on the electrochemical behavior and charge transfer kinetics at the electrode–electrolyte interface [101, 103].

Computational modeling and simulation offer a powerful tool for elucidating the underlying photocatalytic mechanisms. Density Functional Theory (DFT) calculations, employing appropriate exchange–correlation functionals and pseudopotentials, accurately predict electronic structure, band structure, density of states (DOS), and optical properties, providing a theoretical framework for understanding the electronic behavior [104–106]. Molecular dynamics (MD) simulations, utilizing reactive force fields, delve into reactant interactions, adsorption energies, reaction pathways, and the overall catalytic mechanism at the atomic level. Time-Dependent Density Functional Theory (TD-DFT) calculations can simulate UV–Vis absorption spectra, bridging the gap between theoretical predictions and experimental observations. By integrating these advanced characterization techniques with computational modeling, a comprehensive understanding of the structure–property-performance relationships in CuO-based composites can be achieved, paving the way for the rational design of highly efficient photocatalysts.

2.3 Mechanisms of Enhanced Photocatalysis by CuO-Based Composites

Optimizing the photocatalytic performance of CuO composites requires a sophisticated understanding of the intricate mechanisms governing light-matter interactions, charge carrier dynamics, and interfacial phenomena. Figure 6 shows a summary of the mechanisms responsible for the enhanced activity of copper oxide-based composites.

Boosting Light Absorption: Incorporating non-metal dopants like nitrogen or sulfur into the CuO lattice can narrow the band gap, extending light absorption into the visible region [104, 107, 108]. Surface plasmon resonance arising from noble metal nanoparticles deposited on CuO can also significantly enhance visible light absorption [108–110]. Furthermore, forming heterojunctions with semiconductors possessing complementary absorption spectra broadens the overall range of light harvested by the composite photocatalyst. The increased light absorption translates to a higher generation rate of electron–hole pairs, ultimately boosting photocatalytic efficiency.

Increasing Charge Carrier Mobility: Metal dopants can introduce additional energy levels within the CuO band gap, acting as stepping-stones for charge carrier transport and enhancing their mobility. The high electrical conductivity of carbon materials like graphene and CNTs, when integrated with CuO, provides efficient pathways for electron transport, further improving charge carrier mobility. This enhanced mobility reduces the probability of charge recombination, allowing more charge carriers to participate in redox reactions at the photocatalyst surface.

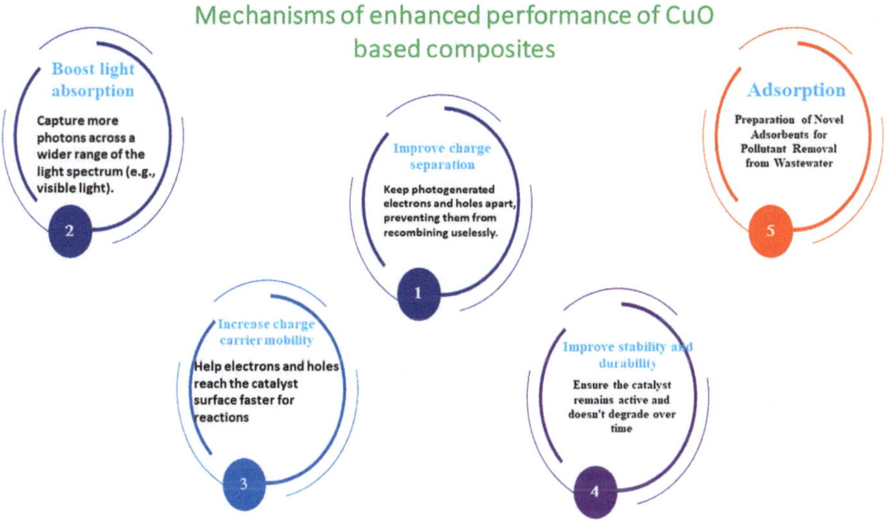

Fig. 6 Summary of the mechanisms responsible for the enhanced activity of copper oxide-based compounds

Improving Charge Separation: The creation of heterojunctions between CuO and other semiconductors establishes an internal electric field that drives photogenerated electrons and holes towards opposite sides of the interface [111–113]. Similarly, metal nanoparticles deposited on CuO act as electron sinks, effectively separating photogenerated charge carriers. Doping CuO with specific metal or non-metal ions can introduce trapping sites for electrons or holes, further promoting charge separation and inhibiting recombination. This improved charge separation maximizes the availability of charge carriers for photocatalytic reactions.

Enhancing Adsorption Capacities: Incorporating carbon materials like graphene, CNTs, or chitosan in the CuO composites significantly increases the surface area and porosity, leading to enhanced adsorption of pollutants onto the photocatalyst surface [114–120]. Chitosan, with its functional groups, also introduces specific adsorption sites, improving the selectivity of the composite towards certain pollutants [121–123]. This increased adsorption capacity ensures a higher concentration of reactants at the photocatalyst surface, promoting more efficient photocatalytic degradation.

Improving Stability and Durability: Doping CuO with certain metal or non-metal ions can stabilize the crystal structure, preventing photocorrosion and aggregation. The integration of carbon materials can provide a protective layer around CuO nanoparticles, enhancing their resistance to mechanical degradation and harsh reaction conditions. Heterojunction formation can also improve the stability of CuO by facilitating charge transfer and reducing the accumulation of reactive oxygen species that can damage the photocatalyst, ensuring long-term photocatalytic performance.

3 Application of CuO-Based Photocatalysts in Wastewater Remediation

3.1 Degradation of Organic Pollutants

AOPs heterogeneous photocatalysis has proved as a promising approach for the degradation and mineralization of organic pollutants in wastewater. This approach leverages the power of semiconductor photocatalysts to harness solar energy generating electron–hole pairs and ROS that drive redox reactions and transform harmful organic molecules into benign substances. This direct oxidation pathway can involve the abstraction of a hydrogen atom from the organic molecule, initiating a radical chain reaction that ultimately mineralizes the pollutant into CO_2, and H_2O. Alternatively, ˙OH radicals can then attack the organic pollutants non-selectively, leading to their degradation via a series of radical reactions. Recently, copper oxide-based photocatalysts have become a focal point in the study of the degradation of organic pollutants in wastewater. Combining CuO in nanostructured materials has shown promise for wastewater treatment due to enhanced visible light absorption. Thus, the researchers created a hybrid material by integrating low-bandgap copper oxide nanosheets with high-bandgap zinc oxide nanowires on various substrates [124]. This structure demonstrated superior pollutant removal compared to single metal oxides when tested for the degradation of methyl orange and rhodamine B dyes. The enhanced performance stemmed from staggered p-n junctions at the oxide interface, suppressing electron–hole recombination [124]. Another study investigated Congo red dye degradation using Cu_2O/titanium dioxide (CT), and Cu_2O/zinc oxide (CZ) hybrid photocatalysts synthesized via a solvothermal method [125]. While both CT and CZ showed strong adsorption and rapid activity, an optimized CZ composition completely degraded 30 ppm Congo red in 10 min. This impressive result was attributed to the p–n heterojunction facilitating efficient charge separation [125]. A different approach involved depositing Cu_2O nanoparticles onto hydrothermally grown zinc oxide nanorods [126]. The resulting bicatalysts exhibited a 50% improvement in conversion rate and quantum yield compared to bare nanorods. This enhancement resulted from the electric field at the Cu_2O/ZnO interface, promoting charge separation upon light absorption, as evidenced by suppressed photoluminescence and energy shifts. This effect was most prominent on the nanorod side facets, the primary sites for dye adsorption [126]. Researchers also developed visible-light-responsive p–n heterojunction photocatalysts based on CuO/MoO_3 with varying CuO ratios [127]. A 5% CuO/MoO_3 nanocomposite exhibited the best photocatalytic degradation of Rhodamine B and Alizarine Yellow dyes, maintaining excellent stability over five reusability cycles. This enhanced performance was attributed to increased visible light absorption, favorable surface morphology, and improved charge separation, inhibiting electron–hole recombination [127]. Another study synthesized Cu/Cu_2O/CuO nanocomposites using a wet chemical method for dye degradation [128]. Analysis confirmed the nanocomposite formation with an average particle size of approximately 17 nm and band gaps of 1.71 eV (CuO) and 1.92 eV (Cu_2O). The presence of oxygen vacancies was confirmed spectroscopically. The nanocomposite's enhanced performance was attributed to heterojunction formation and oxygen vacancies. It achieved 80 and 60% degradation of Congo red and Malachite green, respectively, after 120 min of light exposure [128]. A microwave method was employed to synthesize TiO_2–CuO systems [129]. Microscopy (TEM and HRTEM) confirmed the presence of octahedral and rod-shaped TiO_2 particles, sheet-like CuO particles, and the formation of a heterojunction between them. X-ray diffraction and Raman spectroscopy revealed anatase TiO_2 and monoclinic CuO. The binary oxides, particularly the 9:1 and 7:3 TiO_2:CuO ratios, exhibited significantly higher photocatalytic activity for tetracycline decomposition (95% degradation in 90 min) compared to pure TiO_2, suggesting a Z-scheme heterojunction mechanism [129]. A 0D/3D CuO/ZnO heterojunction was fabricated via in situ deposition of CuO nanoparticles onto flower-like ZnO structures using a wet chemistry method [130]. This heterojunction demonstrated improved visible-light absorption and charge carrier separation compared to pure ZnO. The CuO/ZnO composite achieved 78% phenol removal under light irradiation, approximately 2 and 4 times higher than pristine ZnO and CuO, respectively. Furthermore, the composite exhibited good stability and durability. Trapping experiments confirmed superoxide and hydroxyl radicals as the primary reactive species in the photocatalytic degradation process [130]. Finally, a 2D/2D CuO/g-C_3N_4 (Cu/CN) heterojunction photocatalyst was designed for peroxymonosulfate (PMS)-activated degradation of oxytetracycline (OTC) under visible light [131]. The optimized Cu/CN catalyst achieved near-complete OTC degradation within 10 min and effectively degraded other antibiotics like tetracycline, ciprofloxacin, levofloxacin, and sulfadiazine. Electron spin resonance and quenching experiments identified multiple reactive oxygen species involved in the degradation process, with hydroxyl and superoxide radicals playing the most significant roles [131].

4 Conclusion and Perspectives

This chapter has highlighted the potential of CuO-based photocatalysts for wastewater treatment, showcasing their effectiveness in degrading a wide range of organic pollutants under UV and visible light irradiation. CuO's narrow band gap, cost-effectiveness, and facile synthesis methods make it an attractive alternative to more expensive photocatalytic materials. Addressing this limitation has driven research towards modifying CuO nanostructures through various strategies. Elemental doping with suitable elements can introduce trapping sites for charge carriers, effectively prolonging their lifespan and promoting ROS generation. Coupling CuO with other semiconductors to form heterojunctions offers another avenue for enhancing charge separation. Specifically, Z-scheme and S-scheme heterojunctions are advantageous as they maintain the strong redox potential of the composite system while efficiently separating electrons and holes, unlike type-II heterojunctions, which can suffer from reduced redox potential due to inter-semiconductor charge transfer. Furthermore, optimizing the morphology, surface area, and dispersibility of CuO nanostructures plays a critical role in maximizing active sites and facilitating efficient interaction with pollutants. Beyond material modifications, environmental factors, including the nature and level of the target pollutant, solution pH, light intensity, the presence of other species in the wastewater matrix, and temperature, significantly influence the photocatalytic activity. Understanding the interplay of these factors is crucial for achieving optimal pollutants' degradation efficiency. While the pseudo-first-order kinetic model often describes the removal of organic pollutants, the actual reaction mechanism can be complex, involving multiple intermediate steps and competing reactions. While significant progress has been made, challenges remain in translating laboratory successes to practical applications, including:

1. Material Design and Synthesis: Developing scalable, environmentally friendly synthesis methods for producing CuO nanostructures with tailored morphologies, controlled doping, and optimized heterojunction interfaces remains a priority. Exploring novel materials and composite structures can further enhance photocatalytic performance and expand the range of treatable pollutants.
2. Enhancing stability and reusability: Addressing the photocorrosion and leaching issues of CuO is crucial for long-term application. Surface modifications, protective coatings, and the development of robust composite materials should be investigated to improve stability and recyclability.
3. Mechanistic understanding: A deeper understanding of the photocatalytic mechanisms involved in pollutant's degradation is essential for rational design and optimization. Advanced spectroscopic techniques and theoretical calculations can provide insights into charge transfer processes, reactive oxygen species generation, and degradation pathways.
4. Real wastewater application: Most studies have focused on model pollutants in controlled laboratory settings. Evaluating the performance of CuO-based photocatalysts in real wastewater matrices containing complex mixtures of pollutants is crucial for practical implementation. Pilot-scale studies and cost-effectiveness analyses are needed to assess the feasibility of large-scale applications.
5. Toxicity assessment: While CuO is considered relatively non-toxic, the potential release of Cu ions during photocatalysis and the formation of byproducts during pollutant degradation warrant further investigation. Toxicity studies should be conducted to ensure the environmental safety of CuO-based photocatalytic systems.

By addressing these challenges, CuO-based photocatalysts can emerge as a viable and sustainable approach for wastewater treatment, contributing to a cleaner and healthier environment.

References

1. Adel M et al (2022) Removal of heavy metals and dyes from wastewater using graphene oxide-based nanomaterials: a critical review. Environ Nanotechnol Monit Manage 18:100719

2. Krishnamoorthy K et al (2024) Selective synthesis of copper oxide nanocube for the electrochemical sensing of environmental aquatic pollutant hydroxylamine. Ionics 30(1):553–562
3. Ahmed A et al (2024) Synthesis of visible-light-responsive lanthanum-doped copper ferrite/graphitic carbon nitride composites for the photocatalytic degradation of toxic organic pollutants. Diam Relat Mater 141:110630
4. Ahmed MA, Amin S, Mohamed AA (2024) Current and emerging trends of inorganic, organic and eco-friendly corrosion inhibitors. RSC Adv 14(43):31877–31920
5. Ahmed MA, Mahmoud SA, Mohamed AA (2024) Unveiling the complexities of microbiologically induced corrosion: mechanisms, detection techniques, and mitigation strategies. Front Environ Sci Eng 18(10):120
6. Okpara EC et al (2024) An overview of advanced oxidation processes using copper-based catalytic degradation of organic pollutants in water. Appl Mater Today 36:102053
7. Wang X et al (2024) Insight into the inhibition mechanisms of Ca-, N-, S-containing compounds on persistent organic pollutant formation from the thermal reaction of anthracene on copper chloride. Chem Eng J 497:154824
8. Xu J et al (2024) Photodegradation of polychlorinated biphenyls (PCBs) on suspended particles from the Yellow River under sunlight irradiation: QSAR model and mechanism analysis. Water Res 267:122547
9. Ren Y et al (2024) Research progress on remediation of organochlorine pesticide contamination in soil. Environ Geochem Health 46(1):25
10. Yao C et al (2024) Formation mechanisms and degradation methods of polycyclic aromatic hydrocarbons in biochar: a review. J Environ Manage 357:120610
11. Ahmed MA, Mahmoud SA, Mohamed AA (2025) Insights into MXene-based materials for environmental applications: performance, mechanisms, and future directions. FlatChem 100825
12. Ahmed MA, Ahmed MA, Mohamed AA (2023) Adsorptive removal of tetracycline antibiotic onto magnetic graphene oxide nanocomposite modified with polyvinylpyrroilidone. React Funct Polym 191:105701
13. Huang X, Wen D, Wang J (2024) Radiation-induced degradation of sulfonamide and quinolone antibiotics: a brief review. Radiat Phys Chem 215:111373
14. Ahmed MA, Ahmed MA, Mohamed AA (2022) Facile adsorptive removal of dyes and heavy metals from wastewaters using magnetic nanocomposite of zinc ferrite@ reduced graphene oxide. Inorg Chem Commun 144:109912
15. Adel M, Ahmed MA, Mohamed AA (2021) Synthesis and characterization of magnetically separable and recyclable crumbled $MgFe_2O_4$/reduced graphene oxide nanoparticles for removal of methylene blue dye from aqueous solutions. J Phys Chem Solids 149:109760
16. Adel M, Ahmed MA, Mohamed AA (2021) Effective removal of indigo carmine dye from wastewaters by adsorption onto mesoporous magnesium ferrite nanoparticles. Environ Nanotechnol Monit Manage 16:100550
17. Adel M, Ahmed MA, Mohamed AA (2021) A facile and rapid removal of cationic dyes using hierarchically porous reduced graphene oxide decorated with manganese ferrite. FlatChem 26:100233
18. Adel M, Ahmed MA, Mohamed AA (2020) Effective removal of cationic dyes from aqueous solutions using reduced graphene oxide functionalized with manganese ferrite nanoparticles. Compos Commun 22:100450
19. Ahmed MA, Mohamed AA (2023) A systematic review of layered double hydroxide-based materials for environmental remediation of heavy metals and dye pollutants. Inorg Chem Commun 148:110325
20. Ahmed MA, Mohamed AA (2023) The use of chitosan-based composites for environmental remediation: a review. Int J Biol Macromol 242:124787
21. Noor MHM, Ngadi N (2024) Global research landscape on coagulation-flocculation for wastewater treatment: a 2000–2023 bibliometric analysis. J Water Process Eng 64:105696
22. Şenol ZM et al (2024) Removal of food dyes using biological materials via adsorption: a review. Food Chem 139398

23. Adel M et al (2022) Characterization of fouling for a full-scale seawater reverse osmosis plant on the Mediterranean sea: membrane autopsy and chemical cleaning efficiency. Groundw Sustain Dev 16:100704
24. Ahmed MA, Ahmed MA, Mohamed AA (2023) Synthesis, characterization and application of chitosan/graphene oxide/copper ferrite nanocomposite for the adsorptive removal of anionic and cationic dyes from wastewater. RSC Adv 13(8):5337–5352
25. Ahmed MA, Mahmoud SA, Mohamed AA (2024) Nanomaterials-modified reverse osmosis membranes: a comprehensive review. RSC Adv 14(27):18879–18906
26. Ahmed MA, Mohamed AA (2024) Evaluation and optimization of antiscalant substances for enhanced reverse osmosis performance. J Saudi Chem Soc 101923
27. Tang Y et al (2024) Different mobile counter ion types of magnetic ion-exchange resin coupling with ozonation treatment of the organic matter, antibiotic resistance genes, and pathogens in secondary effluent. Chem Eng J 485:149962
28. Ahmed MA, Amin S, Mohamed AA (2023) Fouling in reverse osmosis membranes: monitoring, characterization, mitigation strategies and future directions. Heliyon 9(4)
29. Ahmed MA, Mohamed AA (2024) Advances in ultrasound-assisted synthesis of photocatalysts and sonophotocatalytic processes: a review. iScience 27(1)
30. Xie Z-H et al (2023) Review of characteristics, generation pathways and detection methods of singlet oxygen generated in advanced oxidation processes (AOPs). Chem Eng J 468:143778
31. Angelé-Martínez C et al (2017) Reactive oxygen species generation by copper (II) oxide nanoparticles determined by DNA damage assays and EPR spectroscopy. Nanotoxicology 11(2):278–288
32. MuthuKathija M et al (2024) Photocatalytic degradation of methylene blue dye using biogenic copper oxide nanoparticles and its degradation pathway analysis. Inorg Chem Commun 161:111929
33. Bouzayani B et al (2024) Advancements in copper-based catalysts for efficient generation of reactive oxygen species from peroxymonosulfate. Appl Sci 14(17):8075
34. Li X et al (2024) A review of metallurgical slags as catalysts in advanced oxidation processes for removal of refractory organic pollutants in wastewater. J Environ Manage 352:120051
35. Koundle P et al (2024) Ozone nanobubble technology as a novel AOPs for pollutants degradation under high salinity conditions. Water Res 263:122148
36. Koundle P et al (2024) Tetracycline degradation for wastewater treatment based on ozone nanobubbles advanced oxidation processes (AOPs)–Focus on nanobubbles formation, degradation kinetics, mechanism and effects of water composition. Chem Eng J 501:156236
37. Mróz K et al (2024) Copper oxide thin films as components for heterojunction formation with TiO_2 for photocatalytic CO_2 reduction. Catal Today 114890
38. Nassar AA et al (2024) Sustainable green synthesis and characterization of nanocomposites for synergistic photocatalytic degradation of Reactive Orange 16 in textile wastewater using CuO@ A-TiO_2/Ro-TiO_2. Sci Rep 14(1):16188
39. Saraev AA et al (2024) Cu/TiO_2 photocatalysts for CO_2 reduction: structure and evolution of the cocatalyst active form. Trans Tianjin Univ 30(2):140–151
40. Danish MSS et al (2021) Photocatalytic applications of metal oxides for sustainable environmental remediation. Metals 11(1):80
41. Bai L et al (2022) Role of transition metal oxides in g-C_3N_4-based heterojunctions for photocatalysis and supercapacitors. J Energy Chem 64:214–235
42. Ahmed MA, Ahmed MA, Mohamed AA (2024) Fabrication of NiO/g-C_3N_4 Z-scheme heterojunction for enhanced photocatalytic degradation of methylene blue dye. Opt Mater 151:115339
43. Ahmed MA, Mahmoud SA, Mohamed AA (2024) Unveiling the photocatalytic potential of graphitic carbon nitride (gC_3N_4): a state-of-the-art review. RSC Adv 14(35):25629–25662
44. Cheng C et al (2014) Enhanced photocatalytic performance of TiO_2-ZnO hybrid nanostructures. Sci Rep 4(1):4181
45. Xiao F-X (2012) Construction of highly ordered ZnO–TiO_2 nanotube arrays (ZnO/TNTs) heterostructure for photocatalytic application. ACS Appl Mater Interfaces 4(12):7055–7063

46. Matar SF, Campet G, Subramanian MA (2011) Electronic properties of oxides: chemical and theoretical approaches. Prog Solid State Chem 39(2):70–95
47. Pastor E et al (2022) Electronic defects in metal oxide photocatalysts. Nat Rev Mater 7(7):503–521
48. Woodward P et al (2005) The electronic structure of metal oxides. Metal oxides. CRC Press, pp 155–216
49. Siddique AB et al (2024) Photocatalytic and biological efficacy of Carissa macrocarpa fruit extract-mediated one-pot synthesized ternary metal oxide dual S-scheme heterojunction. J Environ Chem Eng 12(3):112725
50. Sajid M et al (2024) Progress in the development of copper oxide-based materials for electrochemical water splitting. Int J Hydrogen Energy 62:209–227
51. Chokejaroenrat C et al (2024) Decomposition of microplastics using copper oxide/bismuth vanadate-based photocatalysts: insight mechanisms and environmental impacts. Mar Pollut Bull 201:116205
52. Jaf ZN et al (2024) DFT+ U investigation on high pressure properties of monoclinic CuO. Can J Phys 102(5):316–323
53. Seling TR et al (2024) Transition metal-doped CuO nanosheets for enhanced visible-light photocatalysis. J Photochem Photobiol, A 448:115356
54. Yu Q et al (2024) The study of optical-electrical properties of ZnO/Cu_2O (CuO)/Si heterojunctions. Physica B 690:416253
55. Abdelfattah I, El-Shamy A (2024) A comparative study for optimizing photocatalytic activity of TiO_2-based composites with ZrO_2, ZnO, Ta_2O_5, SnO, Fe_2O_3, and CuO additives. Sci Rep 14(1):27175
56. Lin C et al (2024) Electrocatalytic conversion of nitrate Into ammonia through heterogeneous catalysis of $NiMoO_4$ and Cu/Cu_2O. Adv Funct Mater 2401287
57. Ahmed MA, Mohamed AA (2023) Recent progress in semiconductor/graphene photocatalysts: synthesis, photocatalytic applications, and challenges. RSC Adv 13(1):421–439
58. Baran T et al (2021) Copper oxide-based photocatalysts and photocathodes: fundamentals and recent advances. Molecules 26(23):7271
59. Wang W et al (2022) Photocatalytic CO_2 reduction over copper-based materials: a review. J CO_2 Utilization 61:102056
60. Singhal A et al (2013) Copper (I) oxide nanocrystals–one step synthesis, characterization, formation mechanism, and photocatalytic properties. Eur J Inorg Chem 2013(14):2640–2651
61. Xu L et al (2017) Improved photocatalytic activity of nanocrystalline ZnO by coupling with CuO. J Phys Chem Solids 106:29–36
62. Kumari V et al (2021) Hydrothermal synthesis conditions effect on hierarchical ZnO/CuO hybrid materials and their photocatalytic activity. J Mater Sci: Mater Electron 32:9596–9610
63. Karuppiah C et al (2015) A simple hydrothermal synthesis and fabrication of zinc oxide–copper oxide heterostructure for the sensitive determination of nonenzymatic glucose biosensor. Sens Actuators, B Chem 221:1299–1306
64. Shammi ZM, Kianfar A, Momeni M (2020) Hydrothermal synthesis and characterization of CuO–CoO/TiO_2 for photocatalytic degradation of methylene blue under visible light and catalytic reduction of P-nitrophenol. J Mater Sci: Mater Electron 31(17):14810–14822
65. Li D et al (2021) Preparation and photocatalytic performance of TiO_2-RGO-CuO/Fe_2O_3 ternary composite photocatalyst by solvothermal method. Mater Res Express 8(1):015025
66. Mousavi-Kamazani M, Rahmatolahzadeh R, Beshkar F (2017) Facile solvothermal synthesis of CeO_2–CuO nanocomposite photocatalyst using novel precursors with enhanced photocatalytic performance in dye degradation. J Inorg Organomet Polym Mater 27:1342–1350
67. Jiang W et al (2024) Solvothermal preparation of CuO/g-C_3N_4 and photocatalytic degradation of ciprofloxacin. J Dispersion Sci Technol 1–12
68. Van Thanh H et al (2024) Photocatalytic activities of the ZnO: CuO nanocomposites synthesized by microwave-assisted hydrothermal under UV-Vis irradiation. Adv Compos Mater 1–20

69. Mariadhas J et al (2023) Microwave assisted green synthesis of Ag doped CuO NPs anchored on GO-sheets for high performance photocatalytic and antimicrobial applications. J Ind Eng Chem 128:383–395
70. Alsulmi A et al (2024) Sonochemical synthesis of S-scheme CuO/CeO_2 heterojunction for enhanced photocatalytic degradation of rhodamine B dye under solar radiations. Ceram Int 50(19):36349–36362
71. Alsulmi A et al (2023) Engineering S-scheme CuO/ZnO heterojunctions sonochemically for eradicating RhB dye from wastewater under solar radiation. RSC Adv 13(19):13269–13281
72. Hendry T, Taufik A, Saleh R (2016) Fe_3O_4/CuO/ZnO/Nano graphene platelets (Fe_3O_4/CuO/ZnO/NGP) composites prepared by sol-gel method with enhanced sonocatalytic activity for the removal of dye. AIP Conf Proc
73. Moosavi F, Bahrololoom ME, Kamjou R (2016) Effects of cu doping on nano structure, morphology and photocatalytic activity of zno thin film synthesized by sol-gel method. Studia Univ Babes-Bolyai Chem 1:79
74. Tju H et al (2017) Synthesis and characterization of noble metal nanocomposites: Ag/Fe_3O_4/ZnO and Ag/Fe_3O_4/CuO/ZnO for better photocatalytic activity under visible light irradiation. IOP Conf Ser: Mater Sci Eng
75. Kabir MH et al (2022) Structural, nonlinear optical and antimicrobial properties of sol-gel derived, Fe-doped CuO thin films. Heliyon 8(9)
76. Kumar N et al (2021) Structural and optical properties of sol–gel derived CuO and Cu_2O nanoparticles. Mater Today: Proc 41:237–241
77. Tseng I-H, Wu JC, Chou H-Y (2004) Effects of sol–gel procedures on the photocatalysis of Cu/TiO_2 in CO_2 photoreduction. J Catal 221(2):432–440
78. Yang J-Z et al (2021) Oxidation efficiency and reaction mechanisms of gaseous elemental mercury by using CeO_2/TiO_2 and CuO/TiO_2 photo-oxidation catalysts at low temperatures in multi-pollutant environments. J Taiwan Inst Chem Eng 125:413–423
79. Lillo-Ramiro J et al (2021) Optical and microstructural characteristics of CuO thin films by sol gel process and introducing in non-enzymatic glucose biosensor applications. Optik 229:166238
80. Sivayogam D et al (2022) Study on structural, electro-optical and optoelectronics properties of CuO nanoparticles synthesis via sol gel method. Mater Today: Proc 48:508–513
81. Wang Y-T et al (2021) Fabrication and characterization of Al–CuO nanocomposites prepared by sol-gel method. Defence Technol 17(4):1307–1312
82. Thakar MA et al (2022) X ray diffraction (XRD) analysis and evaluation of antioxidant activity of copper oxide nanoparticles synthesized from leaf extract of Cissus vitiginea. Mater Today: Proc 51:319–324
83. Sharma A et al (2015) XANES, EXAFS and photocatalytic investigations on copper oxide nanoparticles and nanocomposites. RSC Adv 5(28):21762–21771
84. Kandasamy M et al (2024) Fabrication of ternary zinc-nickel-copper oxide microsphere nanocomposite photoanode material for boosting dye-sensitized solar cell efficiency. J Alloys Compd 175039
85. Arulkumar E, Thanikaikarasan S (2024) Synthesis of CuO@ NiO/rGO nanocomposite with enhanced electrochemical and photocatalytic properties in energy and environmental applications. J Alloys Compd 175008
86. Dharmalingam K et al (2024) Direct Z-scheme heterojunction impregnated MoS_2–NiO–CuO nanohybrid for efficient photocatalyst and dye-sensitized solar cell. Sci Rep 14(1):14518
87. Arulkumar E, Thanikaikarasan S (2024) Structure, morphology, composition, optical properties and catalytic activity of nanomaterials CuO, NiO CuO/NiO using methylene blue. Optik 302:171685
88. Arulkumar E, Thanikaikarasan S, Siddhardhan E (2024) Synthesis and characterization of CuO@ NiO/g-C_3N_4 nanocomposite for photocatalytic and electrochemical application. Res Chem 7:101439
89. Basit A et al (2024) Fabrication of CuO derived reduced graphene oxide photocatalyst for the strategic decolorization of Congo red from an aqueous environment. Diam Relat Mater 142:110839

90. Jeevarathinam M, Asharani I (2024) Synthesis of CuO, ZnO nanoparticles, and CuO–ZnO nanocomposite for enhanced photocatalytic degradation of Rhodamine B: a comparative study. Sci Rep 14(1):9718
91. Song MS et al (2024) Microwave-assisted CuO-modified porous ZnO nanosheet for photocatalytic degradation of organic pollutants and antibacterial inactivation. J Environ Chem Eng 12(6):114453
92. Amrollahi R (2024) Comparison of photocatalytic activity of Cu/TiO_2, Cu/NiO and Cu/ZnO nanocomposites for the degradation of organic dyes. Appl Catal O: Open 195:207010
93. Akter N et al (2024) XPS valence band observable light-responsive system for photocatalytic acid Red114 dye decomposition using a ZnO–Cu_2O heterojunction. Heliyon 10(10)
94. Vo TM, Nguyen TMH, Bark CW (2024) Reduced graphene oxide-supported copper (I) oxide composites for the degradation of methylene blue: exploring the capacity of RGO as an electron capturer for achieving highly stable photocatalytic activity. ACS Appl Electron Mater
95. Kota R et al (2024) The synthesis of A copper (Ii) oxide (Cuo)/reduced graphene oxide (Rgo) nanocomposite using a hydrothermal method for use in biological applications. Rasayan J Chem 17(3)
96. Dustgeer MR et al (2024) Reduced graphene oxide supported polyaniline/copper (II) oxide nanostructures for enhanced photocatalytic degradation of Congo red and hydrogen production from water. J Water Process Eng 59:105053
97. Andronic L, Abreu-Jaureguí C, Silvestre-Albero J (2024) Construction of TiO_2@ Cu_2O-CuS heterostructures integrating RGO for enhanced full-spectrum photocatalytic degradation of organic pollutants. J Alloy Compd 994:174682
98. Hosseinian E et al (2024) Synthesis of rGO/$CuBi_2O_4$ nanocomposite as an effective photocatalyst in the reduction of nitroaromatic compounds to corresponding amines under visible light. J Chem Technol Biotechnol 99(11):2464–2481
99. Bhatia N et al (2024) Green synthesis of chitosan-encapsulated CuO nanocomposites for efficient degradation of cephalosporin antibiotics in contaminated water. Environ Sci Pollut Res 31(23):33638–33650
100. Sarengan N et al (2024) Exploring the n–p type zinc oxide/copper oxide nanocomposite under Xenon light irradiation with enhanced photocatalytic activities for norfloxacin and methyl orange. J Mater Sci: Mater Electron 35(32):1–18
101. Li X-F, Feng X-Q (2024) Construction of Chitosan-biochar/ZnO/NiO composites for efficient photocatalytic removal of Congo red and tetracycline. J Alloy Compd 1008:176765
102. Mahmoud ZH et al (2024) $CdIn_2Se_4$@ chitosan heterojunction nanocomposite with ultrahigh photocatalytic activity under sunlight driven photodegradation of organic pollutants. Int J Biol Macromol 267:131465
103. Ramírez O et al (2024) Hydrogel composites based on chitosan and $CuAuTiO_2$ photocatalysts for hydrogen production under simulated sunlight irradiation. Int J Biol Macromol 132898
104. Dat NM et al (2024) Non-metallic X (X= C, N, S, and P) co-doped copper oxide derived from Mangifera indica leaf extract: synthesis, characterization, density functional theory simulation of structure, and photoactivities. Colloids Surf, A 687:133393
105. Boucherdoud A et al (2024) Experimental exploration and DFT analysis of the kinetics and mechanism of malachite green photodegradation catalyzed by polyaniline-copper oxide nanocomposite. J Mol Model 30(7):235
106. Asadinamin M et al (2024) Role of interfacial morphology in Cu_2O/TiO_2 and band bending: insights from density functional theory. ACS Appl Mater Interfaces 16(27):35781–35792
107. Zhou J et al (2024) Copper-zinc nanoparticle-decorated nitrogen-doped carbon composite for electrochemical determination of triclosan. Microchim Acta 191(3):155
108. Naggar AH et al (2024) Hybrid mesoporous microbeads based on sheets-like sulfur-doped copper oxide embedded in calcium alginate-derived carbon for batch and columnar adsorption of cationic dye. Microporous Mesoporous Mater 378:113253
109. Rathi VH, Jeice AR, Jayakumar K (2023) Green synthesis of Ag/CuO and Ag/TiO_2 nanoparticles for enhanced photocatalytic dye degradation, antibacterial, and antifungal properties. Appl Surf Sci Adv 18:100476

110. Manikandan DB et al (2023) Sustainable fabrication of hybrid silver-copper nanocomposites (Ag-CuO NCs) using Ocimum americanum L. as an effective regime against antibacterial, anticancer, photocatalytic dye degradation and microalgae toxicity. Environ Res 228:115867
111. Faisal AD, Aljubouri AA, Khalef WK (2022) Photodetector fabrication based on heterojunction of $CuO/SnO_2/Si$ nanostructures. Bull Mater Sci 45(2):84
112. Khan MA et al (2021) Interface study of hybrid CuO nanoparticles embedded ZnO nanowires heterojunction synthesized by controlled vapor deposition approach for optoelectronic devices. Opt Mater 117:111132
113. Lopez R et al (2021) Cupric oxide (CuO)/zinc oxide (ZnO) heterojunction diode with low turn-on voltage. Results Phys 22:103891
114. Kohnehsari AA et al (2025) Anchorage of ZnO quantum dots and CuO on graphene for sonophotocatalytic treatment of pharmaceutical effluent: from experimental data and prediction by advanced machine learning algorithms. Colloids Surf, A 705:135724
115. Nazeer M et al (2024) CuO@ 3D graphene modified glassy carbon electrode towards the detection of Orange II and Rhodamine B. Mater Sci Eng, B 310:117770
116. Altin I (2022) $CuO\text{-}TiO_2$/graphene ternary nanocomposite for highly efficient visible-light-driven photocatalytic degradation of bisphenol A. J Mol Struct 1252:132199
117. Yadav A, Hunge Y, Kang S-W (2021) Spongy ball-like copper oxide nanostructure modified by reduced graphene oxide for enhanced photocatalytic hydrogen production. Mater Res Bull 133:111026
118. Alomayrah N et al (2024) Fabrication of a highly efficient $CuO/ZnCo_2O_4$/CNTs ternary composite for photocatalytic degradation of hazardous pollutants. RSC Adv 14(34):24874–24897
119. Duan J et al (2020) Enhanced photocatalytic degradation of organic pollutants using carbon nanotube mediated CuO and Bi_2WO_6 sandwich flaky structures. Nanotechnology 31(42):425202
120. Abo-Dief HM et al (2022) Ternary metal oxide WO3. NiO. ZnO nanoparticles and their composite with CNTs for organic dye photocatalytic degradation. Ceram Int 48(15):22228–22236
121. Li Y et al (2021) Copper oxide functionalized chitosan hybrid hydrogels for highly efficient photocatalytic-reforming of biomass-based monosaccharides to lactic acid. Appl Catal B 291:120123
122. Narayanan VS et al (2020) Schiff-base derived chitosan impregnated copper oxide nanoparticles: an effective photocatalyst in direct sunlight. Mater Sci Semicond Process 119:105238
123. Gaol JML, Imae T (2023) CuO/ZnO catalysts in chitosan film for CO_2 reduction. J Environ Chem Eng 11(6):111333
124. Pal S et al (2015) Low temperature solution processed ZnO/CuO heterojunction photocatalyst for visible light induced photo-degradation of organic pollutants. CrystEngComm 17(6):1464–1476
125. Mohammed AM et al (2021) Ultrafast degradation of Congo Red dye using a facile one-pot solvothermal synthesis of cuprous oxide/titanium dioxide and cuprous oxide/zinc oxide pn heterojunction photocatalyst. Mater Sci Semicond Process 122:105481
126. Montero J et al (2021) Copper–zinc oxide heterojunction catalysts exhibiting enhanced photocatalytic activity prepared by a hybrid deposition method. RSC Adv 11(17):10224–10234
127. Hussain MK et al (2022) Fabrication of CuO/MoO3 pn heterojunction for enhanced dyes degradation and hydrogen production from water splitting. Int J Hydrogen Energy 47(34):15491–15504
128. Kumar S et al (2023) New insights into $Cu/Cu_2O/CuO$ nanocomposite heterojunction facilitating photocatalytic generation of green fuel and detoxification of organic pollutants. J Phys Chem C 127(15):7095–7106
129. Kubiak A et al (2020) Microwave-assisted synthesis of a TiO_2–CuO heterojunction with enhanced photocatalytic activity against tetracycline. Appl Surf Sci 520:146344
130. Zhu L et al (2018) Synthesis of the 0D/3D CuO/ZnO heterojunction with enhanced photocatalytic activity. J Phys Chem C 122(17):9531–9539

131. Wang M et al (2021) CuO/g-C_3N_4 2D/2D heterojunction photocatalysts as efficient peroxymonosulfate activators under visible light for oxytetracycline degradation: characterization, efficiency and mechanism. Chem Eng J 416:128118
132. Ahmed, M. A., Mahmoud, S. A., & Mohamed, A. A. (2025). Interfacially engineered metal oxide nanocomposites for enhanced photocatalytic degradation of pollutants and energy applications. RSC Adv 15(20):15561–15603
133. Enhancing the adsorption capacity of SnS2-g-C3N4 through chitosan integration: Insights from structural and optical characterization
134. Ahmed MA, Mahmoud SA, Mohamed AA (2025) Advances in metal oxide composites for sustainable environmental remediation. Inorg Chem Commun 114641
135. Ahmed MA, Mohamed AA (2025) Graphene–metal oxide hybrid composites: promising adsorbents for the purification of water

Sol–Gel and Hydrothermal Synthesis of Zirconium-Based Photocatalysts for Environmental Applications

Emad Elshehy, M. F. Cheira, Wagih Eldesouky, and Abdelaal S. A. Ahmed

Abstract Water cleaning solutions that are efficient, cost-effective, and non-invasive are urgently needed due to the increasing contamination of water bodies caused by industrial and agricultural activities. Photocatalysis is one of these methods that has recently gained attention as a potential strategy for degrading organic contaminants in water. This method leverages semiconductor nanocrystals that, upon light irradiation, generate electron–hole pairs, triggering chemical reactions that lead to the breakdown of pollutants. Zirconium oxide (ZrO_2), a semiconductor with a wide band gap, has gotten noteworthy consideration as a potential photocatalyst for environmental remediation due to its thermal stability, mechanical strength, and phase stability under various conditions. However, its limited absorption of visible light and high electron recombination rate hinders its photocatalytic performance. Recent advancements in synthesis methods, such as hydrothermal techniques, have improved photocatalytic efficiency of ZrO_2. Strategies such as doping with transition metals, the creation of oxygen vacancies, and hybridizing ZrO_2 with other materials have shown promise in narrowing the band gap, enhancing light absorption, and easing electron–hole recombination. This chapter reviews the various methods of synthesizing ZrO_2-based photocatalysts, their structural characteristics, and their applications in removing organic contaminants from water. Special attention is given to how modifications, including the introduction of dopants and the development of composite materials, enhance the photocatalytic activity of ZrO_2, making it a viable candidate for large-scale environmental remediation. The potential for further improvements in the performance of ZrO_2-based photocatalysts is discussed alongside their practical implications for water treatment technologies.

Keywords Zirconium · Sol–gel synthesis · Hydrothermal synthesis · Environmental remediation · Photocatalysis

E. Elshehy · M. F. Cheira · W. Eldesouky
Nuclear Materials Authority, El Maadi, Cairo 13759, Egypt

A. S. A. Ahmed (✉)
Chemistry Department, Faculty of Science, Al-Azhar University, Assiut 71524, Egypt
e-mail: abdelaalsaiyd@azhar.edu.eg

A. Ahmad et al. (eds.), *Metal Oxide Based Nanophotocatalyst for Wastewater Purification*, Smart Nanomaterials Technology,
https://doi.org/10.1007/978-981-96-7683-5_10

1 Introduction

The rapid technological advancements of recent years have contributed significantly to the increasing pollution of water systems. Both freshwater and wastewater require treatment to remove contaminations of organic foundations, such as hydrocarbons, phenols, organic dyes, surfactants, aromatic amines, inorganic substances, and pesticides, like acids, bases, and heavy metals. As the concentrations of these hazardous substances rise, there is an urgent need for cost-effective, highly efficient, and non-invasive water purification methods. Adsorption and catalysis-based techniques have emerged as promising solutions for this challenge [38, 78, 100]. One intriguing approach to water purification is photocatalysis, which uses semiconductor nanocrystals that produce electron–hole pairs when exposed to light with an energy level equivalent to or greater than the band gap. Hole generation in the valence band and electron formation in the conduction band are the outcomes of this process. Although most charge carriers undergo recombination, a small number of electron–hole pairs can get to the surfaces of nanocrystals. They engage in chemical interactions with adsorbed molecules or ions, facilitating photocatalytic reactions. Under ambient conditions, photocatalysis has emerged as a highly esteemed technique for organic pollutant removal from water. Titanium dioxide (TiO_2) is among the most commonly used photocatalysts, though its requirement for UV light for activation limits its practical use [50]. The increasing interest in photocatalysis can be attributed to the improved availability of photocatalysts, advancements in synthesis techniques, and their potential for large-scale applications [48, 62]. An assortment of variables, such as the reaction medium, the existence of inhibitor ions, the quantity of photoactive substance, and the type, intensity, and duration of radiation, affect the efficacy of photocatalytic reactions [39, 75]. The synthesis procedure determines the crystallite size, area of surface, and energy gap to achieve efficient photocatalysis. Among various semiconductors, TiO_2 is most frequently employed due to its excellent properties and popularity as a photocatalyst [97]. TiO_2 can be synthesized through various methods [29], and other commonly used photocatalysts include oxides such as ZnO, CeO_2, SnO_2, ZrO_2, WO_3, and Fe_2O_3, as well as sulfides and selenides like CdS, ZnS, PbS, CdSe, and CdTe [17, 66, 67, 85]. Among these, zirconium oxide (ZrO_2) is a notable semiconductor with a wide stretch gap (5.0–5.5 eV) [53, 54]. ZrO_2 is widely used in many industries for ceramics, gas sensors, fuel cells, catalysis, solid-state electrolytes, and optical devices [89]. When ZrO_2 is in its monoclinic phase at 1000 °C, it changes from n-type to p-type conductivity close to an oxygen partial pressure of 10–16 atm, indicating that it is polymorphic. At elevated temperatures and pressures, completely ionized zirconium vacancies develop, generating positive holes but with limited mobility [82]. Because of its high oxygen vacancy density, the p-type semiconductor ZrO_2 has garnered much attention for use in various catalytic applications [61]. Depending on the temperature and atmospheric conditions, ZrO_2 can exist in three polymorphic forms: cubic (c-ZrO_2), tetragonal (t-ZrO_2), and monoclinic (m-ZrO_2). Up to 1100 °C, the m-ZrO_2 phase remains stable; between 1100 and 2370 °C, the t-ZrO_2 phase is stable; and above 2370 °C, the c-ZrO_2 phase remains stable. Recent examination has also shown that the *t*-ZrO_2 phase can remain at lower temperatures for extended periods. The TEM and HRTEM images of all these ZrO_2 phases are presented in Fig. 1a–f [16]. The p-type semiconducting materials known as ZrO_2 nanoparticles typically have a large band gap, high energy consumption, and enormous oxygen vacancies on their outermost layer. More and more catalytic applications are looking to ZrO_2 for its high electron transfer capacity and redox activity. This includes fuel cell devices made of solid oxide and sensors for oxygen and nitrogen gas [91]. Additionally, ZrO_2 is known for its thermal stability, which is essential for both oxidation and reduction reactions, as well as its ability to exhibit acidic or basic properties depending on the crystalline phase [2]. By employing different preparatory methods and thermal treatments, tetragonal and monoclinic phases of ZrO_2 can be obtained at temperatures as low as 425 °C. Hydrothermal managements at 750 °C yield pure monoclinic zirconia, with the addition of noble metals further stabilizing the material [18]. Improving visible light absorption by adjusting band gap of ZrO_2 is essential for increasing its photocatalytic efficiency. Several methods have been devised, incorporating carbonaceous materials or doping agents. The light-absorbing capability of ZrO_2 materials are greatly enhanced by doping with lower band gap agents as well as introducing new hybrid states. Under exposed to visible light, sensitizers directly absorb electrons, which are subsequently transferred to ZrO_2 and used for driving oxidation reactions [16].

Another approach to improving photocatalytic performance involves hybrid systems, where two different oxides, such as TiO_2–ZrO_2, ZnO–SiO_2, or TiO_2–SiO_2, are combined. These hybrids offer enhanced properties over individual oxides, leading to improved photocatalytic activity. The photocatalytic efficiency of these hybrid systems depends on the molar ratio of oxides [1, 7, 72]. Additionally, doping with alkaline or transition metals can further enhance the photocatalytic performance of the material [8, 45]. The significance of oxygen vacancies in zirconate dioxide as mobile electrical charge carriers is crucial in photocatalysis. These empty spaces are created when tetravalent Zr^{4+} ions are substituted with divalent or trivalent cations, like Ca^{2+}, Mg^{2+}, or Y^{3+}. Depending on the valence state, concentration, and crystallite size of the dopants, the structural and defect association energies of ZrO_2 are affected by the insertion of oxygen vacancies [6]. Additionally, the band gap is efficiently narrowed by doping, which improves the photocatalytic activity. Due to its high redox potential and photogenerated charge carriers, ZrO_2 exhibits substantial photocatalytic activity under UV light despite its broadband gap of about 5 eV [26]. There are many desirable properties of ZrO_2 make it a valuable material. These include a high melting point, low thermal conductivity, high mechanical strength, resistance to wear, and resistance to high temperatures and corrosion. A further advantage of a photocatalyst is its capacity to demonstrate acid–base and redox catalytic activities [25].

Great research have focused on ZrO_2 because of its nontoxic and sustainable features and its good chemical and mechanical characteristics, including superior strength, fracture resistance, and phase stability [60]. ZrO_2 is used in diverse applications, including thermal barrier coatings [42, 76], refractory materials [64], gas sensors [74], bioceramics [71], biological labeling [52], radiation detectors [73],

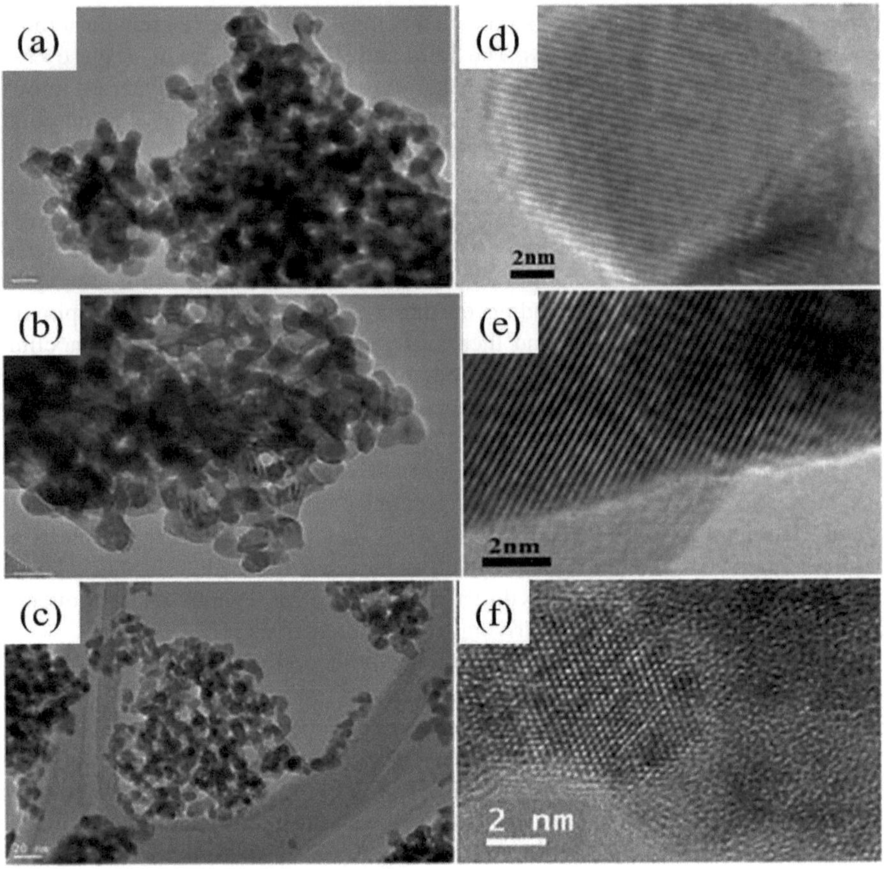

Fig. 1 TEM images of **a** m-ZrO_2, **b** t-ZrO_2, **c** c-ZrO_2, and HR-TEM of **d** m-ZrO_2, **e** t-ZrO_2, **f** c-ZrO_2. Reproduced from Ref. [16] with permission, Copyright 2015 Springer Nature

and solid oxide fuel cells [102]. ZrO_2 is an excellent material for catalytic applications when exposed to UV and visible light because it is a semiconductor with a wide band gap of 5.0 eV, excellent thermal stability, and a high surface area [77]. However, despite its advantages, ZrO_2 typically experiences high electron recombination, which results in low efficiency in organic pollutant degradation under normal conditions. The introduction of suitable dopants, however, can prevent electron–hole pair recombination and significantly improve the photocatalytic efficiency of ZrO_2.

Nanoparticles of ZrO_2 have been produced by a number of different techniques, such as sol–gel, spray pyrolysis, hydrolysis, and microwave plasma [46]. Hydrothermal synthesis stands out among these methods for producing ZrO_2 nanoparticles because it is both efficient and inexpensive. The fabrication of ZrO_2 monoclinic and tetragonal crystalline states using various procedures has been documented in a large number of investigations. For instance, m-ZrO_2 nanorods have been synthesized using a hydrothermal method [58], and m-ZrO_2 nanopowder has been synthesized using a ball-milling-assisted hydrothermal methodology [30]. Subcritical hydrothermal conditions were used to produce single-phase m-ZrO_2 nanocrystals by Taguchi et al. [92], while tetragonal ZrO_2 nanoparticles were reported by Dwivedi et al. [31] by microwave-aided sol–gel synthesis.

2 Methods for Synthesis Zirconium-Based Materials

2.1 Sol–Gel Method

The sol–gel method is a popular choice for synthesizing ZrO_2-based compounds and other materials since it is easy to use, does not cost a lot of money, and can create high-purity products with little equipment. It involves the hydrolysis of precursor solutions, leading to the formation of a gel that can be processed into a solid material. This method is particularly advantageous for synthesizing nanoparticles, as it allows for precise control over material properties such as homogeneity, purity, and physical characteristics at low temperatures [68]. Several factors influence the composition and assets of the substances during the sol–gel process. For example, doping ZrO_2 with metals like Ce, Cu, or Fe can reduce its energy gap, enhancing its catalytic efficiency [23]. In their study on Cu-doped ZrO_2, Sharaf et al. [88] found that photocatalytic degredation of methylene blue (MB) and rhodamine B (RhB) dyes was improved due to a reduction in the band gap and a rise in crystallite size caused by increasing copper concentration. Additionally, it was demonstrated that Cu doping enhances photocatalytic activity by promoting the development of oxygen vacancies. Composites of ZrO_2 and Ti, synthesized using the sol–gel technique, show improved photocatalytic activity when exposed to both UV and visible light [36]. Additionally, sol–gel synthesis has been applied to develop materials like Fe-doped $ZrTiO_4$ for photocatalysis under visible light [56] and Zr/Ti/CHT hybrid catalysts for dye degradation [28]. Incorporating carbon materials, such as mesoporous carbon or carbon quantum dots, into ZrO_2-based catalysts can further improve photocatalytic activity by reducing electron–hole recombination and enhancing light absorption [15, 83].

ZnO–ZrO_2 nanocomposites were fabricated via the sol–gel method using contrasting quantities of zinc (II) acetylacetonate and ZnO (13, 25, 50, and 75 mol%). The addition of ZnO significantly influenced the properties of nanocomposites, particularly in the structural composition, where the materials exhibited a tetragonal ZrO_2 structure and hexagonal ZnO. The external area of the composites expanded with ZnO content, reaching 46 m^2/g for pure ZrO_2. ZnO incorporation also lowered the band gap of ZrO_2, enhancing photocatalytic phenol degradation. The 75% ZnO–ZrO_2 composite showed superior photocatalytic performance in phenol degradation, outperforming pure ZnO in terms of mineralization and reduced intermediates [96].

Bensmaine et al. [19] synthesized TiO_2, TiO_2–ZrO_2, and ZrO_2–TiO_2 composites utilizing the sol–gel approach and reported that mixing ZrO_2 with TiO_2 reduced the band gap and altered. Their photocatalytic tests below UV and blue LED irradiation showed that TiO_2–ZrO_2 composites achieved 100% sulfamethoxazole (SMX) conversion under UV light. Furthermore, Pt-ZrO_2-HAP (hydroxyapatite) nanocomposites demonstrated high photocatalytic activity for easing m-phenylenediamine, particularly in visible light [47]. Akşit et al. [4] investigated the photocatalytic degradation of paracetamol using sol–gel synthesized Bi_2O_3, ZnO, TiO_2, and ZrO_2, finding that Bi_2O_3 exhibited the highest degradation rate, despite its low surface area. Additionally, a ZrO_2/carbon xerogel composite showed enhanced photocatalytic efficiency, achieving a 67% higher degradation rate than pure ZrO_2 under UV light [27]. TiO_2/ZrO_2 nanocomposites synthesized via sol–gel showed reduced recombination rates and enhanced photocatalytic activity for Ponceau BS dye degradation, especially with 6.0% ZrO_2 loading [79]. Cu-doped ZrO_2 nanoparticles demonstrated improved photocatalytic efficiency in degrading Alizarin Yellow, with the 0.05 wt% Cu-doped sample showing the best performance [13]. ZrO_2 mixed with SnO_2 or CeO_2 through sol–gel led to effective photocatalytic degradation of 2-nitrophenol, with the ZrO_2/SnO_2 mixture achieving complete degradation in 45 min when combined with hydrogen peroxide [81]. Ternary ZrO_2/TiO_2/Fe_3O_4 nanocomposites were synthesized for naproxen removal, showing 100% degradation under optimized conditions with high stability over seven cycles [103]. Al_2O_3/ZrO_2 composites synthesized using a sol–gel method demonstrated superior photocatalytic efficiency in dye degradation in visible light, with over 90% of Reactive Yellow 145 degradation and Reactive Blue 222 [101]. Mesoporous Al_2O_3/ZrO_2 composites showed effective degradation of 2,4,6-trichlorophenol, with photocatalytic activity linked to the defects present in the materials [98]. Karunakaran et al. [49] synthesized Al_2O_3–TiO_2 and ZrO_2–TiO_2 binary oxide nanocomposites utilizing the sol–gel approach with PVP-PEG as templating mediators. The Al_2O_3–TiO_2 composite displayed a finish-centered monoclinic crystal configuration, while the ZrO_2–TiO_2 composite exhibited both monoclinic and tetragonal phases, with anatase TiO_2 present in both composites. Energy dispersive X-ray (EDX) analysis showed Al: Ti and Zr: Ti ratios of 1:12 and 1:1, respectively. SEM analysis revealed particle sizes of 30–77 nm for Al_2O_3–TiO_2 and 38–57 nm for ZrO_2–TiO_2. Both composites showed UV-A excitation, photoluminescence similar to TiO_2, and lower charge-transfer resistance in ZrO_2–TiO_2. Photocatalytic degradation of RhB dye and MB dye was effective under UV-A light; nonetheless, methyl orange (MO) degradation was slower. Both composites also degraded RhB dye under visible light through a dye-sensitized photocatalytic mechanism. Liu et al. [63] developed a composite catalyst using the sol–gel technique by combining Fe_2O_3, Al_2O_3, and ZrO_2. This catalyst displayed high catalytic performance, achieving over 93% phenol conversion under optimal conditions. The catalyst maintained its activity after two regeneration cycles, confirming its stability and potential for degrading persistent organic pollutants. Wahba et al. used a modified sol–gel Pechini approach to synthesize Zr-doped ZnO and ZrO_2/ZnO heterogeneous structures [99]. Characterization revealed that pure ZnO and Zr-doped ZnO (5 wt%) had a hexagonal structure, while higher Zr concentrations resulted in multi-phase composites. The ZrO_2/ZnO (50:50) nanomaterials achieved 97% degradation of MB and 99% degradation of indigo carmine. The improved photocatalytic activity was attributed to better charge separation caused by strong contacts at the interface. Matějová and colleagues [69] used a sol–gel technique and $AuCl_3$ impregnation to create TiO_2–ZrO_2 composites and TiO_2-gold-enriched TiO_2 composites. Composites of TiO_2 and ZrO_2 showed more photocatalytic activity when exposed to UV light than TiO_2 alone. On the other hand, the larger gold particles in gold-doped composites made them less efficient because they impeded the absorption of light. Amorphous TiO_2–ZrO_2 outperformed nanocrystalline TiO_2 partly because it had a larger surface area.

A sol–gel nanocomposite of cellulose, γ-Fe_2O_3, and ZrO_2, was created by Helmiyati et al. [43]. They discovered that adding ZrO_2 decreased the spectral band gap of γ-Fe_2O_3, which improved its photocatalytic degradation. The composite of cellulose, γ-Fe_2O_3, and ZrO_2, showed exceptional photocatalytic activity when exposed to visible light, degrading 98.5% of Congo Red at 30 min. With the use of surfactants, a sol–gel process was used to create TiO_2–ZrO_2 binary production oxides with different ZrO_2 ratios. The anatase phase was preserved up to 40 mol% ZrO_2, but an amorphous phase was produced at 50 mol% composition. With a rise in ZrO_2 content, these binary oxides showed a blue shift in UV–visible spectra and a greater surface area. They improved solar energy conversion when utilized as electrodes in solar cells sensitive to dyes, outperforming pure TiO_2 [51]. Researchers Mbiri et al. [70] evaluated sol–gel and evaporation-induced self-assembly (EISA) nanocomposites for the photocatalytic breakdown of chloridazon. For maximum photocatalytic activity, composites made by EISA should contain 0.14 mol% zirconium, and sol–gel samples should have 0.05 mol% zirconium. The phase transition from anatase to rutile was blocked by zirconium up to 600 °C. Shao et al. [87] created TiO_2 and ZrO_2–TiO_2 composites using a more straightforward sol–gel technique. According to photocatalytic testing, composites have a Zr/Ti ratio of < 2.2 decolorized MB dye more effectively than pure TiO_2.

Nanoparticles of TiO_2, ZrO_2, and TiO_2–ZrO_2 were produced by Kumari et al. [55] utilizing sol–gel and liquid combustion techniques. Nanoparticles of mixed-phase TiO_2–ZrO_2 showed excellent photocatalytic activity, degrading Eosin Yellow efficiently, especially when exposed to visible light. Nanoparticles of manganese and zinc oxide were produced by López et al. [65], who found that the addition of manganese increased the photocatalytic effectiveness. This composite showed promise for environmental purification due to its remarkable improvement in photocatalytic degradation. Co_3O_4–ZrO_2 nanocomposites exhibited improved photocatalytic activity when produced by sol–gel synthesis using acetyl trimethylammonium bromide as a template. Complete cyanide degradation was obtained in 30 min, and stability was maintained across five cycles with a 3 wt% Co_3O_4-ZrO_2 combined [5]. Fakhri et al. [33] produced t-ZrO_2 nanoparticles doped with carbon for photocatalytic and antibacterial uses. Zinc zirconate ($ZnZrO_3$) was also studied by Habibi [44] for dye-sensitized solar cells and by Zhu et al. [104] for its optical characteristics in perovskite nanoparticles. Aronne et al. [14] added to our knowledge of yttria-doped zirconia materials by synthesizing ZrO_2–Y_2O_3 (5%) powder using sol–gel and studying its thermal and structural characteristics. The hydrolysis approach was used by Esquivias et al. [32] to investigate the crystalline process of yttria-doped zirconia. The improved photocatalytic characteristics of sol–gel-produced nanocomposites, especially those containing ZrO_2 and other metal oxides, make them useful for ecological tasks such as degrading dyes, removing pollutants, and converting energy.

2.2 *The Hydrothermal Synthesis*

One of the most important ZrO_2-based materials for photocatalysis, ZrO_2 nanoparticles, may be efficiently and sustainably synthesized using hydrothermal processes. The chemical reactions that produce ceramic sols are known as hydrothermal synthesis, and they take place in an aqueous or organo-aqueous solution at elevated temperatures (> 100 °C) and pressures (> 1 atm). The size and phase of the materials can be controlled using this method by altering synthesis variables such as pH, temperature, reaction time, and surfactants [80]. One "soft chemical" method that can produce phase-pure compounds at low temperatures is the hydrothermal approach [90].

The hydrothermal synthesis of zirconia-based materials has attracted much interest because of their photocatalytic capabilities. The breakdown of organic contaminants has been investigated, for instance, in relation to the photocatalytic power of nanoparticles made from ZrO_2 [86]. Over a range of 110–240 min, Basahel et al. [16] discovered that cubic ZrO_2 catalysts degraded MO dye with an efficiency of 80%, tetragonal ZrO_2 catalysts at 99%, and pure monoclinic ZrO_2 catalysts at 90%. Pure tetragonal nanoparticles of ZrO_2 degraded MO dye by 99% in 50 min when exposed to UV radiation.

Additionally, there has been great interest in studying how to dope ZrO_2 nanoparticles to improve their photocatalytic capabilities. As an example, Reddy et al. [84] reported that yttrium-doped tetragonal ZrO_2 particles have demonstrated better photocatalytic activity than pure ZrO_2. The TEM images of the undoped and yttrium-doped ZrO_2 nanoparticles (Fig. 2a, d) show that the prepared nanoparticles were uniformly sized and spherical in shape. As seen by the HR-TEM images (Fig. 2b, e), the prepared nanoparticles show the well-ordered lattice fringe of the ZrO_2 lattice. Similar to Fig. 2b, the undoped sample has two lattice planes separated by 0.35 nm, which is consistent with the (101) crystallite planes of tetragonal ZrO_2. In contrast, Fig. 2e shows both ZrO_2 and yttrium lattice planes which confirm that yttrium ions have been incorporated into the ZrO_2 host lattice. The selected area electron diffraction (SAED) patterns (Fig. 2c, f) also confirmed the crystallinity of the undoped and yttrium-doped samples. The higher photocatalytic degradation mainly assigned to that the modified nanoparticles have a smaller optical band gap, a larger specific surface area, and a higher photocurrent. As a result, RhB can be degraded more efficiently with faster charge transfer.

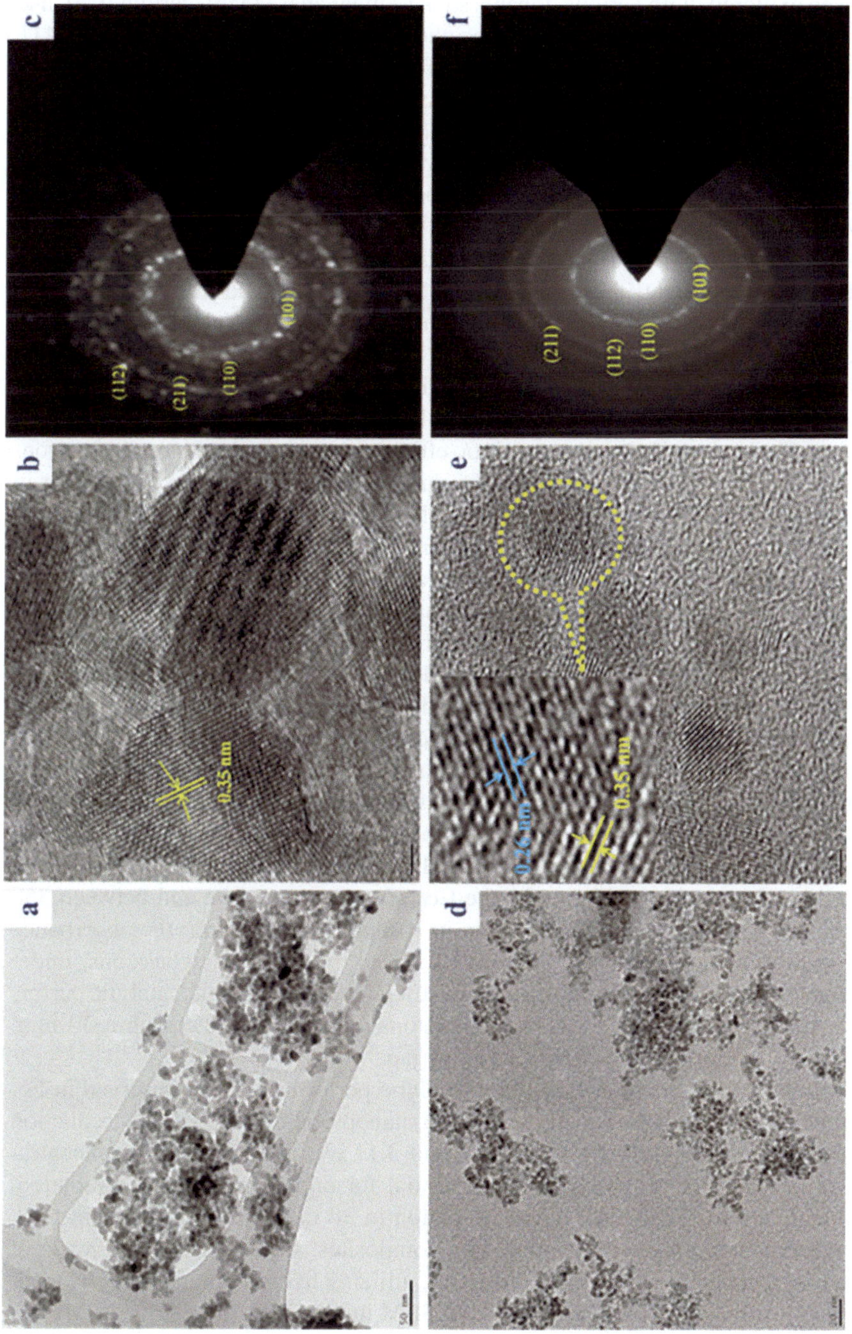

Fig. 2 **a, d** TEM, **b, e** HR-TEM, and **c, f** SAED images of the undoped and yttrium-doped ZrO_2 nanomaterials. Reproduced from Ref. [84] with permission, Copyright 2018 Elsevier

The influence of doping on photocatalytic and electrochemical properties has also been explored, with various dopants such as manganese (Mn), cerium (Ce), and erbium (Er) being used to further enhance photocatalytic performance of ZrO_2 [41]. The hydrothermal method has also been used to synthesize various composites, such as $ZnZrO_3/ZnO/ZrO_2$ perovskite nanocomposites, which demonstrated remarkable photocatalytic implementation for the breakdown of Congo Red (CR) dye in sunlight irradiation [3]. Additionally, Cu-doped $Ce_2Zr_2O_7$ photocatalysts synthesized via a hydrothermal approach exhibited excellent performance in degrading MB dye with a 96% degradation efficiency within 50 min [9]. The use of ZrO_2-based composites, such as CuO/ZrO_2, in environmental cleanup has also demonstrated encouraging results. Under UV light, these composites efficiently degrade drugs like tetracycline. Variables such as nanocomposite concentration and solution pH affect the functionality of these materials [95].

Many zirconium-based composites with improved photocatalytic efficiency for organic pollutant degradation have been developed using the hydrothermal synthesis approach. To increase ZrO_2 photocatalytic efficiency and broaden its use in environmental cleanup, several dopants and material composites have been added to it in recent studies. As an example, in order to test their photocatalytic activity in degrading sulfonamides, researchers created cerium-zirconium oxide ($Ce_xZr_yO_2$) composites (Fig. 3) [59]. As in Fig. 3a, after four sporadic on–off cycles of photoirradiation at $\lambda > 420$ nm, all samples showed quick and steady photocurrent responses. The $Ce_{0.9}Zr_{0.1}O_2$ sample had the largest photocurrent among all, indicating the maximum efficiency in the separation of photogenerated electron–hole pairs under visible light. From Fig. 3b, the $Ce_{0.9}Zr_{0.1}O_2$ sample showed the smallest diameter of the semicircle, indicating the lowest charge-transfer resistance (R_{ct}). It can be seen from Fig. 3c that the internal resistance (R_s) of all catalysts was very close, wherein the $Ce_{0.9}Zr_{0.1}O_2$ sample showed a smaller value. This suggested faster separation of electrons and holes, due to faster electron transition from the valence band to the conduction band. From UV–vis (Fig. 3d), it can be seen that all catalysts showed a well-defined photo absorption edge at about 550 nm for the $Ce_{0.9}Zr_{0.1}O_2$ composite and between 350 and 400 nm for the others. The addition of a small amount of Zr, the absorbance of the catalysts was significantly enhanced. From the above disscoussions, under solar light, a $Ce_{0.9}Zr_{0.1}O_2$ composite demonstrated the highest photocatalytic performance, with a degradation efficiency of sulfonamide was 91.33% within 30 min. This composite also had the narrowest band gap.

A comparable improvement in photocatalytic performance was observed in Fe-doped zirconium vanadate (Fe-doped ZrV_2O_7) nanoparticles, with a drastic reduction in band gap between 4.02 eV (undoped) and 3.11 eV at a 9 mol% Fe concentration. As an example of the material's potential for organic pollutant degradation, this change led to a 72% MB dye degredation in 30 min under visible light [10]. Other investigations have concentrated on composites, such as nanowires made of zirconium manganese oxide (ZrO_2–Mn_3O_4), utilizing hydrothermal synthesis techniques and surfactants (Fig. 4a) [34]. From SEM image (Fig. 4b), the ZrO_2–Mn_3O_4 displyed a clear nanowires with a diameter of nanowires less than 30 nm, and the

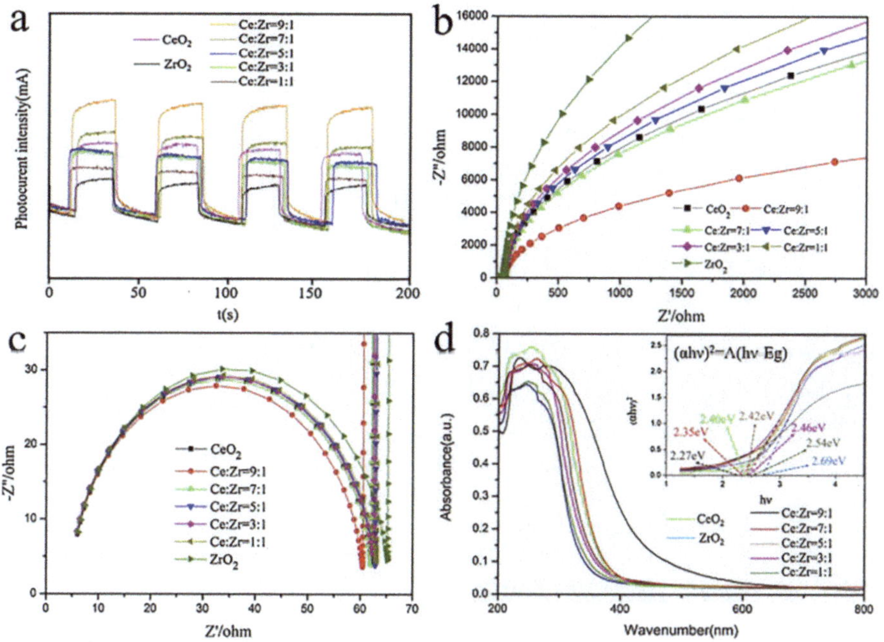

Fig. 3 **a** Photocurrent response of $Ce_xZr_yO_2$, CeO_2, and ZrO_2 in visible light ($\lambda > 420$ nm). **b, c** Nyquist plots of $Ce_xZr_yO_2$, ZrO_2, and CeO_2. **d** UV–vis spectra and $(\alpha h\nu)^2$ versus $h\nu$ for $Ce_xZr_yO_2$, CeO_2, and ZrO_2. Reproduced from Ref. [59] with permission, Copyright 2019 Elsevier

BET surface area was found to be 65 m^2/g. In comparison with commercial photocatalysts, these nanowires showed superior photocatalytic conversion of CO_2 to CH_4 when exposed to UV light, with an estimated rate of 1.46 μmol/g.h. Hydrothermally produced ZrO_2 nanopowders doped with Si^{4+} also demonstrated encouraging results in photocatalytic degradation of pollutants like trichlorophenol and RhB [21] due to stabilization of the tetragonal crystalline state of ZrO_2 and reduction of particle size.

The hydrothermal synthesis process was used to create bi and ternary composites with different sulfur contents, such as ZrO_2–$CdZrO_3$ and ZrO_2–$CdZrO_3$–S_{xy}. Adding sulfur to the ZrO_2–$CdZrO_3$–S_{20} composite lowered its band gap over 4.99 eV to 2.00 eV, moving its optical absorption edge into the visible spectrum. Crystal Violet (CV) degradation efficiency was greatly enhanced by this alteration; it achieved 95% degradation in 30 min and remained reusable for five cycles [93]. Figure 5 shows the combined sonication and hydrothermal synthesis results of g-C_3N_4/ZrO_2 composites. The photocatalytic activity for RhB dye degradation under sunshine was superior in these composites compared to pure g-C_3N_4 and ZrO_2. Due to their uniform distribution throughout the g-C_3N_4 layer, the ZrO_2 nanoparticles improved the photocatalytic activity by reducing resistance to charge transfer and blocking electron–hole recombination [22].

In other studies, the synthesis of Cu-NiO-ZrO_2 composites was examined. These composites degraded dyes such as MO and MB dye and showed high photocatalytic

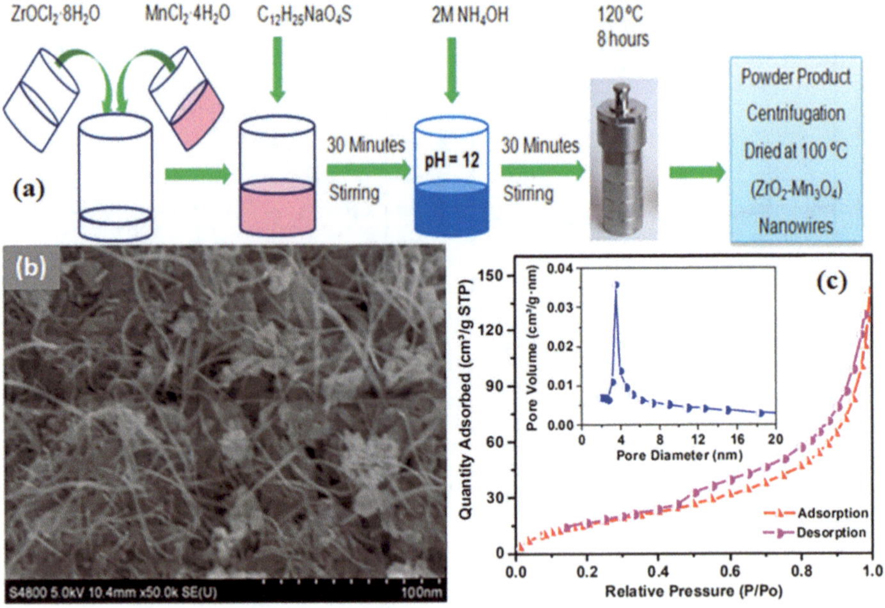

Fig. 4 **a** Scheme for preprtion of zirconium manganese oxide nanowires, **b** SEM image, and **c** Nitrogen adsorption–desorption for Zr, Mn oxide nanowires. Reproduced from Ref. [34] with permission, Copyright 2022 Springer Nature

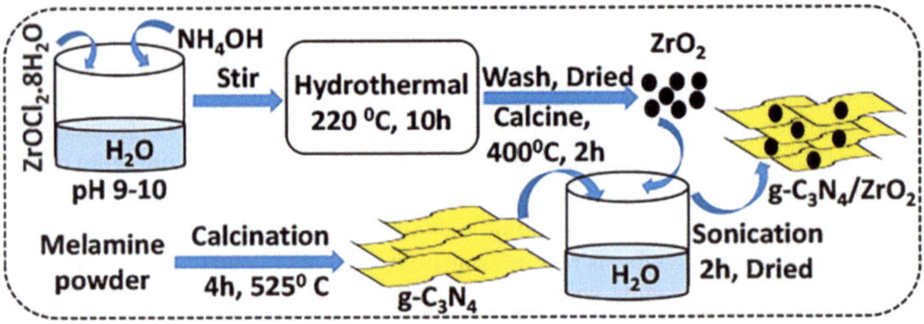

Fig. 5 Scheme for the preparation of g-C_3N_4/ZrO_2 composite. Reproduced from Ref. [22] with permission, Copyright 2023 Elsevier

capabilities for reducing 4-nitrophenol. The degradation efficiency was found to be influenced by parameters such as initial dye concentration, photocatalyst dosage, and pH [12]. Silver-doped zirconia (Ag-ZrO_2) nanoparticles prepared by hydrothermal and precipitation methods improved the direct band gap and photocatalytic action, particularly in damaged Rhodamine 6G dye [20]. The development of advanced zirconium-based photocatalysts through doping and composite formation has significantly enhanced the photocatalytic performance for a wide range of applications, including nitrogen fixation, pollutant degradation, and CO_2 reduction. Several studies have utilized the hydrothermal synthesis approach to modify ZrO_2 and improve its photocatalytic efficiency. For instance, a Fe_2O_3/ZrO_2 photocatalyst with oxygen vacancies was developed to enhance nitrogen fixation. The introduction of oxygen vacancies improved N_2 adsorption and activation, broadened the light absorption spectrum, and significantly increased ammonia production compared to pure ZrO_2. This modification boosted the photocatalytic performance, making it a promising material for efficient nitrogen fixation [57]. Other studies have focused on metal-doped ZrO_2 composites. ZrO_2@Ru and ZrO_2@Au nanostructures, incorporating ruthenium and gold nanoparticles, were synthesized for pesticide degradation. These materials exhibited excellent efficiency in degrading diuron and methyl parathion, with hydroxyl radicals playing a key task in the photocatalytic mechanism. The reusability of the ZrO_2@Ru composite was also confirmed, highlighting its potential for environmental applications [24]. Carboxymethyl chitosan (CMCS) was used as a template in a simple and mild microwave hydrothermal process to create a range of TiO_2/ZrO_2 composites with different molar ratios of ZrO_2:TiO_2. [94] Compared to pure ZrO_2, the TiO_2/ZrO_2 composites with heterogeneous structure had particles with a greater specific surface area, a smaller energy band gap, and superior regularity and uniformity. The particles had a diameter of approximately 800 nm. The morphologies of TiO_2/ZrO_2-CMCS-5 and TiO_2/ZrO_2-5 composites were investigated by SEM (Fig. 6). As presented in Fig. 6b, the agglomerated particles of TiO_2/ZrO_2-CMCS-5 show an average diameter of 800–900 nm, and the morphology of sample is similar to the coral rock. Figure 6d–f shows the images of TiO_2/ZrO_2-5 at different magnifications. Compared with TiO_2/ZrO_2-CMCS-5 in Fig. 6a, the morphology of TiO_2/ZrO_2-5 in Fig. 6d, e has no significant change. Meanwhile, the roughness degree of the sample surface after calcination (Fig. 6f, g) increases significantly, which can be attributed to release of gas generating by organic templates in the process of calcination. Due to the loss of the CMCS templates, the volume of particle (about 800 nm) decreases slightly. The disappearance of the CMCS leads to the obviously improved particle dispersity. The combination of TiO_2 and ZrO_2 has also been explored. TiO_2/ZrO_2 composites templated with chitosan derivatives were manufactured via a microwave hydrothermal approach, showing enhanced photocatalytic assets compared to pure ZrO_2. The 5% TiO_2/ZrO_2 composite exhibited the best performance, demonstrating elevated efficiency in pollutant breakdown and good reusability in multiple cycles.

The photocatalytic characteristics of ZrO_2 materials have been greatly improved by doping it with elements including yttrium, copper, nickel, silver, and iron and mixing it with various semiconductors like TiO_2. These doped materials successfully degrade dyes, reduce pollutants, and fix nitrogen because of their enhanced transmission of charge, light absorption, and reusability. For example, Mo-modified ZrO_2 photocatalysts (Fig. 7) were prepared by one-step hydrothermal process. The results showed improved CO_2 reduction activity, increased absorption of visible light, and a CO_2 output of 27.3 $\mu mol.g^{-1}$ [35].

To remove 4-nitrophenol, a heterojunction photocatalyst was created using reduced-graphene oxide (rGO), ZrO_2, and silver phosphate (Ag_3PO_4) (Fig. 8). From

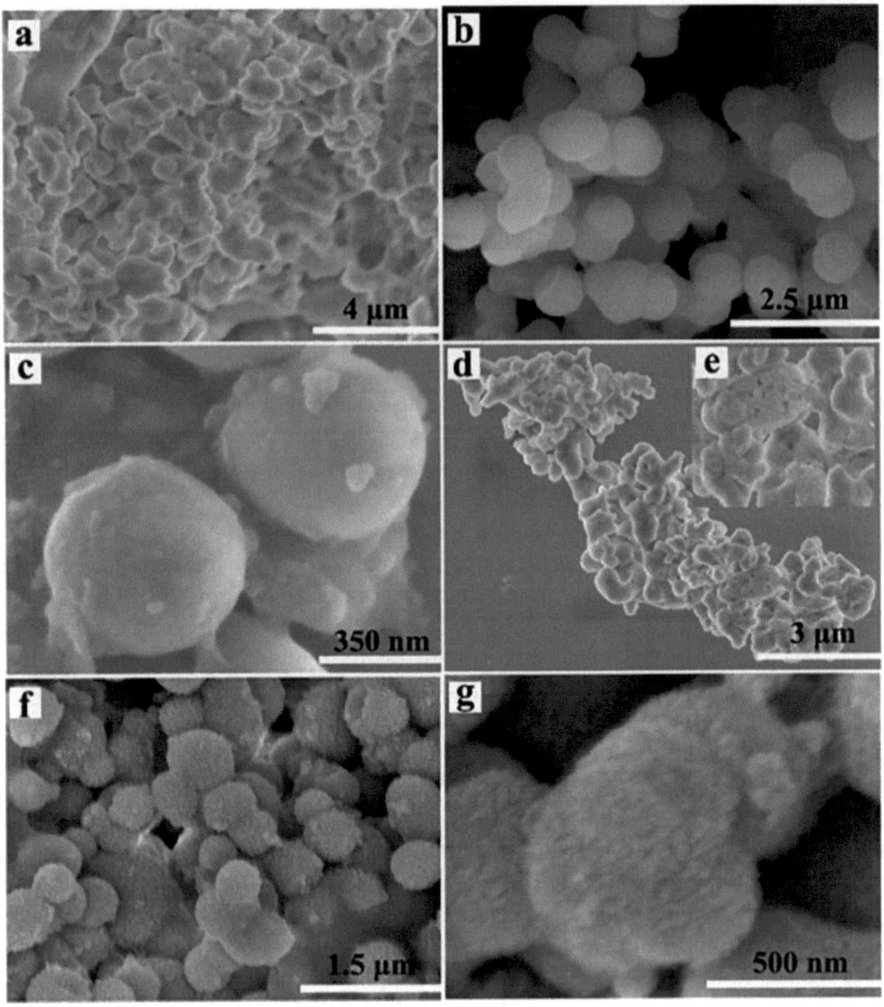

Fig. 6 SEM images of **a–c** TiO_2/ZrO_2-CMCS-5 and **d–f** TiO_2/ZrO_2-5. Reproduced from Ref. [94] with permission, Copyright 2019 Elsevier

Fig. 8a, pure ZrO_2 shows strong absorption in the UV region due to its large band gap energy, while Ag_3PO_4 shows characteristic absorption between 300 and 500 nm. The ZrO_2/Ag_3PO_4 composite displayed strong UV and moderate visible absorption with an absorption edge at about 440 nm. From Tauc plot in Fig. 8b, the calculated band gap energies of ZrO_2 and Ag_3PO_4 are 4.91 and 2.38 eV, respectively. Incorporation of rGO and Ag_3PO_4 as dopants lowered the band gap energy of $rGO/ZrO_2/Ag_3PO_4$ (2.3 eV). This low band gap energy implies that the nanocomposite can also exhibit photocatalytic activity under visible light. As presented in Fig. 8c, the red shift in

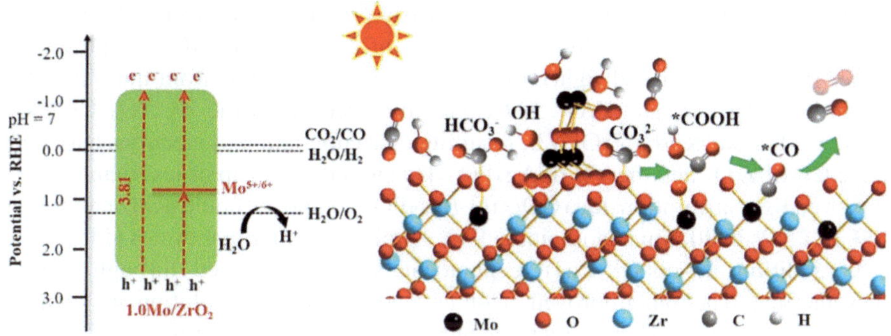

Fig. 7 Diagram depicting the photocatalytic degredation of CO_2 using ZrO_2 modified with molybdenum. Reproduced from Ref. [35] with permission, Copyright 2023 Elsevier

the PL spectra of the synthesized rGO/ZrO_2/Ag_3PO_4 spectrum could be attributed to interfacial charge transfer from ZrO_2/Ag_3PO_4 to rGO.

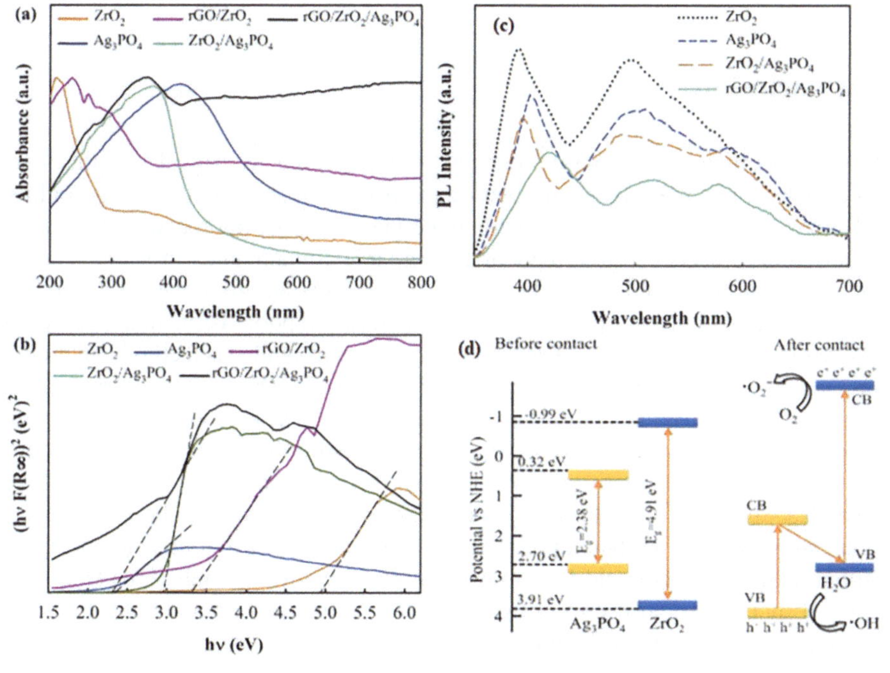

Fig. 8 **a** UV–Vis. spectra of the composites, **b** Tauc scenario to establish band gap energies, and **c** photoluminescence spectra of the nanocomposites, **d** band-edge positions of Ag_3PO_4 and ZrO_2 before and and after irraiation by UV light. Reproduced from Ref. [11] with permission, Copyright 2018 Elsevier

The PNP degradation scheme was proposed as presented in Fig. 8d. Before contact (Fig. 8d left), the VB and CB edge positions of Ag_3PO_4 and ZrO_2 are presented. Under irradiation (Fig. 8d right), the excited electrons moved from the VB to the CB in both Ag_3PO_4 and ZrO_2. Subsequently, electrons in the Ag_3PO_4 CB were transferred to the nearby ZrO_2 VB, resulting in accumulation of electrons and holes in the CB of ZrO_2 and the VB of Ag_3PO_4. Due to the band gap of less than one and improved light absorption, the rGO/ZrO_2/Ag_3PO_4 composite could achive removing 97% of PNPs in 90 min and retaining 94% elimination after five cycles [11].

In addition, Zr-doped $CaTiO_3$ photocatalysts, synthesized through a mild hydrothermal reaction, improved photocatalytic hydrogen evolution by creating nano-step structures. These structures increased active site exposure, surface area, and oxygen vacancies, which enhanced carrier separation and overall photocatalytic activity [37]. Lastly, ZrO_2 doped with rare earth oxides (CeO_2 and Er_2O_3) produced by a hydrothermal process demonstrated enhanced photocatalytic activity in the degradation of RhB dye when exposed to artificial sunlight. Because of a direct relationship between their physical–chemical characteristics and photoactivity, the doped ZrO_2 samples outperformed the pure ZrO_2 in terms of photocatalytic performance [40].

3 Conclusion

Developing efficient, long-term strategies to purify water is critical in light of the growing water contamination problem caused by human activities in agriculture and industry. Utilizing semiconductor materials as ZrO_2, photocatalysis presents an intriguing prospect for the reduction of organic contaminants in water. Though mechanical strength and thermal stability make ZrO_2 an attractive photocatalyst, issues including high electron rate of recombination and poor visible light absorption have reduced its effectiveness. Recent advancements in synthesis methods, including doping, the creation of oxygen vacancies, and hybridization with other materials, have significantly enhanced its photocatalytic performance. These modifications not only improve the ability of ZrO_2 to absorb light but also reduce electron–hole recombination, making it a more effective solution for environmental remediation. As research continues to refine these methods and explore novel approaches, ZrO_2-based photocatalysts hold significant promise for large-scale water treatment applications. Further advancements could lead to even more efficient materials, broadening the scope of their practical use in mitigating water contamination and contributing to cleaner, safer water sources worldwide.

References

1. Abdi J, Yahyanezhad M, Sakhaie S, Vossoughi M, Alemzadeh I (2019) Synthesis of porous TiO_2/ZrO_2 photocatalyst derived from zirconium metal organic framework for degradation of organic pollutants under visible light irradiation. J Environ Chem Eng 7(3):103096. https://doi.org/10.1016/j.jece.2019.103096
2. Aghabeygi S, Khademi-Shamami M (2018) ZnO/ZrO_2 nanocomposite: sonosynthesis, characterization and its application for wastewater treatment. Ultrason Sonochem 41:458–465. https://doi.org/10.1016/j.ultsonch.2017.09.020
3. Aghili SA, Hajiaghababaei L, Aberoomand Azar P, Badiei A, Saber-Tehrani M (2025) Fast preconcentration and determination of Zn(II) and Fe(III) using phenyl isothiocyanate functionalized nanoporous Silica. Int J Nano Dimension 16(1):1–10. https://doi.org/10.57647/j.ijnd.2025.1601.08
4. Akşit D, Soylu GSP (2022) Photocatalytic degradation of paracetamol by semiconductor oxides under UV and sunlight illumination. Turk J Chem 46(6):1866–1874
5. Al-Namshah KS, Mohamed RM (2019) Co_3O_4-ZrO_2 nanocomposites: simple preparation and enhanced photocatalytic performance for cyanide degradation under visible light. Desalin Water Treat 145:318–325. https://doi.org/10.5004/dwt.2019.23659
6. Aldebert P, Traverse J-P (1985) Structure and ionic mobility of zirconia at high temperature. J Am Ceram Soc 68(1):34–40. https://doi.org/10.1111/j.1151-2916.1985.tb15247.x
7. Ali AM, Ismail AA, Najmy R, Al-Hajry A (2014) Preparation and characterization of ZnO–SiO_2 thin films as highly efficient photocatalyst. J Photochem Photobiol, A 275:37–46. https://doi.org/10.1016/j.jphotochem.2013.11.002
8. Aloui Z (2024) Application of silver-iron co-doped ZrO_2 and TiO_2 bimetallic catalysts in the photodegradation of pharmaceutics pollutants and in inhibition of corrosion of commercial X65 steel. JOM 76(8):4713–4723. https://doi.org/10.1007/s11837-024-06636-w
9. Alshgari RA, Abdullah M, Abbas Shah SI, Abid AG, Mohammad S, Ehsan MF, Ashiq MN, Allakhverdiev SI (2024) Enhanced photocatalytic property of Cu doped $Ce_2Zr_2O_7$ toward photodegradation of methylene blue under visible light. Heliyon 10(14):e34266. https://doi.org/10.1016/j.heliyon.2024.e34266
10. Alyousef HA, Hassan AM, Ali AS, Issa SAM, Zakaly HMH (2024) Enhancing photocatalytic efficiency through iron doping: a comprehensive study on zirconium vanadate synthesis and characterization. J Mater Sci: Mater Electron 35(36):2258. https://doi.org/10.1007/s10854-024-13967-6
11. Anwer H, Park J-W (2018) Synthesis and characterization of a heterojunction $rGO/ZrO_2/Ag_3PO_4$ nanocomposite for degradation of organic contaminants. J Hazard Mater 358:416–426. https://doi.org/10.1016/j.jhazmat.2018.07.019
12. Areeb A, Yousaf T, Murtaza M, Zahra M, Zafar MI, Waseem A (2021) Green photocatalyst Cu/NiO doped zirconia for the removal of environmental pollutants. Mater Today Commun 28:102678. https://doi.org/10.1016/j.mtcomm.2021.102678
13. Arjun A, Dharr A, Raguram T, Rajni K (2020) Study of copper doped zirconium dioxide nanoparticles synthesized via sol–gel technique for photocatalytic applications. J Inorg Organomet Polym Mater 30:4989–4998
14. Aronne A, Marotta A, Pernice P, Catauro M (1996) Sol-gel processing and crystallization of yttria-doped zirconia. Thermochim Acta 275(1):75–82. https://doi.org/10.1016/0040-6031(95)02715-7
15. Bailón-García E, Elmouwahidi A, Carrasco-Marín F, Pérez-Cadenas AF, Maldonado-Hódar FJ (2017) Development of carbon-ZrO_2 composites with high performance as visible-light photocatalysts. Appl Catal B 217:540–550
16. Basahel SN, Ali TT, Mokhtar M, Narasimharao K (2015) Influence of crystal structure of nanosized ZrO_2 on photocatalytic degradation of methyl orange. Nanoscale Res Lett 10(1):73. https://doi.org/10.1186/s11671-015-0780-z

17. Baudys M, Krýsa J, Zlámal M, Mills A (2015) Weathering tests of photocatalytic facade paints containing ZnO and TiO_2. Chem Eng J 261:83–87. https://doi.org/10.1016/j.cej.2014.03.112
18. Behbahani A, Rowshanzamir S, Esmaeilifar A (2012) Hydrothermal synthesis of zirconia nanoparticles from commercial zirconia. Proc Eng 42:908–917. https://doi.org/10.1016/j.proeng.2012.07.483
19. Bensmaine Z, El Korso S, Moral-Rodríguez AI, Bedrane S, Ziani-Cherif C, Pérez-Cadenas AF, Carrasco-Marín F, Bailón-García E (2024) Enhanced photodegradation of sulfamethoxazole through cutting-edge titania-zirconia-based materials. Catalysts 14(11):784
20. Bharathi E, Sivakumari G, Karthikeyan B, Senthilvelan S (2020) Hydrothermal implement with supporting of semiconductor ZrO_2 (ZO), Ag doped ZrO_2 (AZO) nanomaterial and its astrophysical, UV photocatalytic employment on Rh6G dye. Appl Nanosci 10(9):3491–3502. https://doi.org/10.1007/s13204-020-01453-4
21. Carević MV, Abazović ND, Novaković TB, Pavlović VB, Čomor MI (2016) Zirconium dioxide nanopowders with incorporated Si^{4+} ions as efficient photocatalyst for degradation of trichlorophenol using simulated solar light. Appl Catal B 195:112–120. https://doi.org/10.1016/j.apcatb.2016.05.005
22. Chand S, Mondal A (2023) G-C_3N_4/ZrO_2 composite material: a pre-eminent visible light-mediated photocatalyst for rhodamine B degradation in the presence of natural sunlight. Ceram Int 49(3):5419–5430. https://doi.org/10.1016/j.ceramint.2022.10.065
23. Chang S-M, Doong R-A (2004) The effect of chemical states of dopants on the microstructures and band gaps of metal-doped ZrO_2 thin films at different temperatures. J Phys Chem B 108(46):18098–18103
24. Chauhan A, Dhenadhayalan N, Yeh J-C, Lin K-C (2022) Photocatalytic degradation-based efficient elimination of pesticides using ruthenium/gold metal nanoparticle-anchored zirconium dioxide. New J Chem 46(47):22561–22573. https://doi.org/10.1039/D2NJ03361E
25. Chu H-L, Hwang W-S, Du J-K, Chen K-K, Wang M-C (2016) Crystallization behavior of $ZrO_{23}Y_2O_{3x}$SrO precursor powders synthesized by a coprecipitation process. J Alloy Compd 678:518–526
26. Davar F, Loghman-Estarki MR (2014) Synthesis and optical properties of pure monoclinic zirconia nanosheets by a new precursor. Ceram Int 40(6):8427–8433
27. de Moraes NP, de Azeredo CASH, Bacetto LA, da Silva MLCP, Rodrigues LA (2018) The effect of C-doping on the properties and photocatalytic activity of ZrO_2 prepared via sol-gel route. Optik 165:302–309
28. Demircivi P, Simsek EB (2018) Fabrication of Zr-doped TiO_2/chitosan composite catalysts with enhanced visible-light-mediated photoactivity for the degradation of Orange II dye. Water Sci Technol 78(3):487–495
29. Dong H, Zeng G, Tang L, Fan C, Zhang C, He X, He Y (2015) An overview on limitations of TiO_2-based particles for photocatalytic degradation of organic pollutants and the corresponding countermeasures. Water Res 79:128–146. https://doi.org/10.1016/j.watres.2015.04.038
30. Duran C, Sato K, Hotta Y, Göçmez H, Watari K (2015) Ball milling assisted hydrothermal synthesis of ZrO_2 nanopowders. Ceram Int 41(4):5588–5593
31. Dwivedi R, Maurya A, Verma A, Prasad R, Bartwal K (2011) Microwave assisted sol–gel synthesis of tetragonal zirconia nanoparticles. J Alloy Compd 509(24):6848–6851
32. Esquivias L, Barrera-Solano C, Piñero M, Prieto C (1996) Short-range order of yttria doped zirconia powders studied by X-ray absorption (II). J Alloy Compd 239(1):71–76. https://doi.org/10.1016/0925-8388(96)02267-0
33. Fakhri A, Behrouz S, Tyagi I, Agarwal S, Gupta VK (2016) Synthesis and characterization of ZrO_2 and carbon-doped ZrO_2 nanoparticles for photocatalytic application. J Molec Liq 216(C):342–346. https://doi.org/10.1016/j.molliq.2016.01.046
34. Farid MA, Ijaz S, Ashiq MN, Ehsan MF, Gul F, Batool SR, Athar M, ul Hassan S (2022) Synthesis of mesoporous zirconium manganese mixed metal oxide nanowires for photocatalytic reduction of CO_2. J Mater Res 37(2):522–532. https://doi.org/10.1557/s43578-021-00463-4

35. Feng S, Zhao J, Liang X, Li H, Wang C (2023) Mo-modified (t-m)-ZrO_2 with narrowed band gap and strong ability to activate reactants for photocatalytic CO_2 reduction. Molec Catal 544:113205. https://doi.org/10.1016/j.mcat.2023.113205
36. Figini Albisetti A, Biffi CA, Tuissi A (2016) Synthesis and structural analysis of Copper-Zirconium oxide. Metals 6(9):195
37. Gao Q, Meng J, Yang Y, Lin Q, Lu Y, Wei X, Li J, Han G, Zhang Z (2021) Zirconium doping in calcium titanate perovskite oxides with surface nanostep structure for promoting photocatalytic hydrogen evolution. Appl Surf Sci 542:148544. https://doi.org/10.1016/j.apsusc.2020.148544
38. Gao Y, Guo Y, Zhang H (2016) Iron modified bentonite: enhanced adsorption performance for organic pollutant and its regeneration by heterogeneous visible light photo-Fenton process at circumneutral pH. J Hazard Mater 302:105–113. https://doi.org/10.1016/j.jhazmat.2015.09.036
39. Gionco C, Fabbri D, Calza P, Paganini MC (2016) Synthesis, characterization, and photocatalytic tests of n-doped zinc oxide: a new interesting photocatalyst. J Nanomater 1:4129864. https://doi.org/10.1155/2016/4129864
40. Gionco C, Hernández S, Castellino M, Gadhi TA, Muñoz-Tabares JA, Cerrato E, Tagliaferro A, Russo N, Paganini MC (2019) Synthesis and characterization of Ce and Er doped ZrO_2 nanoparticles as solar light driven photocatalysts. J Alloy Compd 775:896–904. https://doi.org/10.1016/j.jallcom.2018.10.046
41. Gionco C, Paganini MC, Giamello E, Sacco O, Vaiano V, Sannino D (2017) Rare earth oxides in zirconium dioxide: how to turn a wide band gap metal oxide into a visible light active photocatalyst. J Energy Chem 26(2):270–276. https://doi.org/10.1016/j.jechem.2016.07.006
42. Habibi M, Wang L, Liang J, Guo S (2013) An investigation on hot corrosion behavior of YSZ-Ta_2O_5 in $Na_2SO_4 + V_2O_5$ salt at 1100 °C. Corros Sci 75:409–414
43. Helmiyati H, Fitriana N, Chaerani ML, Dini FW (2022) Green hybrid photocatalyst containing cellulose and γ–Fe_2O_3–ZrO_2 heterojunction for improved visible-light driven degradation of Congo red. Opt Mater 124:111982. https://doi.org/10.1016/j.optmat.2022.111982
44. Hossein Habibi M, Askari E, Habibi M, Zendehdel M (2013) Novel nanostructure zinc zirconate, zinc oxide or zirconium oxide pastes coated on fluorine doped tin oxide thin film as photoelectrochemical working electrodes for dye-sensitized solar cell. Spectrochim Acta Part A Mol Biomol Spectrosc 104:197–202. https://doi.org/10.1016/j.saa.2012.11.055
45. Ilkhechi NN, Kaleji BK (2014) High temperature stability and photocatalytic activity of nanocrystalline anatase powders with Zr and Si co-dopants. J Sol-Gel Sci Technol 69(2):351–356. https://doi.org/10.1007/s10971-013-3224-1
46. Joo J, Yu T, Kim YW, Park HM, Wu F, Zhang JZ, Hyeon T (2003) Multigram scale synthesis and characterization of monodisperse tetragonal zirconia nanocrystals. J Am Chem Soc 125(21):6553–6557
47. Kadi MW, Mohamed RM (2018) Fabrication of Zirconium oxide-Hydroxyapatite nanocomposite for green preparation of m-phenylenediamine. J King Abdulaziz Univ 30(2):1–9
48. Kappadan S, Gebreab TW, Thomas S, Kalarikkal N (2016) Tetragonal $BaTiO_3$ nanoparticles: an efficient photocatalyst for the degradation of organic pollutants. Mater Sci Semicond Process 51:42–47. https://doi.org/10.1016/j.mssp.2016.04.019
49. Karunakaran C, Magesan P, Gomathisankar P, Vinayagamoorthy P (2013) Photocatalytic degradation of dyes by Al_2O_3-TiO_2 and ZrO_2-TiO_2 nanocomposites. Mater Sci Forum 734:325–333. https://doi.org/10.4028/www.scientific.net/MSF.734.325
50. Karunakaran C, Rajeswari V, Gomathisankar P (2010) Antibacterial and photocatalytic activities of sonochemically prepared ZnO and Ag–ZnO. J Alloy Compd 508(2):587–591. https://doi.org/10.1016/j.jallcom.2010.08.128
51. Kitiyanan A, Sakulkhaemaruethai S, Suzuki Y, Yoshikawa S (2006) Structural and photovoltaic properties of binary TiO_2–ZrO_2 oxides system prepared by sol–gel method. Compos Sci Technol 66(10):1259–1265. https://doi.org/10.1016/j.compscitech.2005.10.035
52. Kumar S, Kumar S, Tiwari S, Srivastava S, Srivastava M, Yadav BK, Kumar S, Tran TT, Dewan AK, Mulchandani A (2015) Biofunctionalized nanostructured zirconia for biomedical application: a smart approach for oral cancer detection. Adv Sci 2(8):1500048

53. Kumari L, Du GH, Li WZ, Vennila RS, Saxena SK, Wang DZ (2009) Synthesis, microstructure and optical characterization of zirconium oxide nanostructures. Ceram Int 35(6):2401–2408. https://doi.org/10.1016/j.ceramint.2009.02.007
54. Kumari L, Li WZ, Xu JM, Leblanc RM, Wang DZ, Li Y, Guo H, Zhang J (2009) Controlled hydrothermal synthesis of zirconium oxide nanostructures and their optical properties. Cryst Growth Des 9(9):3874–3880. https://doi.org/10.1021/cg800711m
55. Kumari P, Saha R, Saikia G, Bhujel A, Choudhury MG, Jagdale P, Paul S (2023) Synthesis of mixed-phase TiO_2–ZrO_2 nanocomposite for photocatalytic wastewater treatment. Toxics 11(3):234
56. Kurniawan R, Sudiono S, Trisunaryanti W, Syoufian A (2019) Synthesis of iron-doped zirconium titanate as a potential visible-light responsive photocatalyst. Indonesian J Chem 19(2):454–460
57. Lazuli ARSC, Ramalingam V, Neppolian B (2024) Promoting nitrogen photofixation for the synthesis of ammonia using oxygen-vacant Fe_2O_3/ZrO_2 visible light photocatalyst with straddling heterojunction and enhanced charge transfer. J Clean Prod 451:142095. https://doi.org/10.1016/j.jclepro.2024.142095
58. Lee WS, Kim SW, Koo BH, Bae DS (2008) Synthesis and microstructure of Y_2O_3-doped ZrO_2–CeO_2 composite nanoparticles by hydrothermal process. Colloids Surf, A 313:100–104
59. Li P, Guo M, Wang Q, Li Z, Wang C, Chen N, Wang C-C, Wan C, Chen S (2019) Controllable synthesis of cerium zirconium oxide nanocomposites and their application for photocatalytic degradation of sulfonamides. Appl Catal B 259:118107. https://doi.org/10.1016/j.apcatb.2019.118107
60. Limaye AU, Helble JJ (2004) Secondary atomization as a mechanism for controlling the size of ceramic nanoparticles produced by combustion aerosol synthesis. J Aerosol Sci 35(5):599–610
61. Liu C, Hajagos TJ, Chen D, Chen Y, Kishpaugh D, Pei Q (2016) Efficient one-pot synthesis of colloidal zirconium oxide nanoparticles for high-refractive-index nanocomposites. ACS Appl Mater Interfaces 8(7):4795–4802. https://doi.org/10.1021/acsami.6b00743
62. Liu H-Y, Wang G-H, Yang J, Liu Y-Y, Ma J-F (2014) An unusual Cd-substituted sandwich-type polyoxomolybdate cluster $Mo_{20}Cd_2$ for photocatalysis of organic pollutant. Inorg Chem Commun 50:92–96. https://doi.org/10.1016/j.inoche.2014.10.031
63. Liu W-J, Zeng F-X, Jiang H, Zhang X-S, Li W-W (2012) Composite Fe_2O_3 and ZrO_2/Al_2O_3 photocatalyst: preparation, characterization, and studies on the photocatalytic activity and chemical stability. Chem Eng J 180:9–18. https://doi.org/10.1016/j.cej.2011.10.085
64. López-Romero S, García-Hipólito M, Aguilar-Castillo A (2013) Bright green luminescence from zirconium oxide stabilized with Tb^{3+} ions synthesized by solution combustion technique. World J Condensed Matter Phys 2013
65. López T, Alvarez M, Tzompantzi F, Picquart M (2006) Photocatalytic degradation of 2,4-dichlorophenoxiacetic acid and 2,4,6-trichlorophenol with ZrO_2 and Mn/ZrO_2 sol-gel materials. J Sol-Gel Sci Technol 37(3):207–211. https://doi.org/10.1007/s10971-005-6630-1
66. Luo L, Li Y, Hou J, Yang Y (2014) Visible photocatalysis and photostability of Ag_3PO_4 photocatalyst. Appl Surf Sci 319:332–338. https://doi.org/10.1016/j.apsusc.2014.04.154
67. Ma G, Yan H, Shi J, Zong X, Lei Z, Li C (2008) Direct splitting of H_2S into H_2 and S on CdS-based photocatalyst under visible light irradiation. J Catal 260(1):134–140. https://doi.org/10.1016/j.jcat.2008.09.017
68. Mahmood Q, Afzal A, Siddiqi HM, Habib A (2013) Sol–gel synthesis of tetragonal ZrO_2 nanoparticles stabilized by crystallite size and oxygen vacancies. J Sol-Gel Sci Technol 67:670–674
69. Matějová L, Kočí K, Reli M, Čapek L, Matějka V, Šolcová O, Obalová L (2013) On sol–gel derived Au-enriched TiO_2 and TiO_2-ZrO_2 photocatalysts and their investigation in photocatalytic reduction of carbon dioxide. Appl Surf Sci 285:688–696. https://doi.org/10.1016/j.apsusc.2013.08.111
70. Mbiri A, Wittstock G, Taffa DH, Gatebe E, Baya J, Wark M (2018) Photocatalytic degradation of the herbicide chloridazon on mesoporous titania/zirconia nanopowders. Environ Sci Pollut Res Int 25(35):34873–34883. https://doi.org/10.1007/s11356-017-1023-x

71. Molaei M, Attarzadeh N, Fattah-Alhosseini A (2021) Tailoring the biological response of zirconium implants using zirconia bioceramic coatings: a systematic review. J Trace Elem Med Biol 66:126756
72. Nabih S, Shalan AE, Serea ESA, Goda MA, Sanad MF (2019) Photocatalytic performance of TiO_2@SiO_2 nanocomposites for the treatment of different organic dyes. J Mater Sci: Mater Electron 30(10):9623–9633. https://doi.org/10.1007/s10854-019-01296-y
73. Naik R, Nagaswarupa H, Darukesha B, Tejashwini D (2024) Integration of ZrO_2 nanomaterials into scintillation detectors. In: Advances in space radiation detection: novel nanomaterials and techniques. Springer, pp 65–79
74. Nikolaou P (2023) Drive up electrochemical and electrochemiluminescence signal through biosensing mechanism optimization
75. Nitoi I, Oancea P, Cristea I, Constsntin L, Nechifor G (2015) Kinetics and mechanism of chlorinated aniline degradation by TiO_2 photocatalysis. J Photochem Photobiol, A 298:17–23. https://doi.org/10.1016/j.jphotochem.2014.10.005
76. Padture NP, Gell M, Jordan EH (2002) Thermal barrier coatings for gas-turbine engine applications. Science 296(5566):280–284. https://doi.org/10.1126/science.1068609
77. Pham Dinh K (2022) Functionalization of TiO_2 inverse opal structures for tailored photocatalytic activity. Itä-Suomen yliopisto
78. Pirilä M, Saouabe M, Ojala S, Rathnayake B, Drault F, Valtanen A, Huuhtanen M, Brahmi R, Keiski RL (2015) Photocatalytic degradation of organic pollutants in wastewater. Top Catal 58(14):1085–1099. https://doi.org/10.1007/s11244-015-0477-7
79. Pirzada BM, Mir NA, Qutub N, Mehraj O, Sabir S, Muneer M (2015) Synthesis, characterization and optimization of photocatalytic activity of TiO_2/ZrO_2 nanocomposite heterostructures. Mater Sci Eng, B 193:137–145
80. Piticescu R, Monty C, Millers D (2005) Hydrothermal synthesis of nanostructured zirconia materials: present state and future prospects. Sens Actuators, B Chem 109(1):102–106. https://doi.org/10.1016/j.snb.2005.03.092
81. Pouretedal H, Tofangsazi Z, Keshavarz M (2012) Photocatalytic activity of mixture of ZrO_2/SnO_2, ZrO_2/CeO_2 and SnO_2/CeO_2 nanoparticles. J Alloy Compd 513:359–364
82. Rabah B, Hanene B (2012) Synthesis, Characterization and properties of zirconium oxide (ZrO_2)-doped titanium oxide (TiO_2) thin films obtained via sol-gel process. In: Frank C (ed) Heat treatment. IntechOpen, Rijeka, p Ch 10. https://doi.org/10.5772/51155
83. Ratnayake S, Mantilaka M, Sandaruwan C, Dahanayake D, Murugan E, Kumar S, Amaratunga G, de Silva KN (2019) Carbon quantum dots-decorated nano-zirconia: a highly efficient photocatalyst. Appl Catal A 570:23–30
84. Reddy CV, Babu B, Reddy IN, Shim J (2018) Synthesis and characterization of pure tetragonal ZrO_2 nanoparticles with enhanced photocatalytic activity. Ceram Int 44(6):6940–6948. https://doi.org/10.1016/j.ceramint.2018.01.123
85. Reli M, Edelmannová M, Šihor M, Praus P, Svoboda L, Mamulová KK, Otoupalíková H, Čapek L, Hospodková A, Obalová L, Kočí K (2015) Photocatalytic H_2 generation from aqueous ammonia solution using ZnO photocatalysts prepared by different methods. Int J Hydrogen Energy 40(27):8530–8538. https://doi.org/10.1016/j.ijhydene.2015.05.004
86. Renuka L, Anantharaju KS, Sharma SC, Nagabhushana H, Vidya YS, Nagaswarupa HP, Prashantha SC (2017) A comparative study on the structural, optical, electrochemical and photocatalytic properties of ZrO_2 nanooxide synthesized by different routes. J Alloy Compd 695:382–395. https://doi.org/10.1016/j.jallcom.2016.10.126
87. Shao GN, Imran SM, Jeon SJ, Engole M, Abbas N, Salman Haider M, Kang SJ, Kim HT (2014) Sol–gel synthesis of photoactive zirconia–titania from metal salts and investigation of their photocatalytic properties in the photodegradation of methylene blue. Powder Technol 258:99–109. https://doi.org/10.1016/j.powtec.2014.03.024
88. Sharaf IM, Laifi J, Alraddadi S, Saad M, Koubesy M, Elewa NN, Almohiy H, Ismail YM, Soldatov A (2024) Unraveling the effect of Cu doping on the structural and morphological properties and photocatalytic activity of ZrO_2. Heliyon 10(1)

89. Singh AK, Nakate UT (2014) Microwave synthesis, characterization, and photoluminescence properties of nanocrystalline Zirconia. Scient World J 2014(1):349457. https://doi.org/10.1155/2014/349457
90. Sōmiya S, Roy R (2000) Hydrothermal synthesis of fine oxide powders. Bull Mater Sci 23(6):453–460. https://doi.org/10.1007/BF02903883
91. Sudrajat H, Babel S (2016) Comparison and mechanism of photocatalytic activities of N-ZnO and N-ZrO_2 for the degradation of rhodamine 6G. Environ Sci Pollut Res Int 23(10):10177–10188. https://doi.org/10.1007/s11356-016-6191-6
92. Taguchi M, Nakane T, Matsushita A, Sakka Y, Uchikoshi T, Funazukuri T, Naka T (2014) One-pot synthesis of monoclinic ZrO_2 nanocrystals under subcritical hydrothermal conditions. J Supercrit Fluids 85:57–61
93. Tavakoli-Azar T, Mahjoub A, Seyed Sadjadi M, Ghaznavi-Ghoushchi MB (2022) Enhanced photocatalytic activity of ZrO_2-$CdZrO_3$-S nanocomposites for degradation of Crystal Violet dye under sunlight. J Photochem Photobiol, A 426:113746. https://doi.org/10.1016/j.jphotochem.2021.113746
94. Tian J, Shao Q, Zhao J, Pan D, Dong M, Jia C, Ding T, Wu T, Guo Z (2019) Microwave solvothermal carboxymethyl chitosan templated synthesis of TiO_2/ZrO_2 composites toward enhanced photocatalytic degradation of Rhodamine B. J Colloid Interface Sci 541:18–29. https://doi.org/10.1016/j.jcis.2019.01.069
95. Turki Jalil A, Emad Al. Qurabiy H, Hussain Dilfy S, Oudah Meza S, Aravindhan S, M. Kadhim M, Aljeboree A (2021) CuO/ZrO_2 nanocomposites: facile synthesis, characterization and photocatalytic degradation of tetracycline antibiotic. J Nanostruct 11(2):333–346. https://doi.org/10.22052/JNS.2021.02.014
96. Uribe López M, Alvarez Lemus M, Hidalgo M, López González R, Quintana Owen P, Oros-Ruiz S, Uribe López S (2019) Acosta J (2019) Synthesis and characterization of ZnO-ZrO_2 nanocomposites for photocatalytic degradation and mineralization of phenol. J Nanomater 1:1015876
97. Verbruggen SW (2015) TiO_2 photocatalysis for the degradation of pollutants in gas phase: from morphological design to plasmonic enhancement. J Photochem Photobiol, C 24:64–82. https://doi.org/10.1016/j.jphotochemrev.2015.07.001
98. Vulić TD, Carević MV, Abazović ND, Novaković TB, Mojović ZD, Čomor MI (2022) Application of zirconia/alumina composite oxide ceramics as photocatalysts for removal of 2,4,6-trichlorophenol from water. Photochem 2(4):905–917. https://doi.org/10.3390/photochem2040058
99. Wahba MA, Yakout SM, Mohamed WAA, Galal HR (2020) Remarkable photocatalytic activity of Zr doped ZnO and ZrO_2/ZnO nanocomposites: structural, morphological and photoluminescence properties. Mater Chem Phys 256:123754. https://doi.org/10.1016/j.matchemphys.2020.123754
100. Wang J, Chen B (2015) Adsorption and coadsorption of organic pollutants and a heavy metal by graphene oxide and reduced graphene materials. Chem Eng J 281:379–388. https://doi.org/10.1016/j.cej.2015.06.102
101. Yaghoubi A, Ramazani A, Taghavi Fardood S (2020) Synthesis of Al_2O_3/ZrO_2 nanocomposite and the study of its effects on photocatalytic degradation of reactive blue 222 and reactive yellow 145 dyes. ChemistrySelect 5(32):9966–9973
102. Zakaria Z, Kamarudin SK (2021) Advanced modification of Scandia-stabilized zirconia electrolytes for solid oxide fuel cells application—a review. Int J Energy Res 45(4):4871–4887
103. Zare MH, Mehrabani-Zeinabad A (2022) Photocatalytic activity of ZrO_2/TiO_2/Fe_3O_4 ternary nanocomposite for the degradation of naproxen: characterization and optimization using response surface methodology. Sci Rep 12(1):10388
104. Zhu X, Zhou J, Zhu J, Liu Z, Li Y, Al-Kassab T (2014) Structural characterization and optical properties of perovskite 3 nanoparticles. J Am Ceram Soc 97(6):1987–1992. https://doi.org/10.1111/jace.12883

Bimetallic Oxide-Based Photocatalysts in Wastewater Treatment

Esra Yılmaz Mertsoy and Burcu Palas

Abstract Bimetallic oxide-based photocatalysts hold significant promise for revolutionizing wastewater treatment technologies. By addressing the outlined challenges and exploring new avenues for application, these materials could play a pivotal role in achieving cleaner water resources and a more sustainable future. Bimetallic oxides work together to improve electron transport, reduce recombination rates, and broaden light absorption spectra, all of which enhance photocatalytic activity. The development and use of photocatalysts based on bimetallic oxides containing bismuth, as well as those derived from metal organic frameworks (MOFs) and biomass were investigated in this chapter. Bismuth-based bimetallic oxides offer enhanced photocatalytic activities due to absorption of visible light, efficient charge separation, and tunable band gaps. They exhibit high chemical and thermal stability, improved surface area, and porosity for better pollutant interaction. Their environmentally friendly nature, coupled with scalable and cost-effective synthesis, makes them ideal for sustainable wastewater treatment applications. MOF-based bimetallic oxide photocatalysts increase the degradation efficiency of pollutants by utilizing the tunable topologies and high porosity of metal–organic frameworks. On the other hand, biomass-based bimetallic oxide photocatalysts offer an environmentally friendly and sustainable alternative in accordance with the principles of green chemistry, while maintaining good catalytic performance. This comprehensive strategy helps develop sustainable environmental solutions and demonstrates the potential of these innovative materials in the removal of complex pollutants in wastewater.

Keywords Photocatalytic oxidation · Bismuth based bimetallic oxides · MOF derived bimetallic oxides · Bimetallic oxides derived from natural materials · Organic pollutant removal

E. Y. Mertsoy
Faculty of Engineering, Chemical Engineering Department, Çankırı Karatekin University, Çankırı, Turkey

B. Palas (✉)
Faculty of Engineering, Chemical Engineering Department, Ege University, Bornova, İzmir, Turkey
e-mail: burcu.palas@ege.edu.tr

A. Ahmad et al. (eds.), *Metal Oxide Based Nanophotocatalyst for Wastewater Purification*, Smart Nanomaterials Technology,
https://doi.org/10.1007/978-981-96-7683-5_11

1 Introduction

Water contamination and scarcity have become a serious environmental problem due to the fast population growth and increasing industrialization [9, 20]. Recently, water pollution due to the release of toxic and hazardous chemicals including dyestuffs, pesticides, pharmaceuticals, phenols, heavy metals, and surfactants threaten the environmental health. For instance, the excess use of the carcinogenic and mutagenic chemicals may affect both the aquatic life and human health negatively. The persistent organic pollutants cannot be degraded completely by the application of traditional wastewater treatment methods such as chlorination, flocculation, and filtration techniques. These conventional methods have many disadvantages including requirement for multi-stage sequential processes, high operating cost, and incomplete removal of the contaminates which decrease the treatment efficiency and performance of the treatment system [29, 38].

On the other hand, with the development and application of advanced wastewater treatment technologies, attempts are being made to overcome the bottlenecks of traditional treatment methods. Photocatalysis occurring in the presence of semiconductor catalysts is a promising treatment method for the elimination of organic contaminants. Photocatalytic oxidation has attracted great interest as an advantageous method for degradation of toxic organic and inorganic pollutants in wastewaters. What is more, photocatalysis is a favorable technique, which utilizes sunlight for the conversion of pollutants into carbon dioxide and water. Key benefits of photocatalysis include the absence of solid waste and sludge generation, the achievement of overall mineralization efficiency, mild operating conditions, and the ability to be applied to complex and low-biodegradable chemicals even at high concentrations [21].

Photocatalysts play a significant role in accelerating the oxidation and reduction reactions under light irradiation. The general photocatalytic reactions include three main steps. In the first step, electron (e^-) and hole (h^+) pairs are generated by the absorption of light energy, which is greater than the bang gap energy of the photocatalyst. In the second step, electron and hole pairs are separated and diffused to the surface of the catalyst to initiate the redox reactions. In the third step, photo-oxidation and photo-reduction reactions occurred on the active sites of the photocatalytic surface. The photo-generated h^+ in the valence bands reacts with water to generate hydroxyl radicals and the photo-generated e^- react with O_2 to form superoxide reactive oxygen species. These reactive oxygen species have high oxidation and reduction potentials for the degradation of pollutants in wastewater. The general photocatalysis mechanism is illustrated in Fig. 1, and the main photocatalytic reactions are shown in Eqs. 1–4 [23, 24, 41]:

$$\text{Photocatalyst} + hv \rightarrow e^-_{(CB)} + h^+_{(VB)} \tag{1}$$

$$H_2O + h^+_{(VB)} \rightarrow H^+ + {}^{\cdot}OH \tag{2}$$

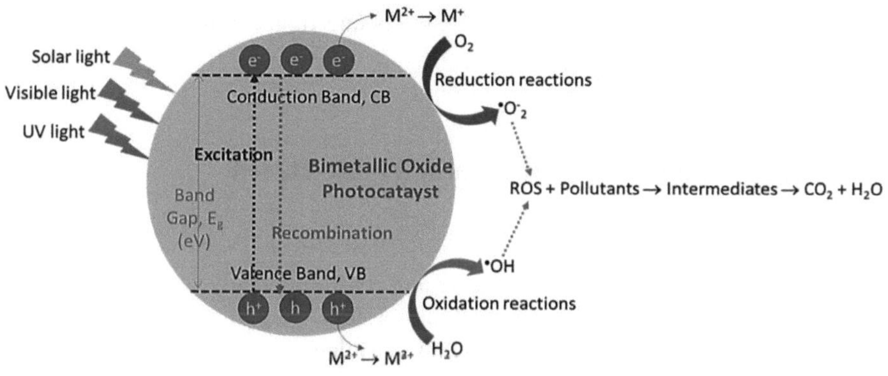

Fig. 1 Basic photocatalytic mechanism

$$O_2 + e^-_{(CB)} \to O_2^{\cdot -} \quad (3)$$

$$\text{Pollutants} + ^{\cdot}OH/O_2^{\cdot -} \to \text{Degradation products} \quad (4)$$

Bimetallic oxide photocatalysts have been attracting attention by both technological and scientific point of view due to potential distinctive optic, electronic, and photocatalytic features which are absent in the corresponding monometallic catalysts. Bimetallic catalysts exhibit not only the combined properties of its individual metals components but also newly acquired features due to the synergistic effect between two metals [49]. Bimetallic metal oxide photocatalysts are very useful recoverable heterogeneous catalysts with enhanced semiconducting efficacy. Conventional metal oxide catalysts such as TiO_2 and ZnO have wide band energies of exceeding 3.0 eV needing UV radiation in photocatalytic processes. Additionally, a single metal oxide has disadvantages of rapid electron hole recombination leading to decrease in photocatalytic efficiency. In order to activate the molecules under solar light, the band gap of the photocatalyst should be around 2.0 eV or below. Therefore, bimetallic photocatalysts containing multiple or hetero-metals have been developed due to their lower band gaps, higher surface areas, and improved catalytic activities. The decrease in the band gap by metal doping provides the ability to operate under visible light which favors environmental applicability in turn [33]. Besides the transition metal-based classical bimetallic oxide photocatalysts, bismuth-based bimetallic oxides, metal–organic framework-derived bimetallic oxides, and natural material-based bimetallic oxides have been extensively studied in in very recent researches.

In this chapter, the recent developments in the field of bimetallic oxide catalyst synthesis and their applications in photocatalytic degradation of various industrial pollutants including the dyes, pharmaceuticals, and phenolic compounds are summarized. Evaluation of photocatalytic reaction mechanisms and comparison of pollutant removal efficiencies of bismuth-containing bimetallic oxides, metal–organic framework-derived bimetallic oxides, and the bimetallic oxides obtained by using various natural materials such as plant extracts and lignocellulosic wastes are the original aspects of this chapter.

2 Bismuth-Containing Bimetallic Oxide-Based Photocatalysts

Bismuth-based compounds are affordable and environmentally friendly and have a potential for a wide variety of chemical and biological applications since they are generally insoluble in biological fluids and considered as nontoxic for organisms. Bismuth-containing photocatalysts have been widely used for wastewater remediation and to reduce air pollution besides water oxidation and H_2 evolution [2]. The recent literature studies on bismuth-containing bimetallic oxide-based catalysts are summarized in Table 1.

Metal oxide semiconductor materials composed of 3*d*, 4*d*, and 5*d* series elements have been utilized in photocatalytic processes due to their valence electrons that have the ability to be excited with solar light. Metal oxides containing higher d and f orbital elements possess extended conduction and valence bands, which increase the photo-generated carrier life time and favor the redox reactions occurring on the catalyst surface.

Photocatalytic performances of metal oxide catalysts consisting of rare earth metals, alkali metals, and transition metals have been investigated widely for the removal of organic contaminants [43]. Among the examined various photocatalysts, bismuth-based metal oxides have gained importance recently due to their high chemical stability, biocompatible properties, and narrow bandgaps below (< 3.0 eV). The recent literature studies on bismuth-containing bimetallic oxide-based catalysts are summarized in 1. It is thought that the valence bands of Bi^{3+} containing materials are formed by hybridization of O 2*p* and Bi 6*s*2 orbitals, thus upshifting the valence bands. $BiFeO_3$, Bi_2WO_6, $BiVO_4$, and Bi_2MoO_6 are $CuBi_2O_4$ are among the most commonly used bismuth-based bimetallic oxide photocatalysts [3]. Bismuth-containing bimetallic oxides have been utilized in photocatalytic processes for the degradation of organic pollutants as well as hydrogen production. Additionally, due to the anti-binding energy between Bi^{3+} and the electrons in the O 2*p* orbit, the Bi-based photocatalysts generally have a narrow band gap for light absorption and improved catalytic efficiency, which makes them good alternatives for high-performance photocatalytic processes [19].

2.1 Bismuth Ferrite-Based Bimetallic Oxides

Bismuth ferrite ($BiFeO_3$) has received significant attention as a class of layered nanocatalysts, and perovskite materials are employed in different fields because of

Table 1 Use of bismuth-based bimetallic oxides in photocatalytic processes for the removal of various industrial pollutants

Bimetallic oxide catalyst	Irradiation source	Target pollutant	Reaction time	Results	Dominant radical type	Reference
$BiFeO_3$–Fe_2O_3	Visible light	Methyl orange (MO) and methylene blue (MB)	300 min	94% dye degradation	$^{\bullet}OH$	[27]
$BiVO_4$	Blue LED light	Tetracycline	30 min	96% degradation	$O_2^{\bullet-}$	[28]
$BiFeO_3$	Solar light	Methylene blue (MB)	80 min	Complete dye degradation	$O_2^{\cdot-}$ and $^{\cdot}OH$	[30]
PVDF-β-Bi_2O_3@$BiFeO_3$	Visible light	Methylene blue (MB), rhodamine B (RhB)	90 min	95% MB degradation and 83% for RhB degradation	$^{\cdot}OH$	[40]
$BiFeO_3$	Visible light	Ofloxacin	180 min	80% ofloxacin degradation,64% TOC reduction	e^-, h^+ and $^{\cdot}OH$	[6]
$BiVO_4$/$BiFeO_3$	Visible light	Rhodamine B	120 min	98% degradation	$O_2^{\cdot-}$ and $^{\cdot}OH$	[13]
Bi_2WO_6/WO_3	Natural sunlight	Tetracycline, methyl orange (MO), rhodamine B (RhB)	50 min for dye removal 150 min for antibiotic removal	74.3 and 54.2% degradation on sunny and overcast days 63.7% MO and 99.0% RhB removal	(Not defined)	[52]
$BiVO_4$	Sunlight	Tetracycline, oxytetracycline	240 min	72% Tetracycline degradation and 83% oxytetracycline degradation	e^- and h^+	[7]
$BiFeO_3$ nanoflakes	Visible light and ultrasound	Rhodamine B	60 min	97% degradation	$O_2^{\cdot-}$ and $^{\cdot}OH$	[47]
Bi_2WO_6 nanoplates	Visible light	Rhodamine B	40 min	100% degradation	$O_2^{\cdot-}$ and $^{\cdot}OH$	[12]

its unique properties including biosafety, abundance, and high photocatalytic efficiency [45]. $BiFeO_3$ crystals have recently emerged as a noticeable photocatalyst for advanced oxidation of pollutants. Its narrower bandwidth between 2.0 and 2.7 eV compared to other multiferroic catalysts makes it a good candidate for visible light absorption and decomposition of organic substances. During the photocatalytic reactions, the electrons react with hydroxide ions to generate $^{\cdot}OH$ radicals in the presence of $BiFeO_3$ leading to the degradation to pollutants [54].

In the presence of H_2O_2 oxidant and bismuth ferrite-based bimetallic oxides, the following reactions (Eqs. 5–7) occur in addition to Eqs. 1–4 to produce reactive oxygen species capable of elimination of complex pollutants [47]:

$$Fe^{3+} + O_2^{\cdot -} \rightarrow Fe^{2+} + O_2 + e^- \quad (5)$$

$$Fe^{3+} + e^- \rightarrow Fe^{2+} \quad (6)$$

$$H_2O_2 + Fe^{2+} \rightarrow {}^{\cdot}OH + Fe^{3+} + OH^- \quad (7)$$

Bismuth ferrite with a rhombohedrally distorted perovskite structure is a novel type of photocatalyst promising for visible light-driven photocatalytic degradation of organic contaminants due to its perfect chemical stability. In addition, the weak ferromagnetic order of $BiFeO_3$ facilitates catalyst recycling from treated wastewaters. The environmental applications of $BiFeO_3$ in the field of advanced wastewater treatment technologies have gained a lot of attention of scientific community [17].

For instance, Parida et al. [30] synthesized $BiFeO_3$ nanoparticles by using leaf extract of Strychnos nux-vomica and tested the photocatalytic performance of the bimetallic oxide catalyst for the removal of methylene blue. At acidic reaction medium and 50 mg catalyst dosage, complete dye degradation was achieved in 80 min under solar light irradiation [30]. Yan et al. [47] achieved 97% rhodamine B degradation efficiency by using $BiFeO_3$ nanoflakes within 60 min in an ultrasound-assisted photo Fenton oxidation process [47]. Senthilkumar et al. [40] investigated the photocatalytic performance of PVDF-Bi_2O_3/$BiFeO_3$ nanocomposite materials. It is reported that 95% methylene blue and 83% rhodamine B degradation was obtained in 90 min [40]. Mehrdadian et al. [27] used $BiFeO_3$-Fe_2O_3 nanocomposites for the removal of dyes in photocatalytic oxidation process under visible light irradiation. Within 300 min 94% methylene blue and methyl orange degradation efficiencies were accomplished in alkaline and acidic medium, respectively [27]. Gupta et al. [6] utilized $BiFeO_3$ with a narrow bandgap of 2.1 eV in photocatalytic oxidation for the mineralization of ofloxacin. At pH 8, 0.5 g/L catalyst loading and 10 mg/L initial ofloxacin concertation, the antibiotic degradation and TOC removal efficiencies were evaluated as 80% and 64%, respectively [6].

2.2 Bismuth Vanadate-Based Bimetallic Oxides

Bismuth vanadate ($BiVO_4$) is one of the most widely examined photocatalysts reported in recent literature. It has several significant advantages as photocatalyst including low toxicity and cost effectiveness, high resistance to photo-corrosion, narrow bandwidth, and good visible light response. Moreover, recent studies have revealed that the fabrication of a n-p type heterojunction between n-type semiconductor $BiVO_4$ and p-type semiconductor $BiFeO_3$ remarkably improves the photocatalytic efficiency of the catalyst [13].

In literature, Obaid Qamar et al. [28] studied on photocatalytic removal of tetracycline in the presence of $BiVO_4$ catalyst in a slurry bubble column reactor. Under blue LED light irradiation, 96% antibiotic degradation was achieved in 30 min by using $BiVO_4$ photocatalyst [28]. Hemavibool et al. [7] studied on photocatalytic decomposition of tetracycline and oxytetracycline by using of $BiVO_4$ under sunlight and visible light. They reported that the degradation efficiencies were calculated as 72% and 83% for tetracycline and oxytetracycline, respectively [7]. Jing et al. [13] used $BiVO_4/BiFeO_3$ heterojunctions for the removal of rhodamine B. Using visible light irradiation, 98% dye degradation was accomplished within 120 min [13].

2.3 Bismuth Tungstate-Based Bimetallic Oxides

The application of bismuth tungstate (Bi_2WO_6) in photocatalytic processes has become an important research topic due to its ability to convert photocatalytic energy and environmental remediation. As a type of Aurivillius materials, Bi_2WO_6 is composed of $[Bi_2O_2]^{2+}$ and $[WO_4]^{2-}$ layers and has several distinctive features such as layered structure, visible light response, high photostability and thermal stability, and low toxicity. Bi_2WO_6 has been used for carbon dioxide reduction, water splitting, and water and air pollution treatment [4, 44]. Additionally Bi_2WO_6 has some interesting features including piezoelectric and ferroelectric characteristics, and it is sensitive to volatile organic compounds [39].

In literature, Jiang et al. [12] studied on organic pollutant removal by piezo-photocatalysis in the presence of Bi_2WO_6 nanoplates. Rhodamine B removal efficiency reached to approximately 100% in 40 min by combining the coupling photocatalytic and the piezoelectric effects [12]. Zhang et al. [52] tested the photocatalytic performance of Bi_2WO_6/WO_3 heterojunctions by utilizing natural sunlight. The tetracycline removal efficiencies were evaluated as 74.3 and 54.2% on sunny days and overcast days, respectively, within 150 min. 63.7% methyl orange degradation and 99.0% rhodamine B degradation efficiencies were achieved in 50 min [52].

3 Metal-Organic Framework-Derived Bimetallic Oxides-Based Photocatalysts

Metal–organic frameworks (MOFs) consisting of core metal ions and organic ligands as. The development of metal–organic frameworks (MOFs) offers a promising approach for separation, adsorption, and degradation water pollutants [1, 16]. Due to their high surface area, structural diversity, and high porosity, MOFs are highly suitable candidates for catalytic applications. However, the fast recombination of photo-generated electron–hole pairs in transition metal-containing MOF materials significantly hampers their photocatalytic performance [48]. To address this, MOFs can be transformed into derivatives, such as metal oxides or composites with porous carbon materials, through various treatment methods, which greatly enhance their conductivity and functionality.

Metal–organic frameworks have also been employed as precursors for the fabrication of metal oxide crystals with tailored morphologies. Several metal oxides derived from MOFs, such as CuO, Co_3O_4, Fe_2O_3, and ZnO, have demonstrated promising catalytic and photocatalytic properties [50, 51].

The greater chemical stability of MOF-derived metal oxides distinguishes them from their parent MOFs. They can also provide structural continuity while preserving some of the shape and porosity of the original MOFs.

Pyrolysis of MOFs in air at high temperatures is one of the most practical ways to obtain these properties by providing metal oxides. Metal oxides produced from MOFs partially inherit the porosity and shape of their parent MOFs while preserving the chemical stability of the oxide form. Furthermore, the properties of the resulting metal oxides can be customized by varying the pyrolysis conditions. Metal oxides produced from MOFs are interesting options for heterogeneous catalysis due to these advantages. Furthermore, due to the synergistic interaction between the two components, bimetallic MOF-derived oxides offer enhanced catalytic activity and significant photocatalytic potential.

Bimetallic MOFs have attracted considerable attention as a way to improve the performance of monometallic MOFs. To further enhance the performance of single-metallic MOFs, bimetallic MOFs have gained significant attention [31, 42].

Bimetallic MOF-based composite materials offer several advantages over their monometallic counterparts: (i) Monometallic MOFs exhibit diverse metallic elements, organic ligands, morphologies, and structures. By combining these distinct characteristics through tailored strategies, bimetallic MOFs with unique compositions, structures, and functionalities can be synthesized; (ii) While many monometallic MOFs possess limited porosity, the development of bimetallic MOFs effectively enhances this property. Moreover, the synthesis process for bimetallic MOFs is typically straightforward and occurs under mild conditions; (iii) The orderly arrangement of metal ions and organic ligands within bimetallic MOF frameworks facilitates the stabilization and dispersion of metals (or metal oxides), thereby enhancing both the stability and catalytic performance of these materials. The recent literature studies on metal–organic framework-derived bimetallic oxide-based catalysts are summarized in Table 2.

3.1 Bimetallic Oxides Derived from MIL Series

Recent research underscores the promise of bimetallic materials derived from MOFs for cutting-edge photocatalytic applications.

A significant instance is the fabrication of mesoporous materials, as showcased by Kohantorabi et al. [15], who synthesized a mesoporous $ZnO@NiFe_2O_4$ nanorod catalyst from a bimetallic Fe_2Ni-MIL-88 MOF.

The catalytic performance of the $ZnO@NiFe_2O_4$ structure was greatly improved by the addition of 1.0% w/w Ag. When exposed to UVA-LED light, the resultant Ag/$ZnO_{0.04}@NiFe_2O_4$ nanocomposite showed remarkable efficacy in PMS activation for the breakdown of acetaminophen (ACT).

Within 15 min, the $NiFe_2O_4$/PMS and 1.0% w/w Ag/$ZnO_{0.04}@NiFe_2O_4$/PMS systems had 62.2% and 100% ACT removal efficiencies, respectively. The Fe^{3+} sites on the surface of $NiFe_2O_4$ play a critical role in activating PMS, leading to the generation of SO_5^{-} and $SO_4^{\cdot-}$ radicals. The $NiFe_2O_4$/PMS system primarily facilitates two key reactions, as outlined below (Eqs. 8 and 9):

$$HSO_5^- + Fe^{3+} \rightarrow Fe^{2+} + SO_5^{\cdot-} + H^+ \quad (8)$$

$$HSO_5^- + Fe^{2+} \rightarrow Fe^{3+} + SO_4^{\cdot-} + HO^- \quad (9)$$

In another study, Zheng et al. [53] synthesized mixed bimetallic Ti/Zr MOFs with varying Ti/Zr molar ratios by incorporating a Zr source into the NH_2-MIL-125 preparation system under hydrothermal conditions. These MOF precursors were then calcined at 600 °C, yielding N self-doped TiO_2/ZrO_2 composites that retained the MOF structure. The influence of the Ti/Zr molar ratio on the photocatalytic degradation efficiency of methylene blue (MB) was systematically investigated. This method successfully addressed the challenges of uneven nitrogen distribution and harsh doping conditions. The resulting N self-doped TiO_2/ZrO_2 composites demonstrated significantly higher MB photodegradation rates compared to N-doped TiO_2 and N-doped ZrO_2. Moreover, the composites exhibited high degradation efficiencies for tetracycline and phenol. Reusability tests across eight photocatalytic cycles confirmed the excellent stability and reusability of these photocatalysts, highlighting their potential for sustainable applications in pollutant degradation. The possible mechanism of photocatalytic degradation of methylene blue (MB) under UV light irradiation using N-TiO_2/ZrO_2-5 expressed by Eqs. 10–14:

$$TiO_2 \rightarrow^{UV} TiO_2(e^-) + TiO_2(h^+) \quad (10)$$

Table 2 Various metal oxide nanostructures synthesized from MOF precursors

MOF precursor	Bimetallic oxide catalyst	Irradiation Source	Target pollutant	Reaction time	Results	Dominant radical type	Reference
Fe_2Ni-MIL-88 MOF	Ag/ $ZnO_{0.04}$@$NiFe_2O_4$	Visible light	Acetaminophen	60 min	100% acetaminophen degradation	$SO_4^{\cdot-}$ and $^{\cdot}OH$	[15]
Cr-Fe BTC	Cr_2O_3/Fe_2O_3	Visible light	Acid Blue 92	30 min	Complete degradation,	$O_2^{\cdot-}$ and h^+	[37]
NH_2-MIL-125	N-TiO_2/ZrO_2-5	UV light	Methylene blue, phenol, tetracycline	80 min for MB, 120 min for phenol and tetracycline	93.2% MB degradation, 86.3% tetracycline degradation, 72.4% phenol degradation	$^{\cdot}OH$	[53]
Fe-MIL-88B	AgPt@γ-Fe_2O_3/CuO	Visible light	Anionic and cationic dyes	160 min	Complete degradation	$O_2^{\cdot-}$ and $^{\cdot}OH$	[14],
MIL-101(Fe/Ni)	NiS/γ-Fe_2O_3/C	Visible light	Cr(VI)	35 min	83.85% Cr(VI) reduction	h^+ and $^{\cdot}OH$	[35]
MIL-125(Ti)	Ti^{3+}-TiO_2/Ar-Fe_2O_3	Visible light	Norfloxacin	5 min	97.80% degradation	$^{\cdot}OH$	[11]
ZIF-8	CuS/ZnO	Visible light	Methylene orange (MO), ciprofloxacin (CIP)	40 min	99.76% MO degradation, 94.59% CIP degradation	$O_2^{\cdot-}$	[26]
ZIF-67	$Bi_{12}O_{17}Cl_2$/GO/ Co_3O_4	Visible light	Levofloxacin (LVX)	120 min	91.2% degradation	h^+ and $^{\cdot}OH$	[22]

$$ZrO_2 \rightarrow^{UV} ZrO_2(e^-) + ZrO_2(h^+) \quad (11)$$

$$e^- + O_2 \rightarrow O_2^{\cdot -} \quad (12)$$

$$h^+ + H_2O \rightarrow {}^{\cdot}OH + H^+ \quad (13)$$

$$h^+/O_2^{\cdot -}/{}^{\cdot}OH + MB \rightarrow CO_2 + H_2O + \text{Other products} \quad (14)$$

Rashid et al. [35] address the urgent need for efficient removal and recycling of Cr(VI), a hazardous pollutant. The magnetic NiS/γ-Fe_2O_3/C type-II heterojunction synthesized via a one-pot calcination method using a type of bimetallic MOF, MIL-101(Fe/Ni) nanoflakes as a precursor. This synthesis leverages the unique dual functionality of MIL-101(Fe/Ni): as a metallic source for constructing NiS nanoparticles on a γ-Fe_2O_3 matrix and as a carbon source for forming an encapsulating C matrix.

The resultant NiS/γ-Fe_2O_3/C heterojunction exhibits remarkable photocatalytic reduction performance under visible light, achieving an 83.85% Cr(VI) reduction efficiency with optimal conditions of 20 mg/L Cr(VI), 0.003 g catalyst, and 35 min of irradiation. This efficiency is driven by enhanced interfacial charge separation and reduced recombination rates of photo-generated carriers within the heterojunction structure.

Jia et al. [11] synthesized a magnetic Ti^{3+}-TiO_2/Ar-Fe_2O_3 type-II heterojunction under Argon atmosphere. The enhanced light-harvesting ability of Ti^{3+}-TiO_2/Ar-Fe_2O_3 makes it an effective candidate for the peroxydisulfate (PDS) activation under visible light. By using Ti^{3+}-TiO_2/Ar-Fe_2O_3 and peroxydisulfate 97.80% degradation of norfloxacin (NOR) was achieved within just 5 min under visible light irradiation. The possible degradation mechanism of NOR involves the following reactions (Eqs. 15–21):

$$Ti^{3+} - TiO_2\backslash Ar - Fe_2O_3 + hv \rightarrow e^- + h^+ \quad (15)$$

$$Fe^{III} + e^- \rightarrow Fe^{II} \quad (16)$$

$$Fe^{II} + S_2O_8^{2-} + e^- \rightarrow Fe^{III} + SO_4^{\cdot -} + SO_4^{2-} \quad (17)$$

$$S_2O_8^{2-} + e^- \rightarrow 2SO_4^{\cdot -} \quad (18)$$

$$SO_4^{\cdot -} + OH^- \rightarrow SO_4^{2-} + {}^{\cdot}OH \quad (19)$$

$$h_{vB}^+ + OH^- \rightarrow {}^{\cdot}OH \quad (20)$$

$$h^{+}_{vB} + H_2O \rightarrow H^{+} + {}^{\cdot}OH \quad (21)$$

The introduction of noble metals into MOFs offers several advantages, including promoting electron transfer across heterojunctions, enhancing catalytic activity by providing thermal catalytic sites for intermediates, and acting as co-catalysts to suppress electron–hole recombination. Additionally, the porous structure of MOFs has been utilized as a template for the synthesis of porous carbon materials, multi-component nanoparticles, and metal oxide frameworks. Among these, metal oxides with efficient porosity, high surface area, controlled pore size, excellent cycle stability, and enhanced rate capability have shown significant potential for decomposing organic contaminants in wastewater, particularly when prepared via pyrolysis.

Khasevani et al. [14] incorporated the noble metals Ag, Pt, and Ag–Pt onto the surface of the γ-Fe_2O_3/CuO by in situ pyrolysis of Fe-MIL-88B and Cu-MOF (Cu (tpa)). The well-dispersed noble metal nanoparticles within the binary nanocomposite effectively enhance both catalytic and photocatalytic reactions. The AgPt@γ-Fe_2O_3/CuO nanocomposite exhibits exceptional performance in degrading both anionic and cationic dyes, as well as catalyzing the reduction of 4-Nitrophenol (4-Nip) to 4-Aminophenol (4-Amp). It can degrade organic dyes within 160 min. The reusability of these nanocomposites highlights their potential for sustainable green chemical synthesis, suitable for both laboratory-scale and industrial applications. Reactive species analysis identified $OH^{\cdot}$ and ${}^{\cdot}O_2^{-}$ radicals as the key contributors to the photodecomposition of organic pollutants. Kinetic studies confirmed that the dye degradation follows pseudo-first-order kinetics, while the 4-Nip reduction proceeds via first-order reaction kinetics.

3.2 *Bimetallic Oxides Derived from ZIF Series*

Mariappan et al. [26] developed ZIF-8-derived CuS/ZnO (CZ). Exceptional photocatalytic performance was achieved with the developed CuS/ZnO catalyst (CZ-2), and this catalyst showed significant improvement in the charge separation of photogenerated electrons. Using the CZ heterostructure, the degradation efficiencies of ciprofloxacin (CIP) and methylene orange (MO) were 94.59% and 99.76% in 40 min, respectively, with rate constants of 0.0312 and 0.0695 min^{-1}. ${}^{\cdot}O_2^{-}$ was found to be the dominant radical involved in the photocatalytic reactions.

For levofloxacin (LVX) degradation, Lu et al. [22] synthesized a $Bi_{12}O_{17}Cl_2$/GO/$Co_3O_4Cl_2$ using graphene oxide (GO), Co_3O_4 derived from zeolitic imidazole framework-67 (ZIF-67) and $Bi_{12}O_{17}Cl_2$. According to the characterization results, the combination of GO and Co_3O_4 increased the visible light response range, improved the LVX degradation efficiency, and enhanced the dissociation efficiency of photo-generated electron–hole pairs (e^{-}-h^{+}). Under ideal conditions, the $Bi_{12}O_{17}Cl_2$/1.5GO/7.5Co_3O_4 composite had the highest reaction rate constant (0.0151 min^{-1})

and degraded LVX by 91.2% in 120 min. This performance indicated the superior photocatalytic activity, which was 2.17, 13.14, and 1.53 times higher than that of $Bi_{12}O_{17}Cl_2$, Co_3O_4, and $Bi_{12}O_{17}Cl_2/7.5Co_3O_4$, respectively. Trapping experiments and electron spin resonance (EPR) studies revealed that superoxide radicals ($O_2^{\cdot-}$) and photo-generated holes (h^+) were present. Additionally, HPLC–MS was used to clarify the photocatalytic breakdown mechanism of LVX. The theoretical basis for creating $Bi_{12}O_{17}Cl_2$-based photocatalysts derived from ZIF-MOFs with enhanced performance is provided by this work.

3.3 Bimetallic Oxides Derived from Other MOFs

Salari [37] derived Cr_2O_3/Fe_2O_3 from a bimetallic MOF, as a promising photocatalyst with high surface area and enhanced photocatalytic performance. Using Acid Blue 92 (AB92), a commonly used anionic azo dye in the textile industry, as a model pollutant, the Cr_2O_3/Fe_2O_3 heterostructure was synthesized via a Cr-Fe-MOF template approach. This facile synthesis resulted in a material with a narrow band gap and high surface area, optimizing its photocatalytic efficiency under visible light irradiation.

Among the tested compositions, the Cr_2O_3:Fe_2O_3 ratio of 1:20 exhibited the highest activity. The pseudo-first-order kinetic rate constant for AB92 degradation was determined to be 0.078 min^{-1}, highlighting the material's potential for practical applications in dye degradation. Electrons in the conduction band can react with adsorbed O_2 molecules, leading to the formation of superoxide species. These reactive oxygen species are capable of degrading organic compounds, facilitating their decolorization. Equations 22–29 outline the general mechanistic pathways of these reactions.

$$Cr_2O_3/Fe_2O_3 + hv\ (\geq Eg) \rightarrow Cr_2O_3/Fe_2O_3\left(e^-_{CB} + h^+_{vB}\right) \tag{22}$$

$$OH^-_{ads} + h^+_{vB} \rightarrow {}^{\cdot}OH \tag{23}$$

$$O_{2\,ads} + e^-_{CB} \rightarrow O_2^{\cdot-} \tag{24}$$

$$H_2O_2 + e^-_{CB} \rightarrow OH^- + {}^{\cdot}OH \tag{25}$$

$$H_2O_2 + h^+_{vB} \rightarrow H^+ + {}^{\cdot}OH \tag{26}$$

$$\text{Decolorization reaction: Dye} + {}^{\cdot}OH \rightarrow \text{Products} \tag{27}$$

$$\text{Oxidation reaction: Dye} + h^+_{vB} \rightarrow \text{Products} \tag{28}$$

$$\text{Reduction reaction: Dye} + e^-_{CB} \rightarrow \text{Products} \tag{29}$$

4 Bimetallic Oxides-Based Photocatalysts Derived from Natural Materials

The management of agricultural waste poses a significant challenge due to costly and complex disposal processes. Natural lignocellulosic materials, primarily composed of cellulose, lignin, and hemicellulose, represent a major portion of this waste. Annually, around 200 billion tons of lignocellulosic biomass are generated from sources such as forestry, agriculture, industry, and domestic waste. Over time, the utilization of lignocellulosic biomass has evolved beyond traditional fuel applications to include advanced materials, pharmaceuticals, various chemicals, and synthesis of nanomaterials.

In particular, lignocellulose-derived nanomaterials have gained significant attention due to their unique properties. These materials exhibit a porous structure, high thermal stability and chemical stability, and tunable structural and morphological chemistry, making them ideal for use in biomaterials, electrodes, separators, and binders. Additionally, carbonization of lignocellulose generates efficient, porous, and electrically conductive catalysts, expanding their applications in electrochemical and photochemical processes and opening new avenues in carbon catalyst chemistry.

Due to their stability, ideal band gap, excellent charge carrier generation, and high surface-to-volume ratio, transition metal oxides (TMOs) have demonstrated efficacy in photocatalytic destruction of organic pollutants, particularly in wastewater treatment. These properties facilitate the formation of reactive oxygen species (ROS), which in turn decompose organic pollutants into less harmful forms. In photocatalytic applications, common TMOs such as TiO_2, ZnO, Fe_2O_3, Mn_2O_3, ZrO_2, V_2O_5, Co_3O_4, and WO_3 have been widely used [5, 18, 55]. The combination of two metal oxides in a single photocatalyst can further enhance performance by improving charge separation, reducing recombination, and creating synergistic effects, making bimetallic oxides highly valuable in advanced photocatalytic systems.

The recent literature studies on bimetallic oxides-based photocatalysts derived from natural materials catalysts are summarized in Table 3.

4.1 Bimetallic Oxides Incorporating Transition Metals: Photocatalysts Derived from Natural Materials

Nickel oxide (NiO), with its semiconducting behavior, stands out as a photocatalyst and antibacterial agent due to its ability to disrupt cell membranes and enzymes. Rani et al. [34] demonstrated a sustainable synthesis of a CaO@NiO nanocomposite using

Table 3 Bimetallic oxides-based photocatalysts derived from natural materials

Bimetallic oxide catalyst	Irradiation source	Target pollutant	Reaction time	Efficiency with time, results	Dominant radical type	Reference
Egg shell- and *S. mukorossi* seed extract derived CaO@NiO	Sunlight	3-Amino phenol (3-AP), lindane (γ-HCH)	5 h for 3-AP and 6 h for γ-HCH	97% 3-AP degradation 91% γ-HCH degradation	$^{\cdot}O^{2-}$	[34]
Egg shell–derived $CaMn_2O_4$/fly ash	Sunlight	Ofloxacin	30 min	88.57% degradation	$O_2^{\cdot-}$ and $^{\cdot}OH$	[46]
Egg shell–derived CaO–MgO	Sunlight	Methylene blue	20 min	Complete MB removal	$\cdot O_2-$ and h^+	[36]
$CuFe_2O_4$, $CuMn_2O_4$, and $MnZn_2O_4$ were synthesized using *Azadirachta indica* leaf extract	Sunlight	3-amino phenol (3-AP) and eriochrome black T (EBT)	5 h for 3-AP and 3 h for EBT	3-AP removal: 93% ($CuFe_2O_4$), 90% ($CuMn_2O_4$), 85% ($MnZn_2O_4$) EBT removal, 97% ($CuFe_2O_4$), 95% ($CuMn_2O_4$), 80% ($MnZn_2O_4$)	$^{\cdot}OH$ and $O_2^{\cdot-}$	[32]
A.P leaves and metal salts derived NiO-PdO-Pd	Solar light	Methyl orange	15 min	96% degradation	–	[10]

waste eggshells for CaO nanoparticles and *S. mukorossi* seed extract for NiO via a co-precipitation method. The resulting nanocomposite, CaO@NiO, characterized by high crystallinity, showed exceptional photocatalytic activity, achieving 91% degradation of pollutants under sunlight and 97% for 3-amino phenol (3-AP), following Langmuir adsorption and first-order kinetics.

With stability and reusability across six cycles, the CaO@NiO nanocomposite presents itself as a promising, eco-friendly material for practical applications, reducing environmental stress.

Hosny et al. [8] utilized the extract of *Persicaria salicifolia*, a hydrophyte from Egypt's Nile Delta, for the first time in the green synthesis of AgNPs, ZnONPs, and a biochar-based Ag/ZnO@BC nanocomposite. This study introduced an eco-friendly and cost-effective method, highlighting its novelty. The research aimed to explore the synergistic effects of AgNPs and ZnONPs on biochar for photocatalytic degradation of tetracycline (TC) under various optimized conditions.

The Ag/ZnO@BC nanocomposite demonstrated significant efficiency in photocatalytic degradation of TC, achieving 70.3% under optimized conditions: TC concentration of 50 ppm, pH 6, catalyst dose of 0.01 g, temperature at 25 °C, and H_2O_2 concentration of 100 mM. Additionally, the nanocomposite showed acceptable reusability, retaining 53% efficiency after six reuse cycles, with a rate constant (*k*) of 0.0182 min^{-1} [8].

Yadav et al. [46] highlighted the successful synthesis of $CaMn_2O_4$ (CMFA) composites on hydrothermally pretreated fly ash using an innovative co-precipitation method, emphasizing the efficiency of oxide formation. The fly ash substrate was hydrothermally tailored to enhance surface characteristics, while calcium precursors were derived from waste eggshells, reducing chemical dependency and promoting sustainability. Using CMFA composites, ofloxacin degradation efficiency of 88.57 $\pm$ 1.0% surpassed pure fly ash (78.2 $\pm$ 1.2%) in terms of photocatalytic activity. The reduction in band gap energy (1.9 eV for CMFA compared to 2.9 eV for fly ash) enhanced sunlight absorption, and photoluminescence analysis revealed reduced charge recombination rates and improved electron–hole separation.

Rani and Shanker [33] reported that single metal oxide nanomaterials often face limitations such as a large bandgap (> 3 eV) and charge recombination issues. To overcome these drawbacks, heterometallic oxides (HMOs) like $CuMn_2O_4$, $MnZn_2O_4$, and $CuFe_2O_4$ were prepared using a green approach with *Azadirachta indica* leaf extract. These HMO nanocatalysts were employed as photocatalysts for the degradation of 3-amino phenol (3-AP) and eriochrome black T (EBT) in water under sunlight. Significant contaminant degradation (80–95%) was achieved at neutral pH using 80 mg of HMOs for 50 mg/L pollutants under sunlight. The superior performance of $CuFe_2O_4$ was attributed to its large surface area of 35.7 m^2/g, low bandgap (2.4 eV), enhanced stability, and reduced photoluminescence intensity.

4.2 *Bimetallic Oxides Incorporating Noble Metals: Photocatalysts Derived from Natural Materials*

Coupling NiO with noble metal oxides further enhances its catalytic performance. As an example, MariaJoseph et al. [25] developed a versatile strategy utilizing carbon-supported NiO/PdO (NiO/PdO@C) nanocomposites derived from waste coconut coir for managing lignocellulose waste, degrading toxic dyes. The study highlights the synergistic effects of NiO and PdO integration in achieving superior photocatalytic and antibacterial performance. Individually, NiO@C and PdO@C nanostructures exhibited dye removal efficiencies of 44% and 34%, respectively, for rhodamine 6G. However, the heterostructured NiO/PdO@C composite achieved an impressive 94% dye degradation within 20 min. This significant improvement is attributed to the heterojunction formed at the NiO/PdO interface, which creates optimal chemical potential differences. These differences induce band bending, promoting enhanced photo-generated electron transfer and hole migration while effectively suppressing electron–hole recombination.

Iram et al. synthesized [10] inorganic–organic patterned synthesis of nanoscaled mixed metal oxides by using metal salts and *A.P* leaves. The synthesized nanocomposite of NiO-PdO-Pd was examined for of methyl orange (MO) degradation both in darkness and under solar light. By the use of organic compound incorporated NiO-PdO-Pd catalyst, 96 and 76% methyl orange removal was achieved under light as well as in darkness, in 15 min.

4.3 Bimetallic Oxides Incorporating Other Metals: Photocatalysts Derived from Natural Materials

Reyes-Vallejo et al. [36] presented the synthesis of Ca–Mg oxide using ball milling with eggshells as the CaO source, followed by annealing at temperatures between 450 and 900 °C. The material was tested for methylene blue (MB) photodegradation (10 ppm, 100 mL) under LED light (200 W) and sunlight. The highest performance was achieved with the catalyst annealed at 750 °C, attributed to its homogeneous size distribution, larger surface area, narrower band gap, and less surface defects, enabling efficient light absorption and charge transport. Complete MB removal occurred in 20 min under sunlight, reduced to 12 min using 400 mg catalyst. Reactive species such as superoxide radicals ($^{\cdot}O_{2-}$) and holes (h^+) were identified as key parameters in the photodegradation process, which followed pseudo 1st kinetics. These results surpass previous reports for Ca-Mg oxide prepared by combustion or ultrasound-based methods.

These studies underscore the importance of oxide formation from wastes in enhancing photocatalytic activity and present a low-cost, environmentally friendly approach for pollutant degradation in wastewater treatment.

5 Conclusion and Future Perspective

Bimetallic oxides offer significant advantages over monometallic systems, including improved surface area, better charge separation, and enhanced porosity. These properties result from the synergistic interaction between the two metallic species, which also increases the stability and distribution of active catalytic sites. Such benefits make bimetallic oxides more efficient for a wide range of applications in wastewater treatment, including the removal of refractory organic compounds and heavy metals.

Bimetallic oxide-based photocatalysts, synthesized from MOFs or biomass-derived precursors, have emerged as promising materials for addressing environmental challenges, particularly in wastewater treatment. These catalysts exhibit excellent photocatalytic performance due to their high surface area, tunable pore structures, and synergistic interactions between the metallic components. The incorporation of bimetallic oxides into the photocatalyst framework enhances electron mobility, reduces charge recombination, and broadens the absorption spectrum, enabling efficient utilization of solar energy for degrading complex organic pollutants and contaminants of emerging concern. Additionally, the use of sustainable precursors such as biomass and MOFs aligns with global green chemistry and circular economy principles.

Future studies should focus on fine-tuning the synthesis parameters to achieve precise control over the composition, shape, and crystallinity of bimetallic oxides. This will provide better stability and photocatalytic activity in various wastewater scenarios. To translate laboratory-scale procedures to industrial applications, it is necessary to focus on scalable and economical production techniques, including using abundant biomass waste as precursors. The diverse requirements of wastewater treatment can be met by creating photocatalysts specifically designed to degrade specific pollutants, such as heavy metals and pharmaceutical residues. The overall pollutant removal efficiency can be increased by combining bimetallic oxide-based photocatalysts with other state-of-the-art oxidation techniques such as ozonation or Fenton-like systems.

References

1. Bai Y, Nie G, He Y, Li C, Wang X, Ye L (2022) Cu-MOF for effectively organic pollutants degradation and E. coli inactivation via catalytic activation of peroxymonosulfate. J Taiwan Inst Chem Eng 132:104154. https://doi.org/10.1016/J.JTICE.2021.11.021
2. Balakumar S, Mahesh N, Kamaraj M, Shyamalagowri S, Manjunathan J, Murugesan S, Aravind J, Babu PS (2022) Outlook on bismuth-based photocatalysts for environmental applications: a specific emphasis on Z-scheme mechanisms. Chemosphere 303:135052. https://doi.org/10.1016/j.chemosphere.2022.135052
3. Chen Q, Cheng X, Long H, Rao Y (2020) A short review on recent progress of Bi/semiconductor photocatalysts: the role of Bi metal. Chin Chem Lett 31:2583–2590. https://doi.org/10.1016/j.cclet.2020.08.018
4. Chen T, Liu L, Hu C, Huang H (2021) Recent advances on Bi_2WO_6-based photocatalysts for environmental and energy applications. Chin J Catal 42:1413–1438. https://doi.org/10.1016/S1872-2067(20)63769-X
5. Delsouz Khaki MR, Shafeeyan MS, Raman AAA, Daud WMAW (2018) Evaluating the efficiency of nano-sized Cu doped TiO_2/ZnO photocatalyst under visible light irradiation. J Mol Liq 258:354–365. https://doi.org/10.1016/J.MOLLIQ.2017.11.030
6. Gupta G, Kansal SK, Umar A, Akbar S (2023) Visible-light driven excellent photocatalytic degradation of ofloxacin antibiotic using $BiFeO_3$ nanoparticles. Chemosphere 314:137611. https://doi.org/10.1016/j.chemosphere.2022.137611
7. Hemavibool K, Sansenya T, Nanan S (2022) Enhanced photocatalytic degradation of tetracycline and oxytetracycline antibiotics by $BiVO_4$ photocatalyst under visible light and solar light irradiation. Antibiotics 11:761. https://doi.org/10.3390/antibiotics11060761

8. Hosny M, Fawzy M, Eltaweil AS (2022) Green synthesis of bimetallic Ag/ZnO@Biohar nanocomposite for photocatalytic degradation of tetracycline, antibacterial and antioxidant activities. Sci Rep 12:1 12:1–17. https://doi.org/10.1038/s41598-022-11014-0
9. Huang L, Huang X, Yan J, Liu Y, Jiang H, Zhang H, Tang J, Liu Q (2023) Research progresses on the application of perovskite in adsorption and photocatalytic removal of water pollutants. J Hazard Mater 442:130024. https://doi.org/10.1016/j.jhazmat.2022.130024
10. Iram S, Ahmad KS, Shaheen I, Aljuwayid AM, Malik MA (2024) Phytogenic synthesis of facile NiO-PdO-Pd nanocomposite: green photocatalyst for fast degradation of organic pollutant. Bionanoscience 14:4644–4652. https://doi.org/10.1007/S12668-024-01321-3/FIGURES/6
11. Jia M, Yang Z, Xiong W, Cao J, Xiang Y, Peng H, Jing Y, Zhang C, Xu H, Song P (2021) Magnetic heterojunction of oxygen-deficient Ti3+-TiO_2 and Ar-Fe_2O_3 derived from metal-organic frameworks for efficient peroxydisulfate (PDS) photo-activation. Appl Catal B 298:120513. https://doi.org/10.1016/J.APCATB.2021.120513
12. Jiang Z, Tan X, Xu J, Huang Y (2022) Piezoelectric-induced internal electric field in Bi_2WO_6 nanoplates for boosting the photocatalytic degradation of organic pollutants. ACS Appl Nano Mater 5:7588–7597. https://doi.org/10.1021/acsanm.2c01698
13. Jing Q, Liu Z, Cheng X, Li C, Ren P, Guo K, Yue H, Xie B, Li T, Wang Z, Shu L (2023) Boosting piezo-photocatalytic activity of $BiVO_4/BiFeO_3$ heterojunctions through built-in polarization field tailoring carrier transfer performances. Chem Eng J 464:142617. https://doi.org/10.1016/j.cej.2023.142617
14. Khasevani SG, Taheri M, Gholami MR (2021) Green synthesis of Ag–Pt bimetallic nanoparticles supported on the metal-organic framework (MOF)–derived metal oxides (γ-Fe_2O_3/CuO) nanocomposite as a reusable heterogeneous nanocatalyst and nanophotocatalyst. Mater Chem Phys 261:124218. https://doi.org/10.1016/J.MATCHEMPHYS.2021.124218
15. Kohantorabi M, Moussavi G, Mohammadi S, Oulego P, Giannakis S (2021) Photocatalytic activation of peroxymonosulfate (PMS) by novel mesoporous Ag/ZnO@NiFe2O4 nanorods, inducing radical-mediated acetaminophen degradation under UVA irradiation. Chemosphere 277:130271. https://doi.org/10.1016/J.CHEMOSPHERE.2021.130271
16. Konnerth H, Matsagar BM, Chen SS, Prechtl MHG, Shieh F, Wu KC (2020) Metal-organic framework (MOF)-derived catalysts for fine chemical production. Coord Chem Rev 416:213319. https://doi.org/10.1016/j.ccr.2020.213319
17. Lam S-M, Sin J-C, Mohamed AR (2017) A newly emerging visible light-responsive $BiFeO_3$ perovskite for photocatalytic applications: a mini review. Mater Res Bull 90:15–30. https://doi.org/10.1016/j.materresbull.2016.12.052
18. Li Y, Liu K, Zhang J, Yang J, Huang Y, Tong Y (2020) Engineering the band-edge of Fe_2O_3/ZnO nanoplates via separate dual cation incorporation for efficient photocatalytic performance. Ind Eng Chem Res 59:18865–18872. https://doi.org/10.1021/ACS.IECR.0C03388/ASSET/IMAGES/LARGE/IE0C03388_0006.JPEG
19. Liu D, Jin Z, Li H, Lu G (2017) Modulation of the excited-electron recombination process by introduce g-C3N4 on Bi-based bimetallic oxides photocatalyst. Appl Surf Sci 423:255–265. https://doi.org/10.1016/j.apsusc.2017.06.156
20. Long Z, Li Q, Wei T, Zhang G, Ren Z (2020) Historical development and prospects of photocatalysts for pollutant removal in water. J Hazard Mater 395:122599. https://doi.org/10.1016/j.jhazmat.2020.122599
21. Lotfi S, El OM, Ahsaine HA, Assani A (2024) Recent progress on the synthesis, morphology and photocatalytic dye degradation of $BiVO_4$ photocatalysts: a review. Catal Rev 66:214–258. https://doi.org/10.1080/01614940.2022.2057044
22. Lu H, Wu X, Zhu P, Liu M, Li X, Xin X (2025) A novel $Bi_{12}O_{17}Cl_2$/GO/Co_3O_4 Z-type heterojunction photocatalyst with ZIF-67 derivative modified for highly efficient degradation of antibiotics under visible light. J Colloid Interface Sci 677:1052–1068. https://doi.org/10.1016/J.JCIS.2024.08.032
23. de Macedo OB, de Oliveira ALM, dos Santos IMG (2022) Zinc tungstate: a review on its application as heterogeneous photocatalyst. Cerâmica 68:294–315. https://doi.org/10.1590/0366-69132022683873265

24. Malathi A, Madhavan J, Ashokkumar M, Arunachalam P (2018) A review on BiVO 4 photocatalyst: activity enhancement methods for solar photocatalytic applications. Appl Catal A Gen 555:47–74. https://doi.org/10.1016/j.apcata.2018.02.010
25. MariaJoseph A, Okhawilai M, Rajendran S, Pattananuwat P (2024) Waste lignocellulosic biomass-derived graphitic carbon encased bimetallic nickel-palladium oxide nanofibers for efficient organic dye pollutant removal and antibacterial actions. Int J Biol Macromol. https://doi.org/10.1016/J.IJBIOMAC.2024.137655
26. Mariappan A, Mannu P, Thiruppathiraja T, Nga TTT, Lakshmipathi S, Dong CL, Dharman RK, Oh TH (2023) Interfacial oxygen vacancy modulated ZIF-8-derived ZnO/CuS for the photocatalytic degradation of antibiotic and organic pollutants: DFT calculation and degradation pathways. Chem Eng J 476:146720. https://doi.org/10.1016/J.CEJ.2023.146720
27. Mehrdadian E, Sheibani S, Ataie A (2024) Outstanding photocatalytic activity of a mechano-thermally synthesized Z-scheme $BiFeO_3$-Fe_2O_3 heterostructure. J Alloys Compd 997:174900. https://doi.org/10.1016/j.jallcom.2024.174900
28. Obaid Qamar M, Balu K, Ahn Y-H (2024) $BiVO_4$-Driven photocatalytic degradation of pharmaceutical in slurry bubble column reactor: influencing factors and toxicological profiling. Chem Eng J 496:153526. https://doi.org/10.1016/j.cej.2024.153526
29. Pal Singh B, Tyagi L, Vikal S, Tyagi S, Tyagi D, Rani M, Sharma K, Shukla G, Shanker U, Gautam YK, Sillanpää M (2024) Spotlighting graphene-based nanomaterials for the mitigation of hazardous water pollutants: a review. Inorg Chem Commun 166:112618. https://doi.org/10.1016/j.inoche.2024.112618
30. Parida S, Nanda J, Sarangi B, Behera R (2024) Plant extract mediated synthesis of $BiFeO_3$ nanoparticles for photocatalytic degradation of methylene blue dye. Biomass Convers Biorefin. https://doi.org/10.1007/s13399-024-06100-4
31. Raja DS, Chuah X, Lu S (2018) In situ grown bimetallic MOF-based composite as highly efficient bifunctional electrocatalyst for overall water splitting with ultrastability at high current densities. https://doi.org/10.1002/aenm.201801065
32. Rani M, Keshu, Shanker U (2021) Sunlight-induced photocatalytic degradation of organic pollutants by biosynthesized hetrometallic oxides nanoparticles. Environ Sci Pollut Res 28:43 28:61760–61780. https://doi.org/10.1007/S11356-021-15003-0
33. Rani M, Shanker U (2018) Photocatalytic degradation of toxic phenols from water using bimetallic metal oxide nanostructures. Colloids Surf A Physicochem Eng Asp 553:546–561. https://doi.org/10.1016/j.colsurfa.2018.05.071
34. Rani M, Sharma S, Keshu SU (2024) Biowaste-derived nanocomposite of calcium oxide incorporated in nickel oxide for efficient removal of organic pollutants. Biomass Convers Biorefin. https://doi.org/10.1007/S13399-024-05438-Z/FIGURES/9
35. Rashid M, Mowla D, Esmaeilzadeh F, Dashtian K, Bahmani M (2021) Boosted Cr(VI) clean up over magnetically recoverable NiS/γ-Fe_2O_3/C type-II heterostructure derived from bimetal (Fe/Ni)-organic framework under visible light. J Clean Prod 317:128471. https://doi.org/10.1016/J.JCLEPRO.2021.128471
36. Reyes-Vallejo O, Sánchez-Albores R, Maldonado-Álvarez A, Ashok A, Duran-Alvarez JC, Subramaniam V (2023) Calcium-Magnesium oxide by the ball-milling method using eggshell as calcium source: its study for photodegradation of methylene blue. J Mater Sci: Mater Electron 34:1–12. https://doi.org/10.1007/S10854-023-10163-W/TABLES/5
37. Salari H (2019) Kinetics and mechanism of enhanced photocatalytic activity under visible light irradiation using Cr_2O_3/Fe_2O_3 nanostructure derived from bimetallic metal organic framework. J Environ Chem Eng 7:103092. https://doi.org/10.1016/J.JECE.2019.103092
38. Samsudin MFR, Sufian S, Hameed BH (2018) Epigrammatic progress and perspective on the photocatalytic properties of $BiVO_4$-based photocatalyst in photocatalytic water treatment technology: a review. J Mol Liq 268:438–459. https://doi.org/10.1016/j.molliq.2018.07.051
39. Segovia-Sandoval SJ, Ojeda-Galván HJ, Moral-Rodríguez AI, Rodríguez-Hernández J, Peralta-Rodríguez RD, Gómez-Villegas KJ, Hernández-Juárez NI, Mendoza-Mendoza E (2022) A novel green synthesis of Bi_2WO_6-based photocatalysts for efficient pollutants degradation using low-power UV-A LEDs. J Alloys Compd 911:165018. https://doi.org/10.1016/j.jallcom.2022.165018

40. Senthilkumar N, Selvaraj Y, Eswaramoorthy N, Pandiaraj S, Alibrahim KA, Alodhayb AN (2024) Optimizing photocatalytic and supercapacitive performance by β-Bi_2O_3@$BiFeO_3$ modification with PVDF polymer based nanocomposites. Colloids Surf A Physicochem Eng Asp 695:134189. https://doi.org/10.1016/j.colsurfa.2024.134189
41. Sinar Mashuri SI, Ibrahim ML, Kasim MF, Mastuli MS, Rashid U, Abdullah AH, Islam A, Asikin Mijan N, Tan YH, Mansir N, Mohd Kaus NH, Yun Hin T-Y (2020) Photocatalysis for organic wastewater treatment: from the basis to current challenges for society. Catalysts 10:1260. https://doi.org/10.3390/catal10111260
42. Sun H, Ren D, Kong R, Wang D, Jiang H, Tan J, Wu D, Chen S, Shen B (2019) Tuning 1-hexene/n-hexane adsorption on MOF-74 via constructing Co-Mg bimetallic frameworks. Microporous Mesoporous Mater 284:151–160. https://doi.org/10.1016/j.micromeso.2019.04.031
43. Ullah R, Sun H, Ang HM, Tadé MO, Wang S (2012) Visible light photocatalytic degradation of organics on nanoparticles of bi-metallic oxides. Sep Purif Technol 89:98–106. https://doi.org/10.1016/j.seppur.2012.01.014
44. Wang Z, Cai X, Xie X, Li S, Zhang X, Wang Z (2021) Visible-LED-light-driven photocatalytic degradation of ofloxacin and ciprofloxacin by magnetic biochar modified flower-like Bi2WO6: the synergistic effects, mechanism insights and degradation pathways. Sci Total Environ 764:142879. https://doi.org/10.1016/j.scitotenv.2020.142879
45. Wu Y, Shangs H, Pan X, Zhou G (2024) Recent advances of $BiFeO_3$-based catalysts based on the piezoelectric built-in electric field. Mol Catal 556:113820. https://doi.org/10.1016/j.mcat.2024.113820
46. Yadav G, Yadav N, Gadore V, Mishra SR, Gupta SV, Ahmaruzzaman M (2023) Biogenic growth of egg shell–derived $CaMn_2O_4$ over tailored FLY ASH surface for synergistically photodegradation of ofloxacin: materialistic and chemical studies. Biomass Convers Biorefin 14:29923–29940. https://doi.org/10.1007/S13399-023-04978-0/FIGURES/19
47. Yan Y, Wang R, Maimaitituersun M, Sun H, Liu X (2022) Built-in electric field enhanced $BiFeO_3$ photo-Fenton degradation rhodamine B solution. J Mater Sci 57:6900–6913. https://doi.org/10.1007/s10853-022-06916-3
48. Yang C, You X, Cheng J, Zheng H, Chen Y (2017) A novel visible-light-driven In-based MOF/graphene oxide composite photocatalyst with enhanced photocatalytic activity toward the degradation of amoxicillin. Appl Catal B 200:673–680. https://doi.org/10.1016/j.apcatb.2016.07.057
49. Zaleska-Medynska A, Marchelek M, Diak M, Grabowska E (2016) Noble metal-based bimetallic nanoparticles: the effect of the structure on the optical, catalytic and photocatalytic properties. Adv Colloid Interface Sci 229:80–107. https://doi.org/10.1016/j.cis.2015.12.008
50. Zeng D, Li Y, Xia T, Cui F, Zhang J (2022) MOF-derived Co_3O_4 nanoparticles catalyzing hydrothermal deoxygenation of fatty acids for alkane production. https://doi.org/10.1021/acsomega.2c04382
51. Zhang Y, Luo B, Ai C, Li J, Jing D, Ma L (2022) MOF-derived non-noble metal CoP nanoparticle modified TiO_2 for enhanced photocatalytic hydrogen production. Ind Eng Chem Res 61:16653–16661. https://doi.org/10.1021/ACS.IECR.2C02304/ASSET/IMAGES/LARGE/IE2C02304_0009.JPEG
52. Zhang Y, Tong C, Shi L, Li X, Zhang X, Sun X, Wu J, Shi Y, Huang Z, Zhang D (2023) Enhanced natural sunlight driven photocatalytic degradation under outdoor inconstant weather by Bi_2WO_6/WO_3 heterojunctions. Mater Res Bull 167:112404. https://doi.org/10.1016/j.materresbull.2023.112404
53. Zheng J, Sun L, Jiao C, Shao Q, Lin J, Pan D, Naik N, Guo Z (2021) Hydrothermally synthesized Ti/Zr bimetallic MOFs derived N self-doped TiO_2/ZrO_2 composite catalysts with enhanced photocatalytic degradation of methylene blue. Colloids Surf A Physicochem Eng Asp 623:126629. https://doi.org/10.1016/J.COLSURFA.2021.126629
54. Zhou T, Zhai T, Shen H, Wang J, Min R, Ma K, Zhang G (2023) Strategies for enhancing performance of perovskite bismuth ferrite photocatalysts ($BiFeO_3$): a comprehensive review. Chemosphere 339:139678. https://doi.org/10.1016/j.chemosphere.2023.139678

55. Zyoud AH, Zubi A, Hejjawi S, Zyoud SH, Helal MH, Zyoud SH, Qamhieh N, Hajamohideen AR, Hilal HS (2020) Removal of acetaminophen from water by simulated solar light photodegradation with ZnO and TiO_2 nanoparticles: catalytic efficiency assessment for future prospects. J Environ Chem Eng 8:104038. https://doi.org/10.1016/J.JECE.2020.104038

Graphene Oxide-Based Metal Oxide Composites for Wastewater Purification

Nibedita Nayak, Prakash Chandra Sahoo, Dipanjali Swain, Tapan Dash, Subash Ch Mallick, and Surendra Kumar Biswal

Abstract Water is the essential cause of all life on earth. But the needy of purified water has been come to a challenging stage and it is needy of the present time. Wastewater generated from many sources. Disposal in its original form produces concern about environmental sustainability. The wastewater recycling and its reuse are highly attractive area for the researchers. So these problems invite the implementation of novel materials and technologies that are cost and energy efficient. In this background, graphene and its metal oxide-based composites exhibit remarkable physical and chemical properties, including a high surface area and significant surface activity. These characteristics make them highly suitable for applications in adsorbents, catalysts, and membranes used in water and wastewater treatment. Graphene oxide-metal oxide composites, in particular, have shown great potential for the large-scale removal of toxic heavy metals such as arsenic (As), chromium (Cr), and lead (Pb), as well as organic contaminants like aldehydes, dyes, and phenols from wastewater. This review aims to explore the advancements in materials and technologies for water purification, with a specific focus on graphene oxide-based metal oxide composites.

Keywords Wastewater · Graphene oxide · Removal · Toxic metals · Water purification

N. Nayak · P. C. Sahoo · D. Swain · T. Dash (✉)
Centurion University of Technology and Management, Bhubaneswar, Odisha, India
e-mail: tapanphy@gmail.com

T. Dash · S. K. Biswal
International PranaGraf Mintech Research Centre (IGMRC), Bhubaneswar, Odisha, India

S. C. Mallick
LN Indtech Services Pvt Ltd., Bhubaneswar, Odisha, India

A. Ahmad et al. (eds.), *Metal Oxide Based Nanophotocatalyst for Wastewater Purification*, Smart Nanomaterials Technology,
https://doi.org/10.1007/978-981-96-7683-5_12

1 Introduction

1.1 Overview of Water Pollution

Water pollution is a significant global challenge with severe impacts on both human health and ecological systems. It stems from various sources, including industrial, agricultural, and household activities. Pollutants can be categorized as inorganic, organic, pathogenic, thermal, and radioactive [43]. Chemical contaminants like heavy metals and pesticides disrupt aquatic food chains and pose significant health risks. In India, water pollution-related diseases are responsible for the deaths of approximately 580 people each day [32]. The World Health Organization reports that 1.1 billion people lack access to safe water globally. Recent emerging concerns include microplastics, thermal pollution, and suspended solids [43]. To address these issues, many countries have implemented regulations and treatment methods to minimize water contamination [6]. Ongoing evaluation and revision of water resource policies at all levels are necessary to combat this pervasive problem [32]. Water pollution from industrial, urban, and agricultural sources poses serious health risks, especially in areas with poor sanitation and contaminated drinking water [54]. The study of Graphene Oxide-Based Metal Oxide Composites for wastewater purification is highly important due to several critical reasons, primarily related to environmental protection, sustainability, and the growing global demand for effective water treatment technologies. This chapter has focused to analyze and discuss these points.

1.2 Importance of Wastewater Treatment

Wastewater treatment is essential for safeguarding the environment, protecting public health, and ensuring the sustainable management of water resources [67]. The process involves removing pollutants, including organic matter, nutrients, pathogens, and emerging contaminants from sewage and industrial effluents. Various treatment methods, such as physical, chemical, and biological processes, are employed to achieve regulatory compliance and enable water reuse [74]. Effective wastewater treatment not only reduces pressure on natural water sources but also contributes to energy recovery through bio-methane production and provides nutrients for agriculture. Recent advancements in treatment technologies have focused on addressing emerging contaminants and improving nutrient removal efficiency [51]. Implementing appropriate wastewater management strategies is essential for maximizing the benefits of treatment processes and ensuring environmental sustainability [67]. Wastewater treatment is vital for protecting water resources and preserving the environment, using conventional and AI-based methods.

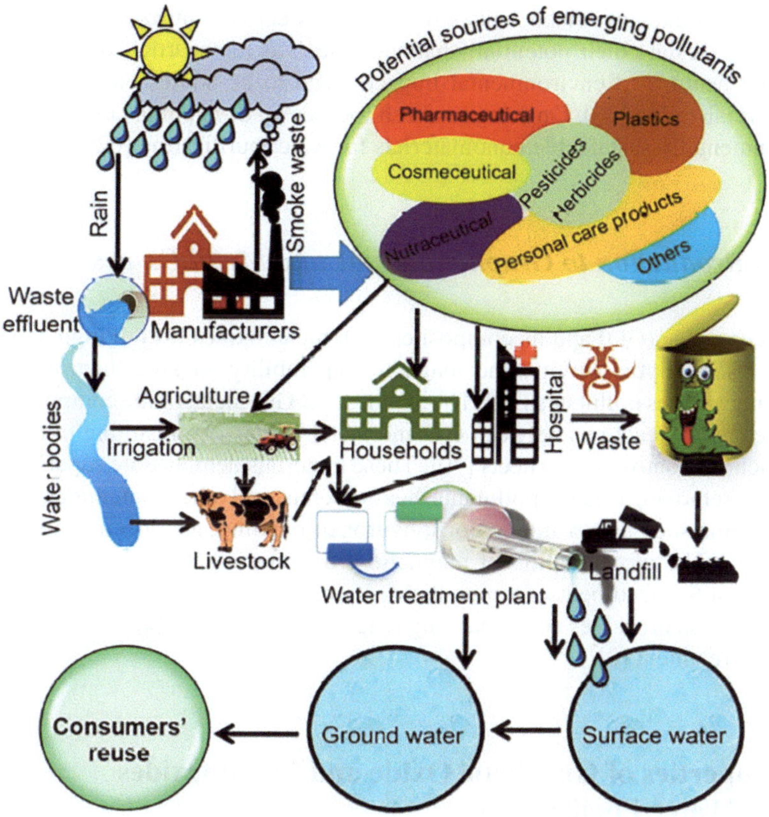

Fig. 1 Schematic representation of life-cycle distribution of pollutants from sources to receptors (Rasheed et al. [58], reproduced with permission of copyright © 2019 Elsevier Ltd.)

Effective wastewater treatment and management is necessary to reduce contaminants and protect the environment [56]. Schematic representation of life-cycle distribution of emerging (micro)-pollutants from sources to receptors viewpoint is shown in Fig. 1 [58].

1.3 Role of Nanomaterials in Water Purification

Various nanomaterials, including graphene oxide, carbon nanotubes, nanosorbents, and nanoparticles, have proven effective in eliminating contaminants such as heavy metals, organic pollutants, and pathogens from water [10, 33]. Nanofiltration techniques and nano-membranes are widely used for water treatment, while nanosorbents serve as effective separation media [10] Titanium dioxide nanoparticles have demonstrated significant potential as photocatalysts for water purification, while nanoagglomerates of mixed oxides have been effectively utilized in groundwater treatment. Despite their potential, challenges such as nanoparticle aggregation and potential health and environmental impacts need to be addressed for widespread commercialization [33]. Ongoing research focuses on developing cost-effective and environmentally sustainable nanomaterials for water purification [62].

1.4 Introduction to Graphene Oxide and Metal Oxides

Graphene oxide (GO) and its composites with metal oxides exhibit high adsorption capacities, catalytic activity, and magnetic separability, making them effective in removing various pollutants [79]. The synthesis of GO-metal oxide composites often involves redox reactions between GO and metal precursors, resulting in nanoparticles anchored onto carbon sheets [70]. These materials demonstrate superior performance in removing organic pollutants, heavy metals, and other contaminants through mechanisms such as π-π interactions, hydrogen bonding, and electrostatic interactions [73]. The incorporation of magnetic properties enables easy separation and recovery of the adsorbents. Despite their potential, challenges remain in optimizing the fabrication processes and enhancing long-term stability for practical applications in water treatment [8].

2 Properties of Graphene Oxide and Metal Oxides for Water Purification

2.1 Structural Characteristics of Graphene Oxide

Graphene oxide's (GO) high surface area, good selectivity, and low separation cost make it suitable for removing various pollutants, including heavy metals, dyes, and pesticides [44]. GO's sheet-like structure, with a polyaromatic π-system, facilitates π–π interactions, hydrogen bonding, and electrostatic interactions with organic pollutants [73]. Two-dimensional nanomaterial, graphene oxide produced from graphite via chemical route is presented in Fig. 2 [27]. The presence of oxygen moieties on GO's surface enhances its adsorption capacity and allows for tunable physicochemical properties. GO-based nanocomposites, including reduced GO (rGO) and magnetic GO (MGO), have shown optimal performance in water purification due to their high mechanical strength, hydrophilicity, and compatibility with other materials [8]. These structural features make GO and its derivatives highly effective in removing a wide range of contaminants from wastewater, including heavy metals, dyes, and agricultural pollutants. Recent advancements in GO synthesis and surface modification have also enabled enzyme immobilization, creating new possibilities for environmental remediation using nano biocatalysts [57].

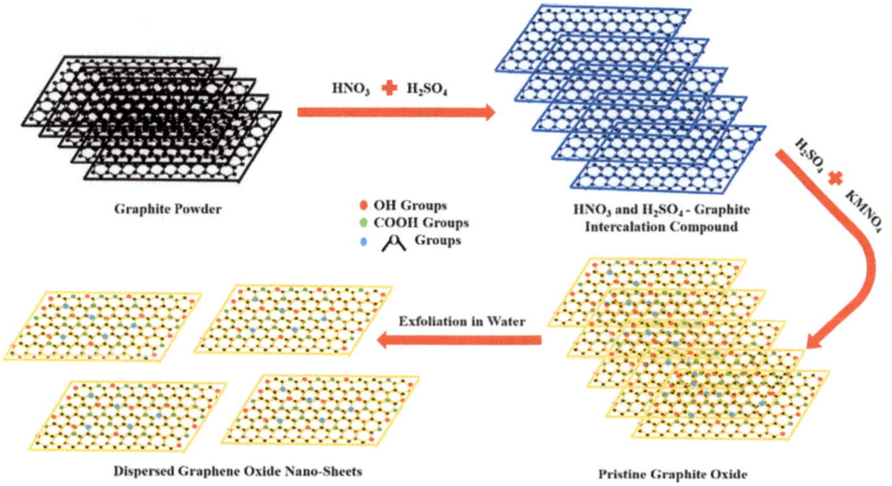

Fig. 2 Schematic diagram of the fabrication mechanism of GO nanosheets from graphite (Hegab and Zou [27], reproduced with permission of copyright © 2015 Elsevier Ltd.)

2.2 Unique Properties of Metal Oxides

2.2.1 Titanium Dioxide (TiO_2)

Titanium dioxide (TiO_2) possesses unique properties that make it highly suitable for wastewater treatment, including low cost, non-toxicity, biocompatibility, and high chemical stability. TiO_2-based photocatalytic membrane reactors (PMRs) have gained significant interest due to their ability to remove organic, inorganic, and microbial pollutants from water [64]. The large surface area of TiO_2 nanoparticles provides ample active sites for reactions, enhancing catalytic activity [60]. However, challenges such as particle aggregation, mass transfer limitation, and difficulty of recovery need to be addressed. Optimizing operational parameters and exploring new engineering approaches can improve the efficiency of TiO_2-based PMRs for water purification [60]. Future research should focus on magnetized TiO_2-based membrane technology to mitigate fouling and improve photocatalyst recoverability, as well as conducting life-cycle assessments to promote large-scale applications.

2.2.2 Zinc Oxide (ZnO)

Zinc oxide (ZnO) nanoparticles have emerged as promising materials for wastewater treatment due to their unique properties. ZnO exhibits strong photocatalytic activity, effectively degrading organic pollutants like azo dyes under UV illumination [38]. Its high UV absorption ability, chemical stability, and non-toxicity make it an excellent photocatalyst [77]. Additionally, ZnO nanoparticles demonstrate potent antibacterial activity against various microorganisms commonly found in wastewater, including *Escherichia coli* and *Staphylococcus aureus* [45]. The antibacterial efficacy of ZnO increases with concentration and varies depending on bacterial strain [19]. ZnO nanoparticles can be synthesized through various methods, including green synthesis using plant extracts, which enhances their eco-friendliness [45]. These properties make ZnO nanoparticles suitable for wastewater treatment, potentially serving as both a photocatalyst for pollutant degradation and an antimicrobial agent for disinfection.

2.2.3 Iron Oxides (Fe_2O_3, Fe_3O_4)

Iron oxides, particularly Fe_2O_3 and Fe_3O_4, possess unique properties that make them highly suitable for wastewater treatment. These nanomaterials exhibit extremely small sizes, high surface-area-to-volume ratios, and excellent magnetic properties, enabling their use as nanosorbents and photocatalysts. Their low band gap (2.3–2.4 eV) allows for electron–hole activation under visible light, making them promising alternatives to conventional TiO_2 photocatalysts [65]. Iron oxide nanoparticles are advantageous due to their low cost, easy separation using external magnetic fields, and high adsorption capacity [28].

These materials can be synthesized through various methods, including co-precipitation, and can be functionalized to enhance their performance [48]. Additionally, iron oxide nanoparticles can serve as coagulants in coagulation and flocculation processes for removing oil droplets from produced water [28], further demonstrating their versatility in wastewater treatment applications. Reaction mechanism of photocatalysis, (b) band structures of typical semiconductor photocatalysts, and (c) proposed photothermal catalyst and its components are presented in Fig. 3 [80, 81].

2.3 *Synergistic Effects in Composites*

Graphene oxide (GO) and its composites, particularly those incorporating metal oxides, exhibit synergistic effects in removing various pollutants from water [79]. The synthesis of reduced graphene oxide (RGO)-metal/metal oxide composites has been demonstrated as a versatile and simple method for creating efficient adsorbents [70]. These materials show enhanced performance in removing heavy metals, dyes, and other organic pollutants through mechanisms such as π-π interactions, hydrogen bonding, and electrostatic interactions [73]. Magnetic graphene oxide (MGO) composites have gained attention for their ability to be easily separated from treated water using magnetic fields. Despite their potential, challenges remain in optimizing these materials for large-scale water treatment applications [8].

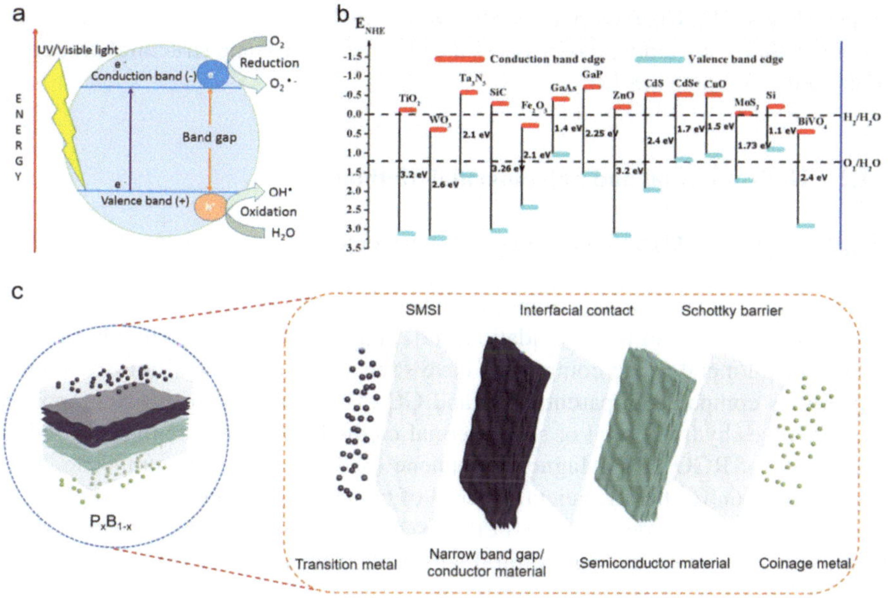

Fig. 3 Reaction mechanism of photocatalysis, **b** band structures of typical semiconductor photocatalysts, and **c** proposed photothermal catalyst and its components (Zhang et al. [80, 81]; adapted on permission of copyright approval for free cultural works—under-creative commons)

3 Synthesis Methods of Graphene Oxide-Based Metal Oxide Composites

3.1 *Chemical Methods*

Graphene oxide (GO)-based metal oxide composites are synthesized through various chemical methods, each tailored to achieve specific structural, morphological, and functional properties of the composites. Here's an overview of the commonly used synthesis techniques:

3.1.1 Sol–Gel Synthesis

Graphene oxide (GO) and its nanocomposites effectively remove various contaminants, such as heavy metals, dyes, and organic pollutants, through adsorption, photocatalytic degradation, and synergistic effects [8, 79]. The synthesis of GO-based metal oxide composites often involves sol–gel methods, utilizing the inherent reduction ability of reduced graphene oxide (RGO) to form metal nanoparticles anchored onto carbon sheets [70]. Characterization techniques, including spectroscopic and microscopic methods, are employed to assess the physiochemical and structural properties of these nanocomposites [8].

3.1.2 Hydrothermal and Solvothermal Methods

Graphene oxide (GO) and its composites have emerged as promising materials for wastewater purification due to their unique properties, including large surface areas, pore volumes, and functional groups [40]. These materials effectively remove various pollutants through adsorption, oxidation, and catalysis [73]. Reduced graphene oxide (RGO)-metal/metal oxide composites have shown enhanced metal ion scavenging capabilities compared to parent RGO and GO. The synthesis of these composites often involves hydrothermal or solvothermal methods, utilizing the inherent reduction ability of RGO [70]. Magnetic graphene oxide (MGO) nanocomposites have gained attention for their efficient removal of contaminants and ease of separation. These materials have demonstrated effectiveness in removing heavy metals, dyes, pharmaceutical waste, and agricultural pollutants from water [8, 73].

3.1.3 Co-precipitation Method

Graphene oxide (GO) and its nanocomposites can be synthesized through various methods, including chemical co-precipitation [78] and hydrothermal decomposition [55]. The resulting materials exhibit excellent adsorption capabilities for removing organic dyes and heavy metals from water [55, 70]. GO-based composites demonstrate high distribution coefficients for pollutant uptake, often surpassing those of pristine GO [70]. These nanocomposites can be further enhanced by incorporation into polymeric membranes, improving fouling resistance, water permeability, and solute selectivity [25]. The synthesis and application of GO-metal oxide composites offer a cost-effective and sustainable alternative to traditional water purification methods, addressing challenges such as the trade-off between permeability and selectivity [25]. Overall, these materials show great potential for advancing water treatment technologies.

3.1.4 Chemical Vapor Deposition (CVD)

Graphene oxide (GO) and its composites effectively remove various pollutants, such as heavy metals, dyes, and pharmaceutical waste, through mechanisms like π–π interactions and hydrogen bonding [73]. Reduced graphene oxide (RGO) composites with metal/metal oxides have shown enhanced performance in water purification, demonstrating higher distribution coefficients for pollutant uptake compared to parent RGO and GO [70]. Metal ferrites and their graphene-based nanocomposites offer additional advantages, including magnetic separation and synergistic adsorption-enhanced degradation of contaminants [79]. Recent research has focused on developing GO-modified nanocomposites, particularly magnetic graphene oxide (MGO), which exhibit optimal performance in water treatment and allow for efficient magnetic separation of pollutants [8]. CVD method is used in preparing such composites for water purification.

3.1.5 In Situ Reduction

The synthesis of GO-metal oxide composites often involves in situ reduction methods, where GO's inherent reduction ability is utilized to form metal nanoparticles anchored onto carbon sheets [70]. These composites offer advantages such as high surface area, thermal and chemical stability, and magnetic properties, enabling easy separation after treatment [79]. Furthermore, GO-based composites can be incorporated into polymeric membranes, enhancing their performance in terms of fouling resistance, water permeability, and solute selectivity [25]. The versatility and effectiveness of GO-metal oxide composites make them promising candidates for addressing various water purification challenges [8].

3.1.6 Microwave-Assisted Synthesis

Microwave-assisted synthesis has emerged as a rapid and efficient method for producing graphene oxide-based metal oxide composites. This approach enables the preparation of nanocomposites with enhanced properties in just a few minutes, often at lower temperatures and with reduced reagent consumption [12]. Microwave-assisted synthesis of graphene oxide (GO) and its metal oxide composites has emerged as a promising approach for wastewater purification. These nanocomposites exhibit high surface areas and excellent adsorption capabilities for heavy metal ions like Pb^{2+} and Cd^{2+} [17, 18]. The synthesis typically involves a modified Hummers' method, resulting in increased interlayer spacing and wrinkled morphology of GO sheets. The composites' effectiveness in water purification is attributed to their large surface area, high stability, and tunable chemical composition [79]. Magnetic nanocomposites, such as GO/Fe_3O_4, offer the additional advantage of easy separation after treatment [17]. These materials demonstrate potential for removing various organic and inorganic contaminants through adsorption and photocatalytic degradation mechanisms [79].

3.2 *Physical Methods*

Physical methods for synthesizing graphene oxide (GO)-based metal oxide composites generally rely on mechanical or thermal processes rather than chemical reactions.

These techniques are useful for producing composites with unique structural features, enhanced dispersion, and desirable physicochemical properties.

3.2.1 Physical Vapor Deposition (PVD)

Graphene oxide (GO) and its metal/metal oxide composites have emerged as promising materials for wastewater purification due to their exceptional properties and versatile applications. These composites can be synthesized through various methods, including redox reactions between GO and metal precursors [70]. The resulting materials exhibit high adsorption capacities for pollutants like mercury, with distribution coefficients exceeding 10 L g^{-1} [70]. GO-based composites have been utilized to remove metal ions, organic pollutants, and dyes from water bodies. Their incorporation into polymeric membranes enhances overall performance, including antibacterial properties and surface hydrophilicity [25]. Metal ferrite-graphene composites offer excellent removal of contaminants through adsorption, photocatalytic degradation, and synergistic effects, with the added benefit of easy magnetic separation [79]. The π-π interactions, hydrogen bonding, and electrostatic interactions between graphene derivatives and organic pollutants contribute to their effectiveness in wastewater treatment [73].

3.2.2 Thermal Evaporation

Graphene oxide-metal oxide composites exhibit enhanced thermal stability and decreased activation energy compared to graphene oxide alone [63]. The large surface area of graphene oxide provides ample space for metal oxide nanoparticles, preventing agglomeration and improving dispersion [46]. Metal ferrites and their graphene-based nanocomposites have shown particular promise in wastewater treatment, offering excellent removal of contaminants through adsorption, photocatalytic degradation, and synergistic effects [79]. GO-metal oxide nanocomposites can also be incorporated into polymeric membranes, enhancing their performance in terms of fouling resistance, water permeability, and solute selectivity [25]. These advancements in GO-based materials present opportunities for developing cost-effective and sustainable solutions for water purification applications.

3.2.3 Mechanical Mixing (Ball Milling)

GO can be synthesized through ball milling, offering advantages in processing time and product quality [11]. Reduced graphene oxide (RGO) composites with metal/metal oxides have shown enhanced metal ion scavenging capabilities compared to pure GO [70]. These composites can be fabricated into various forms, including magnetic graphene oxide (MGO), which allows for efficient magnetic separation in water treatment processes [8]. Metal ferrites and their graphene-based nanocomposites have also demonstrated excellent performance in wastewater treatment through adsorption, photocatalytic degradation, and synergistic effects. The high surface area, stability, and controllable properties of these materials make them ideal candidates for removing a wide range of contaminants, including heavy metals, dyes, and agricultural pollutants from water [79].

3.2.4 Ultrasonic-Assisted Mixing

Ultrasonic-assisted synthesis of graphene oxide (GO) and GO-based metal oxide composites offers several advantages over traditional methods. This approach enables faster, more efficient, and environmentally friendly production of high-quality GO with fewer layers and larger interlayer spacing [21, 68]. The synthesis of GO-based composites often involves ultrasonic-assisted mixing methods, which improve the dispersion and integration of components. Recent advancements include the development of magnetic GO composites for easier separation and the exploration of GO-based materials in advanced oxidation processes like Fenton reactions and photocatalysis [8, 57]. These innovations demonstrate the potential for scaling up GO-based technologies for practical wastewater treatment applications.

3.2.5 Mechanical Exfoliation

The synthesis of reduced graphene oxide (RGO)-metal/metal oxide composites utilizes RGO's inherent reduction ability, resulting in materials with enhanced metal ion scavenging capabilities [70]. Metal ferrites and their graphene-based composites offer additional advantages, such as magnetic separation and photocatalytic degradation of contaminants [79]. The integration of GO with various materials, including magnetic nanoparticles, has led to the development of versatile adsorbents for diverse pollutants, including hazardous metals, radioactive ions, and agricultural contaminants [8]. These advancements highlight the potential of GO-based nanocomposites in addressing global water pollution challenges.

4 Mechanisms of Pollutant Removal

4.1 Adsorption Mechanisms

Graphene oxide (GO) and its composites have emerged as promising adsorbents for removing various pollutants from water due to their large surface area, diverse active sites, and unique physicochemical properties [1, 75]. These materials effectively remove heavy metals, dyes, and other hazardous pollutants through mechanisms

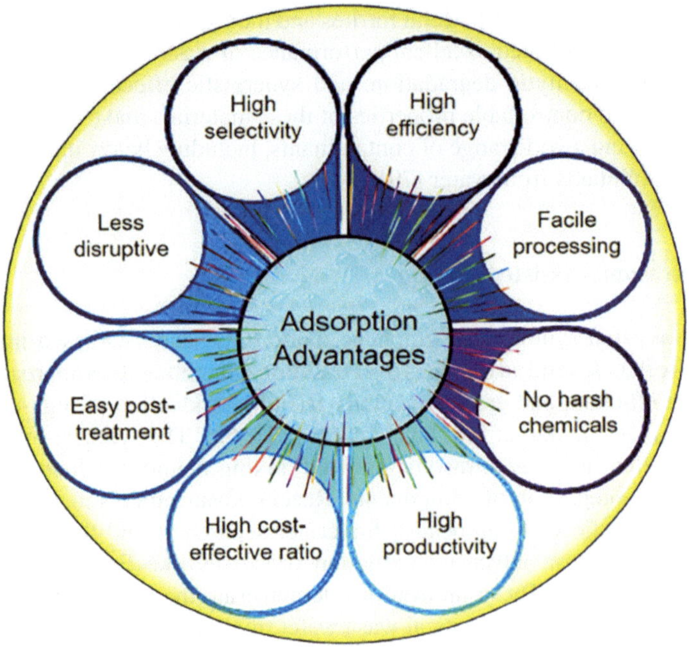

Fig. 4 Illustration of advantageous features of adsorption technique (Rasheed et al. [58], reproduced with permission of copyright © 2019 Elsevier Ltd.)

such as π–π interactions, hydrogen bonding, and electrostatic interactions [73]. The adsorption performance is influenced by factors like pH, agitation, and temperature [1]. GO content plays a crucial role in enhancing adsorption efficiency by increasing active binding sites, reducing agglomeration of nanoparticles, and improving electrostatic interactions between adsorbents and adsorbates [37]. Adsorption behaviors often follow Langmuir or Freundlich isotherm models and pseudo-second-order kinetics [37]. Understanding these mechanisms is essential for designing effective GO-based adsorbents and optimizing their performance in water treatment applications [75]. Graphene oxide-based nanocomposites are efficient adsorbents for removing various water pollutants due to their reactive oxygen functional groups [47]. The advantages of adsorption technique are briefly presented in Fig. 4 [58].

4.1.1 Physical Versus Chemical Adsorption

Graphene oxide (GO) and its composites have emerged as promising adsorbents for removing various pollutants from water due to their large surface area, chemical stability, and high adsorption capacity [61]. These materials effectively remove heavy metals, dyes, and organic pollutants through physical and chemical adsorption mechanisms, including π–π interactions, hydrogen bonding, and electrostatic interactions [73]. The adsorption process is influenced by factors such as pH, agitation, and temperature [1]. GO-based nanocomposites exhibit enhanced stability and adsorption efficacy in aqueous media, making them suitable for multiple adsorption–desorption cycles [47]. Research on these materials focuses on synthesis methods, adsorption behaviors, isotherms, kinetics, and regeneration processes. Despite their potential, challenges remain in scaling up production and improving economic viability for large-scale water treatment applications [1, 61].

4.1.2 Role of Functional Groups

The presence of oxygen-containing functional groups on GO's surface, such as hydroxyl, carboxyl, and epoxy groups, plays a crucial role in the adsorption mechanisms through electrostatic interactions, hydrogen bonding, and π–π interactions. Functionalization of GO with various materials can introduce heteroatoms, enhancing its affinity for specific pollutants and improving its adsorption performance. The adsorption behavior of pollutants on GO-based composites often follows Langmuir or Freundlich isotherm models, depending on the adsorbate. Furthermore, GO content in composites significantly influences the adsorption mechanisms, with higher GO content generally improving removal efficiency for both organic dyes and inorganic ions [37].

4.2 Photocatalytic Mechanisms

Photocatalytic mechanisms for pollutant removal involve the generation of electron–hole pairs in semiconductor materials upon light irradiation [49]. These charge carriers can directly interact with pollutants or form reactive species like hydroxyl radicals for degradation [39]. The process can simultaneously remove multiple pollutants through synergistic oxidation and reduction reactions, enhancing efficiency by inhibiting electron–hole recombination [9]. Various photocatalysts, including binary compounds like TiO_2 and ZnO, and ternary compounds such as tungstates and bismutates, have been employed for treating organic and inorganic pollutants in air and water [72]. The effectiveness of photocatalytic systems is often evaluated using metrics like quantum yield and space time yield, with organic pollutants generally showing higher removal efficiencies compared to inorganic ones [72]. Factors such as catalyst design, pH, and additives can influence the treatment efficacy [9].

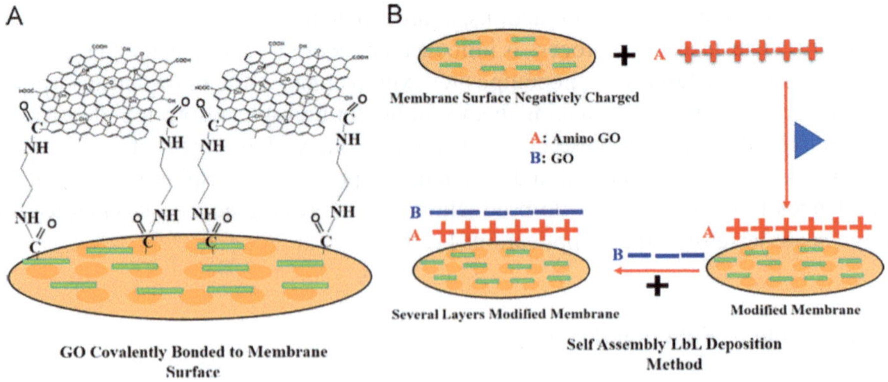

Fig. 5 Schematic diagram showing the preparation mechanism of GO-based membrane surface modification (Hegab and Zou [27], reproduced with permission of copyright © 2015 Elsevier Ltd.)

4.3 *Membrane Filtration Mechanisms*

Membrane filtration technologies have emerged as promising solutions for removing various pollutants from water sources. These techniques can effectively eliminate heavy metals, organic micropollutants, and other contaminants [13, 30, 34, 52]. Different membrane processes, including reverse osmosis, nanofiltration, ultrafiltration, and microfiltration, have been studied for their efficacy in treating diverse organic contaminants [30]. Tight membranes are generally more efficient due to their removal mechanisms, which are influenced by membrane and pollutant properties [52]. Membrane adsorption techniques, combining filtration and adsorption, have shown promise in removing trace amounts of pollutants [34]. Recent research has focused on membrane modifications and the incorporation of nanomaterials to enhance performance [13]. Despite their potential, challenges such as membrane fouling persist, necessitating further research and interdisciplinary collaboration to advance water purification technologies [30]. Functioned graphene oxide-based material is very much effective for water purification by membrane filtration technique.

The fundamental mechanism of GO-based membrane surface modification is shown in Fig. 5 [27]. The coupling technology leveraging the advantages and mitigating the disadvantages of photocatalytic and membrane separation processes for water purification is presented in Fig. 6 [81].

4.4 *Ion Exchange*

Ion exchange filtration is an effective method for removing various pollutants from water and wastewater. It has been successfully used to treat industrial and domestic

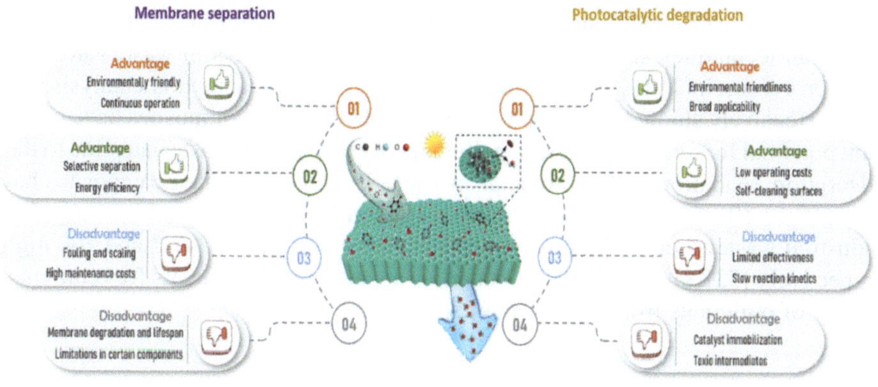

Fig. 6 Schematic illustration of the coupling technology leveraging the advantages and mitigating the disadvantages of photocatalytic and membrane separation processes (Zhang et al. [81] adapted on permission of copyright approval for free cultural works—under-creative commons)

wastewater, demonstrating high removal efficiencies for total suspended solids (95%), total dissolved solids (93.33%), BOD (76.25%), and COD (85%) (Chauhan et al., 2022). Ion exchange resins, which are porous polymers with exchangeable ionic groups, can remove dissolved organic matter, nitrate, heavy metals, and other contaminants from water [4]. The process has shown up to 90% removal rates for organic compounds in petroleum refinery wastewater. Ion exchange technology is particularly advantageous due to its high separation selectivity, simple handling, and reusability of resins, making it a sustainable and cost-effective option for water treatment [4]. Recent bibliometric analyses have highlighted the growing importance of ion exchange research in addressing water pollution challenges worldwide.

5 Application of Graphene Oxide-Based Composites in Wastewater Treatment

5.1 Industrial Wastewater Treatment

Graphene oxide (GO) and its composites have emerged as promising materials for industrial wastewater treatment due to their exceptional properties and versatility. GO-based composites have shown remarkable efficiency, with some materials achieving up to 99–100% removal of pollutants [29]. The integration of GO with metal ferrites has further enhanced their performance through synergistic effects of adsorption and photocatalytic degradation [79]. GO membranes offer advantages such as organic solvent tolerance, oxidation resistance, and reduced fouling compared to traditional organic membranes [7]. Various synthesis methods, characterization techniques, and applications of GO-based materials in wastewater treatment processes, including Fenton-like reactions, ozonation, and photocatalysis, have been extensively studied [3]. Graphene and graphene oxide composites have significant potential for removing organic pollutants from industrial wastewater due to their adsorption, oxidation, and catalytic properties [73]. GO-silica nanocomposites have also proven effective in removing heavy metal ions, such as lead and mercury, from industrial wastewater, with maximum adsorption capacities of 347.2 and 266 mg/g, respectively [53]. Graphene-based materials are effective adsorbents for removing a variety of pollutants from wastewater [5].

5.2 Agricultural Runoff

Graphene oxide (GO) and its composites have emerged as promising materials for wastewater treatment, particularly in removing organic pollutants, heavy metals, and dyes [73]. GO-based nanocomposites have shown excellent performance in water purification, with some achieving up to 99–100% removal efficiency for various pollutants [29]. The fabrication of GO, reduced GO, and magnetic GO nanocomposites has been explored for enhanced contaminant removal, including agricultural runoff [8]. However, challenges remain in optimizing GO-based nanocomposites for specific pollutants and improving their regeneration after multiple adsorption cycles [8]. Chitosan-graphite composites have demonstrated effectiveness in removing nitrates and phosphates from agricultural runoff, with up to 70% removal efficiency at lower concentrations [16].

5.3 Domestic Wastewater

The nanostructured materials of Graphene oxide (GO) and its composites effectively remove organic pollutants, heavy metals, and dyes from water through various mechanisms such as π–π interactions, hydrogen bonding, and electrostatic interactions [29, 73]. GO-based composites can be synthesized and modified through doping, co-doping, and immobilization techniques to enhance their performance [3, 57]. Applications of GO-based materials in wastewater treatment include adsorption, photocatalysis, Fenton-like reactions, ozonation, and photo-Fenton processes [3, 29]. These materials have demonstrated high removal efficiencies, often exceeding 90% for various pollutants [29]. When combined with TiO_2 and silver, these composites demonstrate enhanced photocatalytic activity, effectively degrading pollutants in domestic wastewater. N-doped TiO_2/GO composites show improved visible light photocatalysis, with a 57.69% degradation rate for rhodamine B [41].

6 Challenges and Limitations

6.1 *Stability and Durability of Composites*

Graphene oxide (GO) membranes show promise for wastewater treatment due to their unique properties and selective pollutant removal capabilities [31]. Graphene oxide composite membranes with excellent stability in aqueous solutions were developed using an environmental-friendly electrostatically driven method. [22]. However, stability in aqueous environments remains a significant challenge. GO membranes are prone to swelling and delamination, which reduce selectivity and mechanical stability [22]. The colloidal properties of GO affect its stability, with factors such as pH, salt type, and ionic strength influencing aggregation kinetics [15]. To address these issues, researchers have developed composite hybrid membranes by chemically treating GO layers and decorating them with functional materials, improving properties like permeability, hydrophilicity, and antifouling tendency [31]. An environmentally friendly, electrostatically driven method for preparing GO composite membranes has shown promising results, maintaining stability for over 30 days at 60 °C in water and saline solutions [22]. Despite progress, achieving long-term stability of GO membranes in aqueous environments remains a challenge [76].

6.2 *Economic Considerations*

6.2.1 Cost of Raw Materials

Graphene oxide (GO) and its composites materials effectively remove heavy metals, dyes, and other organic pollutants through adsorption and photocatalytic degradation, with removal efficiencies often exceeding 90% [29]. Adsorption using graphene-based materials is considered a cost-effective and efficient method for large-scale wastewater treatment operations [29]. However, challenges remain in optimizing experimental parameters and improving recyclability [1]. The eco-friendly nature of graphene-based materials makes them attractive for commercial applications in wastewater treatment. Ongoing research focuses on enhancing the properties of graphene-based materials to improve their suitability for wastewater treatment.

6.2.2 Scalability Issues

Graphene oxide (GO) and its composites have shown promise for wastewater treatment due to their large surface area and functional groups [61]. These materials can effectively remove various pollutants, including heavy metals, dyes, and pharmaceuticals. However, scaling up GO production for industrial use faces several challenges, such as sourcing appropriate graphite, safe oxidation processes, and wastewater treatment. The synthesis methods, adsorption behaviors, and regeneration processes of GO-based adsorbents have been extensively studied [61]. Recent research has focused on developing rGO/GO-hybrid nanomaterials for improved photocatalytic degradation and antibacterial activity in wastewater treatment. Despite their potential, GO-based composites still face limitations and challenges in large-scale applications, necessitating further research to optimize their performance and address scalability issues [61].

6.3 Environmental Impact and Toxicity

6.3.1 Nanomaterial Toxicity

Graphene oxide (GO) and its composites have shown effectiveness in removing various pollutants. However, concerns about the toxicity of GO in wastewater treatment systems have been raised. Ahmed and Rodrigues [2] found that GO can have dose-dependent toxic effects on microbial communities in activated sludge, impacting nutrient removal and effluent quality at concentrations between 50 and 300 mg/L. The toxicity mechanism may involve the generation of reactive oxygen species. Despite these challenges, ongoing research focuses on developing novel GO-based nanocomposites to enhance pollutant removal efficiency and address regeneration issues. Further studies are needed to optimize GO-based materials for safe and effective wastewater treatment applications.

6.3.2 Environmental Regulations

Graphene oxide's unique thermal, electrical, and chemical properties make it suitable for various applications, including energy storage and water purification [57]. The adsorption method using graphene-based materials is considered efficient and eco-friendly for wastewater treatment. Recent advances in graphene oxide synthesis and surface modification have enabled enzyme immobilization, creating new possibilities for environmental remediation [57]. While graphene oxide-based adsorbents show great potential, challenges remain in scaling up these technologies for practical applications. Ongoing research focuses on optimizing experimental parameters and improving adsorption–desorption recycling processes to enhance the efficiency of graphene-based materials in wastewater treatment [1].

7 Recent Advances and Innovations

7.1 Functionalization Techniques

7.1.1 Surface Modification

Graphene oxide (GO) and its composites have emerged as promising materials for wastewater treatment due to their unique physicochemical properties and high adsorption capacity for various pollutants. Surface modification techniques, such as plasma treatment and functionalization, can enhance GO's selectivity and efficiency in removing specific contaminants. GO-based nanocomposites have been developed for the removal of heavy metals, dyes, pharmaceuticals, and oil spills from water [57]. These materials can be utilized in various treatment processes, including Fenton reactions, photocatalysis, and ozonation [57]. The π–π interactions, hydrogen bonding, and electrostatic interactions between GO and organic pollutants contribute to their effective removal [73]. Recent research has focused on improving GO-based materials' performance, scalability, and regeneration potential for practical wastewater treatment applications [57].

7.1.2 Hybrid Materials

GO-based hybrid membranes demonstrate improved permeability, porosity, and antifouling properties compared to pristine GO membranes [31]. The removal mechanisms involve size exclusion, electrostatic interactions, and adsorption [31, 73]. GO's large surface area, excellent mechanical strength, and adaptability to different functional groups contribute to its effectiveness in water treatment [29]. Additionally, GO serves as a promising additive in membrane technology for organic–inorganic nanohybrids in separation applications [36]. These materials offer cost-effective, efficient, and reusable solutions for large-scale wastewater treatment operations [29].

7.2 Novel Composite Designs

Graphene oxide (GO)-based nanocomposites, particularly those incorporating metal ferrites, have shown excellent performance in removing both organic and inorganic contaminants through adsorption and photocatalytic degradation [79]. GO-based nanocomposites have demonstrated over 90% removal efficiency for heavy metal ions and dyes, with some materials achieving up to 99–100% decontamination [29]. Recent advancements include the development of magnetic GO nanocomposites, which offer enhanced pollutant removal and easy magnetic separation [8]. Despite their potential, challenges remain in designing efficient GO-based nanocomposites for specific contaminants and ensuring their regeneration after multiple adsorption cycles. Ongoing research focuses on optimizing these materials for various pollutants, including heavy metals, dyes, pharmaceuticals, and oil spills.

7.3 Integration with Other Treatment Technologies

Recent advancements include the development of GO-hydrogel composites for improved adsorption [20] and the immobilization of enzymes on GO-based substrates for environmental remediation. Integration of GO with other treatment technologies, such as Fenton reactions, ozonation, and photo-electrocatalysis, has shown promising results in water purification [57]. Reduced graphene oxide (RGO) composites with metal/metal oxides have shown superior performance in mercury removal compared to pristine GO [70]. Magnetic graphene oxide (MGO) composites offer the additional advantage of easy separation using magnetic fields [8].

8 Real-World Applications of Graphene Oxide-Based Metal Oxide Composites in Wastewater Treatment

Graphene oxide-based metal oxide composites (GO-MOCs) have emerged as efficient materials for wastewater treatment, offering high surface area, unique electronic properties, and a rich surface chemistry ideal for adsorbing and degrading pollutants. The combination of graphene oxide (GO) with various metal oxides enhances the composite's catalytic and adsorptive capabilities, making it a promising candidate for addressing current environmental challenges in wastewater purification.

8.1 Heavy Metal Removal in Mining Effluents

Use of Graphene Oxide-Iron Oxide (GO-Fe_3O_4) Composites in Mining Effluents:

A mining company in China faced severe heavy metal contamination (primarily lead, cadmium, and chromium) in its wastewater. Researchers collaborated with the company to introduce GO-Fe_3O_4 composites into their filtration systems, which utilized the high adsorption capacity of GO for metal ions. The GO-Fe_3O_4 composites successfully removed up to 98% of the lead and 95% of cadmium ions. This cost-effective approach allowed the facility to meet environmental discharge standards, showcasing GO composites' viability for mining wastewater treatment on an industrial scale. GO-Fe_3O_4 composites have demonstrated high adsorption capacities for various pollutants, including organic dyes, pesticides, and pharmaceutical drugs.

8.2 Dye Removal in Textile Industry Wastewater

Photocatalytic Degradation of Dye-Contaminated Wastewater Using GO-TiO_2 Composites in an Indian Textile Mill:

A textile manufacturing plant in India struggled with dye-laden wastewater containing complex organic dyes such as methylene blue and methyl orange. The plant installed a pilot-scale reactor that used GO-TiO_2 composites under UV light exposure to degrade dye molecules. TiO_2's photocatalytic properties, enhanced by the large surface area of GO, facilitated the breakdown of organic dye molecules [35]. The system achieved up to 85% dye removal in the wastewater, reducing the environmental impact and enabling safe discharge of treated water. The success of the pilot led to the implementation of the GO-TiO_2 system in other facilities.

Recent studies have explored the use of graphene oxide-titanium dioxide (GO-TiO_2) composites for photocatalytic degradation of dyes in wastewater. These nanocomposites have shown superior performance compared to pure TiO_2 in degrading various dyes, including azo dyes and methylene blue [23, 35]. The optimal GO content in the composite was found to be 15 wt%, achieving a 95.8% degradation rate for methylene blue [23]. Under optimized conditions, nearly complete removal of methylene blue was achieved within 4 h using a GO-TiO_2 composite, significantly outperforming TiO_2 alone [35]. The composites were characterized using various techniques such as XRD, FTIR, and SEM, revealing their structural and morphological properties [23]. Additionally, the photocatalytic performance of GO-TiO_2-ZnO nanocomposites has been investigated, with varying compositions to optimize dye removal efficiency.

8.3 Organic Pollutant Treatment in Pharmaceutical Industry Wastewater

Degradation of Pharmaceutical Pollutants Using GO-ZnO Nanocomposites in a South Korean Pharmaceutical Plant:

A pharmaceutical company in South Korea was releasing wastewater containing residual antibiotics and other pharmaceutical compounds, posing ecological risks. Researchers collaborated with the plant to install a treatment reactor using GO-ZnO composites. The photocatalytic activity of ZnO, enhanced by GO, produced reactive oxygen species under UV light, which oxidized and decomposed organic pollutants. The GO-ZnO composite system achieved up to 90% removal of pharmaceutical contaminants, helping the plant comply with strict discharge regulations. This cost-effective, eco-friendly system has since become a model for pharmaceutical wastewater treatment.

8.4 Oil and Grease Removal in Oil Refinery Wastewater

Application of GO Composites for Oil Separation in a Middle Eastern Oil Refinery:

An oil refinery in the Middle East faced high concentrations of oil and grease in its wastewater, challenging regulatory compliance. Engineers developed a separation unit using GO-based hydrophobic composites, which were particularly effective at adsorbing and removing oil droplets. This treatment reduced oil content in wastewater by up to 88%, allowing the facility to recover the separated oil for reuse and reduce overall treatment costs. This success has prompted further applications in other oil and gas facilities in the region.

Recent research has explored graphene oxide (GO) composites for efficient oil–water separation. A super-hydrophobic/super-oleophilic UIO-66-F4@rGO composite capable of separating water-in-oil emulsions with high efficiency. Fabricated GO-polyethersulfone membranes, demonstrate enhanced hydrophilicity and flux for oil–water separation. A multifunctional fly ash-based GO/geopolymer composite membrane with superhydrophilic and underwater superoleophobic properties exhibit high water flux and separation efficiencies for oil–water mixtures and emulsions. This membrane also showed stability under harsh conditions and photocatalytic degradation of organic dyes. These studies highlight the potential of GO composites for addressing oil separation challenges in various industrial applications, including oil refineries.

8.5 Pathogen Removal in Municipal Wastewater Treatment

Antimicrobial Graphene Oxide-Silver (GO-Ag) Composites in a Municipal Wastewater Treatment Plant in Spain:

A municipal wastewater plant in Spain required a more efficient solution for pathogen removal to meet stringent European Union discharge standards. GO-Ag composites were introduced in the final filtration stage, where silver nanoparticles embedded in the GO matrix acted as a powerful antimicrobial agent to inactivate bacteria and viruses. The GO-Ag system achieved over 99% reduction in bacterial load, significantly improving the treated water quality. This case demonstrated the effectiveness of GO composites in pathogen control, making it a scalable option for other wastewater treatment plants.

Graphene oxide-silver (GO-Ag) nanocomposites have shown promising antibacterial properties for water disinfection and surface treatment. These composites exhibit strong bactericidal activity against both Gram-negative and Gram-positive bacteria, including *E. coli* and *S. aureus* [69]. The antimicrobial mechanism involves cell membrane destruction and oxidative stress [69]. GO-Ag nanocomposites have demonstrated superior antibacterial efficacy compared to GO or silver nanoparticles alone. They can effectively inhibit biofilm formation on surfaces, making them potential candidates for antibacterial coatings in food packaging and medical devices. The synthesis of GO-Ag composites typically involves the reduction of silver ions on GO sheets, resulting in well-dispersed silver nanoparticles with an average size of 7.5 nm. These nanocomposites show promise for various applications, including water purification and surface disinfection in municipal wastewater treatment plants.

9 Future Perspectives

9.1 Scalability and Cost-effectiveness

While lab-scale studies show promise, there are challenges in scaling GO-MO composites to pilot or industrial-scale applications due to synthesis costs and difficulties in controlling material quality. Production methods like chemical vapor deposition (CVD) and hydrothermal synthesis are often costly and energy-intensive, limiting large-scale applicability. Development of economical, high-yield synthesis methods, such as green synthesis or direct deposition methods, could make GO-MO composites more accessible for widespread use.

9.2 Long-Term Stability and Reusability

GO-MO composites face issues with stability, material degradation, and reduced efficiency over multiple treatment cycles, especially under harsh environmental conditions. The mechanical stability and chemical integrity of GO-MO composites often decrease with use, leading to leaching of metal ions and a decline in performance. Studies focused on enhancing the durability of GO-MO composites, potentially through surface coatings or cross-linking, are essential to improve reusability and long-term stability.

9.3 Environmental and Health Impacts

Potential toxicity and environmental impact of GO-MO composites are not yet well understood, raising concerns about the safety of these materials, especially if they degrade into smaller particles or release metal ions. The leaching of heavy metals from composites can pose risks to aquatic ecosystems and human health. Comprehensive studies on the ecological and health impact of GO-MO composites are needed to develop safe usage protocols and to design more environmentally benign composites.

9.4 Limited Multi-contaminant Removal Studies

Most studies focus on individual contaminants, but real wastewater typically contains a mixture of pollutants (e.g., heavy metals, organic pollutants, nutrients, and pathogens). The efficacy of GO-MO composites in treating complex mixtures is unclear, as interactions between contaminants can affect adsorption and degradation processes. Multi-target studies simulating real wastewater conditions would provide a more realistic assessment of GO-MO composites' performance and identify any potential limitations in mixed-contaminant environments.

9.5 Limited Pilot and Field Studies

Most research on GO-MO composites remains at the laboratory scale, with limited pilot or field-scale studies. The lack of real-world data makes it difficult to evaluate the practical feasibility, operational challenges, and long-term performance of GO-MO composites in industrial settings. Field studies and pilot projects in diverse industrial settings would provide valuable data, enabling better design and optimization for full-scale applications.

10 Conclusions

In this chapter, we have explored the critical role of graphene oxide-based metal oxide composites in wastewater treatment. The growing concern over water pollution and the need for effective wastewater treatment methods make these advanced materials essential. Graphene oxide-based metal oxides, with their unique properties, offer promising solutions for the efficient removal of pollutants through mechanisms such as adsorption, photo catalysis, and ion exchange. Their high surface area, stability, and tenable properties enable these composites to tackle diverse contaminants in various wastewater sources, including industrial, agricultural, and domestic waste.

The synthesis of these composites through methods like sol–gel, hydrothermal, and emerging techniques such as microwave-assisted synthesis and electrochemical methods provide flexibility and scalability for large-scale applications. The performance metrics, including removal efficiency and kinetic studies, highlight their potential, though challenges like stability, cost, and environmental toxicity need careful consideration.

Recent advancements, particularly in functionalization and hybrid material development, are pushing the boundaries of these materials' efficiency and applicability. Moreover, integrating graphene oxide-based composites with other treatment technologies, such as bioremediation and advanced oxidation processes (AOPs), offers a multi-faceted approach to tackling complex wastewater treatment challenges.

While there are challenges regarding the long-term stability, economic feasibility, and potential environmental impact, the future of graphene oxide-based composites looks promising. With continued research and innovation, these materials could revolutionize wastewater treatment, providing sustainable and effective solutions for global water pollution issues. The combination of cutting-edge materials science and practical applications paves the way for a cleaner, healthier environment.

11 Future Perspective

The future perspective of Graphene Oxide-Based Metal Oxide Composites for wastewater purification can be summarized in the following steps:

Enhanced Functionalization: Advances in surface modification and functionalization techniques will further improve the efficiency of graphene oxide-based composites, allowing for better pollutant adsorption and removal.

Sustainable Synthesis Methods: Development of eco-friendly, cost-effective, and scalable synthesis methods will make these materials more viable for large-scale applications.

Integration with Other Technologies: Combining graphene oxide-based composites with other treatment technologies, such as bioremediation and advanced oxidation processes (AOPs), will enable multifunctional treatment systems for complex wastewater.

Improved Durability and Stability: Research into enhancing the stability, recyclability, and long-term performance of these composites will address current limitations, making them more suitable for industrial applications.

Economic Viability: Reducing the cost of raw materials and increasing the scalability of production will make graphene oxide-based composites economically feasible for large-scale wastewater treatment.

Environmental Impact Reduction: Understanding and mitigating the potential toxicity of nanomaterials will be crucial to ensure that these advanced materials are environmentally friendly and safe for widespread use.

Real-World Applications and Commercialization: Continued successful case studies and real-world implementations will lead to the commercialization of graphene oxide-based composites as a leading solution for wastewater purification across various industries.

References

1. Adel M, Ahmed MA, Elabiad MA, Mohamed AA (2022) Removal of heavy metals and dyes from wastewater using graphene oxide-based nanomaterials: a critical review. Environ Nanotechnol Monit Manag 18:100719

2. Ahmed F, Rodrigues DF (2013) Investigation of acute effects of graphene oxide on wastewater microbial community: a case study. J Hazard Mater 256–257:33–39
3. Ajala OJ, Tijani JO, Bankole MT, Abdulkareem AS (2022) A critical review on graphene oxide nanostructured material: properties, synthesis, characterization and application in water and wastewater treatment. Environ Nanotechnol Monit Manag 18:100673
4. Ali A, Sadia M, Azeem M, Ahmad MZ, Umar M, Abbas ZU (2023) Ion exchange resins and their applications in water treatment and pollutants removal from environment: a review: ion exchange resins and their applications. Futuristic biotechnology, 12–19
5. Ali I, Basheer AA, Mbianda XY, Burakov AE, Galunin EV, Burakova IV, Mkrtchyan ES, Tkachev AG, Grachev VA (2019) Graphene based adsorbents for remediation of noxious pollutants from wastewater. Environ Int 127:160–180
6. Alrumman SA, El-kott AF, Keshk SM (2016) Water pollution: source & treatment. Am J Environ Eng 6:88–98
7. An Y, Gao X, Jiang W, Han J, Ye Y, Chen T, Ren R, Zhang J, Liang B, Li Z, Wang A, Ren N (2023) A critical review on graphene oxide membrane for industrial wastewater treatment. Environ Res 115409
8. Asghar F, Shakoor B, Fatima S, Munir S, Razzaq H, Naheed S, Butler IS (2022) Fabrication and prospective applications of graphene oxide-modified nanocomposites for wastewater remediation. RSC Adv 12:11750–11768
9. Aziz FFA, Jalil AA, Hassan NS, Fauzi AA, Azami MS, Jusoh NWC, Jusoh R (2022) A review on synergistic coexisting pollutants for efficient photocatalytic reaction in wastewater remediation. Environ Res 209:112748
10. Bhattacharya S, Saha I, Mukhopadhyay A, Chattopadhyay D, Ghosh UC, Chatterjee D (2013) Role of nanotechnology in water treatment and purification: potential applications and implications. Int J Chem Sci Technol 3(3):59
11. Caicedo FM, López EV, Agarwal A, Drozd V, Durygin A, Hernandez AF, Wang C (2020) Synthesis of graphene oxide from graphite by ball milling. Diam Relat Mater 109:108064
12. Carriazo D, Patiño J, Gutiérrez MC, Ferrer ML, del Monte F (2013) Microwave-assisted synthesis of $NiCo_2O_4$–graphene oxide nanocomposites suitable as electrodes for supercapacitors. RSC Adv 3:13690–13695
13. Cevallos-Mendoza J, Amorim CG, Rodríguez-Díaz JM, Montenegro MDCB (2022) Removal of contaminants from water by membrane filtration: a review. Membranes 12(6):570
14. Chauhan MS, Rahul AK, Shekhar S, Kumar S (2023) Removal of heavy metal from wastewater using ion exchange with membrane filtration from Swarnamukhi river in Tirupati. Mater Today Proceed 78:1–6
15. Chowdhury I, Duch MC, Mansukhani ND, Hersam MC, Bouchard DC (2013) Colloidal properties and stability of graphene oxide nanomaterials in the aquatic environment. Environ Sci Technol 47(12):6288–6296
16. Dongre RS (2018) Phosphate & nitrate removal from agricultural runoff by chitosan-graphite composite. Res Develop Mater Sci 11
17. Dubey A, Dube CL (2023) Microwave assisted synthesized graphene oxide nanocomposites for remediation of toxic metal ions. Nano-Struct Nano-Objects 35:101017
18. Dubey AK, Bhavsar N, Pachchigar V, Saini M, Ranjan M, Dube CL (2021) Microwave assisted ultrafast synthesis of graphene oxide based magnetic nano composite for environmental remediation. Ceram Int 48(4):4821–4828
19. Elmi F, Alinezhad H, Moulana Z, Salehian F, Mohseni Tavakkoli S, Asgharpour F, Fallah H, Elmi MM (2014) The use of antibacterial activity of ZnO nanoparticles in the treatment of municipal wastewater. Water Sci Technol J Int Assoc Water Pollut Res 70(5):763–770
20. Eraković ZB, Stefanović D (2022) Purification of contaminated wastewater with the help of graphene composites with hydrogels. Facta Universitatis, Series, Working and Living Environmental Protection, pp 027–036
21. Farazas A, Mavropoulos A, Christofilos D, Tsiaousis I, Tsipas DN (2018) Ultrasound assisted green synthesis and characterization of graphene oxide. Int J NanoSci Nanotechnol 14:11–17

22. Fontananova E, Tocci E, Abu-Zurayk RA, Grosso V, Meringolo C, Muzzi C, Al Bawab A, Qtaishat M, De Filpo G, Curcio E, Drioli E, Di Profio G (2022) An environmental-friendly electrostatically driven method for preparing graphene oxide composite membranes with amazing stability in aqueous solutions. J Membr Sci 655:120587
23. Fu Z, Zhang S, Fu Z (2019) Preparation of multicycle GO/TiO_2 composite photocatalyst and study on degradation of methylene blue synthetic wastewater. Appl Sci 9(16):3282
24. Garino N, Sacco A, Castellino M, Muñoz-Tabares JA, Chiodoni A, Agostino V, Margaria V, Gerosa M, Massaglia G, Quaglio M (2016) Microwave-assisted synthesis of reduced graphene oxide/SnO_2 nanocomposite for oxygen reduction reaction in microbial fuel cells. ACS Appl Mater Interfaces 8(7):4633–4643
25. Goyat R, Saharan Y, Singh J, Umar A, Akbar S (2022) Synthesis of graphene-based nanocomposites for environmental remediation applications: a review. Molecules 27(19):6433
26. He P, Zhang F, Zhang Y, Chen H (2023) Multifunctional fly ash-based GO/geopolymer composite membrane for efficient oil-water separation and dye degradation. Ceram Int 49(2):1855–1864
27. Hegab HM, Zou L (2015) Graphene oxide-assisted membranes: fabrication and potential applications in desalination and water purification. J Membr Sci 484:95–106
28. Jabbar KQ, Barzinjy AA, Hamad SM (2022) Iron oxide nanoparticles: preparation methods, functions, adsorption and coagulation/flocculation in wastewater treatment. Environ Nanotechnol Monit Manag 17:100661
29. Joshi NC, Gururani P (2022) Advances of graphene oxide based nanocomposite materials in the treatment of wastewater containing heavy metal ions and dyes. Curr Res Green Sustain Chem 5:100306
30. Kafle SR, Adhikari S, Shrestha R, Ban S, Khatiwada G, Gaire P, Tuladhar N, Jiang G, Tiwari A (2024) Advancement of membrane separation technology for organic pollutant removal. Water Sci Technol J Int Assoc Water Pollut Res 89(9):2290–2310
31. Kamran U, Rhee KY, Lee S, Park S (2022) Innovative progress in graphene derivative-based composite hybrid membranes for the removal of contaminants in wastewater: a review. Chemosphere 306:135590
32. Kavitha S, Dhandapani C (2018) Water pollution in India: an overview. Int Rev Bus Econ 1(3):39
33. Khan ST, Malik A (2019) Engineered nanomaterials for water decontamination and purification: from lab to products. J Hazard Mater 363:295–308
34. Khulbe KC, Matsuura T (2018) Removal of heavy metals and pollutants by membrane adsorption techniques. Appl Water Sci 8:1–30
35. Kurniawan T, Mengting Z, Fu D, Yeap SK, Othman MH, Avtar R, Ouyang T (2020) Functionalizing TiO_2 with graphene oxide for enhancing photocatalytic degradation of methylene blue (MB) in contaminated wastewater. J Environ Manage 270:110871
36. Kyzas GZ, Mitropoulos AC, Favvas EP (2020) Nanohybrid graphene-based materials for advanced wastewater treatment: adsorption and membrane technology. In: Membrane desalination. CRC Press, pp 91–123
37. Lan Huong PT, Tu N, Lan H, Thang LH, Van Quy N, Tuan PA, Dinh NX, Phan VN, Le A (2018) Functional manganese ferrite/graphene oxide nanocomposites: effects of graphene oxide on the adsorption mechanisms of organic MB dye and inorganic As(v) ions from aqueous solution. RSC Adv 8(22):12376–12389
38. Lee KM, Lai CW, Ngai KS, Juan JC (2016) Recent developments of zinc oxide based photocatalyst in water treatment technology: a review. Water Res 88:428–448
39. Litter MI (2015) Mechanisms of removal of heavy metals and arsenic from water by TiO_2-heterogeneous photocatalysis. Pure Appl Chem 87:557–567
40. Liu A, Zhou W (2014) Graphene-based metal oxide nanocomposites and their application as adsorbents in water treatment schemes. Chin Sci Bull 59:2039–2054
41. Liu S, Su ZL, Liu Y, Yi LY, Chen ZL, Liu ZZ (2021) Mechanism and purification effect of photocatalytic wastewater treatment using graphene oxide-doped titanium dioxide composite nanomaterials. Water 13(14):1915

42. Lowe ammed MN, Aljibori HS, Jweeg MJ, Oqaili FA, Abdullah TA, Abdullah OI, Meharban F, Rashed RT, Aldulaimi M, Al-Azawi K (2024) A comprehensive review on graphene oxide based nanocomposites for wastewater treatment. Polish J Chem Technol 26:64–79
43. Madhav S, Ahamad A, Singh AK, Kushawaha J, Chauhan JS, Sharma S, Singh P (2020) Water pollutants: sources and impact on the environment and human health. Sensors in water pollutants monitoring: role of material 43–62
44. Maheshwari U, Varshney D, Deshpande D, Ghodke S (2023) Development and utilizing graphene oxide for the wastewater treatment: an experimental evaluation. Mater Today Proceed 78:23–27
45. Mahlaule-Glory LM, Hintsho-Mbita NC (2022) Green derived zinc oxide (ZnO) for the degradation of dyes from wastewater and their antimicrobial activity: a review. Catalysts 12(8):833
46. Majumder P, Dutta K, Dutta P (2019) Synthesis, properties of graphene oxide–metal oxide mixed nanocomposites and their applications-review. Int J Adv Sci Eng 3:1032–1039
47. Malhotra P, Jain A (2021) Graphene oxide-based nanocomposites for adsorptive removal of water pollutants. In: Contamination of water. Academic Press, pp 431–448
48. Matei E, Predescu A, Vasile E (2011) Properties of magnetic iron oxides used as materials for wastewater treatment. In: Journal of physics: conference series, vol 304, No. 1. IOP Publishing, p 012022
49. Molla MAI, Ahmed AZ, Kaneco S (2021) Reaction mechanism for photocatalytic degradation of organic pollutants. In: Nanostructured photocatalysts. Elsevier, pp 63–84
50. Mura S, Jiang Y, Vassalini I, Gianoncelli A, Alessandri I, Granozzi G et al (2018) Graphene oxide/iron oxide nanocomposites for water remediation. ACS Appl Nano Mater 1(12):6724–6732
51. Nishat A, Yusuf M, Qadir A, Ezaier Y, Vambol V, ijaz khan M, ben Moussa S, Kamyab H, Sehgal SS, Prakash C, Yang H, Ibrahim H, Eldin SM (2023) Wastewater treatment: a short assessment on available techniques. Alexandria Eng J 76:505–516
52. Ojajuni O, Saroj DP, Cavalli G (2015) Removal of organic micropollutants using membrane-assisted processes: a review of recent progress. Environ Technol Rev 4(1):17–37
53. Pachiyappan J, Gnanasundaram N (2020) Using graphene oxide-silica [GO-Si] nano composite adsorbent, removal of heavy metal ions (lead and mercury) from industrial wastewater and analysing its performance. Rasayan J Chem 13:2027–2035
54. Pandey S (2006) Water pollution and health. Kathmandu Univ Med J (KUMJ) 4(1):128–134
55. Pourzare K, Farhadi S, Mansourpanah Y (2017) Graphene oxide/Co_3O_4 nanocomposite: synthesis, characterization, and its adsorption capacity for the removal of organic dye pollutants from water. Acta Chim Slov 64(4):945–958
56. Prabu SL, Suriyaprakash TNK, Kandasamy R, Rathinasabapathy T (2020) Effective waste water treatment and its management. In: Waste management: concepts, methodologies, tools, and applications. IGI Global, pp 49–72
57. Priyadharshini SD, Manikandan S, Kiruthiga R, Rednam U, Babu PS, Subbaiya R, Karmegam N, Kim W, Govarthanan M (2022) Graphene oxide-based nanomaterials for the treatment of pollutants in the aquatic environment: recent trends and perspectives—a review. Environ Pollut 306:119377
58. Rasheed T, Bilal M, Nabeel F, Adeel M, Iqbal HMN (2019) Environmentally-related contaminants of high concern: potential sources and analytical modalities for detection, quantification, and treatment. Environ Int 122:52–66
59. Raval HD, Gohil JM (2010) Nanotechnology in water treatment: an emerging trend. Int J Nucl Desalin 4:184–188
60. Riaz S, Park S (2020) An overview of TiO_2-based photocatalytic membrane reactors for water and wastewater treatments. J Ind Eng Chem 84:23–41
61. Rout DR, Jena HM, Baigenzhenov O, Hosseini-Bandegharaei A (2023) Graphene-based materials for effective adsorption of organic and inorganic pollutants: a critical and comprehensive review. Sci Total Environ 863:160871

62. Sadegh H, Ali GA, Gupta V, Makhlouf AS, Shahryari-ghoshekandi R, Nadagouda MN, Sillanpää M, Megiel E (2017) The role of nanomaterials as effective adsorbents and their applications in wastewater treatment. J Nanostruct Chem 7:1–14
63. Sadhasivam T, Gurunathan K (2015) A facile synthesis and thermal properties of graphene oxide–mischmetal oxide nanocomposites. J Nanosci Nanotechnol 15(8):5676–5683
64. Sagadevan S, Fatimah I, Egbosiuba TC, Alshahateet SF, Lett JA, Weldegebrieal GK, Le M, Johan MR (2022) Photocatalytic efficiency of titanium dioxide for dyes and heavy metals removal from wastewater. Bull Chem React Eng Catal 17(2):430–450
65. Shahrodin NS, Jaafar J, Rahmat AR, Yusof N, Othman M, Rahman MA (2020) Superparamagnetic iron oxide as photocatalyst and adsorbent in wastewater treatment—a review. Micro Nanosyst 12(1):4–22
66. Sheshmani S, Amini R (2013) Preparation and characterization of some graphene based nanocomposite materials. Carbohyd Polym 95(1):348–359
67. Silva JA (2023) Wastewater treatment and reuse for sustainable water resources management: a systematic literature review. Sustainability 15(14):10940
68. Soltani T, Lee B (2017) Low intensity-ultrasonic irradiation for highly efficient, eco-friendly and fast synthesis of graphene oxide. Ultrason Sonochem 38:693–703
69. Song B, Zhang C, Zeng G, Gong J, Chang Y, Jiang Y (2016) Antibacterial properties and mechanism of graphene oxide-silver nanocomposites as bactericidal agents for water disinfection. Arch Biochem Biophys 604:167–176
70. Sreeprasad TS, Maliyekkal SM, Lisha KP, Pradeep T (2011) Reduced graphene oxide-metal/metal oxide composites: facile synthesis and application in water purification. J Hazard Mater 186(1):921–931
71. Safeer S, Pandey RP, Rehman B, Safdar T, Ahmad I, Hasan SW, Ullah A (2022) A review of artificial intelligence in water purification and wastewater treatment: recent advancements. J Water Process Eng 49:102974
72. Talaiekhozani A, Rezania S, Kim KH, Sanaye R, Amani AM (2021) Recent advances in photocatalytic removal of organic and inorganic pollutants in air. J Clean Prod 278:123895
73. Thakur KV, Kandasubramanian B (2019) Graphene and graphene oxide-based composites for removal of organic pollutants: a review. J Chem Eng Data 64(3):833–867
74. Topare NS, Attar SJ, Manfe MM (2011) Sewage/wastewater treatment technologies: a review. Sci Revs Chem Commun 1(1):18–24
75. Wang J, Zhang J, Han L, Wang J, Zhu L, Zeng H (2021) Graphene-based materials for adsorptive removal of pollutants from water and underlying interaction mechanism. Adv Coll Interface Sci 289:102360
76. Wang Z, He F, Guo J, Peng S, Cheng XQ, Zhang Y, Drioli E, Figoli A, Li Y, Shao L (2020) The stability of a graphene oxide (GO) nanofiltration (NF) membrane in an aqueous environment: progress and challenges. Mater Adv 1(4):554–568
77. Weldegebrieal GK (2020) Synthesis method, antibacterial and photocatalytic activity of ZnO nanoparticles for azo dyes in wastewater treatment: a review. Inorg Chem Commun 120:108140
78. Yadav P, Bhaduri A (2020) Synthesis and structural characterization of ZnO-graphene nanocomposite by chemical co-precipitation method. Int J Innov Res Phys 1(2):19–24
79. Zahid M, Nadeem N, Hanif MA, Bhatti IA, Bhatti HN, Mustafa G (2019) Metal ferrites and their graphene-based nanocomposites: synthesis, characterization, and applications in wastewater treatment. Magnetic nanostructures: environmental and agricultural applications, pp 181–212
80. Zhang J, Chen H, Duan X, Sun H, Wang S (2023) Photothermal catalysis: from fundamentals to practical applications. Mater Today 68:234–253
81. Zhang J, Wu H, Shi L, Wu Z, Zhang S, Wang S, Sun H (2024) Photocatalysis coupling with membrane technology for sustainable and continuous purification of wastewater. Sep Purif Technol 329:1–21

Doped Metal Oxide-Graphene Oxide Nanocomposites: Advanced Materials for Heavy Metals Removal and Wastewater Purification

Sonal Rajoria, Manish Vashishtha, and Vikas K. Sangal

Abstract The increasing contamination of water resources with heavy metals and organic pollutants poses significant environmental and health challenges. Nanocomposites have emerged as promising materials for wastewater purification due to their high surface area, tunable physicochemical properties, and enhanced catalytic activity. This chapter explores the integration of doped metal oxide with graphene oxide (GO) to develop advanced nanocomposites and electrodes for efficient water treatment. The synergy between doped metal oxide and GO enhances conductivity, chemical stability, and surface reactivity, enabling exceptional performance in photocatalytic and electrochemical processes. These materials demonstrate the effective removal of toxic heavy metals and the degradation of complex organic pollutants through advanced oxidation processes (AOPs) and other catalytic mechanisms. The incorporation of dopant elements in the metal oxide optimizes photocatalytic efficiency by facilitating faster charge transfer and higher reaction rates. Furthermore, this chapter discusses the mechanisms by which organic contaminants are effectively oxidized, leading to their transformation into benign end products such as H_2O and CO_2. This work provides valuable insights into the role of nanocomposites in developing innovative, scalable, and sustainable solutions for addressing industrial effluents and environmental water pollution.

Keywords Doped nanocomposites · Graphene oxide · Heavy metals removal · Industrial effluents · Wastewater purification

1 Introduction

In today's world, water is regarded as a basic necessity due to its environmental significance [8]. Due to human activity, rapid industrialization, population increase, careless exploitation of natural water resources, and unplanned urbanization, the

---

S. Rajoria · M. Vashishtha · V. K. Sangal (✉)
Department of Chemical Engineering, Malaviya National Institute of Technology, Jaipur, India
e-mail: vksangal.chem@mnit.ac.in

A. Ahmad et al. (eds.), *Metal Oxide Based Nanophotocatalyst for Wastewater Purification*, Smart Nanomaterials Technology,
https://doi.org/10.1007/978-981-96-7683-5_13

quality of the water has been significantly impacted [41]. According to a World Health Organization (WHO) report titled "Progress on Sanitation and Drinking Water-2015 Update and MDG Assessment," approximately 2.6 billion people had access to improved drinking water sources in 1990; by 2015, that number could rise to 663 million. The World Water Council estimates that by 2030, around 3.9 billion people will be living in water-scarce places [59]. The primary wellsprings of freshwater contamination can be attributed to releasing untreated clean and poisonous industrial wastes, dumping industrial wastewater, and runoff from agrarian fields. The toxic pollutants released in wastewater can harm aquatic organisms and cause regular waters to be unfit as consumable water sources. Wastewater effluents from industrial processes are contaminated with various toxic heavy metal contaminants, with human and anthropogenic factors being the main causes of increased environmental toxicity [16]. With the rapid development of industries such as metal plating facilities, mining operations, fertilizer industries, tanneries, batteries, paper industries, and pesticides, heavy metals wastewater are directly or indirectly discharged into the environment increasingly, especially in developing countries [12]. Due to the discharge of large amounts of metal-contaminated wastewater, industries bearing heavy metals, such as Cd, Cr, Cu, Ni, As, Pb, and Zn, are the most hazardous among the chemical-intensive industries. Because of their high solubility in the aquatic environments, heavy metals can be absorbed by living organisms. Once they enter the food chain, large concentrations of heavy metals may accumulate in the human body. If the metals are ingested beyond the permitted concentration, they can cause serious health disorders [3]. Unlike other pollutant molecules, heavy metals are excreted slowly through the kidneys, liver, and spleen, and the longer storage leads to damage to the organs by altering their cellular functioning. Continuous exposure to heavy metals may decrease the activity of the central nervous system, change the blood composition, and damage the liver, lungs, kidneys, and other fundamental organs. Therefore, it is necessary to treat metal-contaminated wastewater prior to its discharge into the environment.

We live in the "Pollutant Removal Age," where wastewater treatment has transitioned from a voluntary or optional practice to a multi-stage, essential process aimed at significantly improving the quality of potable water and water discharged into the environment. There is a need to find cost-effective treatment technologies because of the strict laws on the acceptable limits of pollutants in discharged wastewater, which have led to higher disposal costs [24]. While effective for certain applications, traditional wastewater purification methods often struggle with limitations such as high operational costs, inefficiency, and the inability to handle complex and persistent pollutants. To overcome these challenges, researchers are exploring advanced materials and processes to develop innovative solutions for wastewater purification. This has created an urgent need for suitable and sustainable techniques that offer low cost, reduced energy consumption, and high removal efficiency. In the recent years, the focus has shifted toward the use of advanced materials, particularly those derived from nanotechnology, as transformative solutions for wastewater treatment. Nanotechnology-based materials offer unique physicochemical properties such as high surface area, tunable porosity, and enhanced reactivity, making them ideal for addressing the limitations of conventional methods. These advanced materials hold immense potential for tackling complex contaminants and ensuring efficient, cost-effective, and sustainable water purification. Among the various nanomaterials, nanocomposites comprising doped metal oxides and GO have gained significant attention. These hybrid materials combine the advantages of metal oxides known for their catalytic properties with the exceptional electrical conductivity, mechanical strength, and chemical stability of GO. Such synergetic combinations are critical for addressing the limitations of conventional treatment methods, which often suffer from low efficiency, high energy consumption, and secondary pollution.

In wastewater treatment, integrating doped metal oxides with GO has garnered significant attention due to their exceptional performance in removing toxic heavy metals and degrading complex organic pollutants [22]. These nanocomposites leverage synergistic mechanisms such as adsorption, photocatalysis, and electrochemical oxidation, ensuring comprehensive water purification. The novelty of this chapter lies in exploring the unique interplay between doped metal oxides and GO, which enhances material properties. GO improves charge transfer kinetics, stabilizes metal oxide nanoparticles, and supports uniform dispersion, while doped elements in metal oxides increase light absorption, catalytic activity, and selective adsorption of pollutants. This innovative combination enhances the efficiency of heavy metal removal and facilitates the simultaneous degradation of complex organic pollutants, making these materials highly effective for advanced wastewater purification.

This chapter aims to provide a comprehensive understanding of the development, synthesis methods, and functional properties of doped metal oxide-GO nanocomposites, focusing on their role in heavy metal removal and pollutant degradation. The chapter investigates the synergistic interaction between doped metal oxides and GO and seeks to uncover the mechanisms that enable enhanced performance. Additionally, it addresses key challenges such as scalability, stability, and environmental impact, which are essential for transitioning these materials from laboratory-scale research to industrial applications. This work significantly contributes to the field by offering comprehensive insights into the properties, role, and synthesis of doped metal oxide-GO nanocomposites. In particular, the chapter delves into the unique enhanced properties of these nanocomposites, focusing on the synergistic effects of GO and doped metal oxides. It explores various synthesis methods, optimization of parameters, and effective purification performance in wastewater treatment. Practical applications for removing heavy metals and degrading organic pollutants are discussed, alongside challenges in scalability, stability, and industrial use. Finally, the chapter concludes with future perspectives on optimizing these materials for large-scale, sustainable wastewater treatment.

2 Graphene Oxide: Properties, Role, and Synthesis

Synthesis, characterization, investigation, and exploitation of nanomaterials are the main focusses of nanoscience and nanotechnology [2]. Carbon, one of the most prevalent atoms on the planet, may be found in nature in a variety of forms and as a constituent of many substances known as carbon allotropes. The world's thinnest, strongest, and stiffest material is graphene, a "wonder material" that also excels as a heat and electrical conductor [4]. In the recent years, there has been a lot of interest in graphene's two-dimensional conjugated chemical structure, which is responsible for its exceptional conductivity and remarkable chemical resilience. Graphene finds numerous applications in solar cells, energy storage, photocatalysis, and other technologies [19]. GO is a common oxide of graphene, a substance with a distinctive two-dimensional lamellar structure created by adding many oxidation functional groups to the surface or edge of graphene that has specific defects. Because of its massive specific surface area, robust adsorption, broad conversion to graphene, superior charge carrier mobility, and high transparency, GO has garnered a lot of attention. GO flakes are regarded as an insulator material in terms of electricity because of the abundance of oxygen-containing functional groups, such as hydroxyl (˙OH), carboxyl (–COOH), and epoxy (–C–O–C). These functional groups offer active sites for pollutant adsorption and interaction with metal oxides, in addition to improving its dispersibility in aqueous conditions.

Furthermore, the oxygen-containing functional groups in GO can interact with other materials by electrostatic adsorption, covalent bonding, and hydrogen bonding under appropriate circumstances, providing the ideal environment for thin film formation. In addition to making GO chemically stable, the presence of oxygen-containing groups gives GO broad specific surface area and surface modification active spots for the creation of graphene-based and graphene oxide-based products. As a precursor and support carrier for the synthesis of graphene-based composites, GO is highly controlled and easily functionalized. During the composite process with metals, metal oxides, conductive polymers, etc., it can offer a wide specific surface area to efficiently distribute the connected elements and avoid agglomeration. GO is synthesized via various methods, focusing on high yield and oxygen functionality. The Hummer's method oxidizes graphite with potassium permanganate ($KMnO_4$) and concentrated sulfuric acid (H_2SO_4), yielding multilayered GO, while the improved Hummer's method minimizes hazardous by-products. Other techniques include chemical vapor deposition for high-quality GO, though costly, and electrochemical synthesis, a green method using graphite exfoliation. Figure 1 illustrates the basic structure of GO.

GO exhibits excellent physical, chemical, optical, and electrical properties, making it a versatile material with a wide range of applications. GO composites, including those based on metal oxides, have a wide range of applications, making the surface modification of GO a significant research focus. The integration of GO with metal oxides such as titanium oxide (TiO_2) and tin oxide (SnO_2) enhances their functional properties, including catalytic activity, adsorption capacity, and stability. By modifying the surface of GO, researchers can optimize its interaction with metal

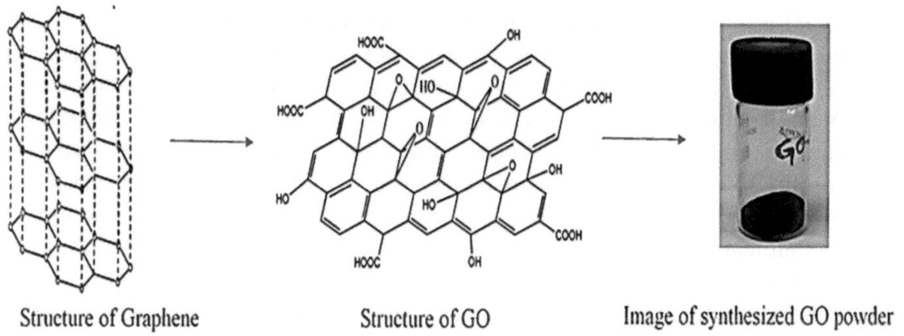

Fig. 1 Structure of graphene and GO [2, 44]

oxides, improve the dispersion of nanoparticles, and enhance the overall performance of the composite in applications like photocatalysis, heavy metals removal, and AOPs.

3 Doped Metal Oxides-GO: Enhanced Properties

Doped metal oxides are modified by introducing different elements (dopants) into their structure to enhance their electrochemical, photocatalytic, and environmental properties. These dopants enhance the stability, activity, and light absorption properties of metal oxides, thereby improving their effectiveness in wastewater treatment applications. This results in superior pollutant degradation, enhanced stability, and greater efficiency compared to individual components.

3.1 Titanium Dioxide (TiO_2)-GO Nanocomposites

Heavy metal reduction has been achieved through the development of GO-based composites including a variety of semiconducting elements, including TiO_2, ZnO, ZnS, CuS, and others.

As a nanocomposite material, the GO/TiO_2 hybrid has garnered some interest for the removal of pollutants such heavy metals, crude oil, and aromatic dyes because of its efficacy and environmental friendliness [23]. In 1972, Fujishima and Honda were the first to achieve photoelectrochemical (PEC) water splitting using TiO_2 as a catalyst [57]. As the most popular semiconductor material, TiO_2 has demonstrated excellent oxidizing power, corrosion resistance, nontoxicity, insoluble in most conditions, and chemical stability. Over the past few decades, TiO_2 has been thoroughly investigated as a potential photocatalyst and supporting substrate for electrocatalysis because of its many advantages. With optical bandgaps (e.g., 1.23 eV < E_g

< 3.26 eV), semiconductor materials can sustain band potentials for water oxidation and reduction. TiO_2-based photocatalysts are the most popular for degradation and water splitting under sunlight due to their superior valence band position at − 4.05 eV and conduction band position around − 7.25 eV [17], which are suitable for photocatalytic redox reactions to proceed, low cost, low ecotoxicity, and high electron mobility [63]. Because of its great chemical stability and corrosion resistance, titanium (Ti) can be used in paints, solar cells, and chemical sensing. The unique physicochemical features of one-dimensional TiO_2 nanotubes (NTs) films have been extensively researched for photocatalysis, making them a promising material for use in a variety of sectors [7, 45, 49]. However, pure TiO_2 nanomaterials have intrinsic problems such as the recombination of photogenerated electrons (e^-) and holes (h^+) and a wide bandgap of 3.2 eV, which limits their sensitivity to the ultraviolet portion of the spectrum [29]. These limitations reduce the efficiency of TiO_2 for PEC water splitting toward H_2 production. A popular and successful method for overcoming these limitations is composite development, in which TiO_2NT is decorated, activated, or merged with nanoparticles of other materials to form heterojunctions. This approach inhibits charge carrier recombination, thereby enhancing device performance. Carbonaceous materials on top of the TiO_2 semiconductor produce a new recombination center for e^--h^+ pairs, which narrows the energy bandgap [50].

3.2 *Mixed Metal Oxides-GO Nanocomposites*

GO is frequently changed by adding dopants like transition metals, non-metals, or different metallic species to enhance the stability, electrocatalytic activity, oxidation potential, and current efficiency of the GO/TiO_2 active layer. It has been demonstrated that metallic species such as IrO_2, RuO_2, SnO_2, and PbO_2 improve electrode durability and performance in wastewater treatment applications. Because of its outstanding electrocatalytic qualities, SnO_2 is the most feasible and efficient option for EO-based degradation of organic contaminants. Increased electrocatalytic activity for producing $^{\cdot}OH$ radicals is the result of higher OEP of SnO_2, which lessens electrode fouling and restores electrode functioning.

The following is a detailed description of the primary reactions occurring at the SnO_2 anode [6, 39]:

$$SnO_2 + xH_2O \rightarrow SnO_2(^{\cdot}OH)_x + xH^+ + xe^- \quad (1)$$

$$R + SnO_2(^{\cdot}OH)_x \rightarrow SnO_2 + CO_2 + xH^+ + xe^- \quad (2)$$

$$SnO_2(^{\cdot}OH)_x \rightarrow SnO_{2+x} + xH^+ + xe^- \quad (3)$$

$$R + SnO_{2+x} \rightarrow SnO_2 \rightarrow RO_x \quad (4)$$

where the organic contaminants are denoted by R. It is demonstrated that when water molecules come into contact with the SnO_2-based metal electrode, physisorbed hydroxyl radicals (SnO_2 $(^{\cdot}OH)_x$) are generated at high anodic potentials [5]. EO eliminates the COD and organic/inorganic effluents when they easily react with organic substrates near the electrode surface to generate CO_2 and H_2O. The anode metal absorbs radicals chemically as a result of the $^{\cdot}$OH radical's strong contact with it. The higher oxide state SnO_{2+x} is produced by this method and may react with organic contaminants. However, the overall oxidation of organic contaminants, as indicated in Eq. (2), proceeds more rapidly than in Eq. (4). It should be mentioned that the OER and the organic oxidation reaction are always in contest in the following ways:

$$SnO_2(^{\cdot}OH)_x \rightarrow SnO_{x+1} + 1/2O_2 + xH^+ + xe^- \tag{5}$$

$$SnO_{x+2} \rightarrow SnO_{1+x} \rightarrow 1/2O_2 \tag{6}$$

The reaction in Eq. (6) is suppressed at higher OEP, which facilitates interactions with organic contaminants.

SnO_2 also exhibits various advantages, including resistance to corrosion, adhesion to the base metal, crystallization in an insulating rutile-type structure, and resistance to chemical etching in acidic or alkaline conditions. H_2O and CO_2 were produced by oxidizing the organic contaminants on the SnO_2 electrode. A variety of metallic species put onto the surface of TiO_2 have shown excellent e^--h^+ separation, significant interfacial charge transfer ability, and photocatalysis initiated by visible light [26]. Thus, the production of materials based on graphene and SnO_2 for electrocatalytic or photocatalytic treatment is a viable promise. As a result, researchers suggest that combining TiO_2, GO, and SnO_2 could be a viable method of achieving exceptional chemical stability, light absorbability, and electron transferability. The metal–carbon/TiO_2 composite can also be employed to enhance noble metals' EO performance and efficient e^--h^+ separation characteristics. A significant amount of research has focused on creating tubular or nanolayer mixed metal oxide–GO composites. It has become essential to employ multistep manufacturing processes, including hydrothermal methods [65], dip-coating [58], chemical vapor deposition [53], and sol–gel methods [33]. These processes are labor-intensive, time-consuming, and frequently require high-temperature and high-pressure processing, as well as the right reaction chamber [11]. Additionally, several researchers created mixed metal oxides-GO nanocomposites using electrochemical techniques, which had straightforward fabrication conditions and good electrochemical catalytic activity [11, 61]. For example, tubular TiO_2 films have been fabricated through Ti anodization combined with electrospinning, electrodeposition [11], anodic layer deposition, and pulse electrodeposition [61], followed by the deposition of SnO_2 or Sn.

4 Synthesis of Doped Metal Oxide-GO Nanocomposites

To synthesize the doped metal oxide-GO nanocomposite, a systematic approach was followed:

(a) **Preparation of GO Powder**
GO was synthesized using the improved Hummer's method [2]. Graphite powder was oxidized under controlled conditions in a solution of $KMnO_4$ and concentrated H_2SO_4. After cleaning with deionized water until the pH became neutral, the final product was dried to obtain GO powder.

(b) **Fabrication of Metal Oxide Nanocomposites**
Ti, renowned for its exceptional stability and cost-effectiveness, is widely employed as a base substrate for nanocomposites. In this study, a seamless Ti plate was used as the base material. The surface of the Ti plate was meticulously prepared by polishing with sandpaper of varying grit sizes, followed by cleaning with acetone and distilled water, and ultrasonic treatment to ensure a pristine surface. Subsequently, the cleaned Ti surface underwent chemical etching using a solution of $HF:HNO_3:H_2O$ in a 1:4:10 (v/v/v) ratio, which facilitated the formation of a porous structure. To fabricate TiO_2NTs, anodization was performed in a 5% HF electrolyte solution at 20 V for 25 min, producing the desired nanostructure.

(c) **Incorporation of Doping Agent**
A dopant (e.g., SnO_2) was incorporated onto the metal oxide nanostructure using electrodeposition. A solution of the dopant precursor (e.g., $SnCl_2$ in deionized water) was prepared, and deposition was performed under controlled voltage (e.g., 15 V for 10 min).

(d) **Integration of GO and Metal Oxide**
GO was dispersed in deionized water (e.g., 0.5 g in 200 ml) using ultrasonication to ensure uniform dispersion. The GO solution was electrodeposited onto the doped metal oxide surface via anodization (e.g., 15 V for 15 min).

(e) **Calcination**
The synthesized nanocomposite was calcined at a specific temperature (e.g., 300–500 °C) in a muffle furnace to enhance structural integrity, crystallinity, and adhesion between the GO and doped metal oxide.

These steps results in a robust and efficient doped metal oxide-GO nanocomposite, suitable for various applications such as photocatalysis, wastewater treatment, and electrochemical processes. The unique characteristics of the doped metal oxide-GO nanocomposite enable it to achieve superior performance compared to individual components, making it a promising material for environmental and energy-related technologies.

5 Applications of Doped Metal Oxide-GO Nanocomposites in Wastewater Purification

Doped metal oxide-GO nanocomposites have emerged as promising candidates for effective wastewater remediation, particularly in the removal of toxic heavy metals. These nanocomposites combine the high surface area, adsorption capacity, and chemical stability of GO with the photocatalytic and electrochemical properties of doped metal oxides, creating a synergistic effect that enhances their overall performance.

5.1 Role in Heavy Metals Removal

Heavy metals, such as Cu(II), Cd(II), Pb(II), Ni(II), and Cr(VI), are persistent and toxic pollutants in industrial wastewater, primarily due to their non-biodegradable nature. These metals do not degrade naturally and can accumulate in the environment, leading to severe health and environmental risks. Their bioaccumulation in aquatic ecosystems poses a significant threat to both human health and ecological systems. Long-term exposure to these metals can result in various health problems, including organ damage, neurological disorders, and carcinogenic effects in humans [28]. Ecologically, these metals disrupt aquatic life and can lead to the bioaccumulation of toxins in the food chain, making their removal a critical environmental priority. The removal of heavy metals from wastewater has become essential, and doped metal oxide-GO nanocomposites have emerged as promising candidates for this task. The combination of GO and metal oxides such as TiO_2, SnO_2, and others provides a synergistic effect that enhances the removal of heavy metals through multiple processes.

These processes include:

(a) **Adsorption**

One of the most common and effective mechanisms for heavy metal removal is adsorption, where contaminants are attracted and bound to the surface of a material. The oxygen-containing functional groups (such as hydroxyl, epoxy, and carboxyl groups) present on the surface of GO play a crucial role in this process [43]. These functional groups provide active binding sites that can specifically interact with and adsorb metal ions from aqueous solutions. The large surface area of GO further enhances its adsorption capacity, making it highly efficient at capturing and retaining heavy metal ions, such as Cu(II), Cd(II), and Pb(II). The adsorbed metal ions are then removed from the wastewater, significantly reducing their concentration and toxicity [9, 64]. According to Agustiono et al. [1], heavy metals such as Cr(III), Cu(II), Cd(II), Cr(VI), Ni(II), and Zn(II) were removed from electroplating wastewater using a variety of low-cost adsorbents made from natural materials, agricultural residues or industrial by-products.

(b) **Photocatalysis**

Photocatalysis is another highly effective mechanism employed by doped metal oxide-GO nanocomposites, particularly under UV or visible light [21]. Doped metal oxides like TiO_2 and SnO_2 have intrinsic photocatalytic properties that are enhanced when combined with GO. When exposed to light, these materials generate electron-hole pairs that can facilitate the breakdown of pollutants in wastewater, including both metal complexes and organic contaminants. The photocatalytic activity enables the degradation of toxic compounds into less harmful by-products, improving the overall quality of the water. The synergy between GO and metal oxides further boosts photocatalytic efficiency by preventing charge carrier recombination, thereby enhancing the removal of pollutants associated with heavy metals. Cr(VI), Hg(II), and Pb(II) are currently the focus of additional study on photocatalytic remediation of heavy metal contaminants in effluent [18, 32]. Various attempts were made by Zhao et al. [67] to degrade Cr(VI) to less harmful and immobile Cr(III) by photocatalytic degradation. For this purpose TiO_2 is used as a photocatalyst along with neodymium, in presence of UV radiation. As a result, the Cr(VI) gets adsorbed on the TiO_2 surface and reduced to Cr(III) [51, 52] used TiO_2 as a photocatalysis to degrade Cu-EDTA and found that about 80–100% of Cu-EDTA degradation and 50–80% of Cu recovery was obtained within a 60-min operation.

(c) **Reduction of Heavy Metal Ions**

The reduction of metal ions is another key mechanism through which doped metal oxide-GO nanocomposites contribute to heavy metal removal. GO, with its excellent electron conductivity, plays a critical role in facilitating the reduction of toxic heavy metal ions like Cr(VI) to their less toxic forms, such as Cr(III) [44, 46]. This transformation is catalyzed by the metal oxide component of the nanocomposite, which acts as a catalyst to drive the redox reaction. The ability to convert Cr(VI) (a known carcinogen) into Cr(III) (a much less harmful form) significantly reduces the toxicity of the wastewater and facilitates the subsequent removal of the metal in its safer state.

(d) **Electrochemical Removal**

Electrochemical removal is a highly efficient method for heavy metal removal that utilizes the electrochemical activity of doped metal oxide-GO nanocomposites. GO's high conductivity, combined with the electrochemical properties of metal oxides, makes these nanocomposites ideal candidates for use in electrochemical setups. In these systems, an applied potential drives electrochemical reactions that facilitate the removal of heavy metals from wastewater. This method is particularly useful for removing metals like Cu(II), Pb(II), and Cr(VI), where electrochemical reduction or oxidation reactions are involved [47]. Electrochemical removal also offers the advantage of regenerating the nanocomposite material for multiple cycles of use, improving the sustainability of the process.

5.2 *Role in Wastewater Purification*

The application of doped metal oxide-GO nanocomposites in wastewater purification is not limited to the efficient removal of heavy metals; these advanced materials are also highly effective in addressing a variety of other contaminants. Their ability to treat organic pollutants, synthetic dyes, pathogens, and other hazardous substances makes them exceptionally versatile for industrial and environmental remediation.

5.2.1 Advanced Purification Capabilities

(a) **Organic Contaminants**

The aquatic environment has become progressively contaminated due to global industrialization and urbanization [62]. Industrial wastewater pollution is a global environmental issue due to the complex composition and other challenging aspects of wastewater produced by numerous industries, including food [36], pesticides [14], tannery [10, 60], electroplating [31, 48, 55], dairy [40], textile [20], pharmaceutical [13, 21, 25], pulp, and paper [56]. The primary constituents of wastewater released by these industries are dyes [30], toxic organic molecules [66], inorganic salts, volatile organic compounds, heavy metals [37, 54], cyanide [27], petroleum constituents, and pharmaceuticals. Due to growing environmental concerns, stringent regulations have been enforced, making it mandatory to treat wastewater before discharging it into natural water sources [34]. The high photocatalytic activity of doped metal oxides, particularly under UV or visible light, plays a significant role in degrading persistent organic pollutants (POPs) [42], such as pesticides, pharmaceuticals, and hydrocarbons. POPs are resistant to natural degradation, posing a critical environmental challenge. When doped metal oxide-GO nanocomposites are exposed to light, they generate reactive oxygen species (ROS) that help break down these harmful pollutants. The synergy between metal oxides like TiO_2 or SnO_2 and GO enhances the separation of e^--h^+ pairs, preventing their recombination and ensuring a more efficient degradation process. This enhanced charge separation improves the photocatalytic efficiency, enabling the transformation of complex organic pollutants into non-toxic end products, such as CO_2 and H_2O, thus contributing to cleaner wastewater and a healthier environment. The ability of these nanocomposites to address a wide range of organic contaminants makes them an invaluable tool in advanced wastewater treatment.

(b) **Dye Degradation**

Industrial dyes, widely used in textiles, plastics, and other sectors, are known for their stability and toxicity, making them challenging to remove from wastewater. The doped metal oxide-GO nanocomposites effectively adsorb and break down these dyes due to the combination of GO's large surface area and the catalytic properties of doped metal oxides [23]. GO's surface provides an abundance of active sites for dye molecules to attach, while the metal oxides enhance the catalytic breakdown of dye molecules into harmless by-products, such as CO_2 and H_2O. The photocatalytic degradation process is particularly effective for the mineralization of complex dye molecules, which are often recalcitrant and difficult to treat by conventional methods. By breaking down these toxic compounds, doped metal oxide-GO nanocomposites help mitigate environmental pollution caused by the textile and dye industries, which are significant contributors to water contamination.

(c) **Pathogen Elimination**

Doped metal oxide-GO nanocomposites exhibit remarkable antimicrobial properties, which are especially important for treating water contaminated with pathogens such as bacteria, viruses, and other microorganisms. The antimicrobial action is primarily driven by ROS generated during photocatalysis. These ROS, including $^{\cdot}OH$ radicals, superoxide anions ($O_2^{\cdot-}$), and hydrogen peroxide (H_2O_2), have strong oxidative capabilities that can damage the cell walls of bacteria and disrupt the cellular functions of both bacteria and viruses. The interaction between ROS and microbial cells leads to oxidative stress, which ultimately results in cell death or inactivation of the pathogens. This makes doped metal oxide-GO nanocomposites an effective tool for pathogen-free water purification. Moreover, the ability of these nanocomposites to eliminate pathogens without the need for harmful chemicals ensures a more sustainable and environmentally friendly approach to disinfection in water treatment applications.

5.2.2 Multifunctionality in Sustainable Wastewater Treatment

(a) **Adsorption and Catalysis**

The oxygen-functional groups (such as hydroxyl, epoxy, and carboxyl groups) present on GO serve as high-affinity adsorption sites, making them highly effective in capturing a wide spectrum of contaminants, including heavy metals, organic pollutants, and dyes. These functional groups, due to their strong affinity for various contaminants, provide localized binding sites, thus enhancing the adsorption capacity of GO. Additionally, the large surface area of GO ensures a higher adsorption efficiency by providing more active sites for the contaminants, facilitating the removal of pollutants from wastewater in an effective and rapid manner. In conjunction with adsorption, the integration of doped metal oxides such as TiO_2 or SnO_2 in the composite material significantly enhances the photocatalytic and redox capabilities. Upon exposure to UV or visible light, the doped metal oxides undergo photoexcitation, generating ROS, including $^{\cdot}OH$ radicals and $O_2^{\cdot-}$ ions. These ROS play a pivotal role in degrading organic pollutants, such as pesticides, pharmaceuticals, and hydrocarbons, into non-toxic by-products. The synergy between the high adsorption capacity of GO and the catalytic properties of the doped metal oxides creates a dual-action purification mechanism. Initially, GO adsorbs the pollutants, concentrating them on its surface. Then, upon exposure to light, the metal oxides catalyze the degradation of these contaminants through photocatalysis and redox reactions. This combined approach ensures a comprehensive and efficient removal of a wide variety of pollutants, resulting in a more thorough purification process compared to traditional methods. Thus, the interaction between GO and doped metal oxides not only improves the removal efficiency but also extends the scope of contaminants that can be effectively treated.

(b) **Electrochemical Activity**

GO is known for its exceptional conductivity, which, when combined with the electrochemical properties of doped metal oxides, significantly enhances the performance of these nanocomposites in electrochemical treatment applications. GO's high conductivity allows for efficient electron transfer, which is crucial for electrochemical processes such as ion removal and redox reactions. When used in electrochemical setups, these nanocomposites can effectively remove dissolved ions, such as heavy metals (e.g., Cr(VI), Cu(II), Pb(II)), from wastewater by applying an electrical potential that drives the movement of ions toward the electrode surface. Additionally, the electrochemical reactions catalyzed by the metal oxides further contribute to the degradation of persistent contaminants. This electrochemical approach is highly efficient, energy-saving, and versatile, as it can treat a wide variety of pollutants without the need for toxic chemicals, making it an ideal solution for sustainable wastewater treatment.

(c) **Environmental Resilience**

Doped metal oxide-GO nanocomposites exhibit remarkable structural stability and reusability, even under harsh environmental conditions. This resilience makes them suitable for large-scale applications in industrial wastewater treatment. These nanocomposites maintain their effectiveness across multiple cycles of pollutant removal, showing minimal loss in performance over time. The stability of GO is enhanced by the metal oxide components, which prevent degradation or leaching during extended use. Additionally, the composites perform well under a range of pH levels, temperatures, and in the presence of various interfering substances commonly found in industrial wastewater. Their ability to retain high performance and catalytic activity over extended periods and under diverse conditions makes them a cost-effective and sustainable solution for large-scale environmental remediation projects. Moreover, their reusability reduces the need for frequent replacements, further lowering operational costs and making them a more economically viable option in long-term wastewater treatment processes.

By integrating properties like high adsorption capacity, enhanced photocatalytic performance, and robust electrochemical reactivity, doped metal oxide-GO nanocomposites play a pivotal role in achieving the goals of sustainable wastewater management. They hold immense potential for replacing conventional treatment technologies, reducing energy consumption, and ensuring the availability of clean water resources. This makes them indispensable in addressing global water scarcity and pollution challenges.

6 Mechanism of Heavy Metals Removal

AOPs include the electrochemical method [38], which uses specific cathodes and anodes and applies an electric field between them to degrade the pollutants. A simple schematic representation of the electrochemical oxidation method is shown in Fig. 2a, which includes the direct and indirect oxidation methods, and Fig. 2b, which shows a potential mechanism for eliminating heavy metals from industrial effluents. Both the cathode and the anode are used in the removal system. Anodic oxidation and cathodic reduction were both involved in the removal reaction because of the presence of both $^{\cdot}OH$ and $H^{\cdot}$. However, either the anode directly or in situ-generated $^{\cdot}OH$ can degrade heavy metal complexes in the anodic region. The anodic oxidation process consists of the following steps: (i) Direct transmission of electrons to anode surface M; (ii) OH^{-} or water in wastewater might produce strong physiosorbed $^{\cdot}OH$ at the anode surface, represented by M($^{\cdot}OH$), using Eq. 7. Equations 8 and 9 show that ROS are oxygen intermediates of water oxidation, while bulk ions electrochemically form other weaker oxidant agents [46].

$$M + H_2O \rightarrow M(^{\cdot}OH) + H^{+} + e^{-} \tag{7}$$

$$2M(^{\cdot}OH) \rightarrow 2MO + H_2O_2 \tag{8}$$

$$3H_2O \rightarrow O_3 + 6H^{+} + 6e^{-} \tag{9}$$

Depending on the type of anode, the interaction between M($^{\cdot}OH$) and the anode surface also produces different effects. For example, in the case of an active anode with low $O_2{}^{-}$ overpotentials, Eq. 10 converts M($^{\cdot}OH$) into MO superoxide, and M serves as a selective intermediate in organic oxidation. (ii) When an anode is not active and has low $O_2{}^{-}$ overpotentials, M($^{\cdot}OH$) is so weakly physisorbed on the surface of doped metal oxide-GO nanomaterial that it can react with organic molecules and cause them to mineralize. A greater potential for O_2 development of the anode material is linked to lower M($^{\cdot}OH$) contact with the doped metal oxide-GO and higher chemical reactivity.

$$M(^{\cdot}OH) + L \rightarrow M + LO + H^{+} + e^{-} \tag{10}$$

As a result, heavy metal complexes are broken down by the powerful oxidant $^{\cdot}OH$ Eq. 11, and free heavy metal ions are deposited on the surface of cathode (Fig. 2).

$$^{\cdot}OH + CA \rightarrow \text{Simple organics} \tag{11}$$

$$\text{Metal} - CA + ^{\cdot}OH/H^{+} + M^{X+} \rightarrow \text{Oxidation products} + nCO_2 \tag{12}$$

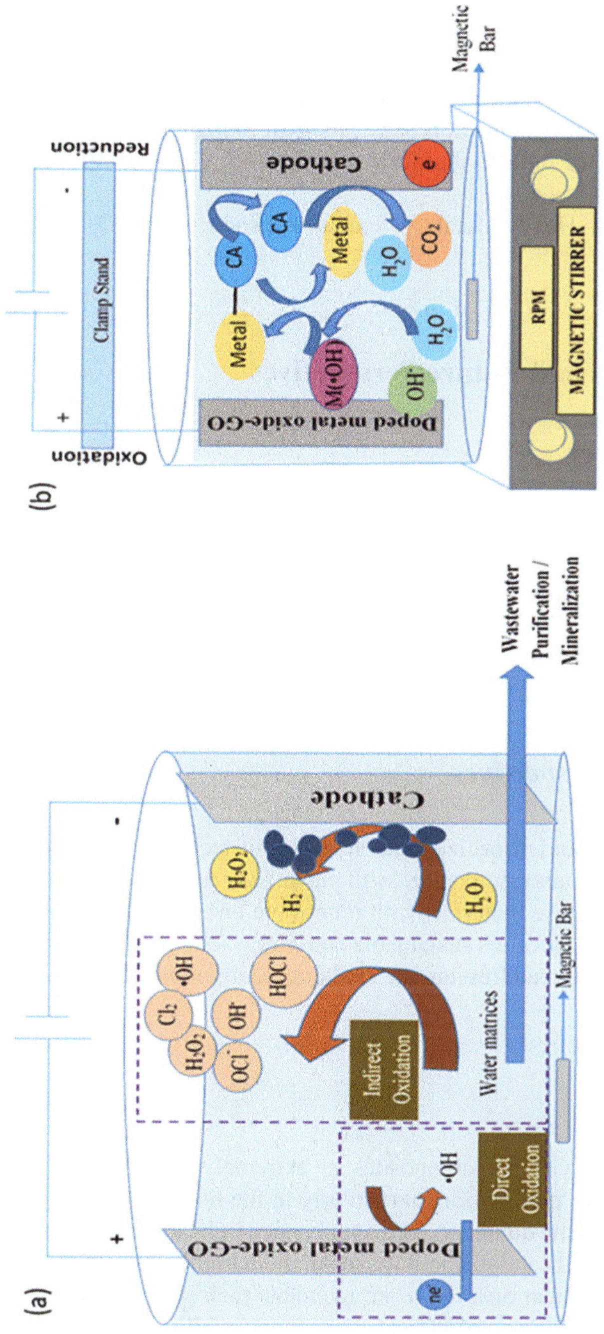

Fig. 2 **a** Schematic representation of the EO process, **b** possible mechanism for the removal of heavy metal complexes from industrial wastewater [46]

The oxidation of organic contaminants in an AOP produces mineral acid, H_2O, and CO_2. Because of its greater removal effectiveness, adaptability, and versatility, the AOPs have attracted a lot of attention [15]. An AOPs provide a straightforward and uncontaminated method of producing $^{\cdot}OH$ radicals electrochemically without the need for further chemical reagents or secondary pollutants. Using $^{\cdot}OH$ generated during the electrochemical process, indirect oxidation was used to rapidly and effectively purify the industrial effluent (Eq. 13) [35, 47].

$$\mathrm{M}\left[{}^{\cdot}\mathrm{OH}\right]_{\mathrm{ads}} + \text{Heavy Metals} \rightarrow \text{oxidation products} + n\mathrm{CO_2} + y\mathrm{H_2O} \quad (13)$$

7 Challenges and Future Perspectives

7.1 Challenges

The major challenges in utilizing nanocomposites for wastewater treatment include scalability issues, as lab-scale synthesis methods are resource-intensive and difficult to industrialize. Stability and recyclability concerns, such as structural degradation and component leaching, hinder long-term applications. Environmental impact and economic feasibility further complicate their large-scale adoption.

7.2 Future Perspectives

Future research should prioritize scalable synthesis techniques and the development of multifunctional nanocomposites with enhanced reactivity, selectivity, and durability. Integrating these materials with renewable energy systems, like solar-driven purification processes, offers sustainable and energy-efficient solutions. These future perspectives, aimed at addressing the challenges, are outlined in Fig. 3.

8 Conclusion

Doped metal oxide-GO nanocomposites are extremely promising as advanced materials for wastewater purification, particularly in the removal of heavy metals. Their remarkable synergistic qualities, such as enhanced light absorption, increased charge transfer kinetics, and catalytic activity, make them highly useful for water treatment and sustainable applications. However, to enable their industrial adoption, practical issues such as environmental effects, long-term stability, and scalability must be

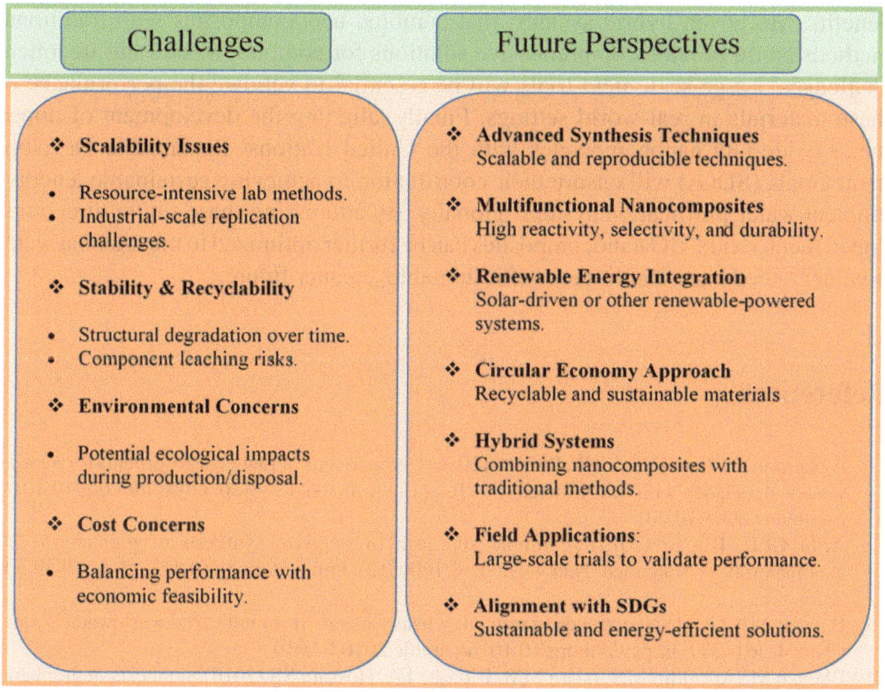

Fig. 3 Challenges and future perspectives of nanocomposites in wastewater treatment

resolved. To maximize their performance, a greater understanding of the underlying mechanisms, functional characteristics, and fabrication methods is necessary. Additionally, their capacity to transform dangerous pollutants into safe by-products aids in the development of environmentally friendly and sustainable water treatment methods. This chapter outlines the vital role that nanocomposites play in improving water purification systems and provides a foundation for further studies to improve performance and optimize material design. By addressing the pressing challenges of water contamination, doped metal oxide-GO nanocomposites pave the way for innovative solutions that align with global goals for sustainable development and environmental conservation.

Looking ahead, several advancements could further enhance the performance and applicability of doped metal oxide-GO nanocomposites in wastewater treatment. Future research should focus on developing advanced, scalable, and reproducible synthesis techniques to enable large-scale production and consistent material quality. Creating multifunctional nanocomposites with high reactivity, selectivity, and durability will improve efficiency in targeting a broader range of contaminants. Additionally, integrating renewable energy sources, such as solar-powered systems, into wastewater treatment processes will help reduce energy consumption and improve sustainability. The adoption of a circular economy approach, where recyclable and sustainable materials are used, will ensure long-term environmental benefits. Moreover, hybrid systems that combine nanocomposites with traditional methods could provide comprehensive solutions for complex wastewater treatment challenges. Large-scale field trials will be essential to validate the performance of these materials in real-world settings. Finally, aligning the development of doped metal oxide-GO nanocomposites with the United Nations' Sustainable Development Goals (SDGs) will ensure their contribution to achieving sustainable, energy-efficient water purification systems globally. By addressing these future directions, doped metal oxide-GO nanocomposites can be further optimized to meet global water treatment needs and contribute to a sustainable, cleaner future.

References

1. Agustiono T, Chan GYS, Lo W, Babel S (2006) Comparisons of low-cost adsorbents for treating wastewaters laden with heavy metals. Sci Total Environ 366:409–426. https://doi.org/10.1016/j.scitotenv.2005.10.001
2. Arthi GPB, BD L, (2015) A simple approach to stepwise synthesis of graphene oxide nanomaterial. J Nanomed Nanotechnol 6:1000253. https://doi.org/10.4172/2157-7439.1000253
3. Barakat MA (2011) New trends in removing heavy metals from industrial wastewater. Arab J Chem 4:361–377. https://doi.org/10.1016/j.arabjc.2010.07.019
4. Bhuyan MSA, Uddin MN, Islam MM, Bipasha FA, Hossain SS (2016) Synthesis of graphene. Int Nano Lett 6:65–83. https://doi.org/10.1007/s40089-015-0176-1
5. Bondarenko EA, Mazanik AV, Streltsov EA, Kulak AI, Korolik OV (2015) SnO_2/reduced graphene oxide composite films for electrochemical applications. Mater Sci Eng, B 202:61–67. https://doi.org/10.1016/j.mseb.2015.10.002
6. Bravo-Yumi N, Pacheco-Álvarez M, Bandala ER, Brillas E, Peralta-Hernández JM (2022) Studying the influence of different parameters on the electrochemical oxidation of tannery dyes using a Ti/IrO_2-SnO_2-Sb_2O_5 anode. Chem Eng Process Process Intensification 181:109173. https://doi.org/10.1016/j.cep.2022.109173
7. Cao X, Wu K, Wang C, Guo Y, Lu R, Wang X, Chen S (2022) Graphene oxide loaded on TiO_2-nanotube-modified Ti regulates the behavior of human gingival fibroblasts. Int J Mol Sci 23:8723. https://doi.org/10.3390/ijms23158723
8. Carolin CF, Kumar PS, Saravanan A, Joshiba GJ, Naushad M (2017) Efficient techniques for the removal of toxic heavy metals from aquatic environment: a review. J Environ Chem Eng 5:2782–2799. https://doi.org/10.1016/j.jece.2017.05.029
9. Chen Q, Yao Y, Li X, Lu J, Zhou J, Huang Z (2018) Comparison of heavy metal removals from aqueous solutions by chemical precipitation and characteristics of precipitates. J Water Process Eng 26:289–300. https://doi.org/10.1016/j.jwpe.2018.11.003
10. Chen X, Ke XX, Liu Y, Weerasooriya R, Li H, Wu YC (2021) Photocatalytically induced Au/mpg-C_3N_4 nanocomposites for robust electrochemical detection of Cr(VI) in tannery wastewater. J Environ Chem Eng 9:104642. https://doi.org/10.1016/j.jece.2020.104642
11. Chen X, Kure-Chu SZ, Matsubara T, Hihara T, Yashiro H, Okido M (2022) Fabrication and electrochemical performance of TiO_2–TiN/Sn–SnO_2 composite films on Ti for LIB anodes with high capacity and excellent conductivity. Adv Mater Interfaces 9:2200956. https://doi.org/10.1002/admi.202200956
12. Fu F, Wang Q (2011) Removal of heavy metal ions from wastewaters: a review. J Environ Manage 92:407–418. https://doi.org/10.1016/j.jenvman.2010.11.011
13. Gong C, Han J, He C, Shi L, Shan Y, Zhang Z, Wang L, Ren X (2023) Insights into degradation of pharmaceutical pollutant atenolol via electrochemical advanced oxidation processes

with modified Ti_4O_7 electrode: efficiency, stability and mechanism. Environ Res 228:115920. https://doi.org/10.1016/j.envres.2023.115920
14. Guo D, Guo Y, Huang Y, Chen Y, Dong X, Chen H, Li S (2021) Preparation and electrochemical treatment application of Ti/Sb–SnO_2-Eu&rGO electrode in the degradation of clothianidin wastewater. Chemosphere 265:129126. https://doi.org/10.1016/j.chemosphere.2020.129126
15. Guo H, Hu W, Xu Z, Guo S, Qiao D, Wang X, Xu H, Yan W (2022) How to improve lead dioxide anodes performance in organic wastewater treatment: review and prospect. Process Saf Environ Prot 164:189–207. https://doi.org/10.1016/j.psep.2022.06.012
16. Hama Aziz KH, Mustafa FS, Omer KM, Hama S, Hamarawf RF, Rahman KO (2023) Heavy metal pollution in the aquatic environment: efficient and low-cost removal approaches to eliminate their toxicity: a review. RSC Adv 13:17595–17610. https://doi.org/10.1039/d3ra00723e
17. Jubu PR, Chahrour KM, Yam FK, Awoji OM, Yusof Y, Bee Choo E (2022) Titanium oxide nanotube film decorated with β-Ga_2O_3 nanoparticles for enhanced water splitting properties. Sol Energy 235:152–162. https://doi.org/10.1016/j.solener.2022.02.033
18. Kabra K, Chaudhary R, Sawhney RL (2008) Solar photocatalytic removal of Cu(II), Ni(II), Zn(II) and Pb(II): speciation modeling of metal-citric acid complexes. J Hazard Mater 155:424–432. https://doi.org/10.1016/j.jhazmat.2007.11.083
19. Karim AV, Selvaraj A (2021) Graphene composites in photocatalytic oxidation of aqueous organic contaminants—a state of art. Process Saf Environ Prot 146:136–160. https://doi.org/10.1016/j.psep.2020.08.042
20. Kaur P, Kushwaha JP, Sangal VK (2018) Transformation products and degradation pathway of textile industry wastewater pollutants in electro-Fenton process. Chemosphere 207:690–698. https://doi.org/10.1016/j.chemosphere.2018.05.114
21. Kaur P, Park Y, Sillanpaa M, Imteaz MA (2021) Synthesis of a novel SnO_2/graphene-like carbon/TiO_2 electrodes for the degradation of recalcitrant emergent pharmaceutical pollutants in a photo-electrocatalytic system. J Clean Prod 313:127915. https://doi.org/10.1016/j.jclepro.2021.127915
22. Kolluru SS, Agarwal S, Sireesha S, Sreedhar I, Kale SR (2021) Heavy metal removal from wastewater using nanomaterials-process and engineering aspects. Process Saf Environ Prot 150:323–355. https://doi.org/10.1016/j.psep.2021.04.025
23. Kong EDH, Chau JHF, Lai CW, Khe CS, Sharma G, Kumar A, Siengchin S, Sanjay MR (2022) GO/TiO_2-related nanocomposites as photocatalysts for pollutant removal in wastewater treatment. Nanomaterials 12:3536. https://doi.org/10.3390/nano12193536
24. Krishna RH, Chandraprabha MN, Samrat K, Krishna Murthy TP, Manjunatha C, Kumar SG (2023) Carbon nanotubes and graphene-based materials for adsorptive removal of metal ions—a review on surface functionalization and related adsorption mechanism. Appl Surf Sci Adv 16:100431. https://doi.org/10.1016/j.apsadv.2023.100431
25. Krishnan RY, Manikandan S, Subbaiya R, Biruntha M, Govarthanan M, Karmegam N (2021) Removal of emerging micropollutants originating from pharmaceuticals and personal care products (PPCPs) in water and wastewater by advanced oxidation processes: a review. Environ Technol Innov 23:101757. https://doi.org/10.1016/j.eti.2021.101757
26. Li D, Tang J, Zhou X, Li J, Sun X, Shen J, Wang L, Han W (2016) Electrochemical degradation of pyridine by Ti/SnO_2-Sb tubular porous electrode. Chemosphere 149:49–56. https://doi.org/10.1016/j.chemosphere.2016.01.078
27. Li M, Li B, Chen J, Cui S, Yang Y, Shen X, Liu K, Han Q (2022) Purifying cyanide-bearing wastewaters by electrochemical precipitate process using sacrificial Zn anode. Sep Purif Technol 284:120250. https://doi.org/10.1016/j.seppur.2021.120250
28. Lim JY, Mubarak NM, Abdullah EC, Nizamuddin S, Khalid M, Inamuddin (2018) Recent trends in the synthesis of graphene and graphene oxide based nanomaterials for removal of heavy metals—a review. J Ind Eng Chem 66:29–44. https://doi.org/10.1016/j.jiec.2018.05.028
29. Liu J, Wang L, Tang J, Ma J (2016) Photocatalytic degradation of commercially sourced naphthenic acids by TiO_2-graphene composite nanomaterial. Chemosphere 149:328–335. https://doi.org/10.1016/j.chemosphere.2016.01.074

30. Liu X, Chen Z, Du W, Liu P, Zhang L, Shi F (2022) Treatment of wastewater containing methyl orange dye by fluidized three dimensional electrochemical oxidation process integrated with chemical oxidation and adsorption. J Environ Manage 311:114775. https://doi.org/10.1016/j.jenvman.2022.114775
31. Liu Y, Liu G, Wang H, Wu P, Yan Q, Vayenas DV (2021) Elongation the duration of steel anode with polypyrrole modification during the electrocoagulation treatment process of electroplating wastewater. J Environ Chem Eng 9:104969. https://doi.org/10.1016/j.jece.2020.104969
32. Luo S, Qin F, Ming Y, Zhao H, Liu Y, Chen R (2017) Fabrication uniform hollow Bi_2S_3 nanospheres via Kirkendall effect for photocatalytic reduction of Cr(VI) in electroplating industry wastewater. J Hazard Mater 340:253–262. https://doi.org/10.1016/j.jhazmat.2017.06.044
33. Maver K, Arčon I, Fanetti M, Al Jitan S, Palmisano G, Valant M, Lavrenčič Štangar U (2022) Improved photocatalytic activity of SnO_2-TiO_2 nanocomposite thin films prepared by low-temperature sol-gel method. Catal Today 397–399:540–549. https://doi.org/10.1016/j.cattod.2021.06.018
34. Mei R, Zhu C, Wei Q, Ma L, Li W, Zhou B, Deng Z, Tong Z, Ouyang G, Jiang C (2018) The dependence of oxidation parameters and dyes' molecular structures on microstructure of boron-doped diamond in electrochemical oxidation process of dye wastewater. J Electrochem Soc 165:H324–H332. https://doi.org/10.1149/2.0221807jes
35. Moradi M, Vasseghian Y, Khataee A, Kobya M, Arabzade H, Dragoi EN (2020) Service life and stability of electrodes applied in electrochemical advanced oxidation processes: a comprehensive review. J Ind Eng Chem 87:18–39. https://doi.org/10.1016/j.jiec.2020.03.038
36. Nayak A, Bhushan B (2019) An overview of the recent trends on the waste valorization techniques for food wastes. J Environ Manage 233:352–370. https://doi.org/10.1016/j.jenvman.2018.12.041
37. Nazaripour M, Reshadi MAM, Mirbagheri SA, Nazaripour M, Bazargan A (2021) Research trends of heavy metal removal from aqueous environments. J Environ Manage 287:112322. https://doi.org/10.1016/j.jenvman.2021.112322
38. Öztürk H, Barışçı S, Turkay O (2021) Paracetamol degradation and kinetics by advanced oxidation processes (Aops): electro-peroxone, ozonation, goethite catalyzed electro-fenton and electro-oxidation. Environ Eng Res 26:180332. https://doi.org/10.4491/eer.2018.332
39. Pahlevani L, Mozdianfard MR, Fallah N (2020) Electrochemical oxidation treatment of offshore produced water using modified Ti/Sb-SnO_2 anode by graphene oxide. J Water Process Eng 35:101204. https://doi.org/10.1016/j.jwpe.2020.101204
40. Parihar RK, Bhandari K, Burnwal PK, Ghosh S, Chaurasia SP, Midda MO (2024) Advancing dairy wastewater treatment: exploring two-stage fluidized bed anaerobic membrane bioreactor for enhanced performance, fouling, and microbial community analysis. J Water Process Eng 58:104917. https://doi.org/10.1016/j.jwpe.2024.104917
41. Parihar RK, Ghosh S, Chaurasia SP, Midda MO (2024) Anaerobic treatment of dairy wastewater: impact of substrate-to-inoculum ratio on biomethane production. J Environ Eng 150:04023090. https://doi.org/10.1061/joeedu.eeeng-7396
42. Rai D, Sinha S (2024) Characterization and electrochemical properties of TiO_2-rNTs/SnO_2–Sb/PbO_2 electrodes for the mineralization of persistent organic pollutants using anodic oxidation coupled electro-Fenton treatment: effect of precursor selection. Chemosphere 352:141307. https://doi.org/10.1016/j.chemosphere.2024.141307
43. Rajivgandhi G, RTV V, Raju N, Sevanan M, Alharbi NS, Kadaikunnan S, Khaled JM, Alanzi KF, Li W-J (2021) Adsorption of nickel ions from electroplating effluent by graphene oxide and reduced graphene oxide. Environ Res 199:111322. https://doi.org/10.1016/j.envres.2021.111322
44. Rajoria S, Vashishtha M, Sangal VK (2023) Preparation and characterization of novel SnO_2-GO/TiO_2NTs electrode for the degradation of electroplating effluents. J Environ Chem Eng 11:110679. https://doi.org/10.1016/j.jece.2023.110679
45. Rajoria S, Vashishtha M, Sangal VK (2024) Oxidative degradation of electroplating wastewater by an electro-Fenton process using GO/TiO_2NTs electrode. Environ Eng Res 29:230056. https://doi.org/10.4491/eer.2023.056

46. Rajoria S, Vashishtha M, Sangal VK (2024) Electroplating wastewater treatment by electro-oxidation using synthesized new electrode: experimental, optimization, kinetics, and cost analysis. Process Saf Environ Prot 183:735–756. https://doi.org/10.1016/j.psep.2024.01.023
47. Rajoria S, Vashishtha M, Sangal VK (2023) Electrochemical treatment of electroplating wastewater using synthesized GO/TiO_2 nanotube electrode. Environ Sci Pollut Res 30:71226–71251. https://doi.org/10.1007/s11356-023-27366-7
48. Rajoria S, Vashishtha M, Sangal VK (2021) Review on the treatment of electroplating industry wastewater by electrochemical methods. Mater Today Proceed 47:1472–1479. https://doi.org/10.1016/j.matpr.2021.04.165
49. Rajput H, Changotra R, Kumar Sangal V, Dhir A (2021) Photoelectrocatalytic treatment of recalcitrant compounds and bleach stage pulp and paper mill effluent using Au-TiO_2 nanotube electrode. Chem Eng J 408:127287. https://doi.org/10.1016/j.cej.2020.127287
50. Rajput H, Dhir A, Sangal VK (2018) GO mediated TiO_2 nanotube electrode for the photoelectrocatalytic degradation of pentachlorophenol. J Electrochem Soc 165:H16–H26. https://doi.org/10.1149/2.0871802jes
51. Rengaraj S, Venkataraj S, Yeon JW, Kim Y, Li XZ, Pang GKH (2007) Preparation, characterization and application of Nd-TiO_2 photocatalyst for the reduction of Cr(VI) under UV light illumination. Appl Catal B 77:157–165. https://doi.org/10.1016/j.apcatb.2007.07.016
52. Rhoads KR, Davis AP (2004) Metal recovery and catalyst reuse from the photocatalytic oxidation of copper-ethylenediaminetetraacetic acid. J Environ Eng 130:425–431. https://doi.org/10.1061/(asce)0733-9372(2004)130:4(425)
53. Scott JI, Martinez-Gazoni RF, Allen MW, Reeves RJ (2019) Optical and electronic properties of high quality Sb-doped SnO_2 thin films grown by mist chemical vapor deposition. J Appl Phys 126:135702. https://doi.org/10.1063/1.5116719
54. Shrestha R, Ban S, Devkota S, Sharma S, Joshi R, Tiwari AP, Kim HY, Joshi MK (2021) Technological trends in heavy metals removal from industrial wastewater: a review. J Environ Chem Eng 9:105688. https://doi.org/10.1016/j.jece.2021.105688
55. Singh P, Itankar N, Patil Y (2021) Biomanagement of hexavalent chromium: current trends and promising perspectives. J Environ Manage 279:111547. https://doi.org/10.1016/j.jenvman.2020.111547
56. Soloman PA, Ahmed Basha C, Velan M, Balasubramanian N, Marimuthu P (2009) Augmentation of biodegradability of pulp and paper industry wastewater by electrochemical pre-treatment and optimization by RSM. Sep Purif Technol 69:109–117. https://doi.org/10.1016/j.seppur.2009.07.002
57. Song H, Zhang J, Sun Z, Sun T, Han Y, Meng H, Zhang X (2020) A novel hybrid electrode of zeolitic imidazolate framework–derived carbon encapsulated in reduced-TiO_2 nanotube arrays: Fabrication and photoelectrocatalytic activity. Appl Organomet Chem 34:1–12. https://doi.org/10.1002/aoc.5791
58. Sun Y, Cheng S, Yu Z, Li L, Li C, Yang J (2020) Elucidating deactivation mechanisms of Pd-doped and un-doped Ti/SnO_2-Sb electrodes. J Alloy Compd 834:155184. https://doi.org/10.1016/j.jallcom.2020.155184
59. Vardhan KH, Kumar PS, Panda RC (2019) A review on heavy metal pollution, toxicity and remedial measures: current trends and future perspectives. J Mol Liq 290:111197. https://doi.org/10.1016/j.molliq.2019.111197
60. Villaseñor-Basulto DL, Picos-Benítez A, Pacheco-Alvarez M, Pérez T, Bandala ER, Peralta-Hernández JM (2022) Tannery wastewater treatment using combined electrocoagulation and electro-Fenton processes. J Environ Chem Eng 10:107290. https://doi.org/10.1016/j.jece.2022.107290
61. Wang HP, Guan ZC, Shi HY, Wang X, Jin P, Song GL, Du RG (2021) Ag/SnO_2/TiO_2 nanotube composite film used in photocathodic protection for stainless steel. J Photochem Photobiol A 417:113353. https://doi.org/10.1016/j.jphotochem.2021.113353
62. Wang K, Wu J, Zhu M, Zheng YZ, Tao X (2020) Highly effective pH-universal removal of tetracycline hydrochloride antibiotics by UiO-66-$(COOH)_2$/GO metal–organic framework composites. J Solid State Chem 284:121200. https://doi.org/10.1016/j.jssc.2020.121200

63. Wu Y, Du X, Pan J, Zhou Q, Zhang J, Li M, Fu D (2023) Study on the efficiency improvement, photocatalytic kinetics and mechanism of trace GO on TiO_2 for removing Cr(VI) under visible light conditions. Research Square 1–28. https://doi.org/10.21203/rs.3.rs-3049370/v1
64. Yan FL, Wang Y, Wang WH, Zhao JX, Feng LL, Li JJ, Zhao JC (2020) Application of biochars obtained through the pyrolysis of Lemna minor in the treatment of Ni-electroplating wastewater. J Water Process Eng 37:101464. https://doi.org/10.1016/j.jwpe.2020.101464
65. Zamiri G, Haseeb ASMA, Jagadish P, Khalid M, Kong I, Krishnan SG (2022) Three-dimensional graphene-TiO_2-SnO_2 ternary nanocomposites for high-performance asymmetric supercapacitors. ACS Omega 7:43981–43991. https://doi.org/10.1021/acsomega.2c05343
66. Zhang Z, Xiao Q, Du X, Xue T, Yan Z, Liu Z, Zhang H, Qi T (2022) The fabrication of Ti_4O_7 particle composite modified PbO_2 coating electrode and its application in the electrochemical oxidation degradation of organic wastewater. J Alloy Compd 897:162742. https://doi.org/10.1016/j.jallcom.2021.162742
67. Zhao Z, An H, Lin J, Feng M, Murugadoss V, Ding T, Liu H, Shao Q, Mai X, Wang N, Gu H, Angaiah S, Guo Z (2019) Progress on the photocatalytic reduction removal of chromium contamination. Chem Rec 19:873–882. https://doi.org/10.1002/tcr.201800153

Historical Development of Photocatalyst and Its Application in Water Purification

Shivam Pandey, Manisha Lingwal, and Ritu Painuli

Abstract Despite showing significant promises in the last few decades for the removal and cleaning of wastewater from manufacturing and town waste sources, photocatalysis has yet to make the leap from research in laboratories to real-world, large-scale applications. A significant portion of this cold progress can be attributed to the inherent physicochemical features of the majority of photocatalytic materials, which have a tendency to have poorer photoefficiency when exposed to visible light, in addition to a variety of engineering obstacles. Water treatment with photocatalysts is a sustainable solution to water contamination. Photocatalysis activates a semiconductor material with visible or UV light to break down organic, viral, and inorganic impurities in water. High efficiency, low energy usage, and sustainability are benefits of this technique. Photocatalysts work under sunlight, making them suitable for dispersed water filtration systems. This chapter reviews the history and activity of the photocatalyst in water purification. The chapter also discusses various types of photocatalysts along with its mechanism. The efficiency of these photocatalytic materials is something that is being more investigated, and thus, it is necessary to have a grasp of the function that these materials play in photocatalytic processes in order to handle the challenges that are associated with wastewater treatment.

Keywords Photocatalysis · Photocatalytic activity · Ultraviolet (UV) · Ultrafiltration (UF) · Reverse osmosis (RO) · Volatile organic compound (VOC)

1 Introduction

The terms "catalysis" and "photo" are combined to form the term "photocatalysis." In general, "catalysis" refers to the process of altering a chemical reaction when a catalyst is present, whereas "photo" denotes light. It is referred to as a photocatalyst

S. Pandey (✉) · M. Lingwal · R. Painuli
Department of Chemistry, School of Applied and Life Sciences, Uttaranchal University, Dehradun, Uttarakhand 248007, India
e-mail: pandeyshivam547@gmail.com

A. Ahmad et al. (eds.), *Metal Oxide Based Nanophotocatalyst for Wastewater Purification*, Smart Nanomaterials Technology,
https://doi.org/10.1007/978-981-96-7683-5_14

when it absorbs photons and converts them into chemical energy [1]. When photocatalysts stimulate and absorb photons of a specific wavelength, an electron–hole pair forms on the catalyst surface, producing hydroxyl radicals and ultimately destroying contaminants. Photocatalysts break down harmful chemicals when exposed to UV radiation from the sun. Among the uses of photocatalysts are drug delivery, water and air purification, hydrocarbon production and generation, conversion of certain molecules, photocatalytic water splitting, and self-cleaning coatings [2].

Due to its shown ability to purify various pollutants, low cost, and great efficiency, photocatalysis is a desirable alternative. Through the stimulation of electron transfer in a semiconductor photocatalyst, photocatalysis can use solar energy to oxidize or decrease contaminants. The photocatalyst, which transforms solar energy into chemical energy and destroys the pollutants, is the most crucial component of this system. Three phases are usually involved in the entire photocatalytic process: charge creation, separation, and consumption. Photocatalysts have gained significant attention in the field of water treatment since the 1970s, when the first evidence of their ability to remove contaminants from water was discovered [3]. It is a promising technique for removing organic pollutants, heavy metals, pathogens, etc., in water or wastewater by utilizing various semiconductors or nanomaterials to harness light energy and induce chemical reactions for breaking down of the contaminants in the water.

For effective treatment of water, there are various methods such as filtration, sedimentation, chlorination, aerobic and anaerobic treatment, ozonation, etc., which can be used separately or in combination with other techniques. These approaches although possess various drawbacks, as well, particularly when it comes to dealing with more recent types of chemicals that are present in water, such as microplastics and everlasting chemicals. These conventional techniques also require sludge control procedures and are frequently costly. Moreover, it has been acknowledged and found that advanced oxidation processes (AOPs) are by far an encouraging approach for removing new pollutants from wastewater effluents. For purifying the wastewater, photocatalysis has been thoroughly investigated within the field of AOPs [4]. Moreover, the problems with different types of wastewater are generation of non-breakable contaminants which can be easily broken down by using the AOPs.

Severe water scarcity affects almost two-thirds of the world's population, with China and India being especially susceptible. The inability of conventional wastewater treatment techniques to efficiently eliminate persistent, harmful, and non-biodegradable organic contaminants emphasizes the need for more sophisticated water treatment technology. The directly creation of extremely reactive radicals, like hydroxyl radicals ($^{\cdot}OH$), which effectively break down organic pollutants into non-toxic molecules, is made possible by advanced oxidation processes (AOPs) [5]. Further development of photocatalysts is impeded by the fact that they can only be activated by UV light without further changes. Therefore, heteroatom doping, vacancy engineering, heterostructure manufacturing, and dye sensitization have been developed to improve the rate of photodegradation by improving the visible light response and boosting charge dynamics [6].

In recent times, a lot of studies have been concerned about the use of photocatalyst for the purification of water and wastewater and it is increasing year after year. A current study by Pandey et al. [7, 8] examined the efficacy of hydrothermally generated zinc oxide (ZnO) and stannous oxide (SnO_2) nanocatalysts using a novel method for the degradation of p-nitrophenol in water. Because of the harmful consequences p-nitrophenol has on the environment and human health, the study focuses on its photocatalytic breakdown. In another study by Chaudhary et al. [9], MnO_2 nanoparticles fabricated with *Withania somnifera* were utilized for breakdown of industrial dyes. Several other studies fabricated different nanoparticles, such as stannous oxide, copper oxide, bismuth oxide, zirconia oxide, etc., for degradation of contaminants in wastewater [7, 8, 10, 11].

2 Processing of Wastewater

The processes and methods used to treat wastewater which is a byproduct of commercial or industrial operations is referred to as industrial wastewater processing. This kind of wastewater must be treated before being returned to the environment [12]. Phase separation techniques like sedimentation are frequently used in wastewater treatment. Another example is chemical and biological processes like oxidation. Polishing is another instance. If anaerobic treatment is used in the process, biogas may be another byproduct. Reclaimed water can be made from treated wastewater. The primary goal of wastewater treatment is to guarantee that the treated water can be safely released or repurposed [13].

There are four stages involved in wastewater treatment, i.e., preliminary processing, initial sedimentation, biological processing, and advanced filtration which can be seen in Table 1.

The process of water purification can be schematically understood by Fig. 1 as well.

Table 1 Stages of water filtration

Preliminary processing	Initial sedimentation	Biological processing	Advanced filtration
Involves filtering of the sewage to remove large debris like sticks, rags, and non-biodegradable objects	Involves holding wastewater in tanks with particles sinking to the bottom and fats and oils rising to the top which removes the lighter materials and collects the settled solids	Involves controlled breakdown of organic material by aerobic bacteria	Involves elimination of nutrients and smaller waste particles, protecting vulnerable ecosystems and preparing the water for safer reuse or disposal

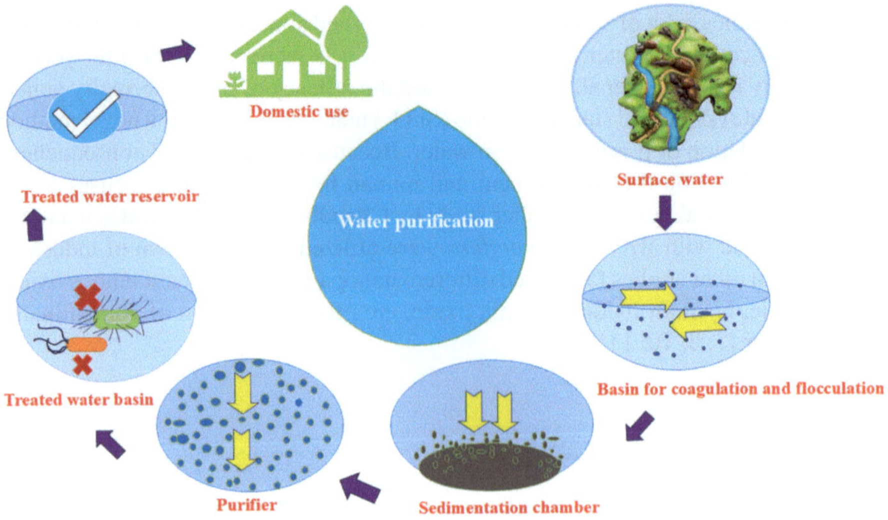

Fig. 1 Process of water purification

2.1 Challenges with Water Purification

Water purification methods ensure that contaminants are eliminated from the raw water. Water purification usually involves the use of physical, biological, and chemical techniques to remove contaminants from raw water and make it drinkable. Providing universal access to pure drinking water necessitates the application of water purification methods, but each method presents its own set of challenges.

Table 2 illustrates different methods and various challenges associated with them.

2.2 Photocatalyst in Water Purification

Heterogeneous photocatalysis has not yet made the leap from initial investigation to practical use despite being acknowledged as a viable method for treating and cleaning industrial and sanitary wastewater over the last generation. Numerous engineering issues and the intrinsic physicochemical characteristics of the majority of photocatalytic materials on the market, which show decreased photoefficiency in the presence of visible radiation, are to blame for this slow progress. Photocatalytic process kinetics, various kinds of solar-driven reactor, and clarifying the importance of the substances used in photocatalytic oxidation are also some factors that affect activity of treatment. Graphitic carbon nitride, zinc oxide, titania, and other commercially used materials for various reactor systems are all methodically examined for effectiveness in this broad evaluation [17].

Table 2 Various methods and challenges

Methods	Challenges
Ultrafiltration (UF) Particles, colloids, and macromolecules (such as proteins) are eliminated from water through the membrane filtration technique known as ultrafiltration. UF membrane pores are normally between 0.01 and 0.1 microns in size	1. Ultrafiltration membranes are extremely vulnerable to fouling by bacteria, particles, and organic compounds, and fouling shortens the membrane's life and raises operating expenses 2. Water treatment facilities may incur greater energy expenditures as a result of UF's requirement for a steady supply of pressure, especially in large-scale operations [14]
Reverse osmosis (RO) It is a membrane filtering technique that pushes water through a semipermeable membrane to eliminate dissolved salts, ions, and a variety of other impurities	1. When organic material, scale-forming salts, or suspended particles are present in excessive amounts, membranes may eventually foul, performance is lowered, and routine maintenance is necessary 2. Significant amounts of wastewater are produced by RO systems, some of which may not be recovered [15]
Sediment filtration The most basic method of purifying water is sediment filtering. To get rid of big particles like sand, mud, silt, and rust, physical barriers like mesh or filter media are used	1. Sediment filters need to be cleaned or replaced frequently because they might clog with suspended particles, decreasing their filtration effectiveness 2. If the filter becomes clogged, the system's pressure may drop, leading to a reduction in the overall filtration efficiency [16]

In photocatalytic water filtration, it is critical to interpret sophisticated processes controlling the production and use of reactive oxygen species. These reactive species which include H_2O_2, hydroxyl radicals ($^{\cdot-}OH$), superoxide radicals ($O_2^{\cdot-}$), electrons (e^-), holes (h^+) and are essential for powering photocatalytic reactions that lead to effective pollutant destruction. The reduction of O_2 molecules by photogenerated electrons produces superoxide radicals ($O_2^{\cdot-}$), which are important intermediates in photocatalytic pathways. Hydroperoxyl radicals ($HO_2^{\cdot}$), which are strong oxidants that can break down organic contaminants, can be created when $O_2^{\cdot-}$ combines with water [18].

2.3 *Types of Photocatalysis*

2.3.1 Homogeneous Photocatalysis

When photocatalysis is homogeneous, both reactants and photocatalysts are in a single state. The process through which the atmosphere removes and breaks down big organic molecules is known as a gas-phase homogeneous photocatalytic reaction. When creating a variety of photocatalysts, the ozone process is frequently mentioned:

$$2O_3(g) + H_2O(g) \xrightarrow{hv} O^2(g) + 2OH \quad (1)$$

Transition-metal complex photocatalysts are used in most aqueous-phase homogeneous photocatalytic techniques. Transition-metal complexes are widely used as photocatalysts due to large band gap and strong species stability. In the manufacture of clean hydrogen fuel, homogeneous photocatalysts are frequently used with cobalt and iron complexes being particularly prominent. The ozone process which is comparable to fenton's reagent method carried out in high hydrogen ion concentration is frequently used in the generation of hydrogen fuel to generate iron complex hydroxy radicals [19].

$$Fe^{2+} + H_2O_2 \xrightarrow{hv} Fe^{3+} + OH^- + HO \quad (2)$$

$$Fe^{3+} + H_2O_2 \xrightarrow{hv} Fe^{2+} + H^+ + HO_2 \quad (3)$$

$$Fe^{2+} + HO \xrightarrow{hv} Fe^{3+} + OH^- \quad (4)$$

2.3.2 Heterogeneous Photocatalysis

When the photocatalyst and reactants are in different phases, heterogeneous catalysis takes place. Metal accumulation, filtration of water, gaseous pollutant removal, moderate to complete oxidations, oxidative reactions, redox reactions, $^{18}O_2$–$^{16}O_2$, and deuterium-alkane isotopic interchange are only a few of the many reactions that fall under the broad category of heterogeneous photocatalysis. Transition metal oxides and semiconductors make up the majority of heterogeneous photocatalysts. Semiconductors feature a void energy region, where no energy levels are available to encourage the recombination of an electron and hole created by photoactivation in the solid, in contrast to metals, which contain a continuum of electronic states. In a semiconductor's MO diagram, a gap known as the 'band gap' is the energy difference between the filled valence band and the empty conduction band [20].

When a photon with energy equal to or greater than the semiconductor's band gap is absorbed, a charged particle in the valence band is stimulated to travel to the conduction band, creating an electron hole in the valence band. These two electrons combine to form an exciton. The excited electron and hole can recombine, releasing the energy that was acquired via the electron's excitation as heat. Such recombination of excitons is more economical and unwanted. By using a range of methods, including phase heterojunctions, noble-metal nanoparticles, silicon nanowires, and substitutional cation doping, the development of functional photocatalysts usually focuses on enhancing electron–hole separation and extending exciton lifetime. Enabling interactions among charged electrons and oxides to produce minimized compounds

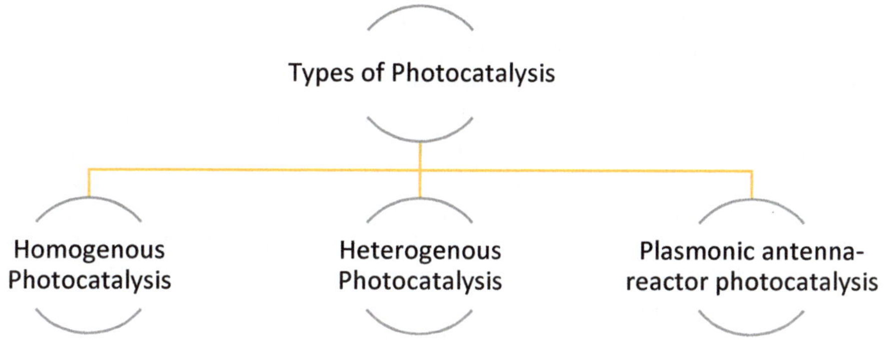

Fig. 2 Types of photocatalysis

or between generated holes and reductants to produce oxidative compounds is the primary objective of developing a photocatalyst.

Positive holes (h^+) and excited electrons (e^-) are produced at the outermost regions of semiconductors exposed to light causing oxidation–reduction processes. A hydroxyl radical is produced when holes react with surface moisture in one oxidative reaction process. The process is started at the metal oxide surface by photo-induced exciton generation followed by photon (hv) absorption [21].

$$MO + hv \rightarrow MO\left(h^+ + e^-\right) \tag{5}$$

Induced oxidative reactions by photocatalysis:

$$h^+ + H_2O \rightarrow H^+ + {}^{\cdot}OH \tag{6}$$

$$2\,h^+ + 2\,H_2O \rightarrow 2\,H^+ + H_2O_2 \tag{7}$$

$$H_2O_2 \rightarrow 2\,{}^{\cdot}OH \tag{8}$$

Ultimately, hydroxyl radicals are produced by both processes.

Figure 2 illustrates types of photocatalysis.

2.3.3 Plasmonic Antenna-Reactor Photocatalysis

It is the kind of photocatalyst that combines a catalyst with an antenna linked to enhance the catalyst's absorption of light and, as a result, its rate of activity. By combining plasmonic nanostructures with conventional photocatalysts, plasmonic antenna-reactor photocatalysis improves photocatalytic processes. Usually composed of noble metals like silver or gold, the plasmonic antennas have the capacity to take in incident light and generate localized surface plasmons which strengthens the electromagnetic fields surrounding the catalyst. Enhanced reaction rates for a range of uses, such as CO_2 reduction and separation of H_2O, result from the photocatalyst's charge carriers being much more excited by the increased light intensity. Higher efficiency and selectivity in photochemical reactions can be achieved by combining plasmonics and photocatalysis as demonstrated by research [22]. Hydrogen sulfide-to-hydrogen reactions were enhanced using a SiO_2 catalyst and a gold (Au) light absorber. The procedure is a substitute for the traditional Claus process, which runs between 800 and 1000 °C (1470 and 1830 °F). At room temperature, visible light can be used to create hydrogen from ammonia (NH_3) using a Fe catalyst and a Cu light absorber. The temperature range for conventional Cu-Ru manufacturing is 650–1000 °C (1202–1832 °F) [23].

2.4 *Mechanism of Photocatalytic Advanced Oxidation Process*

Advanced oxidation process (AOP) may directly use solar radiation to achieve disinfection and chemical detoxification which are effective for treating wastewater. The capacity to create photon, generated using solar energy or to create pairs of electron holes has attracted a lot of interest for photocatalytic degradation techniques [24]. The resulting electron–hole pairs create highly reactive oxygen species (ROSs) which can harm a variety of microorganisms, including bacteria, viruses, spores, protozoa, etc., as well as different other organic and inorganic contaminants. The AOPs provide an excellent water purification system that relies mostly on hydroxyl radicals [25]. Brunet et al. [26] explained the process by which a surface larger than the band gap of a semiconductor promotes the transfer of electrons through the valence band to enter the conduction band, creating gaps in the valence band and initiating the photocatalytic reaction.

With H_2O in the valence region and O_2 in the region of conduction band, the resulting e^-/h^+ pair may interact or recombine [27]. While h^+ has a chemical potential of + 1.0 to + 3 V with respect to electrode, signifying it would react with electron acceptors in a strong oxidative fashion, e^- would react with electron donors in a strong reductive manner due to its chemical potential of + 0.5 to − 1.5 V. These species are extremely reactive and can convert organic contaminants into innocuous metabolic products. The electron can react with molecular oxygen through a reductive mechanism to produce the superoxide anion ($O_2^{\cdot-}$) through an oxidative process. The hole can extract electrons from water or hydroxyl ions to create hydroxyl radicals [28]. Both hydroxyl radical and superoxide anion are potent, nonselective oxidants that can be applied to water, air, and other materials to disinfect and deodorize them. Additionally, they are able to demolish the structures of various biomolecules and organic and inorganic compounds, including proteins, nucleic acids, lipids, and carbohydrates. Oxidizing agents ($OH^{\cdot}$, $O_2^{\cdot-}$, and H_2O_2) generated by photocatalysts

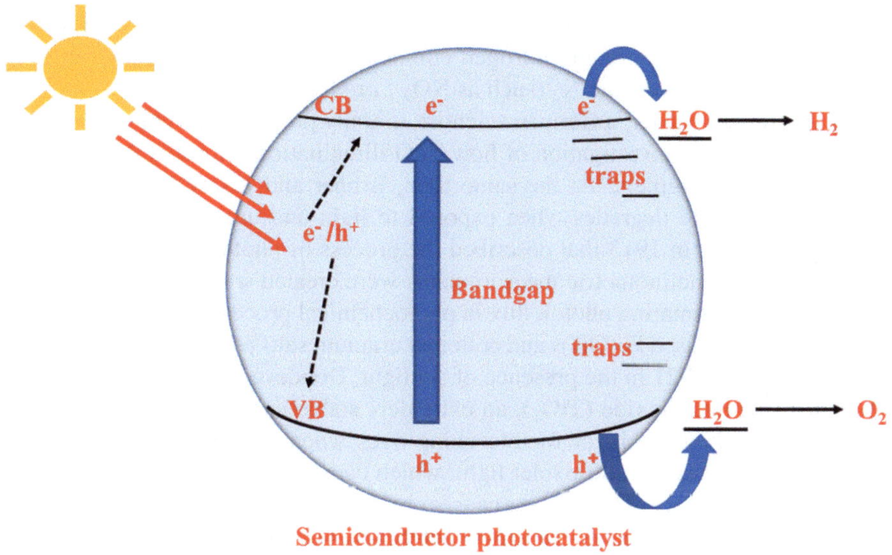

Fig. 3 Mechanism of action of semiconductor photocatalyst

when exposed to sunlight remove bacteria and organic contaminants from water [29]. This can be understood only by looking at Fig. 3.

3 History of Photocatalyst

Since the beginning, the scientific community has held the belief that sunlight may be transformed into more easily used and stored types of energy such as electricity and/or chemical fuels. In 1777, Scheel documented the sunlight induced reduction of AgCl to metallic Ag, marking a pioneering investigation into light-induced chemical reactions in research. The foundational work of Grotthuss (1817) and Draper (1841) introduced the key photochemistry principle that only radiation that penetrates can result in a photochemical transformation, known as the Grotthuss–Draper principle. Early researchers like Ciamician in Bologna relied solely on sunlight to study photochemical reactions, as artificial light sources were unavailable. These limitations posed significant challenges to advancing photochemistry in the nineteenth century [30].

- **Early Mention (1911–1938)**

In the early twentieth century, Plotnikow (1910) and Landau (1912–1913) introduced the concept of performing photochemical reactions with a substance or material that could speed up the process. The first methods that resembled natural photosynthesis appeared in the 1920s, with an emphasis on using the photochemical reaction of CO_2

and H_2O to create organic chemicals like CH_2O and carbohydrates. These methods also included forming organic nitrogen compounds by initiating a photoactivated interaction among inorganic N_2 (such as NO_3^- or NH_3) and CO_2.

Dr. Alexander Eibner, a German scientist, initially proposed the concept in 1911. He utilized it in his investigation of how ZnO illumination impacted the bleaching of Prussian blue pigment. At the same time, Bruner and Kozak released a paper on how oxalic acid degrades when exposed to light and uranyl salts and Landau published a paper in 1913 that described the process of photocatalysis. As a result of their research, actinometric measurements were created so that they serve as the foundation for estimating photon flux in photochemical processes [30]. After a time period, Baly et al. used $Fe(OH)_3$ and colloidal uranium salts as stimulants to generate formaldehyde in 1921 in the presence of sunlight. Doodeve and Kitchener found in 1938 that titanium dioxide (TiO_2), an extremely stable and non-toxic oxide, could function as a photosensitizer for bleaching dyes when oxygen was present. This is because TiO_2 absorbs ultraviolet light, which causes active oxygen species to be produced on its surface. This process causes organic chemicals to be blotched through photooxidation. The essential features of heterogeneous photocatalysis were seen for the first time [2].

- **Medieval Period (1964–2000)**

Research on photocatalysis was again halted up to 1964, when Filimonov examined the photooxidation of isopropanol from ZnO and TiO_2. Doerffler and Hauffe, Ikekawa et al. 1965, and Kato and Mashio 1965 investigated the photooxidation of carbon dioxide and organic solvents from ZnO radiance. Tanaka and Blyholde reported the photocatalytic degradation of N_2O in 1970, whereas Formenti et al. reported the oxidation of several alkenes. When a TiO_2 electrode exposed to ultraviolet radiation was connected electrically to a platinum electrode, Fujishima and Honda made the significant discovery in 1972 that electrochemical photolysis of H_2O took place. H_2 gas was created when electrons migrated from the metal's anode to the platinum cathode as the TiO_2 electrode absorbed the UV radiation [31]. Given that most hydrogen is produced by gasification and natural gas reforming, this was among the original occurrences of H_2 generation from a cost-effective and clean origin. Other advancements resulted from the discoveries made by Fujishima and Honda.

Nozik discovered in 1977 that an external voltage was not necessary to enhance photoactivity through the electrochemical photolysis process when a noble metal, such as gold or platinum, was added. Hydrogen synthesis on the $SrTiO_3$ surface through photogeneration was documented by Wagner and Somorjai [32]. TiO_2 and PtO_2 illumination in ethanol produced hydrogen and methane, respectively. It was years before photocatalysis was developed for commercial usage. Müller and Steinbach [33] showed that ZnO may cause disintegration of isopropanol in the presence of UV light. This finding proved that organic materials in water might be broken down by photocatalysts [3].

- **Latest Development (2001–2024)**

Since 2000, there has been a lot of interest in nanotechnology, and many nanoscale photocatalysts for water purification have been created. This indicated the beginning of a new phase in photocatalyst research. Nano-sized photocatalysts have higher charge transport rates, more exposed active sites, and bigger specific surface areas which all contribute to their increased photocatalytic activity. Creating a photoelectrochemical (PEC) tandem cell that would "replicate natural photosynthesis" at a reasonable cost and with minimal energy consumption is the main problem. Photocatalytic oxidation is a desirable alternative among the several treatment plans for water contamination because of its high efficiency, cheap cost, and shown ability to purify various contaminants. Through the stimulation of electron transport in a photoactive semiconductor, photocatalysis can use solar energy to oxidize or decrease contaminants [31]. The most crucial component of this technique is the photocatalyst which breaks down the pollutants by producing chemical energy from sun radiation.

Three phases are usually involved in the full photocatalytic process: creation, separation, and consumption of charges. The valence band e^- hole couples split when light strikes the photocatalyst pushing the e^- into the conduction band and exiting the valence band with holes from hydrogen ions. Electrons on the surface of the photoactive catalyst can form ROS by reducing common electron acceptors such as O_2 and H_2O. At the same time, the organic contaminants mineralize as the donor species recombine with the pores on the photocatalyst's surface. Since the 1970s, when the first photocatalytic removal of water contaminants was discovered, photocatalysts have garnered more and more attention in the water treatment industry [3].

4 Applications in Water Purification

Heavy metals, inorganic compounds, organic pollutants, and a number of other complicated molecules have been discovered in drinking water, groundwater, and surface water due to increasing industrialization and population increase [34]. Among some of the significant innovations for preserving human existence and the environment are water treatment, and it is a global goal to create efficient methods of treating wastewater from, for example, industrial activities. Effective techniques for treating persistent organic contaminants are especially needed. A significant threat to achieving equitable growth aims will come from changes in the H_2O cycle, as well as to the well-being of humans, food availability, energy generation, revenue creation, and reducing inequality, according to the 2020 United Nations Goals. Therefore, developing sophisticated, economical, environmentally friendly, and extremely efficient wastewater reclamation techniques is essential. Heterogeneous photocatalysis is being researched and used in a number of fields since 1972 including water splitting, CO_2 reduction, N_2 fixation, and water/air purification. Photocatalysis breaks down organic toxins in effluents into H_2O, CO_2, or other tiny molecules while oxidizing or converting inorganic contaminants to innocuous chemicals through mild temperatures, basic procedure, and environmentally friendly technology [35].

4.1 Elimination of Heavy Metals

In living things, trace metal ions have a unique capacity to bind to proteins, nucleic acids, and tiny molecules. Organic cell death and other severe health problems may result from this because trace metal ions are not biodegradable; they will accumulate in people and animals through their diet and drinking water [36]. Therefore, it's crucial to eliminate these hazardous trace metals which frequently include Ni, Hg, Cr, Mn, Cd, Zn, Hg, and Cd ions prior to release wastewater into the environment [37]. A study created an artificial CH-GEL/ZSPNC hybrid nanocomposite ion exchanger that could extract 90.3% Cd^{2+}, 95% Mg^{2+}, 100% Pb^{2+}, 88.9% Cu^{2+}, 94.9% Zn^{2+}, 90% Ni^{2+}, and 84% Rhodamine dye under sunlight. The photocatalytic oxidation procedure used a manufactured chitosan/silver bio-nanocomposites (chitosan/polyvinylidene chloride/silver) to remove 77% of the dye, 97% of the Cu^{2+}, 88% of the Pb^{2+}, and 89% of the Cd^{2+}. This method had a number of drawbacks even if it generates reactive radicals locally and doesn't call for chemicals or sludge. When there are a lot of metals around, it is still ineffective, pH-sensitive and has low throughput [38]. Figure 4 illustrates the procedure for removal of heavy metals from polluted water.

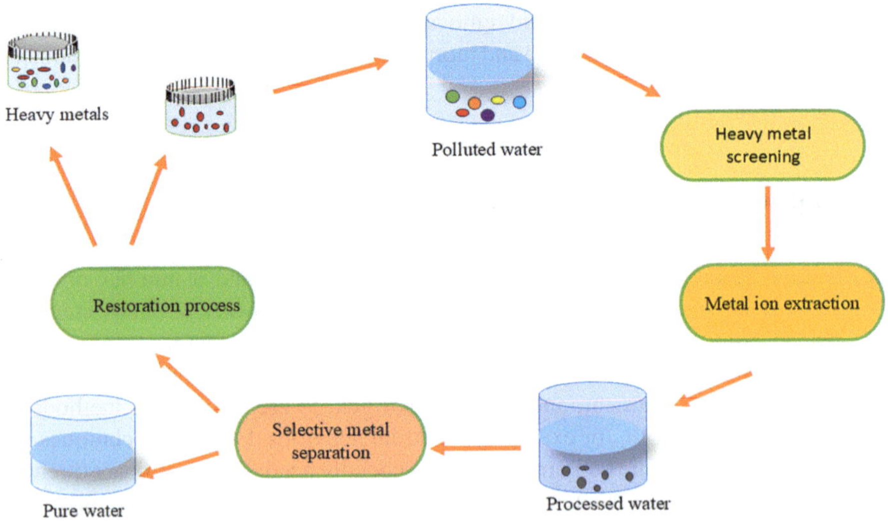

Fig. 4 Process of removal of heavy metals from water

4.2 Elimination of Drugs

Pharmaceutical compounds (PCs) are among the emerging contaminants that have caused concern recently because of their harmful effect on the environment. Heterogeneous photocatalysis is one of the best at degrading dangerous substances like antibiotics [39]. UV rays can break down a number of medications, such as riboflavin, ascorbic acid, folic acid, hydrocortisone, prednisolone, and phenothiazine tranquilizers. Pharmaceuticals are broken down using photocatalytic nanomaterials like TiO_2, ZnO, CuO, doped catalysts, and suspended catalysts [40]. A study by Ahmad et al. [41] emphasized the bio-fabrication of tin oxide nanoparticles (SnO_2 NPs) by utilizing Phyllanthus emblica leaf extract. The investigations focused on the physio-chemical properties of the synthesized material and assessed their efficacy as photocatalysts and antibacterial agents. The photocatalytic effectiveness of the synthesized materials was assessed for the elimination of 2,4-dichlorophenoxyacetic acid (2,4-D); concurrently, the antibacterial efficacy of the developed material was evaluated against the selected bacterial strains.

4.3 Elimination of Lipid Control Agents

Metformin, a lipid-regulating medication, is the most commonly used hypoglycemic drug for managing type 2 diabetes, also referred to as non-insulin-dependent diabetes. Metformin is fully excreted from the body during use and is not metabolized by the body. These pollutants penetrate aquatic habitats and contaminate marine resources in various ways [42]. Synthetic titanium dioxide and zirconium dioxide catalyst were used to remove metformin from sewage, demonstrating the effectiveness of photocatalysis in the elimination of metformin from H_2O. The use of TiO_2 as a photocatalyst for the breakdown of amoxicillin and metformin was examined by Chinnaiyan et al. The experimental analysis found that the maximum elimination rates were achieved by amoxicillin (90%) and metformin (98%) at pH 7.6, 563 mg/L of TiO_2, an initial pollutant concentration of 10 mg/L, and a reaction period of 150 min [43]. Figure 5 demonstrates different applications of photocatalysts.

4.4 Elimination of Organic Substances

Water contains a variety of organic contaminants. Pollutants include plasticizers, organohalides, phenolic compounds, surfactants, hydrocarbons, dyes, and other materials. These organic pollutants are chemically resistant, harmful, and cancer-causing. Additionally, they are difficult to decompose in water [44]. Future research on photocatalysis was made possible by Carey's discovery that TiO_2 catalytically destroys polychlorinated biphenyls in the presence of solar radiation. In general, a

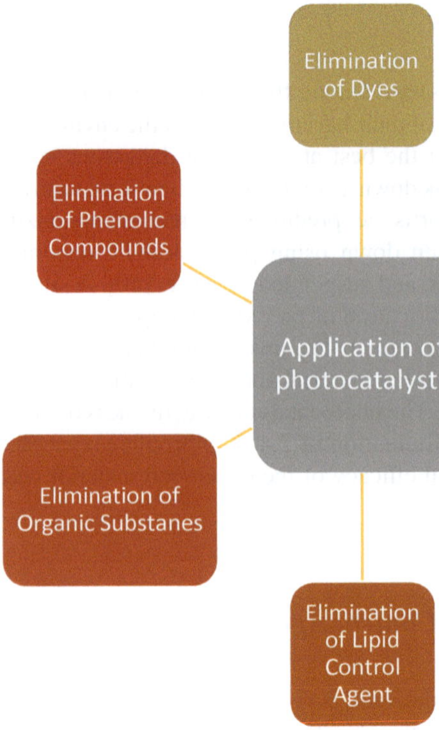

Fig. 5 Application of photocatalysts

variety of semiconductors, including those based on bismuth, ZnO, Fe_2O_3, C_3N_4, and TiO_2, have been strong contenders for breaking down a variety of organic pollutants into less hazardous molecules or easily biodegradable compounds which subsequently undergo additional mineralization to produce harmless carbon dioxide and water [4]. The advancement of a fluorescent pH sensor and the remediation of wastewater using nanoparticles are both essential subjects. The fluctuation in pH influences the form and ensuing properties of nanoparticles, which are employed across several domains, while the latter offers an enhanced treatment method for industrial waste materials [45].

4.5 Elimination of Dyes

In general, a dye is defined as a material that can bind chemically or physically to a substrate to provide color. The creation of color in dyes is explained by the presence of chromophores, to which auxochromes are linked. Dye classification is based on physicochemical characteristics, uses, ecotoxicological consequences, and water solubility. These colors have aromatic molecular structures that come from hydrocarbons including xylene, benzene, toluene, naphthalene, and anthracene. MO, MB, and RhB are among the dyes that are non-decomposable, H_2O soluble, and potentially harmful to the surroundings [46]. It has been noted that photocatalysts have achieved exceptional efficiency of simultaneous dye extraction and great recyclability and revealed that the bismuth ferrite tungstate/α-silver vanadate composite with 1 weight percent bismuth ferrite tungstate may significantly boost RhB removal in comparison with pure α-silver vanadate and bismuth ferrite tungstate [47]. A study by Jaiswal et al. [48] detailed the synthesis of semiconductor selenium nanoparticles/microrods (Se-NP/MRs) via a bio-fabrication method and examined their photocatalytic activity under natural sunlight exposure. Cabbage leaf extract served as a reducing agent for the synthesis of Se-NP/MRs.

4.6 Elimination of Phenolic Compounds

Organic substances with a hydroxyl group joined to one or more aromatic rings in their molecular structure are known as phenols. They are highly soluble in water, highly toxic, and physiologically inert. The photocatalytic process mainly breaks down phenol and its derivatives, nitrophenol, and chlorophenol, in the bulk liquid. A range of oxidation byproducts are produced when hydroxyl radicals approach the cyclic carbon [49]. Phenolic compounds are introduced into water through the breakdown of natural materials (such as the transformation of lignin and tannin, the presence of polyphenolic macromolecules in certain fish and algae species, the decomposition of dead animals, etc.) and human activities, such as industrial discharges, agricultural practices, and human consumption. Chlorophenols are suspected carcinogens and relatively hazardous H_2O contaminants. The primary byproducts identified during its photocatalytic breakdown include hydroquinone, benzoquinone, 4-nitrocatechol, and a few organic acids [50]. A study investigated the TiO_2's capacity for photocatalytic reduction within the breakdown of phenolic chemicals found within refinery effluent. Phenol and its byproducts were shown to degrade most efficiently at optimal conditions of pH-3, 318 K, and 100 mg/l of catalyst. Under these ideal circumstances, phenol degraded 90% of its original material in under two hours. Many studies have also shown that phenolic compounds can be photocatalytically degraded from wastewater utilizing a variety of catalysts, such as CuO, ZnO, Fe_2O_3 decorated on carbon nanotubes, and TiO_2/reduced graphene [51]. A recent study investigated the effectiveness of hydrothermally prepared ZnO and SnO_2 nanocatalysts using a new approach to degrade p-nitrophenol, a toxic compound in water [7, 8].

4.7 Elimination of Petroleum Hydrocarbon

Hydrocarbons that are both aromatic and aliphatic constitute petroleum and its derivatives. The latter is an environmentally persistent, poisonous, carcinogenic, mutagenic, and stable chemical, oil spills, refinery wastewater, water from oil fields, storage tank or pipeline leaks, ballast water, car wash wastewater, construction materials, and runoff from rain-induced repairs or inner-city tunnel washing are some of the sources of oil pollution in aquatic environments [52]. As a result, properly treating petroleum hydrocarbons in water has become crucial under visible light irradiation, K-doped g-C_3N_4 composites supported on biochar demonstrated high-efficiency photocatalytic degradation of naphthalene (82.19%). Due to their abundant surface hydrophilic functional groups, improved light absorption, and reduced charge recombination. After five rounds, rate of photocatalytic breakdown exhibited minimal variation [53]. Organic contaminants from effluent from petroleum refineries could be effectively removed by photocatalysis using titanium dioxide/iron-modified ZSM-5 composite. Following a duration of four hours of exposure to UV radiation at a catalyst proportion of 2.1 g/L, pH of 4, and 45 °C temperature, the maximal COD elimination efficiency was 80% [54]. Petroleum hydrocarbons in aqueous solution can be effectively removed by photocatalytic oxidation using two-dimensional ultrathin g-C_3N_4 nanosheets because of their large specific surface areas, brief carrier transport distances, and controlled electrical structures [55].

4.8 Elimination of Pesticides

During storage or transportation of fruits, pesticides are applied to control leaf shedding, regulate growth, dehydrate, and prevent spoilage. However, pesticides are also a significant source of water pollution. Every pesticide causes cancer and has harmful side effects. Both toxicity and biological resistance are present in pesticides [56]. Even trace levels of pesticides can linger and significantly affect human health and the environment; due to the chemical stability and non-biodegradable of semiconductor-based photocatalysis, it has also been used in the field of pesticide breakdown. Many metal oxides, including TiO_2 and ZnO, are the primary source of semiconductor materials used in photocatalytic pesticide breakdown [57]. In the presence of sunlight, a mesoporous silver/Ag_2O-TiO_2 p-n heterojunction was used to photocatalytically breakdown the herbicide dimethenamid-P. The photoactivity facts demonstrated that, after 180 min, the imazapyr destruction was completely eliminated [58]. Triazophos and profenofos had respective degradation efficiencies of 32.1 and 53.3% after one day, according to the results. Bare TiO_2 and Au-modified TiO_2 were studied for their photocatalytic properties to break down phenoxyacetic acid under ultraviolet radiation and white light [59]. Table 3 presents a summary of various applications of photocatalysts.

Table 3 Photocatalytic elimination of water contaminants

Type of materials	Resources	Light source	Contaminant	Reference
Heavy metals	CuO/g-C_3N_4	Vis light(150 W)	Hg (II)	[33, 37]
	$CoFe_2O_4$/g C_3N_4	Vis light (300 W)		[60]
	ά–Fe_2O_3/g C_3N_4	Vis light (400 W)		[60, 61]
Drugs	TiO_2	UV lamp (125 W)	Amoxicillin metformin	[43]
	TiO_2–ZrO_2	UV light (125 W)	Metformin	[42]
Organic substances (dye)	ZnO/PALFS f	Vis light (300 W)	Congo red	[62]
	SnO_2-MoS_2	Vis light (200 W)	Methylene blue, rhodaminc B	[63]
	$BiFeWO_6$/ ά$AgVO_3$	Vis light	Rhodamine B	[47]
	TiO_2/ZnO/RGO	UV/artificially generated solar light (300 W)	Methylene blue, rhodamine B, methyl orange	[64]
	Ag@AgC@MI L100(Fe)/CCF	Vis light (500 W)	Methylene blue	[46]
Phenolic compounds	TiO_{2-x} @ZIF-67	Vis light	BPA	[65]
	CuO-TiO_2	UV light (96 W)	Phenols	[66]
	Cu-NiO	UV-Vis (150 W)	Phenols	[67]
	Au@Ni/rGO	Sunlight	Phenols	[68]
Petroleum hydrocarbon	g-C_3N_4Hx+	Vis light	Naphthalene	[69]
	K-doped g-C_3N_4	Vis light (200 W)	N-tetradecane	[53]
Pesticides	Ag/$LaTiO_3$	Vis light (300 W)	Atrazine	[70]
	Au/TiO_2	Vis light	Phenoxyacetic acid	[59]
	TiO_2/Ce	Simulated xenon light	Profenofos and triazophos	[71]
	Ag/Ag_2O-TiO_2	Vis light	Imazapyr	[58]

5 Conclusion

One viable and sustainable way to fix the growing issues of aquatic pollution is through the use of photocatalysts in water purification. Through the use of light energy, usually visible or ultraviolet (UV) light, photocatalysis breaks down organic pollutants, viruses, and even inorganic contaminants in water by activating a semiconductor material. This approach has a number of benefits, such as high efficiency, low energy consumption, and sustainability. It is appropriate for decentralized water purification systems since photocatalysts can be utilized in the presence of sunlight.

Numerous organic contaminants, including pesticides, dyes, medicines, and industrial effluents, can be efficiently broken down by photocatalysis, which has also been shown to inactivate dangerous microbes. ROS like hydroxyl radicals and superoxide anions are produced as part of the mechanism and are essential for decomposing the pollutants.

Furthermore, improvements in the creation of new photocatalysts and modification methods have improved the stability, efficiency, and use of UV light in photocatalytic processes, increasing their applicability in actual water treatment situations. The photocatalytic process still has issues, though, such as poor efficiency under visible light, the possibility of catalyst inhibition, as well as the need for modification in large-scale applications. For broad acceptance, it is also necessary to address the energy needed to maintain ideal reaction conditions and the cost of high-performance photocatalysts. The filtration of contaminated water can be achieved by photocatalysis, but further research and development is required to increase its effectiveness, scalability, and affordability. The use of photocatalytic devices in water treatment processes may be crucial to producing safer and cleaner water in places with limited access to conventional water treatment facilities.

6 Future Perspective

Advanced photocatalysts, especially nanoparticles and its doped forms, are regarded as crucial to the future of water purification because they can break down contaminants when exposed to light, providing an energy-efficient and eco-friendly treatment method. Ongoing research is focused on enhancing photocatalysts to function under visible light, as conventional semiconductors are only active under UV light, which restricts its practical use. Future advancements in photocatalysis include hybrid systems that combine photocatalytic methods with adsorption or filtration, forming multi-stage treatment solutions capable of tackling complex pollutants like pharmaceuticals and pesticides. Researchers are also developing nano-engineered photocatalysts with larger surface areas to enable more precise contaminant breakdown and increase efficiency. Additionally, stability and reusability are essential considerations, as many photocatalysts deteriorate over time. Thus, ongoing development aims to create robust, long-lasting materials that can endure extended use. Finally, scaling these technologies for broad application remains challenging; however, ongoing research into cost-effective, efficient photocatalyst materials, and reactor designs holds promise for making large-scale water treatment feasible. As advancements continue, photocatalysis has the potential to become an accessible, sustainable solution for water purification, particularly in areas where traditional purification methods are less accessible.

References

1. Ameta SC, Ameta R (eds) (2018) Advanced oxidation processes for wastewater treatment: emerging green chemical technology. Academic press
2. Fujishima A, Honda K (1972) Electrochemical photolysis of water at a semiconductor electrode. Nature 238(5358):37–38. https://doi.org/10.1038/238037a04
3. Long Z, Li Q, Wei T, Zhang G, Ren Z (2020) Historical development and prospects of photocatalysts for pollutant removal in water. J Hazard Mater 395:122599
4. Ren G, Han H, Wang Y, Liu S, Zhao J, Meng X, Li Z (2021) Recent advances of photocatalytic application in water treatment: a review. Nanomaterials 11(7):1804
5. Zhao C, Meng L, Chu H, Wang JF, Wang T, Ma Y, Wang CC (2023) Ultrafast degradation of emerging organic pollutants via activation of peroxymonosulfate over Fe3C/Fe@ NCx: Singlet oxygen evolution and electron-transfer mechanisms. Appl Catal B 321:122034
6. Zhang J, Wu H, Shi L, Wu Z, Zhang S, Wang S, Sun H (2024) Photocatalysis coupling with membrane technology for sustainable and continuous purification of wastewater. Sep Purif Technol 329:125225
7. Pandey S, Singh A, Kumar A, Tyagi I, Karri RR, Gaur R et al (2024) Photocatalytic degradation of noxious p-nitrophenol using hydrothermally synthesized stannous and zinc oxide catalysts. Phys Chem Earth Parts A/B/C 133:103512
8. Pandey S, Chaudhary J, Sharma H, Pundir S, Rustagi S, Malik S, Choudhary P (2024b) Biosynthesis of zirconia nanoparticles (ZrO_2) by water hyacinth: characterization and its photocatalytic dye degradation activity. Biomass Convers Biorefinery 1–9
9. Chaudhary M, Kumar C, Raghav S, Panwar M, Pandey S, Painuli R (2024) Sunlight-driven photocatalytic degradation of industrial dyes using Withania somnifera decorated MnO_2 nanoparticles. Discover Nano 19(1):206
10. Kandari D, Pandey S (2024) Comparative study of photo catalytic activity of *Tinospora cordifolia* derived nanoparticles for degradation of indigo carmine. Desalin Water Treat 320:100870
11. Mishra M, Urooj T, Singh A, Pandey S (2024) A new approach for synthesis of bismuth oxide derived photocatalyst for degradation of para nitrophenol. Desalination and water treatment 100558
12. Pandey S, Twala B, Singh R, Gehlot A, Singh A, Montero EC, Priyadarshi N (2022) Wastewater treatment with technical intervention inclination towards smart cities. Sustainability 14(18):11563
13. Smarzewska S, Morawska K (2021) Wastewater treatment technologies. In: Handbook of advanced approaches towards pollution prevention and control. Elsevier, pp 3–32
14. Shi X, Tal G, Hankins NP, Gitis V (2014) Fouling and cleaning of ultrafiltration membranes: a review. J Water Process Eng 1:121–138
15. Zhai Y, Liu G, van der Meer WG (2022) One-step reverse osmosis based on riverbank filtration for future drinking water purification. Engineering 9:27–34
16. Abd Wahab N, Kamarudin MKA, Toriman ME, Juahir H, Saad MHM, Ata FM et al (2019) Sedimentation and water quality deterioration problems at Terengganu River Basin, Terengganu, Malaysia. Desalin Water Treat 149:228–241
17. Muscetta M, Ganguly P, Clarizia L (2024) Solar-powered photocatalysis in water purification: applications and commercialization challenges. J Environ Chem Eng 113073
18. Matthew BM, George I, Anastasio C (2003) Hydroperoxyl radical ($HO_2^{\cdot}$) oxidizes dibromide radical anion ($^{\cdot}Br_2^{-}$) to bromine (Br_2) in aqueous solution: implications for the formation of Br_2 in the marine boundary layer. Geophys Res Lett 30(24)
19. He F, Jeon W, Choi W (2021) Photocatalytic air purification mimicking the self-cleaning process of the atmosphere. Nat Commun 12(1):2528
20. Ahmed SN, Haider W (2018) Heterogeneous photocatalysis and its potential applications in water and wastewater treatment: a review. Nanotechnology 29(34):342001
21. Kudo A, Miseki Y (2009) Heterogeneous photocatalyst materials for water splitting. Chem Soc Rev 38(1):253–278

22. Linic S, Aslam U, Boerigter C, Morabito M (2015) Photochemical transformations on plasmonic metal nanoparticles. Nat Mater 14(6):567–576
23. Kale MJ, Avanesian T, Christopher P (2014) Direct photocatalysis by plasmonic nanostructures. ACS Catal 4(1):116–128
24. Duan X, Zhou X, Wang R, Wang S, Ren NQ, Ho SH (2021) Advanced oxidation processes for water disinfection: features, mechanisms and prospects. Chem Eng J 409:128207
25. Colmenares JC, Luque R (2014) Heterogeneous photocatalytic nanomaterials: prospects and challenges in selective transformations of biomass-derived compounds. Chem Soc Rev 43(3):765–778
26. Brunet L, Lyon DY, Hotze EM, Alvarez PJ, Wiesner MR (2009) Comparative photoactivity and antibacterial properties of C60 fullerenes and titanium dioxide nanoparticles. Environ Sci Technol 43(12):4355–4360
27. Liu Y, Chen X, Li J, Burda C (2005) Photocatalytic degradation of azo dyes by nitrogen-doped TiO_2 nanocatalysts. Chemosphere 61(1):11–18
28. Lin S, Huang H, Ma T, Zhang Y (2021) Photocatalytic oxygen evolution from water splitting. Adv Sci 8(1):2002458
29. Chen Y, Xu M, Wen J, Wan Y, Zhao Q, Cao X, Bian Z (2021) Selective recovery of precious metals through photocatalysis. Nat Sustain 4(7):618–626
30. Sordello F, Calza P, Minero C, Malato S, Minella M (2022) More than one century of history for photocatalysis, from past, present and future perspectives. Catalysts 12(12):1572
31. Coronado JM, Fresno F, Hernández-Alonso MD, Portela R (eds) (2013) Design of advanced photocatalytic materials for energy and environmental applications (vol 71). Springer, London
32. Wagner F, Somorjai G (1980) Photocatalytic hydrogen production from water on Pt-free $SrTiO_3$ in alkali hydroxide solutions. Nat 285:559–560. https://doi.org/10.1038/285559a03
33. Müller HD, Steinbach F (1970) Decomposition of Isopropyl Alcohol photosensitized by Zinc Oxide. Nat 225:728–729. https://doi.org/10.1038/225728a02
34. Gómez-Pastora J, Dominguez S, Bringas E, Rivero MJ, Ortiz I, Dionysiou DD (2017) Review and perspectives on the use of magnetic nanophotocatalysts (MNPCs) in water treatment. Chem Eng J 310:407–427
35. Li Z, Meng X, Zhang Z (2019) Fewer-layer BN nanosheets-deposited on Bi_2MoO_6 microspheres with enhanced visible light-driven photocatalytic activity. Appl Surf Sci 483:572–580
36. Khan A, Khan S, Khan MA, Qamar Z, Waqas M (2015) The uptake and bioaccumulation of heavy metals by food plants, their effects on plants nutrients, and associated health risk: a review. Environ Sci Pollut Res 22:13772–13799
37. Wu Y, Pang H, Liu Y, Wang X, Yu S, Fu D, Wang X (2019) Environmental remediation of heavy metal ions by novel-nanomaterials: a review. Environ Pollut 246:608–620
38. Qasem NA, Mohammed RH, Lawal DU (2021) Removal of heavy metal ions from wastewater: a comprehensive and critical review. Npj Clean Water 4(1):1–15
39. Kay P, Hughes SR, Ault JR, Ashcroft AE, Brown LE (2017) Widespread, routine occurrence of pharmaceuticals in sewage effluent, combined sewer overflows and receiving waters. Environ Pollut 220:1447–1455
40. Verma M, Haritash AK (2020) Photocatalytic degradation of amoxicillin in pharmaceutical wastewater: a potential tool to manage residual antibiotics. Environ Technol Innov 20:101072
41. Ahmad W, Ahmad A, Shameem M (2024) Solar light-driven heterogeneous photocatalytic removal of 2, 4 dichlorophenoxy acetic acid from aqueous solution by bio-fabricated SnO_2 nanostructure. Res Chem Intermed 1–23
42. Carbuloni CF, Savoia JE, Santos JS, Pereira CA, Marques RG, Ribeiro VA, Ferrari AM (2020) Degradation of metformin in water by TiO_2–ZrO_2 photocatalysis. J Environ Manage 262:110347
43. Chinnaiyan P, Thampi SG, Kumar M, Balachandran M (2019) Photocatalytic degradation of metformin and amoxicillin in synthetic hospital wastewater: effect of classical parameters. Int J Environ Sci Technol 16(10):5463–5474
44. Ali I, Asim M, Khan TA (2012) Low cost adsorbents for the removal of organic pollutants from wastewater. J Environ Manage 113:170–183

45. Negi P, Rawat BS, Joshi NC, Ahmad W, Kumar N, Upadhyay S et al (2024) Hydrothermal synthesis of nitrogen-doped CQDS of *Rubus niveus* leaves for fluorescent pH sensing and photocatalytic applications. Malays J Sci 13–21
46. Lu W, Duan C, Liu C, Zhang Y, Meng X, Dai L et al (2020) A self-cleaning and photocatalytic cellulose-fiber-supported "Ag@ AgCl@ MOF-cloth" membrane for complex wastewater remediation. Carbohyd Polym 247:116691
47. Senthil RA, Sun M, Pan J, Osman S, Khan A, Sun Y (2019) Facile fabrication of a new BiFeWO6/α-$AgVO_3$ composite with efficient visible-light photocatalytic activity for dye-degradation. Opt Mater 92:284–293
48. Jaiswal KK, Dutta S, Pohrmen CB, Banerjee I, Das HT, Jha N et al (2023) Bio-fabrication of selenium nanoparticles/micro-rods using cabbage leaves extract for photocatalytic dye degradation under natural sunlight irradiation. Int J Environ Anal Chem 103(18):6559–6576
49. Ahmed S, Rasul MG, Martens WN, Brown R, Hashib MA (2010) Heterogeneous photocatalytic degradation of phenols in wastewater: a review on current status and developments. Desalination 261(1–2):3–18
50. Zhong W, Wang D, Wang Z (2018) Distribution and potential ecological risk of 50 phenolic compounds in three rivers in Tianjin, China. Environ Pollut 235:121–128
51. Shahrezaei F, Mansouri Y, Zinatizadeh AAL, Akhbari A (2012) Process modeling and kinetic evaluation of petroleum refinery wastewater treatment in a photocatalytic reactor using TiO_2 nanoparticles. Powder Technol 221:203–212
52. Varjani SJ (2017) Microbial degradation of petroleum hydrocarbons. Biores Technol 223:277
53. Li F, Lin M (2020) Synthesis of biochar-supported K-doped g-C_3N_4 photocatalyst for enhancing the polycyclic aromatic hydrocarbon degradation activity. Int J Environ Res Public Health 17(6):2065
54. Ghasemi Z, Younesi H, Zinatizadeh AA (2016) Preparation, characterization and photocatalytic application of TiO_2/Fe-ZSM-5 nanocomposite for the treatment of petroleum refinery wastewater: optimization of process parameters by response surface methodology. Chemosphere 159:552–564
55. Qiao M, Fu L, Barcelo D (2022) Removal of polycyclic aromatic hydrocarbons by g-C3N4 nanosheets under visible light irradiation and effect of typical co-existence substances in river water. Process Saf Environ Prot 159:376–381
56. Hamza RA, Iorhemen OT, Tay JH (2016) Occurrence, impacts and removal of emerging substances of concern from wastewater. Environ Technol Innov 5:161–175
57. Le TDH, Scharmüller A, Kattwinkel M, Kühne R, Schüürmann G, Schäfer RB (2017) Contribution of waste water treatment plants to pesticide toxicity in agriculture catchments. Ecotoxicol Environ Saf 145:135–141
58. Mkhalid IA, Fierro JLG, Mohamed RM, Alshahri AA (2020) Visible light driven photooxidation of imazapyr herbicide over highly efficient mesoporous Ag/Ag_2O–TiO_2 pn heterojunction photocatalysts. Ceram Int 46(16):25822–25832
59. Lannoy A, Bleta R, Machut-Binkowski C, Addad A, Monflier E, Ponchel A (2017) Cyclodextrin-directed synthesis of gold-modified TiO_2 materials and evaluation of their photocatalytic activity in the removal of a pesticide from water: effect of porosity and particle size. ACS Sustain Chem Eng 5(5):3623–3630
60. Kadi MW, Mohamed RM, Ismail AA, Bahnemann DW (2020) Decoration of g-C_3N_4 nanosheets by mesoporous $CoFe_2O_4$ nanoparticles for promoting visible-light photocatalytic Hg (II) reduction. Colloids Surf, A 603:125206
61. Kadi MW, Mohamed RM, Ismail AA, Bahnemann DW (2020) Soft and hard templates assisted synthesis mesoporous CuO/g-C_3N_4 heterostructures for highly enhanced and accelerated Hg (II) photoreduction under visible light. J Colloid Interface Sci 580:223–233
62. Deebansok S, Amornsakchai T, Sae-Ear P, Siriphannon P, Smith SM (2021) Sphere-like and flake-like ZnO immobilized on pineapple leaf fibers as easy-to-recover photocatalyst for the degradation of congo red. J Environ Chem Eng 9(2):104746
63. Ni X, Chen C, Wang Q, Li Z (2019) One-step hydrothermal synthesis of SnO_2-MoS_2 composite heterostructure for improved visible light photocatalytic performance. Chem Phys 525:110398

64. Nguyen CH, Tran ML, Van Tran TT, Juang RS (2020) Enhanced removal of various dyes from aqueous solutions by UV and simulated solar photocatalysis over TiO_2/ZnO/rGO composites. Sep Purif Technol 232:115962
65. Tang Y, Li X, Zhang H, Ouyang T, Jiang Y, Mu M, Yin X (2020) Cobalt-based ZIF coordinated hybrids with defective TiO_2-x for boosting visible light-driven photo-Fenton-like degradation of bisphenol A. Chemosphere 259:127431
66. Çinar B, Kerimoğlu I, Tönbül B, Demirbüken A, Dursun S, Kaya IC et al (2020) Hydrothermal/electrospinning synthesis of CuO plate-like particles/TiO_2 fibers heterostructures for high-efficiency photocatalytic degradation of organic dyes and phenolic pollutants. Mater Sci Semicond Process 109:104919
67. Ethiraj AS, Uttam P, Varunkumar K, Chong KF, Ali GA (2020) Photocatalytic performance of a novel semiconductor nanocatalyst: copper doped nickel oxide for phenol degradation. Mater Chem Phys 242:122520
68. Darabdhara G, Das MR (2019) Dual responsive magnetic Au@ Ni nanostructures loaded reduced graphene oxide sheets for colorimetric detection and photocatalytic degradation of toxic phenolic compounds. J Hazard Mater 368:365–377
69. Yang X, Yu J, Zhang Y, Peng Y, Li Z, Feng C, Wang Y (2020) Visible-near-infrared-responsive g-C_3N_4Hx+ reduced decatungstate with excellent performance for photocatalytic removal of petroleum hydrocarbon. J Hazard Mater 381:120994
70. Shawky A, Mohamed RM, Mkhalid IA, Youssef MA, Awwad NS (2020) Visible light-responsive Ag/$LaTiO_3$ nanowire photocatalysts for efficient elimination of atrazine herbicide in water. J Mol Liq 299:112163
71. Liu X, Zhan Y, Zhang Z, Pan L, Hu L, Liu K et al (2019) Photocatalytic degradation of profenofos and triazophos residues in the Chinese cabbage, brassica chinensis, using Ce-Doped TiO_2. Catalysts 9(3):294
72. Kadi MW, Mohamed RM, Ismail AA, Bahnemann DW (2020) Performance of mesoporous α-Fe_2O_3/g-C_3N_4 heterojunction for photoreduction of Hg(II) under visible light illumination. Ceram Int 46(14):23098–23106
73. Daneshvar N, Salari D, Khataee AR (2004) Photocatalytic degradation of azo dye acid red 14 in water on ZnO as an alternative catalyst to TiO_2. J Photochem Photobiol A 162(2–3):317–322
74. Fujishima A, Zhang X (2006) Titanium dioxide photocatalysis: present situation and future approaches. C R Chim 9(5–6):750–760
75. Fujishima A, Zhang X (2006) Titanium dioxide photocatalysis: present situation and future approaches. C R Chim 9(5–6):750–760
76. Daneshvar N, Salari D, Khataee AR (2004) Photocatalytic degradation of azo dye acid red 14 in water on ZnO as an alternative catalyst to TiO_2. J Photochem Photobiol A 162(2–3):317–322

Tin Oxide Materials for Water Purification: Synthesis and Applications

Abdelaal S. A. Ahmed, Islam K. M. Thabet, Emad Elshehy, and M. F. Cheira

Abstract Wastewater purification is a critical global challenge due to the increasing contamination of water resources by industrial, agricultural, and domestic pollutants. Tin dioxide (SnO_2)-based materials have emerged as promising candidates for wastewater treatment due to their unique physicochemical properties, including high chemical stability, photocatalytic activity, and tunable surface characteristics. This chapter provides the recent advancements in SnO_2-based materials for wastewater purification. The chapter discusses the most important synthesis methods and applications of SnO_2 materials in wastewater purification specially adsorption and photocatalytic degradation processes. Furthermore, the mechanisms of pollutant degradation, challenges, and future perspectives are included. This chapter aims to provide valuable insights for researchers or readers interested in water treatment.

Keywords Tin oxide · Material synthesis · Wastewater · Adsorption · Photocatalytic degradation

1 Introduction

These days, water scarcity is a major issue that affects everyone. For many uses, such as household, agricultural, industrial, and energy requirements, clean water is essential, especially in emerging nations where the population is growing [17]. More than 1.2 billion people worldwide lack access to clean drinking water [13]. Toxic pollutants in the environment have alarmingly increased in various forms and concentrations due to industrial development, energy plants, mining industries, and environmental disasters. Additionally, the rise in the demand for readily available water is brought on by population growth [28]. It has been reported that approximately

A. S. A. Ahmed (✉) · I. K. M. Thabet
Chemistry Department, Faculty of Science, Al-Azhar University, Assiut 71524, Egypt
e-mail: abdelaalsaiyd@azhar.edu.eg

E. Elshehy · M. F. Cheira
Nuclear Materials Authority, P.O. Box 530, Cairo, Egypt

A. Ahmad et al. (eds.), *Metal Oxide Based Nanophotocatalyst for Wastewater Purification*, Smart Nanomaterials Technology,
https://doi.org/10.1007/978-981-96-7683-5_15

14,000 people die each day due to release of about two million tons of industrial, sewage, and agricultural waste into the world's water sources each day [60]. Typically, organic, inorganic, biological, and macroscopic contaminants are most water pollutants [91]. The most prevalent heavy metals are copper (Cu), nickel (Ni), zinc (Zn), mercury (Hg), etc. [87]. Dyes that are discharged from the textile industries or other related fields usually pollute water and disturb aquatic life as they block sunlight from reaching the deeper areas of water bodies. Additionally, water contaminated by dyes can lead to skin conditions and, over time, cancer in living things [31]. Pesticides (organophosphate, carbamate, organochlorine, pyrethroids, etc.) are another water pollutant that primarily originates from agriculture and irrigation [51]. Rain and other water activities cause the pesticide-treated agricultural areas to leak, eventually reaching freshwater bodies [45]. Aquatic animals may store these materials, which could cause biomagnification when people ingest them. Long-term exposure to pesticide-contaminated water can result in a variety of health problems [63]. These effluents harm both humans and animals by causing pathogenic infections, metal poisoning, and irritations [27]. A healthy environment for sustainable growth, human well-being, and livelihoods all depends on clean water. Thus, it is vitally important to develop new and effective technology for the environmentally sustainable removal of these hazardous substances from various sources. There are many methods that produces a sustainable energy source in addition to treating wastewater including coagulation [48], advanced oxidation processes (AOP) [56], photodegradation [42, 81], biodegradation [54], and adsorption [72] have been considered as most promising chemical techniques extensively used for removal of various pollutants from wastewater. For all these techniques, nanomaterials play a critical role, as these materials are employed as adsorbents or catalysts for the elimination of toxic pollutants from their solutions. The most important advantages of the utilized materials are ecofriendly, effectiveness, as well as recyclability. Till now, a lot of materials including carbon-based materials [5, 57], metal oxides [11, 72], polymers [82], or agricultural waste-derived materials [9], have been utilized as efficient adsorbents and catalysts for removal of pollutants. In the field of industrial wastewater treatment, SnO_2-based materials have emerged as extremely promising candidates for removing different types of pollutants as they possess unique characteristics such as their crystalline structures, high surface area, and variable surface chemistry. Kumar [38] in addition to its non-toxicity, low cost, and theoretically adequate capacity (782 mA/hg) [55].

In adsorption process, SnO_2 nanoparticles (NPs) can effectively adsorb various pollutants from wastewater. This is due to the exceptional surface reactivity which facilitates interactions with contaminants in industrial effluents. The large surface area provides a high adsorption sites, enabling the effective absorption toward different contaminants [39].

For photocatalytic degradation process, SnO_2-based materials show high ability to degrade various organic pollutants under ultraviolet (UV) light irradiation. The degradation of pollutants into smaller and less hazardous chemicals is caused by redox processes, which are triggered by the creation of electron–hole pairs upon exposure to UV radiation. This technique offers an environmentally friendly and energy-efficient way to remove contaminants, demonstrating how these materials have the power to revolutionize wastewater treatment methods [65].

Essentially, the combination of SnO_2 distinct surface characteristics, adaptable chemistry, and reactivity highlights its status as a rising star in industrial wastewater treatment. Thus, this chapter highlights the potential of SnO_2-based materials in addressing the challenges of wastewater purification.

2 Structure and Properties of SnO_2 Relevant to Wastewater Purification

SnO_2 or stannic oxide is an amphotheric colorless solid with stannous oxide with Sn in oxidation state of 2+, sometimes referred to as cassiterite. This inorganic compound exhibits the qualities of good chemical and thermal stability, as well as great optical transparency and high reflectivity in the infrared spectrum [14]. The rutile phase structure of TiO_2 and SnO_2 displayed a similar crystal structure. In a unit cell, Sn (IV) ion is surrounded by six oxygen ions, and each oxygen ion is bound by three Sn (IV) ions as in Fig. 1 [59].

For water purification, SnO_2 possesses high surface area and high electron mobility (100–200 cm^2/Vs) which enhances their adsorption capacity and reaction kinetics for pollutants [68]. Besides the promising surface area, SnO_2 retains structural integrity under harsh conditions, such as high temperatures and acidic/alkaline environments, making it suitable for repeated use in industrial wastewater treatment [6, 62].

For photocatalytic applications, SnO_2 acts as a photocatalyst under UV and visible light, enabling the degradation of organic pollutants. Its wide bandgap (~ 3.6 eV) is

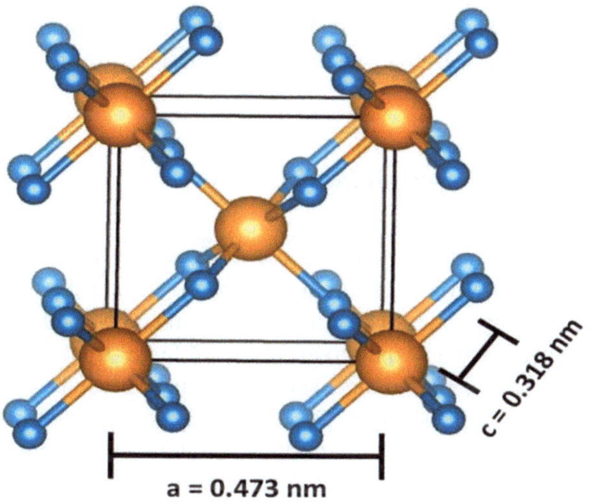

Fig. 1 Unit cell of the tetragonal cassiterite structure of the SnO_2 (Sn is orange spheres and O is blue spheres). Reproduced from Pinto et al. [59] with permission, Copyright 2022 MDPI, Basel, Switzerland

responsible for the photoactivation in the UV range, which responsible for the strong redox potential and generating electron–hole pairs that react with water to produce hydroxyl radicals ($^{\cdot}OH$) and superoxide ions ($^{\cdot}O_2^-$), which break down contaminants like dyes and pharmaceuticals [47, 58, 68]. This allows SnO_2 to prioritize the degradation of organic pollutants via direct electron transfer or reactive oxygen species (ROS) like hydroxyl radicals ($^{\cdot}OH$) and ozone (O_3) [40]. The bandgap of SnO_2 can be adjusted by doping with W or with other oxides (e.g., ZnO, TiO_2), extending light absorption into the visible spectrum. For example, W-doped SnO_2 widened the bandgap to 3.84 eV, improving UV response [67]. Despite the surface modifications via doping can alter the wettability of SnO_2. It has been reported that pure SnO_2 is hydrophilic, and 10% W doping rendered it hydrophobic, aiding in selective pollutant removal [67]. Moreover, SnO_2-functionalized membranes are essential for promoting the pollutant adsorption and antifouling properties [25].

From all the above discussions, the photocatalytic efficiency of SnO_2, structural robustness, and tunable electronic properties make it a cornerstone material in modern water purification systems. Its integration with other nanomaterials (e.g., TiO_2, graphene) and doping strategies further amplifies its utility, addressing challenges in pollutant degradation, disinfection, and sustainable energy use.

3 Synthesis Methods of SnO_2-Based Materials

SnO_2 is a versatile semiconductor used in gas sensors, batteries, solar cells, and photocatalysis. The synthesis method influences its structural, morphological, and functional properties. Various physical and chemical techniques have been utilized to synthesize SnO_2 with specific properties and morphologies. The chemical methods include sol–gel [3, 83], hydrothermal, precipitation [76] mechanochemical method [18], microemulsion [90], and so forth. On the other hand, spray pyrolysis, chemical vapor deposition (CVD) [53], and laser ablation [21] belonging to the physical techniques. Below, we will present the basics of three synthesis methods.

3.1 Sol–gel Method

Sol–gel method is one of the most widely utilized techniques for synthesis nanomaterials. Also, this method helps in varying the morphology of binary and ternary nanocomposites. The materials are first converted to a colloidal suspension which is called a sol and then to an irreversible gel by the hydrolysis and polymerization of compatible precursors. Then, SnO_2 is produced by heat treating the gel at temperatures as high as 800 °C [75]. Typically, to regulate the size and prevent agglomeration of the NPs, certain chemical stabilizers and capping agents, such as oxalic acid or ethylene glycols, can be used during the synthesis of SnO_2 [3, 83]. The size and shape of NPs can also be affected by the pH of a solution, chemical concentration, reaction duration, and calcination temperature [88, 90]. Abdelkader et al. prepared SnO_2 NPs with high crystallinity, tetragonal rutile structure, and average particle size about 13–23 nm. via sol–gel method [2].

3.2 *Hydrothermal Synthesis*

Hydrothermal synthesis method is based on chemical reactions in an aqueous media at elevated temperatures (100–300 °C) and pressures (> 1 atm) within sealed autoclaves. Solvothermal methods are similar but use non-aqueous solvents, enabling better solubility of precursors and tailored nanostructures [20]. These methods are favored for their simplicity, scalability, and eco-friendliness compared to vapor-phase techniques requiring high temperatures (> 700 °C) [12]. Typically, the structural properties of the obtained SnO_2 depend on reaction conditions (temperature, precursors and solvents, additives and templates, pH, and mineralizers). Usually, the higher temperatures (> 120–200 °C) and longer reaction times (5–24 h) promote crystallinity and phase stability. For instance, SnO_2 NPs synthesized at 200 °C for 24 h showed enhanced photocatalytic activity [7]. In this process, both tin chloride ($SnCl_4 \cdot 5H_2O$) and urea are common precursors. Solvent choice (e.g., water, ethanol, or ethylene glycol) influences nucleation rates and morphology. A hydro-solvothermal hybrid approach using mixed solvents can tailor energy gaps (3.64–5.15 eV) [46]. Moreover, some surfactants like polyethylene glycol (PEG) or DL-aspartic acid act as soft templates to control nanostructure shapes (e.g., nanospheres, nanowires) [61]. In addition, the alkalinity (pH ~ 10–12) using NaOH/KOH facilitates SnO_2 formation by promoting hydrolysis and condensation reactions [12]. Till now, great efforts have been made to prepare SnO_2 by this technique. For instance, SnO_2 NPs with an average particle size of 3.0 ± 0.5 nm were prepared by adjusting pH at 12, and temperature of 150 °C for 24h. Thermal treatment of the obtained SnO_2 NPs at 300 °C for 1h under 10% H_2/Ar leads to prepare SnO_2 NPs with an average diameter of 3.3 ± 0.6 nm. The thermally treated SnO2 displayed smaller surface area (92 m^2/g) as compared with that as-synthesized SnO_2 (130 m^2/g). Recently, SnO_2 NPs in nanorods (NRs), NPs, and nanopillars morphologies were prepared by solvothermal process at 180 °C for 6 h [92]. Another study reported the preparation of SnO_2 NPs [64] by various glycol solvents (ethylene glycol, diethylene glycol, tetraethylene glycol, polyethylene glycol 400, and polyethylene glycol 600). As presented in TEM images (Fig. 2a–e), the change the polyols affects the size and shape of the SnO_2 NPs as indicated by the Sn-EG, Sn-DEG and Sn-TEG have a nearly spherical shape with average particles sizes of 2.49, 2.5, and 2.65 nm, respectively, while Sn-PEG4 and Sn-PEG6 have a nearly cubical morphology with sizes of 4.5 and 5 nm, respectively. The XRD patterns (Fig. 2f) show that the rutile SnO_2 NPs were synthesized using a polyol solvent with molecular weights ranging from 2 to 3.5 nm. Three varieties of hierarchical SnO_2 microspheres with one-dimensional nanorods were recently created by Shi et al. [66] using a straightforward hydrothermal approach.

These include urchin-like microspheres (SN-1), flower-like microspheres (SN-2), and hydrangea-like microspheres (SN-3), as shown in Fig. 2.

Composite based on SnO_2 also has been prepared by this technique. For example, Cao et al. [16] used hydrothermal method to synthesis SnO_2/MoO_{3-x} nanocomposites with ultrafine SnO_2 NPs uniformly embedded into an amorphous MoO_{3-x} matrix. Furthermore, Ren et al. [66] prepared three distinct hierarchical SnO_2 microsphere structures known as urchin-like (SN-1), flower-like (SN-2), and hydrangea-like (SN-3) via hydrothermal technique. Figure 2 displays FE-SEM images illustrating the structures and surface morphologies of the three SnO_2 samples. FE-SEM images of the synthesized SN-1 sample are shown in Fig. 3a, b, revealing that it is composed of numerous nanorods arranged in a distinctive radial pattern resembling urchin-like spheres. The average diameter of these spheres and rods are 430 and 20 nm, respectively. Furthermore, each rod of the SN-1 sample features a pointed tip and a smooth surface. The rods are not tightly packed but are interwoven in a way that creates the urchin-like structure, with ample space between them. When the hydrothermal synthesis time is reduced to 12 h, the SN-2 sample is produced. FE-SEM images of the SN-2 sample in Fig. 3c, d reveal a slight change in its structure and morphology. The SN-2 sample still maintains a spherical, flower-like shape made up of one-dimensional nanorods. However, compared to SN-1, the nanorods in SN-2 have a larger diameter (55 nm) and a rectangular cross-section. The nanorods are more tightly packed, and the flower-like spheres appear to be in the bud stage, resembling the early formation of SN-1. This suggests that shorter hydrothermal reaction times lead to a less developed structure, while longer times promote further growth. Both of the above samples were synthesized with the addition of 0.04 g of citric acid, which plays a crucial role in determining the morphological formation. To confirm the significant role of citric acid in the synthesis process, SN-3 sample was prepared by increasing the citric acid amount to 0.08 g, while keeping other conditions identical to those used for the SN-2 sample. FE-SEM images of the SN-3 sample (Fig. 3e, f) reveal that it still consists of one-dimensional nanorods, but the morphology and the core structure of the three-dimensional arrangement have changed. Compared to the SN-2 sample, the nanorods in SN-3 are thicker (145 nm) and resemble hydrangea-like shapes, with numerous nanoparticles on their surface, creating a rough texture. The prepared materials were then tested for their ability to detect indoor formaldehyde (HCHO) gas. Among them, the SN-1-based gas sensor exhibited superior performance, showing higher sensitivity and faster response and recovery times compared to the SN-2 and SN-3 sensors. This enhanced performance is primarily attributed to the finer nanorod structure of the SN-1 sample. These findings highlight the structural advantages of SN-1 and its potential as an effective material for HCHO detection.

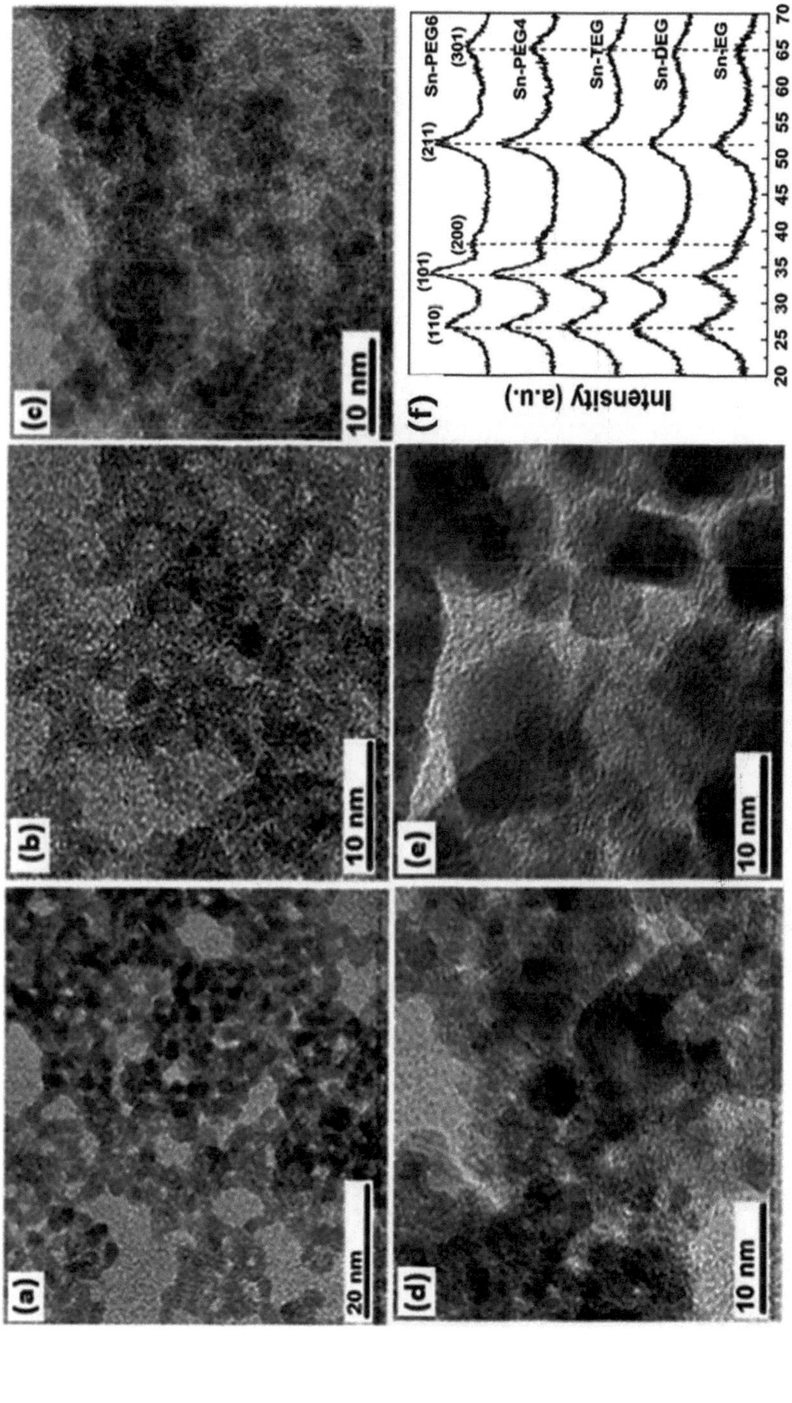

Fig. 2 TEM images of **a** Sn-EG, **b** Sn-DEG, **c** Sn-TEG, **d** Sn-PEG4 and **e** Sn-PEG6 NPs. **f** XRD patterns of the polycl-functionalized SnO_2 NPs. Reproduced from [64] with permission, Copyright 2018 Royal Society of Chemistry

Fig. 3 FE-SEM images of SnO_2 samples: **a**, **b** SN-1, **c**, **d** SN-2, **e**, **f** SN-3. Reproduced from [66] with permission, Copyright 2022 Licensee MDPI, Basel, Switzerland

3.3 Chemical Vapor Deposition (CVD)

CVD is a versatile technique used to produce thin films and coatings of various materials, including SnO_2. This method allows for the precise control of film thickness, composition, and microstructure, making it suitable for applications in electronics, optics, and sensors. In CVD process, gaseous precursors are carried to a reactor chamber where they react with a heated substrate to form a solid coating on the substrate. Both the gas phase above the substrate and the substrate's surface may be the site of the reactions [71]. The ultimate characteristics of the layer are determined by the kind of precursors and the reactions that occur. SnO_2 NPs are frequently deposited in industrial processes using both inorganic and metalorganic precursors in combination with O_2 and/or H_2O because they offer fast enough deposition rates for the process to be profitable [84]. CVD process is important for the continuous process of tin oxide deposition on flat glass, where the speed of the glass traveling along the line limits the deposition time. Usually, a coating with a thickness of 1000–7000 Å may be deposited in 1–3 s. SnO_2 NPs can be deposited from a wide variety of organometallic precursors, although the most widely used ones in industry are monobutyltintrichloride (n-$C_4H_9SnCl_3$, MBTC), dimethyltindichloride ($(CH_3)_2SnCl_2$, DMTC), and tin tetrachloride ($SnCl_4$, TTC) [84]. Recently, Nguyen et al. [55] reported the preparation of SnO_2 NPs attached on carbon fibers via a two-step CVD approach (Fig. 4a). As shown by TEM images (Fig. 4b–d), by adjusting the temperature, the changes in morphologies were observed.

3.4 Electrospinning

Electrospinning is mostly used to prepare 1-D nanorods and nanotubes at a very low cost. This method is a versatile and cost-effective method for synthesizing SnO_2 nanofibers, which are widely used in gas sensing, energy storage, and optoelectronics due to their high surface-to-volume ratio and tunable properties [50]. This procedure makes use of a simple configuration with four key parts known as a high-voltage power supply, a syringe pump, a spinneret (a syringe with a blunt-tip needle), and a fiber collector (Fig. 5). The growth of the nanostructure on the substrate electrode depends on several factors. The applied electric fields should range from 1.25 to 1.4 kV/cm to overcome surface tension and initiate jet formation. And the flow rate should be optimized between 0.2 mL/h and 5 μL/min to control fiber diameter and uniformity. The distance between the needle-to-collector typically should be around 15–40 cm as the longer distances enhance fiber stretching and thinning [70]. Furthermore, it should be adjusting the humidity and temperature as they influence solvent evaporation rates and fiber morphology [4].

The electrospinning is a multistep process [4]. In which the polymer is first dissolved in a highly volatile solvent before being pumped out of a syringe at a controlled rate. The applied electrostatic force leads to separate the positive and

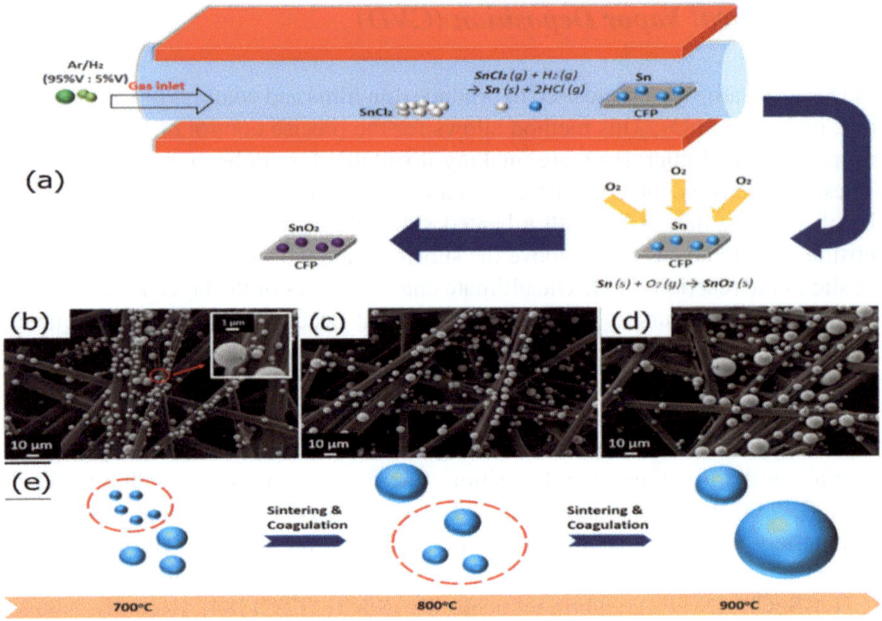

Fig. 4 **a** Schematic diagram of CVD process for SnO_2 NPs growth. The SEM images of SnO_2 NPs at different temperatures: **b** 700 °C, **c** 800 °C, and **d** 900 °C. **e** Schematic illustration of temperature-dependent particle-growth process. Reproduced from [55] with permission, Copyright 2020 Springer Nature

negative charges of the liquid, and charges with the same sign as the needle's polarity migrate toward the surface, creating a charged polymer droplet at the tip of the needle. The surface charge density on the droplet increases as the electrostatic field strength increases, which improves the mutual charge repulsion on the surface of the liquid. Thus, the surface area of the droplet then increases, hence reducing the repulsion. Therefore, the drop becomes a Taylor cone. Surface tension is eventually overcome by electrostatic repulsion, resulting in the emergence of a jet that is rapidly approaching the collector. The solvent evaporates, and the polymer solution elongates and whips as the jet moves closer to the collector. This process leads to prepare very tiny fibers. In general, the electrospinning process includes four sequential steps: the creation of a Taylor cone at the tip of the needle, the straight-line ejection of the charged jet, the expansion of the jet's diameter and the development of physical instability, and, lastly, the solidification and deposition of fibers on the collector [4].

For the electrospinning of SnO_2 nanofibers, a precursor solution is typically prepared by dissolving a tin salt (e.g., $SnCl_4{\cdot}5H_2O$) in a solvent mixture (e.g., ethanol/DMF or water) along with a polymer such as polyvinylpyrrolidone (PVP) or polyvinyl alcohol (PVA) to adjust viscosity and enable fiber formation [70]. The polymer acts as a structural template, ensuring uniform fiber formation during electrospinning. For example, PVP/$SnCl_4$ composites are common, while PVB (polyvinyl

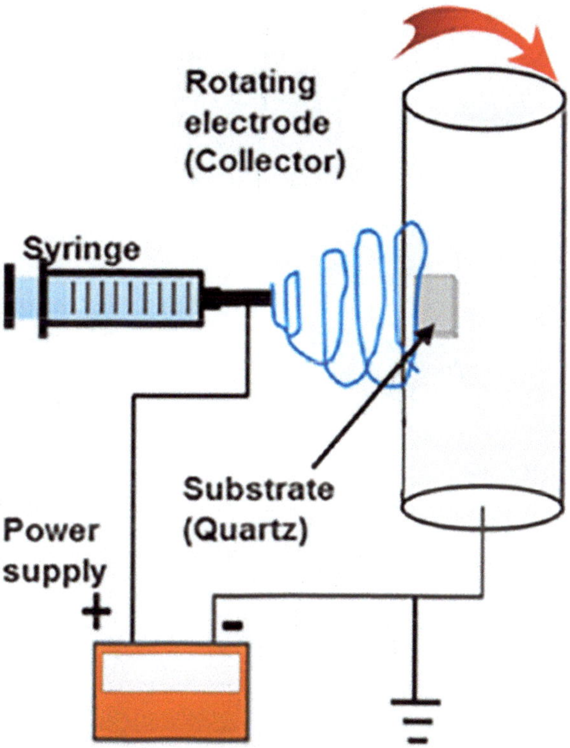

Fig. 5 Schematic representation of electrospinning apparatus. Reproduced from [1] with permission, Copyright 2014 American Chemical Society

butyral) is used for creating hollow nanotubes via Ostwald ripening [78]. Till now, great work has been done in preparing SnO_2 NPs by this technique. For instance, Jiang et al. [32] prepared SnO_2 with tube-like structure via electrospinning at appropriate thermal treatment in atmosphere (Fig. 6a–d). The obtained SnO_2 nanotubes (NTs) consist of 5–10 nm SnO_2 NPs, as Fig. 6e.

Later study by Ab Kadir et al. [1] prepared hollow and filled well-aligned electrospun SnO_2 nanofibers with diameters ranging from 80 to 400 nm at a low temperature. The mechanism for the formation is described in Fig. 7a, b. As in Fig. 7c–h adjusting the concentration of polyacrylonitrile in the solution for electrospinning determines the diameter of the obtained nanofibers.

Jose et al., reported that adjusting the tin precursor concentration successfully prepares multiporous nanofibers (MPNFs), porous nanofibers (PNFs), and nanowires (NWs) of SnO_2. The MPNFs displayed the higher surface area among the other morphologies [85].

Some composites have been prepared by this technique. For example, Lue et al., prepared ZnO-SnO_2 composite nanofibers (Fig. 8a). The SEM (Fig. 8b) and TEM (Fig. 8c) images of the as-prepared ZnO-SnO_2 nanofibers after annealing consist of randomly oriented nanofibers with diameters of 80–150 nm and lengths of 0.5–2 μm.

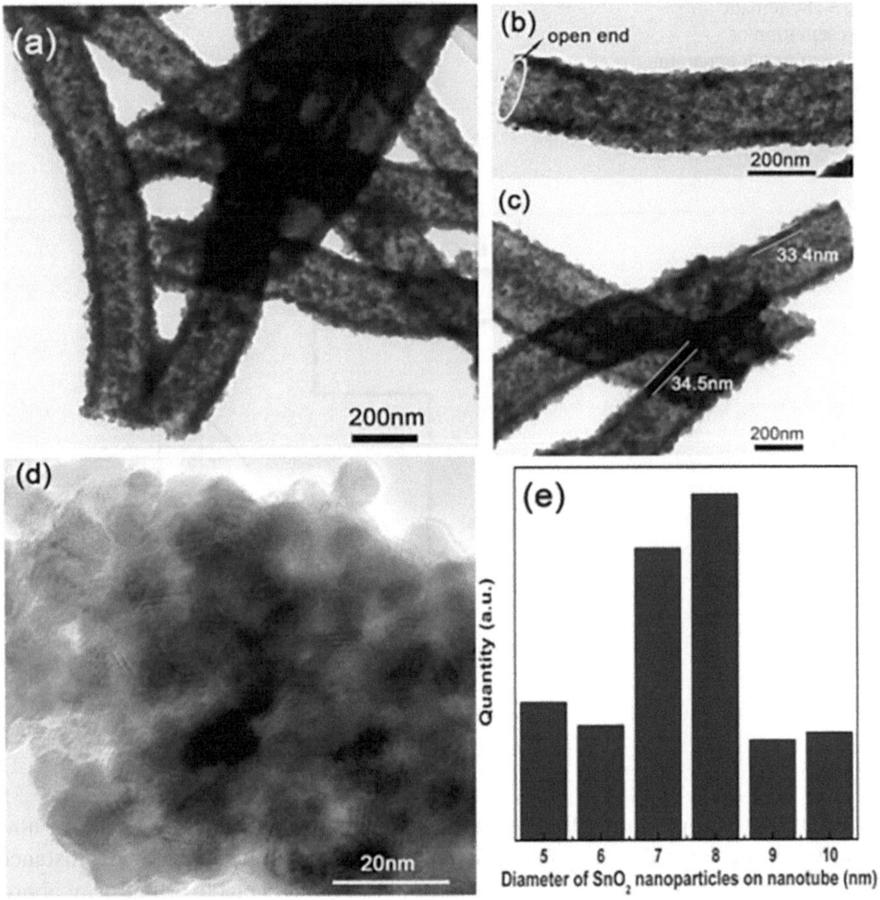

Fig. 6 **a–c** TEM images of TONTs obtained by calcination at 600 °C for 5h, **d** HR-TEM image of SnO_2 NTs, and **e** diameter distribution of SnO_2 NPs on SnO_2 NTs Reproduced from [32] with permission, Copyright 2012 Royal Society of Chemistry

Furthermore, the ZnO-SnO_2 nanofibers are made up of nanoparticles, and their rough surface is a result of the thermal breakdown of PVP brought on by annealing [43].

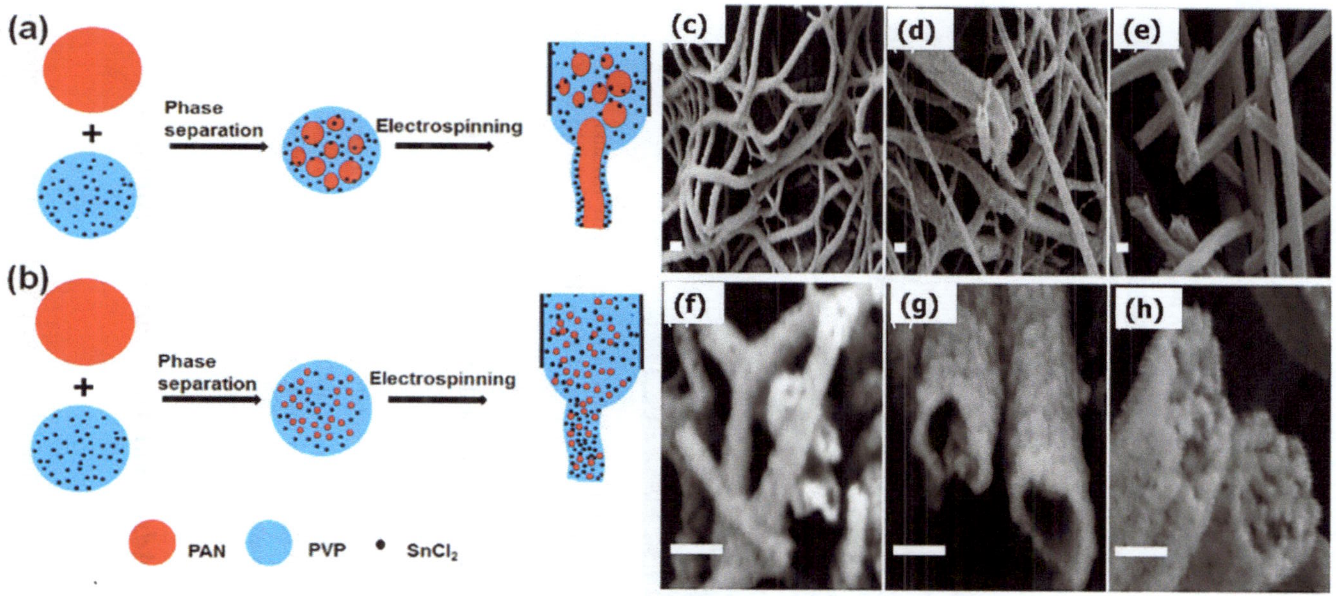

Fig. 7 Scheme for the formation of SnO_2 nanofibers electrospun using solutions of **a** PAN 1 and PAN 2 (hollow structure) and **b** PAN 3 (solid structure). SEM images of the SnO_2 nanofibers electrospun using solutions of **c** PAN 1 (1 mL), **d** PAN 2 (2 mL), and **e** PAN 3 (3 mL). **f**–**h** Magnified SEM of **c**–**e**, respectively. Reproduced from [1] with permission, Copyright 2014 American Chemical Society

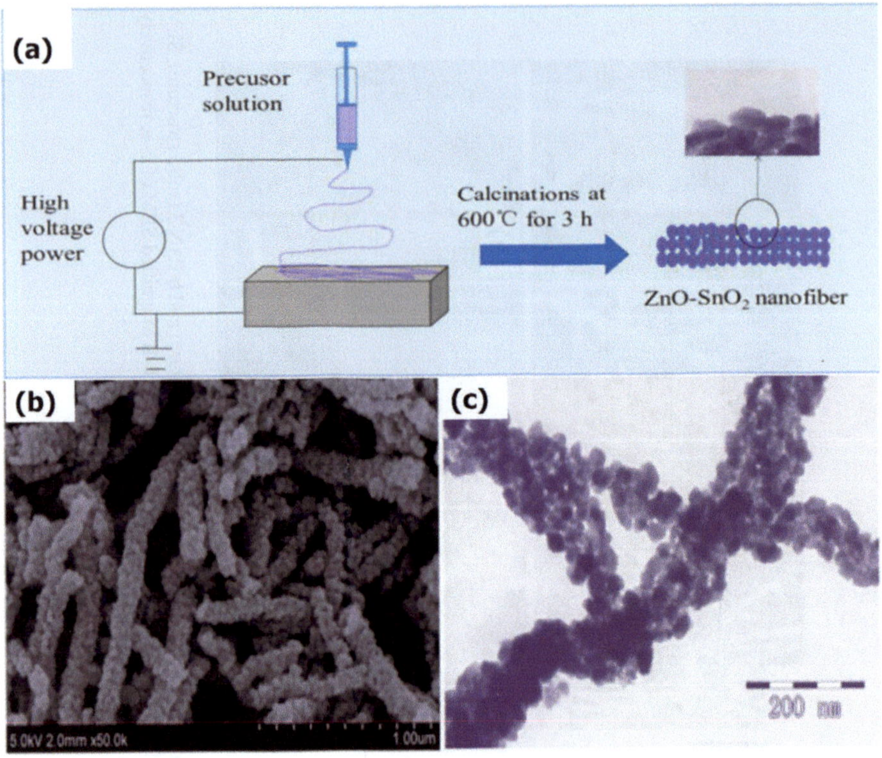

Fig. 8 **a** Scheme for electrospun ZnO-SnO_2 nanofibers. SEM (**b**) and TEM images (**c**) of electrospun ZnO-SnO_2 nanofibers. Reproduced from [43] with permission, Copyright 2018 Frontiers

4 Applications of SnO_2-Based Materials in Wastewater Purification

SnO_2-based materials have been widely used for the adsorption of heavy metals, photocatalytic degradation of organic pollutants, and advanced oxidation processes. Their antibacterial properties also make them effective in eliminating pathogenic microorganisms from wastewater.

4.1 SnO_2-Based Adsorbent Materials in Wastewater Purification

SnO_2 materials display promising adsorptive properties for wastewater purification due to its unique structural, chemical, and catalytic properties. SnO_2-based materials exhibit high surface area, tunable surface chemistry, and active sites (e.g., oxygen vacancies and hydroxyl groups). These features enable both physisorption (via van der Waals forces) and chemisorption (through covalent bonding or electrostatic interactions) [24]. For example, Abdelkader et al. [2] utilized SnO_2 nanoparticles with tetragonal rutile structure and average particle size about 13–23 nm synthesized via sol–gel method as adsorbent toward Congo red (CR) azo dye. The untreated sample SnO_2-80 exhibited high efficiency (84.41%) after 60 min, which was 2 times as much as that of SnO_2-450 sample. The adsorption process was obeyed the pseudo-second-order kinetic model, and the adsorption isotherm follows the Langmuir model.

Further enhancement of adsorption efficiency was achieved by preparing SnO_2-based composites. For inatsnce, Dhal et al. [22] prepared hematite (α-Fe_2O_3) and Fe_2O_3-SnO_2 nanorods by wet chemical method. SEM and TEM results revealed the formation of uniform nanorods with typical diameter of 100–200 nm. The obtained α-Fe_2O_3 and Fe_2O_3-SnO_2 nanorods were used as adsorbents toward CR dye from aqueous solution. The Fe_2O_3-SnO_2 (Fe: Sn = 8:2) nanorod showed adsorption capacity of 182 mg/g. Shrivastava et al. [33] prepared SnO_2 nanorods by co-precipitation technique. The morphology of the prepared SnO_2 nanorods are presented by SEM images (Fig. 9a, b) and TEM images (Fig. 9c, d). The prepared SnO_2 nanorods utilized as adsorbent toward methylene blue (MB) dye from aqueous solution. The maximum removal of MB up to the 66% at pH = 7. The experimental adsorption of MB onto SnO_2 nanorods follows the pseudo-second-order kinetic model and obeys the Langmuir isotherm indicating a chemisorption process. The thermodynamic study indicating that adsorption is spontaneous and favorable endothermic process. Another study by Dutta et al., reported the preparation of mesoporous SiO_2 NPs decorated with SnO_2 quantum dots (QDs-MSN) via simple two-step method [23]. The prepared QDs-MSN at room temperature could adsorb $\sim$ 100% of MB within 5 min. The experimental adsorption data was fitted with Langmuir isotherm model with maximum monolayer adsorption capacity of 73.15 mg/g.

Recently, our colleague prepared spongy sphere-like SnO_2 and 3D cubic-like MgO (SnO_2/MgO) as a promising adsorbent to remove sunset yellow (SSY) dye from real wastewater at neutral pH conditions [52]. The prepared composites could adsorb about 99% of SSY from aqueous solution at pH 7 within 60 min. The adsorption of SSY fitted by pseudo-second order and the Langmuir model, indicating a chemisorption process.

Another composite based on reduced graphene oxide (rGO-SnO_2) with an average particle size of 3.67 nm was prepared by hydrothermal method (Fig. 10a) [89]. The TEM image of the prepared rGO-SnO_2 composites (Fig. 10b) showed that the SnO_2 NPs are uniformly dispersed on the graphene nanosheets. The rGO-SnO_2 composites displayed high adsorptive ability for efficient removal of organic dyes from wastewater. The prepared rGO-SnO_2 composites exhibited fast adsorption rates toward rhodamine B (RhB), MB, methyl green (MG), and methyl red (MR) dyes. The maximum adsorption capability of the prepared rGO-SnO_2 composite toward Rh B, MB, MG, and MR reached 115.4, 72.2, 76.5, and 108.3 mg/g, respectively. Recently, SnO_2/CeO_2 composite was synthesized by sol–gel method to be utilized as adsorbent toward alizarin dyes from aqueous solutions [30]. The BET surface area

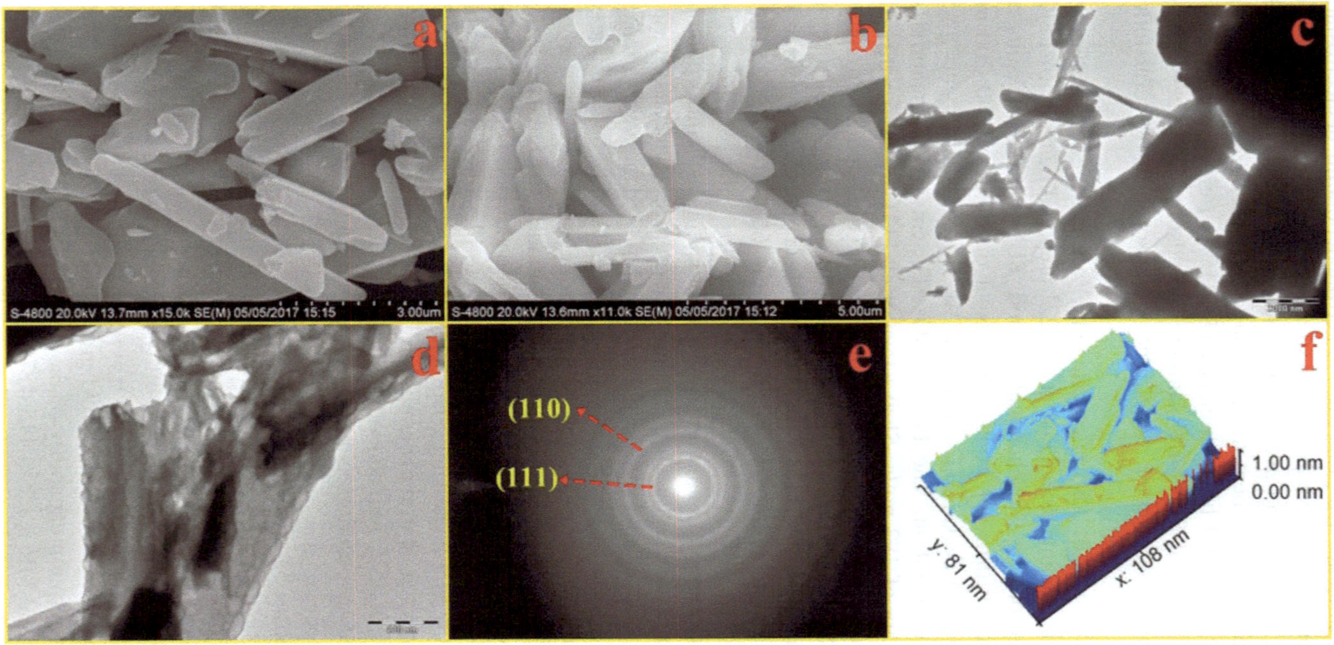

Fig. 9 SEM images of SnO_2 nanorods (**a, b**), TEM images (**c, d**), SAED pattern of SnO_2 nanorods (**e**), and 3D view of SnO_2 nanorods (**f**). Reproduced from [33] with permission, Copyright 2020 Springer Nature

of the prepared SnO_2/CeO_2 composite was 18.970 m^2/g. As indicated from TEM image (Fig. 10c, d), the SnO_2/CeO_2 composite displayed spherical morphology with particle sizes varying from 11.5 to 30 nm. About 96.4, 87.8, and 97.3% of the alizarin yellow, alizarin red, and alizarin-3-methylimino-diacetic acid dyes were eliminated from aqueous solutions, respectively.

Also, SnO_2-based materials have been utilized as adsorbents toward heavy metals from wastewater. For instance, Kumar et al. [39] reported the adsorption of toxic metal ions Cd(II) and Pb(II) onto SnO_2 NPs prepared via precipitation method. The adsorption study confirmed that SnO_2 NPs that the adsorption of heavy metal follows the Langmuir model with a maximum adsorption capacities of 1275.5 and 1265.8 mg/g for Cd(II) and Pb(II), respectively. Further study reported the fabrication of SnO_2 NPs coated with mesoporous silica and further modified with 3-Aminopropyl triethoxysilane to render $SnO_2/PSi/NH_2$ nanocomposite [34]. The maximum adsorption capacity of $SnO_2/PSi/NH_2$ toward Pb (II) ions was 653.62 mg/g. Recently, Nazeeha S. Alkayal reported the preparation of C-PMMA/SnO_2 composites for adsorbents of Cr (III) [10]. The C-PMMA/SnO_2 composite displayed highly efficient adsorption toward Cr (III) ions from wastewater at pH of 6 within 60 min.

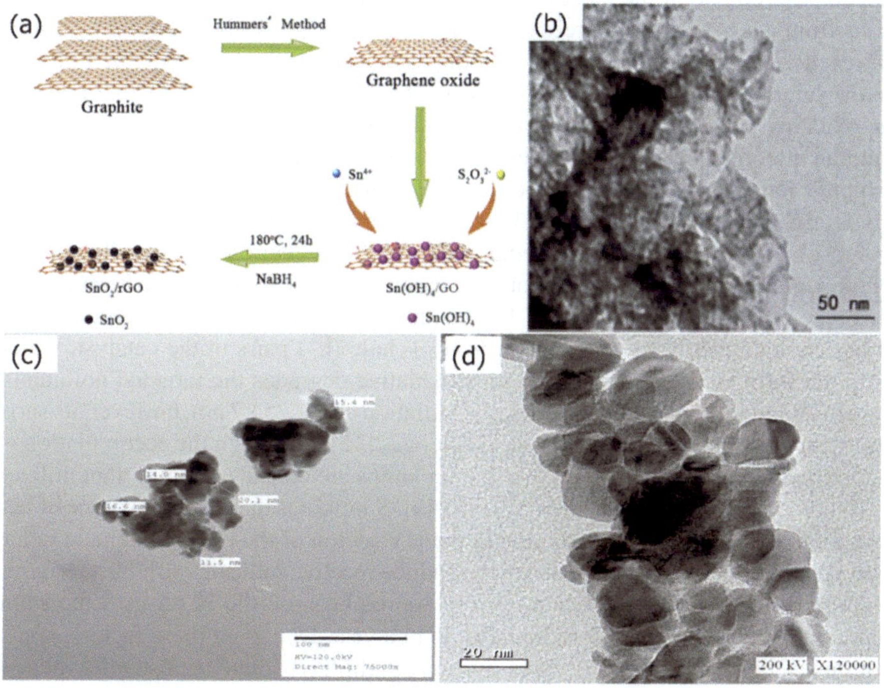

Fig. 10 **a** Scheme for synthesis rGO-SnO_2 composite and **b** TEM image of rGO-SnO_2 composite. Reproduced from [89] with permission, Copyright 2016 Royal Society of Chemistry. **c**, **d** TEM images of SnO_2/CeO_2 composite. Reproduced from [30] with permission, Copyright 2020 MDPI, Basel, Switzerland

The maximum adsorption capacity of the prepared C-PMMA/SnO_2 composite was 1.76 mg/g. The adsorption of Cr (III) ions onto the C-PMMA/SnO_2 composite obeyed the Langmuir model. Sami et al., prepared MgO@SnO_2 composite modified with polyvinylpyrrolidone (PVP) by green technique using Pistacia lentiscus leaf extract to utilize as adsorbent toward heavy metals in petroleum wastewater [69]. Adsorption studies evaluated the nanocomposites' efficiency in eliminating 11 heavy metals from petroleum wastewater. The Cr (III), Mo (II), and Sb (II) ions are completely removed within 20 min with about 99% removal efficiency.

From above discussion, it can be shortly noted that the SnO_2-composite materials are highly efficient adsorbents for organic dyes and heavy metals.

4.2 *SnO_2-Based Photocatalysts in Wastewater Purification*

Although the various techniques that have been utilized for wastewater purifications, most of them do not match the long-term and large-scale usability. Due to the utilization some toxic and expensive reagents [77]. Advanced oxidation processes (AOPs) are considered one of the most promising techniques for eliminating organic pollutants from water or other sources [86]. The working of AOPs is justified by creation of reactive hydroxyl radicals ($OH^{\cdot}$) on the catalyst surface to degrade the pollutant in the sample via oxidation reaction. Typically, the utilized catalysts should display an oxidative behavior under exposed to light. The photocatalytic process is considered an effective process when the OH radical has an oxidation potential of 2.8 V, which is lower than that of fluorine (F2) and higher than that of hydrogen peroxide (H_2O_2). Since this technique have been developed by Fujishima in 1972 [26], great interest has been devoted to remediations of various environmental pollutants via photodegradation technique. Although a variety of light sources have been used in this type of photodegradation, UV light is the commonly used light source because it has enough energy to create electron (e^-)- hole (h^+) pairs in the catalyst, which help the formation of radicals and thus oxidative degrades the targeted pollutants. Usually, mercury vapor lamps with a UV radiation at $\lambda = 253.7$ nm, however, its short life span and high-cost operation are challenges [15]. Although the accumulation of ozone due to the long time uses, the Xenon lamps are considered important utilized UV source with irradiation at $\lambda = 240$ nm [49]. Sunlight is the cheapest source of UV radiation, as $\approx$ 4–5% of its radiation in the UV region of the electromagnetic radiation spectrum with sufficient energy to produce reactive radicals [80]. Furthermore, sunlight is a particularly efficient energy source because the Earth uses the same amount of energy in an hour as humans do in a year. Due to all these factors, applications of photocatalytic processes using sunlight have gained substantial interest [37]. Numerous metal oxides, such as ZnO, SnO_2, and TiO_2 have been studied and assessed for their potential as photocatalysts. SnO_2 materials have potential for application in wastewater treatment due to its unique physicochemical properties such as high chemical stability, non-toxicity, and tunable surface chemistry. Its application in photocatalysis have shown great promise for the treatment of industrial wastewater. SnO_2-based catalysts use their special qualities to help in degradation of many organic contaminants through redox reactions, providing a sustainable and effective method of treating wastewater [74].

SnO_2 is frequently distinguished by oxygen vacancies that improve its n-type semiconductor characteristics. The material is ideal as a photocatalyst due to its 3.6 eV band gap, which corresponds to activation with photons with a wavelength of 350 nm (UV-A range) [8]. When exposed to UV light, the e^-– h^+ couples are generated, which oxidatively degrades the organic molecules. A great research effort on utilizing SnO_2 to eliminate various pollutants from wastewater has been done. The steps of the photocatalytic degradation process are presented in the following equations.

$$SnO_2 + hv \rightarrow SnO_2\left(e^- + h^+\right) \tag{1}$$

$$SnO_2\left(h^+\ VB\right) + H_2O_{ads} \rightarrow SnO_2 + \left(OH^{\cdot}_{ads} + H^+\right) \tag{2}$$

$$SnO_2\left(h^+\ VB\right) + OH^-_{ads} \rightarrow SnO_2 + OH^{\cdot}_{ads} \tag{3}$$

$$SnO_2\left(e^-\ CB\right) + O_{2\,ads} \rightarrow SnO_2 + O_2^{-\cdot} \tag{4}$$

$$O_2^{-\cdot} + H^+ \rightarrow HO_2^{\cdot} \tag{5}$$

$$O_2^{-\cdot} + H_2O \rightarrow HO_2^{\cdot} + OH^- \tag{6}$$

$$\text{Organic pollutant} + OH^{\cdot}/HO_2^{\cdot} \rightarrow CO_2 + H_2O \tag{7}$$

The process starts when the photocatalyst is activated and e^-–h^+ couples are formed (Eq. 1). These couples travel to the surface of the catalysts and produce extremely reactive radicals. The produced h^+ generate hydroxyl radical by the reaction of absorbed water (Eq. 2) or hydroxyl anions (Eq. 3). The e^- react with dissolved oxygen to form peroxy radical anions (Eq. 4) After that, these radicals anions are protonated to produce extremely reactive hydroperoxyl radicals H^+ (Eq. 5) or H_2O (Eq. 6). Then, the organic molecules reacted with the obtained hydroxyl radicals and transformed into CO_2 and H_2O (Eq. 7). The overall process is presented in Fig. 11 [8].

Till now, there are many research groups utilized SnO_2-based materials for water purification. For instance, Koppala et al. [35] reported the synthesis of novel worm-like structured SnO_2 photocatalysts by chemical precipitation route to be utilized as a catalyst for Rhodamine B (RhB) dye. The prepared SnO_2 nanoparticles are amorphous with a tetragonal structure. The prepared SnO2 could degrade about 92% of RhB in 120 min under UV light. Another study by Ma et al. [44] reported

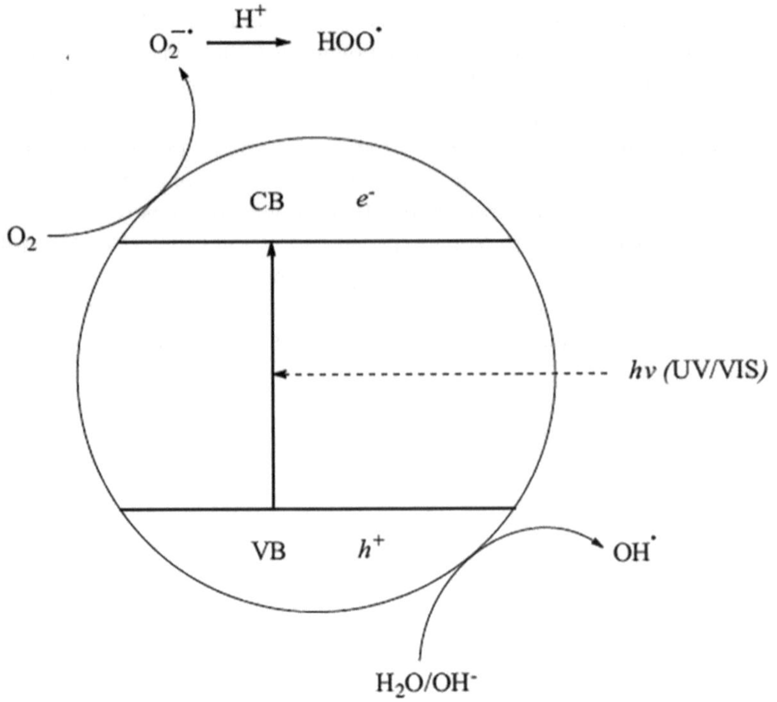

Fig. 11 Scheme for the AOP on the material surface. Reproduced from [8] with permission, Copyright 2017 Elsevier

the utilization of precipitation method to prepare SnO_2 with an average particle size less than 50 nm. The prepared SnO_2 utilized as photocatalysts for the degradation of CR dye from aqueous solution. The prepared SnO_2 successfully degrade about 100% of CR dye within 97 min at 47 °C. Additionally, the prepared SnO_2 displayed excellent stability after five serial photodegradation cycles.

There are some studies that report the synthesis of SnO_2 photocatalysts with an average particle size of 32.1 nm by green methods. For example, Tasisa et al. [79] prepared SnO_2 NPs by using *Croton macrostachyus* leaf extract. The prepared SnO_2 displayed absorbance in visible light, which enhanced the photocatalytic activity in the visible light region. The prepared SnO_2 effectively utilized as photocatalyst toward RhB and MB dyes under visible light. Other work has been done by Haq et al. [29] reported the green synthesis of the NiO-SnO_2 NC using *F. elastic* leaf extract to be utilized toward degradation of MO in aqueous solution under irradiation with full light spectrum.

Despite the successful utilization of pure SnO_2 in the photodegradation process, the UV light must be used due to the large high activation energy results from the large bandgap of pure SnO_2 (3.6 eV), and thus increases the recombination rate of the photogenerated electrons and holes. This restricts the utilization of SnO_2 photocatalyst in large-scale industrial applications. To overcome these limitations and enhance the photocatalytic activity, the recombination of the e^-–h^+ pairs need to be inhibited.

There are many ways to reduce the bandgap of SnO_2 semiconductors, one way is doping SnO_2 with another metal oxide with different band gap. This leads to lowering the bandgap (lowering the activation energy) and thus enhancing the photocatalytic activity [36].

There are many materials that have been successfully dopant for SnO_2. Semiconductors are considered important dopants for SnO_2. For example, Lamba et al. [41] doped SnO_2 by ZnO NPs by hydrothermal technique. The prepared ZnO-doped SnO_2 revealed tetragonal rutile phase of SnO_2 and wurtzite hexagonal phase of ZnO. The photocatalytic activity of prepared photocatalyst was evaluated by the degradation of MB dye. The synthesized ZnO-doped SnO_2 exhibits best photocatalytic performance than those achieved with pure SnO_2.

NiO is also utilized as dopant to enhance the photocatalytic performance of SnO_2 as reported by Sery et al. [73]. In this study, they utilized sol–gel and sonochemical methods to prepare SnO_2 and NiO NPs. The nanoparticles with an average size less than 10 nm was obtained at 250 °C, while at 750 °C, the particle size increased to about 30–40 nm. The estimated band gap energies for the doubly doped tin oxide $Sn_{0.094}Ti_{0.03}Ni_{0.03}O_2$ and nickel oxide $Ni_{0.094}Ti_{0.03}Sn_{0.03}O$ are 3.24 eV and 3.27 eV, respectively, which are lower than those for their corresponding pure oxides. The photodegradation study toward Coomassie Brilliant Blue dye R (CBBR) confirmed that the oxides prepared by sonochemical method are higher than those achieved by the oxides prepared by sol–gel method. TiO_2 also significantly enhanced the overall photocatalytic performance of SnO_2. There are many studies reported the impact of TiO_2 on enhancing the photocatalytic activity of SnO_2 as reviewed by Chen et al. [19]. Many studies on utilizing various dopants such as metals, non-metals, and so on have been reviewed by Al-Hamdi et al. [8].

5 Challenges and Future Directions

Even though SnO_2-based materials show promise in wastewater purification, there are still several obstacles should be taken into consideration for future prospectives. The limited efficiency of pure SnO_2 under solar light is limited by its limited UV absorption range. But improvements in composite synthesis and doping have helped get around this restriction. One significant problem that lowers photocatalytic degradation efficiency is electron–hole recombination. To boost performance, strategies that facilitate charge separation such as the use of co-catalysts and heterostructures are crucial. Large-scale, low-cost production of SnO_2-based materials is still a difficulty. To create synthesis techniques that are affordable and scalable for practical uses, more research is required. The long-term environmental effects of employing nanoparticles in large-scale wastewater treatment systems must be carefully evaluated, even though SnO_2 is widely regarded as non-toxic.

6 Conclusion

SnO_2-based materials are considered promising candidates for wastewater purification, as they possess high adsorptive and photocatalytic degradation toward various pollutants in our environment. Their performance has been greatly enhanced by preparing composites based on SnO_2, doped materials, and surface alterations, which makes them more effective for environmental applications. However, for SnO_2-based photocatalysts to become economically viable, issues including charge recombination, scalability, and UV light dependence must be resolved. There is a lot of promise for creating effective and long-lasting solutions to stop water contamination with more study in this field.

References

1. Ab Kadir R, Li Z, Sadek AZ, Abdul Rani R, Zoolfakar AS, Field MR, Ou JZ, Chrimes AF, Kalantar-zadeh K (2014) Electrospun granular hollow SnO_2 nanofibers hydrogen gas sensors operating at low temperatures. J Phys Chem C 118(6):3129–3139. https://doi.org/10.1021/jp411552z
2. Abdelkader E, Nadjia L, Rose-Noëlle V (2016) Adsorption of Congo red azo dye on nanosized SnO_2 derived from sol-gel method. Int J Ind Chem 7(1):53–70. https://doi.org/10.1007/s40090-015-0061-9
3. Abdelkader E, Nadjia L, Rose-Noëlle V (2016) Adsorption of Congo red azo dye on nanosized SnO_2 derived from sol-gel method. Int J Ind Chem 7:53–70
4. Ahmadi Bonakdar M, Rodrigue D (2024) Electrospinning: processes, structures, and materials. Macromol 4(1):58–103. https://doi.org/10.3390/macromol4010004
5. Ahmed ASA, Sanad MMS, Kotb A, Negm ANRM, Abdallah MH (2023) Removal of methyl red from wastewater using a NiO@Hyphaene thebaica seed-derived porous carbon adsorbent: kinetics and isotherm studies. Mater Adv 4(14):2981–2990. https://doi.org/10.1039/D3MA00226H
6. Akgul FA, Gumus C, Er AO, Farha AH, Akgul G, Ufuktepe Y, Liu Z (2013) Structural and electronic properties of SnO_2. J Alloy Compd 579:50–56. https://doi.org/10.1016/j.jallcom.2013.05.057
7. Akhir MAM, Mohamed K, Lee HL, Rezan SA (2016) Synthesis of tin oxide nanostructures using hydrothermal method and optimization of its crystal size by using statistical design of experiment. Procedia Chem 19:993–998
8. Al-Hamdi AM, Rinner U, Sillanpää M (2017) Tin dioxide as a photocatalyst for water treatment: a review. Process Safety Environ Prot 107:190–205. https://doi.org/10.1016/j.psep.2017.01.022
9. Alhujaily A, Yu H, Zhang X, Ma F (2020) Adsorptive removal of anionic dyes from aqueous solutions using spent mushroom waste. Appl Water Sci 10(7):183. https://doi.org/10.1007/s13201-020-01268-2
10. Alkayal NS (2022) Fabrication of cross-linked PMMA/SnO_2 nanocomposites for highly efficient removal of chromium (III) from wastewater. Polymers 14(10). https://doi.org/10.3390/polym14102101
11. Andrade-Guel M, Cabello-Alvarado C, Bartolo-Pérez P, Medellin-Banda DI, Ávila-Orta CA, Cruz-Ortiz B, Espinosa-Muñoz A, Cadenas Pliego G (2022) Surface modification of TiO_2/ZnO nanoparticles by organic acids with enhanced methylene blue and rhodamine B dye adsorption properties. RSC Adv 12(44):28494–28504. https://doi.org/10.1039/D2RA04961A
12. Barquinha P, Fortunato E, Branquinho R, Martins R, Rovisco A, Vaz Pinto J (2020) Hydrothermal synthesis of zinc tin oxide nanostructures for photocatalysis, energy harvesting

and electronics. In: Krishnamoorthy K (ed) Novel nanomaterials. IntechOpen, Rijeka. https://doi.org/10.5772/intechopen.94294
13. Bolisetty S, Peydayesh M, Mezzenga R (2019) Sustainable technologies for water purification from heavy metals: review and analysis. Chem Soc Rev 48(2):463–487. https://doi.org/10.1039/C8CS00493E
14. Bouras K, Rehspringer JL, Schmerber G, Rinnert H, Colis S, Ferblantier G, Balestrieri M, Ihiawakrim D, Dinia A, Slaoui A (2014) Optical and structural properties of Nd doped SnO_2 powder fabricated by the sol–gel method. J Mater Chem C 2(39):8235–8243. https://doi.org/10.1039/C4TC01202J
15. Brickner PW, Vincent RL, First M, Nardell E, Murray M, Kaufman W (2003) The application of ultraviolet germicidal irradiation to control transmission of airborne disease: bioterrorism countermeasure. Public health reports (Washington, DC: 1974) 118(2):99–114. https://doi.org/10.1093/phr/118.2.99
16. Cao D, Wang H, Li B, Li C, Xie S, Rogach AL, Niu C (2016) Hydrothermal synthesis of sno_2 embedded MoO_{3-x} nanocomposites and their synergistic effects on lithium storage. Electrochim Acta 216:79–87. https://doi.org/10.1016/j.electacta.2016.09.012
17. Chahar M, Khaturia S, Singh HL, Solanki VS, Agarwal N, Sahoo DK, Yadav VK, Patel A (2023) Recent advances in the effective removal of hazardous pollutants from wastewater by using nanomaterials-a review. Front Environ Sci. https://doi.org/10.3389/fenvs.2023.1226101
18. Chakravarty R, Chakraborty S, Shukla R, Bahadur J, Ram R, Mazumder S, Sarma HD, Tyagi AK, Dash A (2016) Mechanochemical synthesis of mesoporous tin oxide: a new generation nanosorbent for 68 Ge/68 Ga generator technology. Dalton Transact 45(34):13361–13372
19. Chen D, Cheng Y, Zhou N, Chen P, Wang Y, Li K, Huo S, Cheng P, Peng P, Zhang R, Wang L, Liu H, Liu Y, Ruan R (2020) Photocatalytic degradation of organic pollutants using TiO_2-based photocatalysts: a review. J Clean Prod 268:121725. https://doi.org/10.1016/j.jclepro.2020.121725
20. Demazeau G (2011) Solvothermal and hydrothermal processes: the main physico-chemical factors involved and new trends. Res Chem Inter 37(2):107–123. https://doi.org/10.1007/s11164-011-0240-z
21. Desarkar HS, Kumbhakar P, Mitra A (2012) Optical properties of tin oxide nanoparticles prepared by laser ablation in water: influence of laser ablation time duration and laser fluence. Mater Charact 73:158–165
22. Dhal JP, Mishra BG, Hota G (2014) Fe_2O_3–SnO_2 composite nanorods: facile synthesis and sorption properties. J Environ Chem Eng 2(4):2188–2198. https://doi.org/10.1016/j.jece.2014.09.016
23. Dutta D, Thakur D, Bahadur D (2015) SnO_2 quantum dots decorated silica nanoparticles for fast removal of cationic dye (methylene blue) from wastewater. Chem Eng J 281:482–490. https://doi.org/10.1016/j.cej.2015.06.110
24. El-Sawaf A, Tolan DA, Abdelrahman MS, Abd El-Hay I, Ismael M, Ahmed ASA, Elshehy EA, Abdu MT (2024) Fast in-situ synthesis of mesoporous Prussian blue-silica nanocomposite for superior silver ions recovery performance. J Chem Tech Biotech 99(9):1941–1954. https://doi.org/10.1002/jctb.7707
25. Elhenawy S, Khraisheh M, AlMomani F, Al-Ghouti M, Selvaraj R, Al-Muhtaseb Aa (2024) Emerging nanomaterials for drinking water purification: a new era of water treatment technology. Nanomaterials 14(21). https://doi.org/10.3390/nano14211707
26. Fujishima A, Honda K (1972) Electrochemical photolysis of water at a semiconductor electrode. Nature 238(5358):37–38. https://doi.org/10.1038/238037a0
27. Gaur VK, Sharma P, Gaur P, Varjani S, Ngo HH, Guo W, Chaturvedi P, Singhania RR (2021) Sustainable mitigation of heavy metals from effluents: toxicity and fate with recent technological advancements. Bioengineered 12(1):7297–7313. https://doi.org/10.1080/21655979.2021.1978616
28. Geise GM, Lee H-S, Miller DJ, Freeman BD, McGrath JE, Paul DR (2010) Water purification by membranes: the role of polymer science. J Poly Sci Part B Polym Phys 48(15):1685–1718. https://doi.org/10.1002/polb.22037

29. Haq S, Sarfraz A, Menaa F, Shahzad N, Din SU, Almukhlifi HA, Alshareef SA, Al Essa EM, Shahzad MI (2022) Green synthesis of NiO-SnO_2 nanocomposite and effect of calcination temperature on its physicochemical properties: impact on the photocatalytic degradation of methyl orange. Molecules 27(23). https://doi.org/10.3390/molecules27238420
30. Hassan SSM, Kamel AH, Hassan AA, Amr AE-GE, El-Naby HA, Elsayed EA (2020) A SnO_2/CeO_2 nano-composite catalyst for alizarin dye removal from aqueous solutions. Nanomaterials 10(2):254
31. Islam T, Repon MR, Islam T, Sarwar Z, Rahman MM (2023) Impact of textile dyes on health and ecosystem: a review of structure, causes, and potential solutions. Environ Sci Pollut Res 30(4):9207–9242. https://doi.org/10.1007/s11356-022-24398-3
32. Jiang C, Zhang G, Wu Y, Li L, Shi K (2012) Facile synthesis of SnO_2 nanocrystalline tubes by electrospinning and their fast response and high sensitivity to NO_x at room temperature. CrystEngComm 14(8):2739–2747. https://doi.org/10.1039/C2CE06405G
33. Khairnar SD, Shirsath DS, Patil PS, Shrivastava VS (2020) Adsorptive and photocatalytic removal of carcinogenic methylene blue dye by SnO_2 nanorods: an equilibrium, kinetic and thermodynamics exploration. Appl Sci 2(5):822. https://doi.org/10.1007/s42452-020-2607-5
34. Kheshtzar I, Ghorbani M, Gatabi MP, Lashkenari MS (2018) Facile synthesis of smartaminosilane modified- SnO_2/porous silica nanocomposite for high efficiency removal of lead ions and bacterial inactivation. J Hazard Mater 359:19–30. https://doi.org/10.1016/j.jhazmat.2018.07.028
35. Koppala S, Balan R, Banerjee I, Li K, Xu L, Liu H, Kishore Kumar D, Raghava Reddy K, Sadhu V (2021) Room temperature synthesis of novel worm like tin oxide nanoparticles for photocatalytic degradation of organic pollutants. Mater Sci Energy Tech 4:113–118. https://doi.org/10.1016/j.mset.2021.03.002
36. Kovács K, He S, Mile V, Csay T, Takács E, Wojnárovits L (2015) Ionizing radiation induced degradation of diuron in dilute aqueous solution. Chem Central J 9(1):21. https://doi.org/10.1186/s13065-015-0097-0
37. Kudo A, Miseki Y (2009) Heterogeneous photocatalyst materials for water splitting. Chem Soc Rev 38(1):253–278. https://doi.org/10.1039/B800489G
38. Kumar S (2023) Smart and innovative nanotechnology applications for water purification. Hybrid Adv 3:100044
39. Kumar KY, Raj TNV, Archana S, Prasad SBB, Olivera S, Muralidhara HB (2016) SnO_2 nanoparticles as effective adsorbents for the removal of cadmium and lead from aqueous solution: adsorption mechanism and kinetic studies. J Water Process Eng 13:44–52. https://doi.org/10.1016/j.jwpe.2016.07.007
40. Kötz R, Stucki S, Carcer B (1991) Electrochemical waste water treatment using high overvoltage anodes. Part I: Physical and electrochemical properties of SnO_2 anodes. J Appl Electroch 21(1):14–20. https://doi.org/10.1007/BF01103823
41. Lamba R, Umar A, Mehta SK, Kansal SK (2015) ZnO doped SnO_2 nanoparticles heterojunction photo-catalyst for environmental remediation. J Alloy Compoun 653:327–333. https://doi.org/10.1016/j.jallcom.2015.08.220
42. Li L, Zhong Y, Hu Y, Bai J, Qiao F, Ahmed ASA, Ali G, Zhao X, Xie Y (2023) Room-temperature synthesis of Ag- and Mn-doped $Cs_2NaBiCl_6$ octahedrons for dye photodegradation. CrystEngComm 25(30):4355–4363. https://doi.org/10.1039/D3CE00372H
43. Lu Z, Zhou Q, Wang C, Wei Z, Xu L, Gui Y (2018) Electrospun ZnO–SnO_2 Composite nanofibers and enhanced sensing properties to sf6 decomposition byproduct H_2S. Front Chem. https://doi.org/10.3389/fchem.2018.00540
44. Ma CM, Hong GB, Lee SC (2020) Facile synthesis of tin dioxide nanoparticles for photocatalytic degradation of congo red dye in aqueous solution. Catalysts 10(7). https://doi.org/10.3390/catal10070792
45. Malla MA, Gupta S, Dubey A, Kumar A, Yadav S (2021) Chapter 7—contamination of groundwater resources by pesticides. In: Ahamad A, Siddiqui SI, Singh P (eds) Contamination of water. Academic Press, pp 99–107. https://doi.org/10.1016/B978-0-12-824058-8.00023-2

46. Mansoor NA, Hussein EH, Abbas KN (2024) Combination of hydro/solvothermal synthesis routes for the enhancement of nanostructures photoactivity. J Cluster Sci 35(6):1933–1939. https://doi.org/10.1007/s10876-024-02621-0
47. Mayedwa N, Mongwaketsi N, Khamlich S, Kaviyarasu K, Matinise N, Maaza M (2018) Green synthesis of zin tin oxide ($ZnSnO_3$) nanoparticles using Aspalathus Linearis natural extracts: structural, morphological, optical and electrochemistry study. Appl Surf Sci 446:250–257
48. McYotto F, Wei Q, Macharia DK, Huang M, Shen C, Chow CWK (2021) Effect of dye structure on color removal efficiency by coagulation. Chem Eng J 405:126674. https://doi.org/10.1016/j.cej.2020.126674
49. Miller SL, Linnes J, Luongo J (2013) Ultraviolet germicidal irradiation: future directions for air disinfection and building applications. Photochem Photobio 89(4):777–781. https://doi.org/10.1111/php.12080
50. Mishra SR, Ahmaruzzaman M (2022) Tin oxide based nanostructured materials: synthesis and potential applications. Nanoscale 14(5):1566–1605. https://doi.org/10.1039/D1NR07040A
51. Mishra S, Sahu A, Dungdung M, Ahmed S, Baitharu I (2023) Chapter 2—pesticide pollution in freshwater and its impact on community health. In: Singh J, Pandey A, Singh S, Garg VK, Ramamurthy P (eds) Current developments in biotechnology and bioengineering. Elsevier, pp 33–52. https://doi.org/10.1016/B978-0-323-91900-5.00005-9
52. Naggar AH, Seaf-Elnasr TA, Thabet M, El-Monaem EMA, Chong KF, Bakr ZH, Alsohaimi IH, Ali HM, El-Nasser KS, Gomaa H (2023) A hybrid mesoporous composite of SnO_2 and MgO for adsorption and photocatalytic degradation of anionic dye from a real industrial effluent water. Environ Sci Pollut Res Int 30(49):108247–108262. https://doi.org/10.1007/s11356-023-29649-5
53. Nagirnyak SV, Lutz VA, Dontsova TA, Astrelin IM (2016) Synthesis and characterization of tin (IV) oxide obtained by chemical vapor deposition method. Nanoscale Res Lett 11:1–7
54. Nelson TF, Baumgartner R, Jaggi M, Bernasconi SM, Battagliarin G, Sinkel C, Künkel A, Kohler H-PE, McNeill K, Sander M (2022) Biodegradation of poly(butylene succinate) in soil laboratory incubations assessed by stable carbon isotope labelling. Nat Commun 13(1):5691. https://doi.org/10.1038/s41467-022-33064-8
55. Nguyen TK, Yu SH, Yan J, Chua DHC (2020) SnO_2-anchored carbon fibers chemical vapor deposition (CVD) synthesis: effects of growth parameters on morphologies and electrochemical behaviors. J Mater Sci 55(32):15588–15601. https://doi.org/10.1007/s10853-020-05121-4
56. Nidheesh PV, Zhou M, Oturan MA (2018) An overview on the removal of synthetic dyes from water by electrochemical advanced oxidation processes. Chemosphere 197:210–227. https://doi.org/10.1016/j.chemosphere.2017.12.195
57. Nizam NUM, Hanafiah MM, Mahmoudi E, Halim AA, Mohammad AW (2021) The removal of anionic and cationic dyes from an aqueous solution using biomass-based activated carbon. Sci Rep 11(1):8623. https://doi.org/10.1038/s41598-021-88084-z
58. Patil YR, Gadave KM, Pathan HM (2025) Stability of aqueous polysulphide electrolyte with SnO_2 photoanode. J Mater Sci Materi Electron 36(1):85. https://doi.org/10.1007/s10854-024-14070-6
59. Pinto AH, Nogueira AE, Dalmaschio CJ, Frigini IN, de Almeida JC, Ferrer MM, Berengue OM, Gonçalves RA, de Mendonça VR (2022) Doped tin dioxide (d-sno_2) and its nanostructures: review of the theoretical aspects, photocatalytic and biomedical applications. Solids 3(2):327–360. https://doi.org/10.3390/solids3020024
60. Programme UNWWA (2003) The world water development report 1: water for people, water for life. UNESCO, Paris, France. https://doi.org/10.1016/j.envadv.2022.100259
61. Qu C, Lv C, Liu J (2024) Preparation and characterization of hollow spherical indium tin oxide with micro size by template assisted hydrothermal method. J Sol-Gel Sci Tech 111(1):38–48. https://doi.org/10.1007/s10971-024-06408-7
62. Rajput RB, Wagh KS, Mane SM, Lee J, Kale RB (2025) Synergistic SnO_2-TiO_2 nanostructures: dual function catalysts as efficient photocatalyst for industrial dye removal and bifunctional electrocatalyst for energy generation through overall water splitting. Sustain Mater Tech 43:e01246. https://doi.org/10.1016/j.susmat.2025.e01246

63. Rajput P, Sinha RK, Devi P (2021) Chapter 8-Current scenario of pesticide contamination in water. In: Ahamad A, Siddiqui SI, Singh P (eds) Contamination of water. Academic Press, pp 109–119. https://doi.org/10.1016/B978-0-12-824058-8.00032-3
64. Rani B, Punniyakoti S, Sahu NK (2018) Polyol asserted hydrothermal synthesis of SnO_2 nanoparticles for the fast adsorption and photocatalytic degradation of methylene blue cationic dye. N J Chem 42(2):943–954. https://doi.org/10.1039/C7NJ03341A
65. Raphulu M, Chauke NM (2024) Tin Oxide materials for industrial wastewater treatment: promising adsorbents and catalyst. In: Kılıç Taşeli B, Jacob-Lopes E, Queiroz Zepka L, Deprá MC (eds) Wastewater treatment and sludge management systems—the gutter-to-good approaches. IntechOpen, Rijeka. https://doi.org/10.5772/intechopen.1004230
66. Ren P, Qi L, You K, Shi Q (2022) Hydrothermal synthesis of hierarchical sno_2 nanostructures for improved formaldehyde gas sensing. Nanomaterials 12(2). https://doi.org/10.3390/nano12020228
67. Roguai S, Djelloul A (2025) Enhanced multifunctionality of SnO_2 and W-doped SnO_2 thin films synthesized via ultrasonic spray pyrolysis: applications in UV photodetectors, photocatalysis, and tunable surface hydrophilicity. Appl Physi A 131(3):167. https://doi.org/10.1007/s00339-025-08256-w
68. Sadik WM, El-Demerdash A, Nashed AW, Mostafa AA, Lamie E (2024) Synthesis and investigation of optical properties and enhancement photocatalytic activity of TiO_2–SnO_2 semiconductor for degradation of organic compounds. Sci Rep 14(1):27846.https://doi.org/10.1038/s41598-024-78755-y
69. Salmi C, Laouini SE, Meneceur S, Mohammed HA (2024) Biosynthesized MgO@SnO_2 nanocomposite and their modification with polyvinylpyrrolidone. Efficiency for removal of heavy metals and contaminants from industrial petroleum wastewater. Clean Tech Environ Policy 26(8):2483–2502. https://doi.org/10.1007/s10098-024-02766-6
70. Santos JP, Fernández MJ, Fontecha JL, Matatagui D, Sayago I, Horrillo MC, Gracia I (2014) Nanocrystalline tin oxide nanofibers deposited by a novel focused electrospinning method. Application to the detection of tatp precursors. Sensors 14(12):24231–24243. https://doi.org/10.3390/s141224231
71. Sayago I, Hontañón E, Aleixandre M (2020) 9—Preparation of tin oxide nanostructures by chemical vapor deposition. In: Orlandi MO (ed) Tin oxide materials. Elsevier, pp 247–280. https://doi.org/10.1016/B978-0-12-815924-8.00009-8
72. Sayed NSM, Ahmed ASA, Abdallah MH, Gouda GA (2024) ZnO@ activated carbon derived from wood sawdust as adsorbent for removal of methyl red and methyl orange from aqueous solutions. Sci Rep 14(1):5384. https://doi.org/10.1038/s41598-024-55158-7
73. Sery AA, Mohamed WAA, Hammad FF, Khalil MMH, Farag HK (2022) Synthesis of pure and doped SnO_2 and NiO nanoparticles and evaluation of their photocatalytic activity. Mater Chem Phys 275:125190. https://doi.org/10.1016/j.matchemphys.2021.125190
74. Sharif F, Roberts EPL (2020) Electrochemical oxidation of an organic dye adsorbed on tin oxide and antimony doped tin oxide graphene composites. Catalysts 10(2). https://doi.org/10.3390/catal10020263
75. Shek C, Lai J, Lin G (1999) Grain growth in nanocrystalline SnO_2 prepared by sol-gel route. Nanostruct Mater 11(7):887–893
76. Song KC, Kang Y (2000) Preparation of high surface area tin oxide powders by a homogeneous precipitation method. Mater lett 42(5):283–289
77. Sun J, Li X, Quan Y, Yin Y, Zheng S (2015) Effect of long-term organic removal on ion exchange properties and performance during sewage tertiary treatment by conventional anion exchange resins. Chemosphere 136:181–189. https://doi.org/10.1016/j.chemosphere.2015.05.002
78. Suzuki T, Cheng J, Qiao L, Xing Y, Zhang MF, Nishijima H, Yano T, Pan W (2020) Preparation of SnO_2 nanotubes via a template-free electrospinning process. RSC Adv 10(37):22113–22119. https://doi.org/10.1039/D0RA01719A
79. Tasisa YE, Sarma TK, Sahu TK, Krishnaraj R (2024) Phytosynthesis and characterization of tin-oxide nanoparticles (SnO_2-NPs) from *Croton macrostachyus* leaf extract and its application under visible light photocatalytic activities. Sci Rep 14(1):10780. https://doi.org/10.1038/s41598-024-60633-2

80. Thompson TL, Yates JT (2006) Surface science studies of the photoactivation of tio$_2$ new photochemical processes. Chem Rev 106(10):4428–4453. https://doi.org/10.1021/cr050172k
81. Tolan D, El-Sawaf A, Ahmed ASA, Nassar A, Mohamed NM, Alhindawy IG, Elshehy EA, Utgikar V (2024) Enhanced photocatalytic activity of (In–Sr–P) tridoped TiO_2/Bi_2O_3 composite loaded on mesoporous carbon: a facile sol-hydrothermal synthesis approach. Mater Chem Phys 322:129570. https://doi.org/10.1016/j.matchemphys.2024.129570
82. Torabinejad A, Nasirizadeh N, Yazdanshenas ME, Tayebi H-A (2017) Synthesis of conductive polymer-coated mesoporous MCM-41 for textile dye removal from aqueous media. J Nanostructu Chem 7(3):217–229. https://doi.org/10.1007/s40097-017-0232-7
83. Ungureanu A-M, Jitaru I, Gosnea F (2013) Mn doped SnO_2 prepared by a sol-gel method. Sci Bull Ser B Chem Mater Sci 75:43–52
84. van Mol AMB, Chae Y, McDaniel AH, Allendorf MD (2006) Chemical vapor deposition of tin oxide: fundamentals and applications. Thin Sol Films 502(1):72–78. https://doi.org/10.1016/j.tsf.2005.07.247
85. Wali Q, Fakharuddin A, Ahmed I, Ab Rahim MH, Ismail J, Jose R (2014) Multiporous nanofibers of SnO_2 by electrospinning for high efficiency dye-sensitized solar cells. J Mater Chem A 2(41):17427–17434. https://doi.org/10.1039/C4TA03056G
86. Wang JL, Xu LJ (2012) Advanced oxidation processes for wastewater treatment: formation of hydroxyl radical and application. Critical Rev Environ Sci Tech 42(3):251–325. https://doi.org/10.1080/10643389.2010.507698
87. Witkowska D, Słowik J, Chilicka K (2021) Heavy metals and human health: possible exposure pathways and the competition for protein binding sites. Molecules 26(19). https://doi.org/10.3390/molecules26196060
88. Yang H, Hu Y, Tang A, Jin S, Qiu G (2004) Synthesis of tin oxide nanoparticles by mechanochemical reaction. J Alloy Comp 363(1–2):276–279
89. Yin W, Cao H (2016) One-step synthesis of SnO_2-reduced graphene oxide (SOG) composites for efficient removal of organic dyes from wastewater. RSC Adv 6(102):100636–100642. https://doi.org/10.1039/C6RA21856C
90. Zamand N, Pour AN, Housaindokht MR, Izadyar M (2014) Size-controlled synthesis of SnO_2 nanoparticles using reverse microemulsion method. Sol State Sci 33:6–11
91. Zhou Q, Sun H, Jia L, Wu W, Wang J (2022) Simultaneous biological removal of nitrogen and phosphorus from secondary effluent of wastewater treatment plants by advanced treatment: a review. Chemosphere 296:134054. https://doi.org/10.1016/j.chemosphere.2022.134054
92. Zong S, Zhang Y, Cao J, Qin C, Bala H, Wang Y (2024) Hydrothermal synthesis of SnO_2 with Different morphologies as sensing materials for HCHO detection. Langmuir 40(20):10814–10824. https://doi.org/10.1021/acs.langmuir.4c01263

MIX
Papier aus verantwortungsvollen Quellen
Paper from responsible sources
FSC® C105338

If you have any concerns about our products,
you can contact us on
ProductSafety@springernature.com

In case Publisher is established outside the EU,
the EU authorized representative is:
Springer Nature Customer Service Center GmbH
Europaplatz 3, 69115 Heidelberg, Germany

Printed by Libri Plureos GmbH
in Hamburg, Germany